Physiology of Growth and Development in Horticultural Plants

The development of a plant is a multifaceted, dynamic phenomenon. Due to their immobility, plants respond not only to internal developmental cues but also to changes in the prevailing environmental conditions. Climate change has increased vulnerability in plants due to increasing concentrations of CO_2 and other pollutants and fluctuations in the growing environment. These changes affect crop growth and productivity, thereby posing a major risk to global food security.

Physiology of Growth and Development in Horticultural Plants contains 22 chapters organized into six parts, beginning with an introduction on basic concepts of plant growth and development, followed by genetic basis of plant development, quantification of growth, and sensing and response of plants to various environmental signals. It also explores plant growth hormones and their role, either singly or in combination, in controlling various aspects of plant growth and development, and hormonal regulation of physiological and developmental processes. The book highlights intricate aspects of growth and development in horticultural plants with classic examples from the real world.

FEATURES

The book:

- presents information on plant growth and development; structure and genetic basis of plant development with quantification of growth; sensing and response of plants to various environmental signals; and various phytohormones and their role in controlling aspects of plant growth and development;
- provides key scientific and technical advances, issues and challenges in various areas of growth and development of horticultural plants;
- demonstrates how the response of various plants to internal and external stimuli can be commercially exploited.

Physiology of Growth and Development in Horticultural Plants encourages the development of new techniques, technologies and innovative practices and is an ideal reference for students of advanced plant sciences courses, researchers and commercial horticultural practitioners.

Physiology of Growth and Development in Horticultural Plants

N. R. Bhat, Arvind Bhatt and M. K. Suleiman

CRC Press
Taylor & Francis Group
Boca Raton London New York

CRC Press is an imprint of the
Taylor & Francis Group, an **informa** business

Designed cover image: Shutterstock

First edition published 2025
by CRC Press
2385 NW Executive Center Drive, Suite 320, Boca Raton FL 33431

and by CRC Press
4 Park Square, Milton Park, Abingdon, Oxon, OX14 4RN

CRC Press is an imprint of Taylor & Francis Group, LLC

Library of Congress Cataloging-in-Publication Data
Names: Bhat, N. R. (Narayana R.) 1953– author. | Bhatt, Arvind, author. | Suleiman, M. K. (Majda Khalil), author.
Title: Physiology of growth and development in horticultural plants / N.R. Bhat, Arvind Bhatt, M.K. Suleiman
Description: First edition | Boca Raton, FL : CRC Press, 2025 | Includes bibliographical references and index
Identifiers: LCCN 2024007939 (print) | LCCN 2024007940 (ebook) | ISBN 9781032429588 (hardback) |
ISBN 9781032447834 (paperback) | ISBN 9781003373889 (ebook)
Subjects: LCSH: Horticultural crops—Physiology. | Plants, Cultivated—Physiology. | Growth (Plants).
Classification: LCC SB319.5 B43 2025 (print) | LCC SB319.5 (ebook) | DDC 635—dc23/eng/20240711
LC record available at https://lccn.loc.gov/2024007939
LC ebook record available at https://lccn.loc.gov/2024007940

ISBN: 978-1-032-42958-8 (hbk)
ISBN: 978-1-032-44783-4 (pbk)
ISBN: 978-1-003-37388-9 (ebk)

DOI: 10.1201/9781003373889

Typeset in Times
by Apex CoVantage, LLC

Contents

PART III Quantitative Aspects of Plant Growth and Development

PART IV Response of Plants to Environmental Stimuli

PART V Hormonal Control of Plant Growth and Development

About the Authors

N. R. Bhat received his PhD degree in horticulture from The Ohio State University in Columbus, Ohio, USA, in 1982, and was a postdoctoral fellow from July 1982 to January 1983 at Paul Ecke Poinsettias in California, USA, and from December 1988 to March 1990 at The Ohio State University in Columbus, Ohio, USA. Dr. Bhat has worked in several universities and research organizations, including Kuwait Institute for Scientific Research (KISR), The Ohio State University, Indo American Hybrid Seeds, Mahatma Phule Agricultural University, Himachal University and Punjab Agricultural University. His research focused on precision agriculture, greenhouse crop production, soilless growing systems, irrigation water and nutrient management, and characterization, documentation, mass propagation and ecophysiology of Kuwait's native plants, and ecological restoration of degraded ecosystems. During his 27-year tenure at KISR, he was the project leader/activity leader/principal investigator of over 55 research projects and mentored several staff members. He also taught in-house and on-the-job training courses and undergraduate and graduate courses in horticulture. He also chaired the Advisory Committee of seven graduate students and served on several important committees. He has published 17 books/book chapters, over 115 peer-reviewed research papers, 170 reports and scientific publications and presented 69 papers in regional and international conferences.

Arvind Bhatt is Research Scientist affiliated with the Desert Agriculture and Ecosystems Program at the Environment and Life Sciences Research Center of the Kuwait Institute for Scientific Research. Dr. Bhatt completed his PhD degree in 2006 from H.N.B. Garhwal University, Uttarakhand, India. With over 17 years of global experience in plant science, he has collaborated with esteemed institutions and universities worldwide, including UKZN in South Africa, Universiti Sains Malaysia, Ghent University in Belgium, Gulf Organization for Research and Development in Qatar, Lushan Botanical Garden in China and the Kuwait Institute for Scientific Research. Dr. Bhatt's scholarly contributions are proved by more than 135 peer-reviewed research publications, which have amassed over 2,800 citations. His remarkable achievements extend to receiving numerous awards and securing research grants from diverse countries.

M. K. Suleiman is working as Research Scientist and Manager of Desert Agriculture and Ecosystems Program at the Kuwait Institute for Scientific Research. She completed her PhD degree in biological sciences (Thesis: Restoration Ecology of *Acacia pachyceras* in the State of Kuwait) from the University of Western Australia. She was part of the team in national projects such as introducing mangroves for protection and enrichment of Kuwait's coastlines and development of Kuwait Greenery Master Plan and Urban Demonstration Garden. Ms. Suleiman led many research projects and authored several research papers in refereed international journals (60), proceedings/abstract books/posters/presentations (62), books (8) and KISR reports (141). Her main areas of research are desert landscaping, rehabilitation, re-vegetation, biodiversity conservation, introduction and mass propagation of ornamental plants suitable for Kuwait's harsh conditions and their use in landscaping, and standardization of propagation techniques for native plants of Kuwait for their utilization in greenery projects. Ms. Suleiman is also involved in Kuwait Environmental Remediation Program (KERP) as a coordinator and as a resource coordinator for Environment and Life Sciences Research Center (ELSRC) staff members. She is a member of several national and international committees in research and development category.

Preface

Higher plants are a group of complex higher organisms with an extraordinary degree of coordination or integration among cells and tissues as well as a distinctive mode of development involving continuous initiation and growth of organs throughout their lifespan for their orderly development. Although the organization of various plant organs is under the genetic control, the overall structure, including the shape, size and pattern of roots, stems, branches, leaves and flowers, is influenced by their growing environment, indicating that the growth and development of various organs can be highly variable. The cells sense and respond to a wide variety of external (light, temperature, moisture, gravity, etc.) and internal (phytohormones, biomolecules) signals, which trigger a series of physiological and developmental processes that enable them to respond to changing conditions in the growing environment. Plants are capable of synthesizing their food through the process of photosynthesis using the energy from sunlight, CO_2 from air and water and minerals from the soil (autotrophs; opposite heterotrophs).

The development of a plant is a complex and multidimensional phenomenon. Immobility of plants enables them to respond not only to developmental cues but also to the changes in the prevailing environmental conditions. The unique facets of plant development are reflected in molecular mechanisms of cell cycle control in a distinct manner, underpinning precise control modules. Phytohormones provide signals that either individually or collectively direct the developmental processes in individual cells or pass the information between and among cells leading to a highly coordinated action.

Recent trends in climate change have exposed the vulnerability of crop plants to increasing levels of CO_2 and pollutants in the growing environment and changing weather patterns marked by sudden, often severe fluctuations in temperatures and rainfall leading to a unique set of biotic and abiotic stresses. For example, rising temperatures not only lead to frequent and severe drought spells and salt concentrations in the soil but also result in modifications in the spread of diseases and insect pest infestations. Significant yield losses caused by these stresses are a major threat to global food security and international food supply chain. The demand for food is expected to grow by almost 70% by the year 2050, but the productivity and production of most crop plants are likely to decline making the situation very gloomy. To address the growing challenges posed by different stresses, there is an urgent need to first advance our understanding of how some plants are able to tolerate stresses better than others and then make concerted efforts to transfer those traits to commercial crop plants. However, the complexity and variability of stress tolerance mechanisms as well as the wide range of different stress conditions experienced by plants make any improvement in stress tolerance a daunting task. Therefore, it is necessary to identify the genes that govern plants' responses to different cues and discover how they function under the given set of conditions and induce directed alterations in developmental process including rooting, flowering, reproduction and plant architecture that are important to the horticultural industry.

This book provides an overview of the key scientific and technical advances, issues and challenges in areas of growth and development of horticultural plants. The book contains 22 chapters, beginning with an introduction of basic concepts and components of plant growth and development, followed by structure and genetic basis of plant development with quantification of growth, sensing and response of plants to various environmental signals (light, temperature, various phytohormones) and their role in controlling various aspects of plant growth and development. This book is intended to serve as a textbook for advanced undergraduate and postgraduate courses, faculty and researchers in horticultural science, plant sciences, forestry and plant developmental biology.

The senior author is grateful to his family members and Dr. B. B. Desai for their support and understanding. He would like to specially thank his younger daughter, Anita Bhat, for her pain-taking efforts in preparing the drawing for the book.

N. R. Bhat
Arvind Bhatt
M. K. Suleiman
August 2023

Part I

Introduction – Definitions, Basic Concepts and Components of Growth and Development

Several definitions have been used to describe "growth." While the botanists employ this term to explain permanent increase in size, volume or mass of a cell, tissue, organ or whole organism when it is accompanied by an increase in dry weight, horticulturists and farmers use it to represent crop growth and yield including biomass. In both the cases, it involves cell division and/or elongation, although it is not necessary for both cell division and cell elongation to precede growth. For example, early growth of an embryo in the flower occurs through an increase in cell number without any significant increase in the size of the embryo. Growth is limited to living cells, occurs at the expense of metabolic energy and involves the synthesis and organization of macromolecules, such as nucleic acid, polysaccharides, proteins and lipids cell organelles. Therefore, the term growth can be more precisely defined as an increase in the size by cell division and cell enlargement, together with the synthesis of macromolecules and organization of cellulose organelles. It is a vital process, which brings about a permanent change in size, form, weight and volume of the whole plant or any of its parts.

Differentiation refers to orderly processes occurring in plants wherein groups of genetically identical cells are distinguished into various tissues and organs to perform a specific function. This specialization occurs due to differential activation of a cell's genome. For example, after fertilization, the fertilized egg undergoes a series of cell divisions to form the radicle and the epicotyl of the embryo. Similarly, the meristematic cells in the apical portion of shoots and roots continue to divide to form various organs such as stem, leaf, flowers, fruits, primary and secondary roots, and root hairs. In plants, the differentiation is not permanent process, because fully differentiated cells may revert to the meristematic state as it happens when a portion of the plant tissue (explant) can be induced to produce a group of undifferentiated cells called callus. The callus can be induced again to differentiate to produce an entire plant with various organs. The ability of the single cell to produce the entire plant is known as "totipotency." The demonstration of totipotency of cells led to the development and commercialization of plant cell and tissue culture, thus providing a major impetus for mass propagation of horticultural plants.

DOI: 10.1201/ 9781003373889-1

In contrast, various qualitative and quantitative changes occurring during the life cycle of the plant are termed as plant "development." This involves both growth and differentiation occurring due to a series of events, such as seed germination, bud sprouting, vegetative growth, maturation, transition to flowering, fruit and seed formation, ripening and senescence. In annual plants, the morphological changes associated with these stages are more distinctive, time bound and easily predictable. However, the developmental stages in trees crops are less predictable, vary from year to year depending on the prevailing environmental conditions and are repeated year after year before they eventually die. Furthermore, in perennial crops like fruits trees, a number of development stages (vegetative growth, leaf abscission, flowering, fruit development, fruit ripening and senescence) may overlap. For example, in mango (*Mangifera indica* L.), the entire tree or only a portion of the tree may produce flowers at a time and the other portion on the same tree produces vegetative flushes. Again, the tree may produce flowers and fruits during alternate years (alternate bearing). Similarly, flowering plants (e.g., rose, chrysanthemum and carnation) produce flowers and vegetative branches simultaneously. In contrast, as in grapes and temperate fruits, the entire plant can be induced to produce flowers at one time. Furthermore, the exact time of commencement and the duration of each developmental stage can vary considerably within the species and under a different set of environmental conditions.

At a cellular level, growth, differentiation and development involve a number of metabolic activities, such as photosynthesis, respiration, biosynthesis of various macromolecules and phytohormones and their organization into cell organelles. Both external (temperature, light, water and nutrient supply, oxygen, carbon dioxide and many others) and internal (phytohormones) factors influence these metabolic processes and, in turn, regulate the growth and developmental process in plants. The exact nature of control by various factors in regulating growth and development has been discussed separately under different chapters in this book.

1 Plant Structure and Development – An Overview

1.1 INTRODUCTION

Plants are immobile living organisms capable of regenerating their cells and various organs and tissues (totipotency or *Latin* word "totipotentia" "ability for all [things]") throughout their lifespan. Unlike animal cells, plant cells are covered by a rigid wall, which protects them from external stresses. The plants with green pigment (chloroplast) in their leaves produce their own food (autotrophs) and directly or indirectly serve as a source of food for most other living organisms on this planet. Although different plants look different from outside (e.g., a cactus plant, aloe, rose and a mango tree), they all are composed of the same basic unit called the cell. These cells are very similar in how they grow, synthesize and transport their food, and their development is regulated in a similar manner. While the basic body pattern of the plant is established very early in their life cycle (embryogenesis), various types of cells, tissues and organs are produced at a later stage of their development. These structures give them their unique shape, size and patterns and allow them to perform unique functions that they are expected to perform.

Plants are classified as vascular (containing xylem and phloem tissues in its body) or nonvascular (without xylem and phloem tissues) depending on the type of tissues present; lower (e.g., bryophytes, pteridophytes) or higher (most cultivated or domesticated plants) based on the level of development or evolution; unicellular (plants with a single cell, e.g., mosses, hornworts, bacteria, liverworts) or multicellular (having several cells, e.g., angiosperms) based on the number of cells present in the body; and either amphibious (those live in water, e.g., mosses, water plants) or land plants based on their natural habitat (Graham et al., 2003).

Vascular plants (derived from the *Latin* word "vasculum" meaning "duct"), also known as tracheophytes, contain lignified tissues (xylem) for conducting water, nutrients and other substances and non-lignified tissues (phloem) to distribute products of photosynthesis. These plants also have group of cells (cambium or stem cells) that continuously divide and produce new cells throughout their lifespan. These newly produced cells are differentiated into xylem and phloem tissues. The xylem, phloem and cambium together are called the vascular bundle. Nonvascular plants do not have vascular bundles. Although nonvascular plants lack vascular bundles, they possess simpler specialized tissues for internal transport of water. Ferns and common liverwort fall into nonvascular plant category. Higher plants are grouped as monocotyledonous (also known as "monocots") or dicotyledonous (also referred to as "dicot") based on the number of leaves (cotyledons) produced during germination. Monocots (e.g., grasses, garlic, onion, lilies, gladiolus, garlic) produce only one seed leaf (cotyledon), whereas dicots (e.g., beans, mango, tree crops and sunflower produce) produce two leaves. Anatomically, these plants differ in their vascular systems and structure of stems, roots, flowers and leaves that are reflected in their growth and development (Table 1.1).

1.2 LIFE CYCLE OF HIGHER PLANTS

Based on the length of the life cycle (from germination to death), vascular plants are classified as annuals, biennials or perennials. Annual plants (e.g., beans, eggplant, tomato, marigold) complete their life cycle within one year or one growing season (Hirasuna, 1985). During this period, the seeds germinate, grow into mature plant, bloom, produce seeds and die. They are further classified as winter and summer annuals. Biennial plants, on the other hand, complete their life cycle in two

DOI: 10.1201/9781003373889-2

TABLE 1.1
Characteristics of Monocot and Dicot Plants

Traits	Monocot plants (grasses, onion, garlic, lilies, etc.)	Dicot plants (beans, tomato, orange, etc.)
Number of leaves in the seed (cotyledon)	One	Two
Seed germination	Hypogeal (belowground)	Hypogeal or epigeal (aboveground)
Vascular bundles in leaves and stems	Xylem and phloem tissues form paired bundle, numerous and scattered throughout leaves and stems	Vascular bundles are fewer in number and arranged in circles or concentric rings.
Cambium	Cambium is absent. No secondary growth.	Cambiums are present. Produce secondary growth.
Root system	Adventitious root system	Tap root system
Annual rings	Not present	Present
Leaves	Leaves have parallel venation	Leaves have reticulate venation.
Stem	Hollow	Usually, solid
Flower	Floral parts are in three or multiples of three.	Floral parts are four or multiples of four.
Seeds	Seeds contain embryonic plant and endosperm (storage tissue).	Seeds contain embryonic plant and food storage tissue or cotyledons.

years or two growing seasons. Their life comprises two phases: first, the vegetative phase, when they produce vegetative structures (stem, leaves etc.) and food storage organs and accumulate sufficient food reserves. This phase is followed by the reproductive phase during which they produce flowers, fruits and seeds (e.g., Swiss chard, onion, carrots, cabbage, cauliflower, beets, parsley). Sometimes, biennials can go from seed germination to seed production in one growing season, a phenomenon referred to as bolting. These plants also bolt instead of going through normal developmental stages if they are exposed to extreme or unfavorable environmental conditions.

Plants that require more than two years to complete their life cycle are called perennials. Perennials could be either herbaceous (with soft or nonwoody stems) or woody plants (with woody stem known as trunk). Herbaceous perennials generally die back to the ground each winter and sprout during the ensuing spring season of each year to produce new stems, flowers and fruits (e.g., Delphinium, aster, canna, hemerocallis, phlox). On the other hand, trees and shrubs with woody stems that have a mechanism to withstand cold temperatures do not die back during winter (e.g., most tree fruits), although they could remain dormant to withstand unfavorable weather conditions (cold or heat).

1.3 ORGANIZATION OF THE PLANT BODY

Cells are the basic structural and physiological units in plants. Similar cells get organized into tissues and work together to perform a specific function. An example would be the xylem tissue that conducts water from roots to the leaf through the stem or the phloem that distributes the food that is produced in leaves to different parts of the plant. Majority of the metabolic processes, such as cell division, photosynthesis and respiration, occur at the cellular level.

Vascular plants comprise a root system and a shoot system (Raven and Johnson, 2001; Figure 1.1). The root system has a network of primary and secondary roots that penetrate to the soil and anchors the plant. They absorb water and nutrients that are crucial for plant growth. The shoot system provides the framework for positioning the branches and leaves to facilitate the absorption of maximum amount of light and carbon dioxide (CO_2) from the atmosphere. The arrangement, size and other

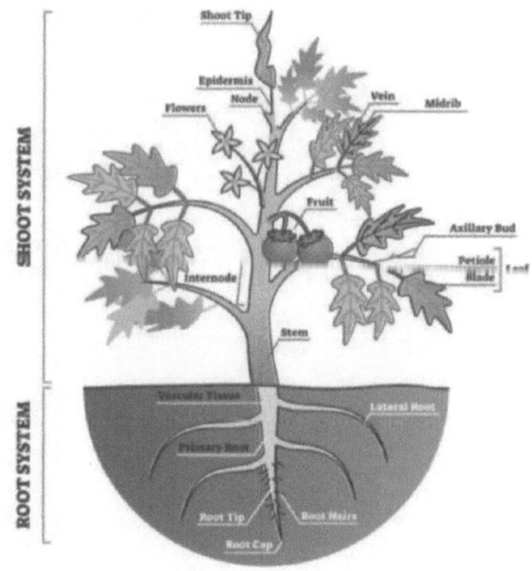

FIGURE 1.1 Structure of a higher plant.

features of the leaves are of critical importance in the plant's ability to produce food. Additionally, the flowers, other reproductive organs and, ultimately, fruits and seeds are also formed on the shoot. The vegetative shoot consists of nodes, internodes, leaves and axillary buds. Axillary buds are the apical meristems derived from the primary apical meristem that allows the plant to branch out or replace the main shoot if it is damaged by any of the several causes and is unable to divide further. A vegetative axillary bud has the capacity to develop the primary shoot. These vegetative axillary buds will produce flowers or floral shoots when the plant has transited to the reproductive phase of development.

The structure of the stem usually determines the overall growth habit (i.e., herbaceous, shrubby or arborescent). Herbaceous plants may live only for one year (annuals) or persist for two (biennials) or more years (perennials). The annual herbaceous plants complete their life cycle in one growing season, thereby avoiding extreme environmental conditions, particularly very high or low temperatures and drought. Suffrutescent (equivalent to Latin suf- = New; Latin frut- = shortened stem; Latin frutex = shrub, bush; Latin ēscent = escent) plants are herbaceous perennials with slightly woody belowground part surviving year after year, but the aboveground parts die back annually to a woody base. In some herbaceous perennials, the aboveground parts persist even during unfavorable weather conditions. The shrubby plants are woody, considerably branched and usually less than 10 meters tall, whereas the arborescent plants are those that are treelike.

The basic structure and shape of the plant are determined very early in their life cycle, usually during embryogenesis (Mauseth, 1988, 2003). At the embryonic stage, one cell divides into two, out of which one remains meristematic, while the other differentiates to form different tissues in the plant body. In this way, the population of meristematic cells is constantly maintained. The cells that continuously divide are called the meristematic ("meristemos" means divisible) tissues (Figure 1.2). During the germination process, the meristematic tissues divide and produce a radicle, which eventually becomes root system, and epicotyl that ultimately becomes the shoot system. The meristematic tissues are called differently on the basis of their location on the plant body. Meristems located at the tips of shoots and roots are called the apical meristems, whereas those distributed within the body are referred to as lateral meristems. Following the seed germination, the cells of the apical meristems divide continuously and add more cells to the tips of a seedling, leading to longitudinal growth of seedlings. The growth brought about by the apical meristems is called the

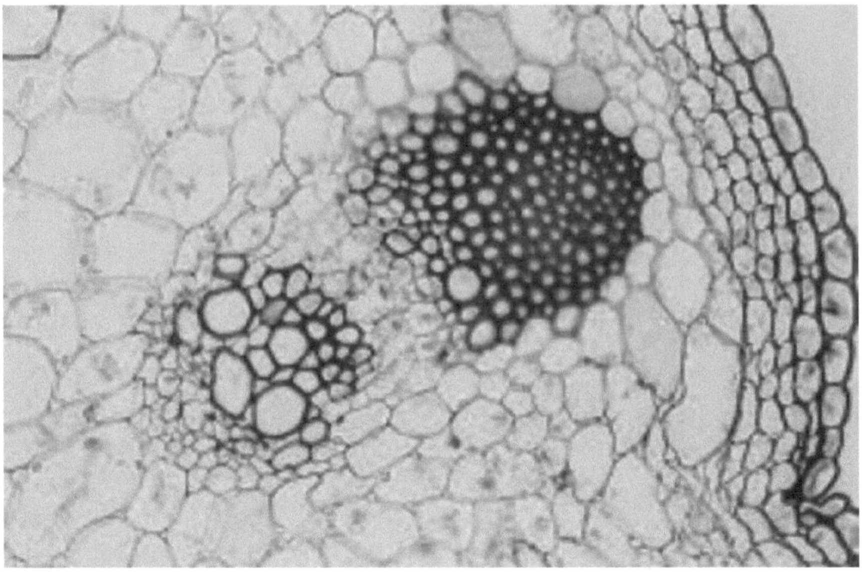

FIGURE 1.2 Cambium tissue.

primary growth. The elongation of the root and stem forms the primary plant growth, producing young shoots and roots. The meristem tissues in roots being delicate are protected by a root cap. These root caps, which are produced by the root meristems, are replaced as the root grows through the soil. In contrast, a variety of adaptive mechanisms, such as bending of epicotyl of the seedling and wrapping of emerging shoot by coleoptile (sheath) in monocot seedlings, protect the shoot apical meristem during germination and early growth. Similarly, the leaf primordia protect the shoot apical meristems at later stages of plant development.

The apical meristem gives rise to three types of embryonic tissue systems called primary meristems: the protoderm, which forms the epidermis; the procambium, which produces primary vascular tissues (primary xylem and primary phloem); and the ground meristem, which differentiates further into ground tissue containing parenchyma cells. In some monocots (e.g., grasses), the intercalary meristems located at the leaf base or in stem internode increase internode length.

Most herbaceous plants exhibit only primary growth, whereas trees, shrubs and some herbaceous plants with lateral meristems also produce secondary growth. The secondary growth increases the stem girth in herbaceous plants with only one lateral meristem. The woody plants, on the other hand, contain two lateral meristems, one within the bark of a woody stem (cork cambium), which is responsible for producing cork cells of the outer bark, and the other just below the bark between the xylem and phloem tissues (vascular cambium) that adds secondary vascular (xylem and phloem) tissues on either side of the vascular cambium. The secondary xylem becomes the wood, whereas the secondary phloem stays very close to the outer surface of a woody stem. Therefore, removal of the bark damages the phloem and may eventually lead to the death of the entire plant. Lateral meristems produce secondary tissues.

The primary and secondary meristematic growth together produces the body parts in an adult plant. Parts such as leaves, roots, branches and flowers all vary both in their size and in number from plant to plant even within a species. The development of the form and structure of plant parts may be fairly flexible. As the plant grows, the number, location, size and even structure of leaves and roots are often influenced by the environmental conditions. The arrangement, size and other features of leaves are of critical importance in plant's ability to synthesize food.

1.3.1 Tissue Types

Three basic types of tissues exist in plants: the ground, dermal and vascular. Each of these tissues has its own unique, functionally related cell types as described in the following.

1.3.1.1 Ground Tissue

Ground tissues make up the bulk of nonwoody plant organs and lie between vascular and dermal tissues (Figure 1.3). Three types of ground tissues (parenchyma, collenchyma and sclerenchyma) differing in form and function are found in plants.

Parenchyma: Parenchyma cells with large vacuoles and thin walls are most abundant in primary tissues, although they may also occur, to a much lesser extent, in secondary tissues (Figure 1.4). Most parenchyma cells have only primary walls that are laid down while the cells are still maturing and are less specialized than other plant cells. They are involved in performing specific functions, such as nectar and resin secretion and storage of latex, proteins and metabolic wastes. Parenchyma cells with functional nuclei are capable of dividing and remaining alive for several years after they mature. The majority of storage cells in fruits are parenchyma. Chloroplast containing parenchyma cells are called the chlorenchyma.

Collenchyma: Like parenchyma, collenchyma tissues have living protoplasts and may live for many years. They differ from parenchyma cells by having walls of varying thickness and being little longer than wide (Figure 1.5). Collenchyma is the primary support tissues in young stems, leaves, stalks and petals. They provide flexibility for plant organs by allowing them to bend without breaking. For example, the strings in the edible portion of celery (leaf stalks) mainly consist of collenchyma and vascular bundles (conducting tissues).

Sclerenchyma: Sclerenchyma cells have thick, lignified walls and are devoid of living protoplasts when they mature (Figure 1.6). Two types of sclerenchyma, namely, fibers and sclereids occur in plants. Fibers are long, slender cells usually grouped together in strands. Sclereids, on the other hand, come in varying shapes and forms, often branched, but usually not elongated, occur either individually or in groups. The long fibers from leaves and stems of many plants are used to make clothes and rope. In contrast, the coarse, grainy texture of flesh in pear fruit is an example of sclereids. Extremely thick secondary wall of sclereids is responsible for hard shells in walnut and

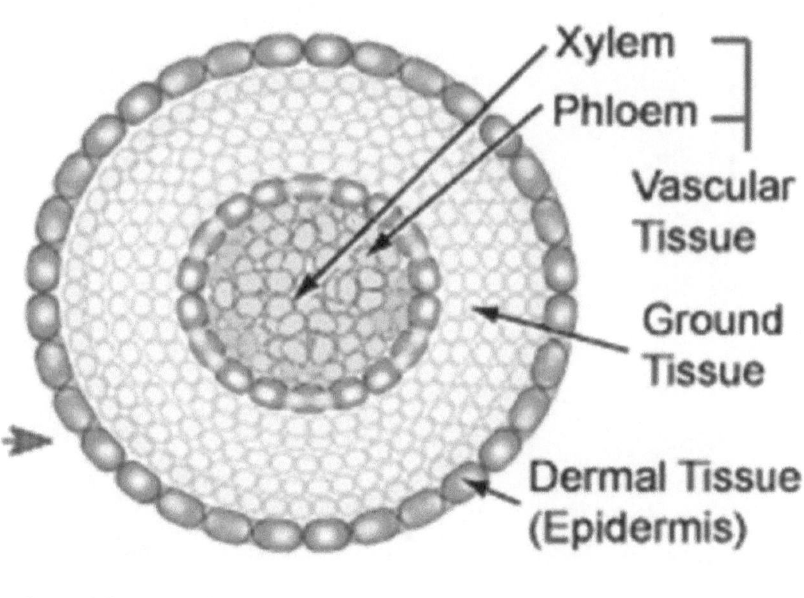

FIGURE 1.3 Ground tissue.

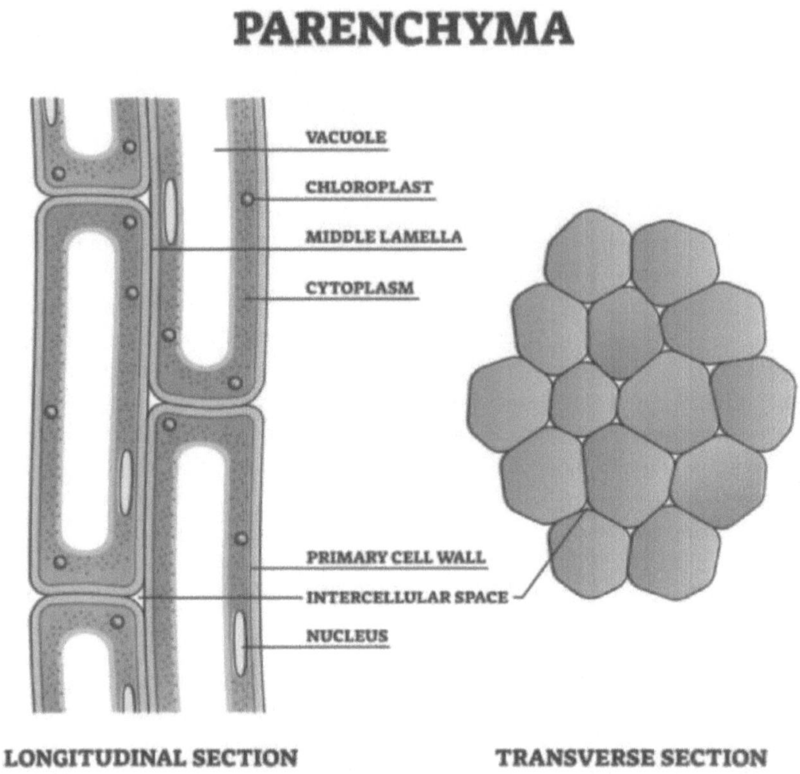

FIGURE 1.4 Plant-parenchyma-structure.

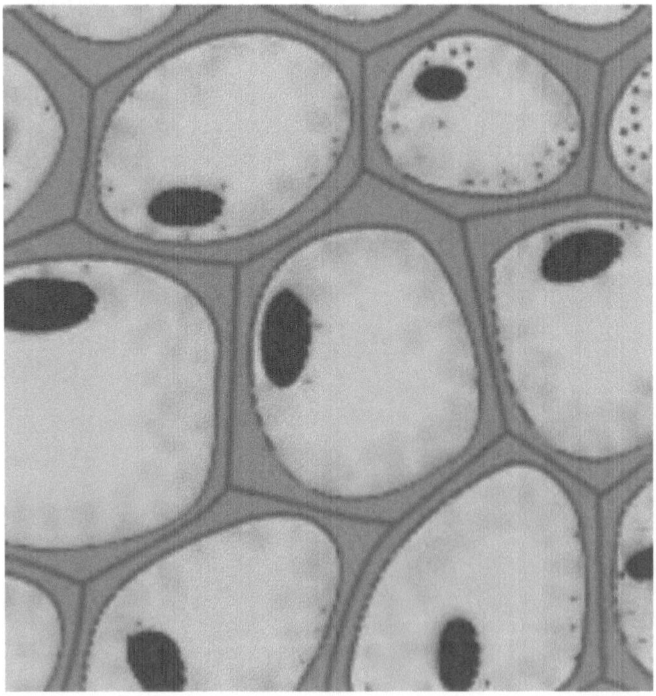

FIGURE 1.5 Transverse view of collenchyma tissue.

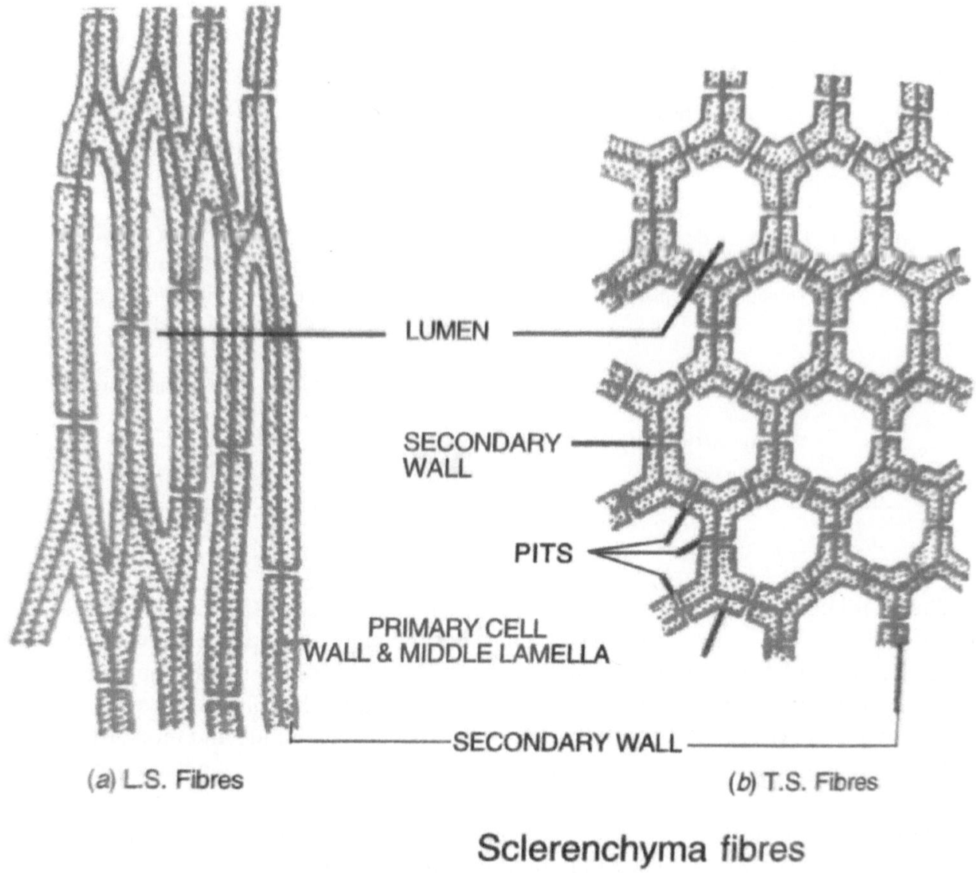

FIGURE 1.6 Transverse view of sclerenchyma tissue.

other fruits. Both fibers and sclereids provide strength to tissues in which they occur and provide mechanical support and protection to plant organs.

The following types of sclereids occur in plants:

- Star shaped – asterosclereids
- Macrosclereids – similar to palisade cells
- Bone-like osteosclereids with enlarged ends
- Isodiametric brachysclereids

The main differences among the above three cell types are shown in Table 1.2.

1.3.1.2 Dermal Tissue

In most plants, the dermal system is one cell thick (epidermis) forming the outer protective covering of the plant. This is probably the earliest tissue system to appear during embryogenesis. The epidermal cells originate from the protoderm and consist of thin-walled parenchyma cells that are initially more or less spherical but may assume other shapes, often ending up with as many as 17 sides. The epidermis in young exposed parts of the plant is covered with a cutaneous layer (cuticle) of various thicknesses depending on the species and environmental conditions. For example, in succulents and desert plants, waxy layer may be added on the cuticle to reduce transpiration. Epidermis may also contain a number of specialized cells, such as guard cells, trichomes and root

TABLE 1.2
Comparison of Ground Tissues in Plants

Features	Parenchyma	Collenchyma	Sclerenchyma
Cell characteristics	• Spherical with thin wall	• Elongated cells with uneven thickening of cell wall (thicker at corners)	• Contain thick, lignified primary and secondary cell walls
	• Living	• Alive at maturity	• Dead at maturity
	• Metabolically active	• Metabolically active	• Metabolically inactive
Presence	• Throughout the plant	• Beneath the epidermis in young stem and below the veins in leaves	• Fibers in wood, bark and stems • Sclereids – in fruits and seeds
Function	• Storage • Regeneration • Metabolism (photosynthesis and respiration)	• Provide flexibility to organs	• Mechanical support

hairs. The guard cells, which are formed as a result of asymmetric cell division, are present in the epidermis of leaves and, sometimes, in other parts of the plant, such as stems or fruits. Although stomata are usually present in greater numbers on the lower side of the leaves, they are also present on the upper surface as in water lilies. The guard cells contain chloroplasts. The stomata close and open in response to external stimuli (light, temperature, carbon dioxide and moisture) and remain open during the periods of active photosynthesis to allow free passage of carbon dioxide into and oxygen out of the leaf.

Trichomes are tiny hair-like structures arising from epidermis that occur frequently on stems, leaves and reproductive organs. They may be unicellular, multicellular or glandular (i.e., secretes substances to deter herbivory) and play an important role in reducing the rate of transpiration.

Root hairs are cylindrical extensions of individual epidermal cells occurring just behind the growing tips of the roots. There is no cross wall between root hairs and the epidermal cells. By increasing the surface area of roots that are in close contact with the surrounding soil particles, root hairs greatly enhance their ability for water and nutrient uptake. The older root hairs normally fall off, and new ones are continuously generated near the growing tip. Therefore, the number of root hairs remains fairly constant as the root grows. The cork cambium replaces the epidermis in secondary growth and produces the bark of a tree trunk or root.

1.3.1.3 Vascular Tissue

Vascular tissue consists of xylem and phloem.

Xylem: Xylem is the main water-conducting tissue in plants. They comprise tube-like vessels, formed from dead, hollow and cylindrical cells (vessel members) that are arranged end to end, and tracheids that taper at the ends and overlap one another (Figure 1.7). Both vessel members and tracheids have thick and lignified secondary walls without living protoplasts when they mature. Lignin is secreted by the cell and deposited on cellulose cell walls before the protoplast dies, leaving only the cell wall. In gymnosperms, water-conducting cells consist of only tracheids in which water passes as continuous stream from the root up through the shoot and then into the leaves. In leaves, much of the water passes in the form of thin film of water outside of the parenchyma cells and then converted into water vapor to diffuse into the intercellular spaces and, ultimately, decipate into the surrounding air through stomata. The process of diffusion of water vapor from a plant surface (leaves, stems and fruits) into the atmosphere is known as transpiration. In addition to serving as a conducting tissue, xylem provides support to the plant body.

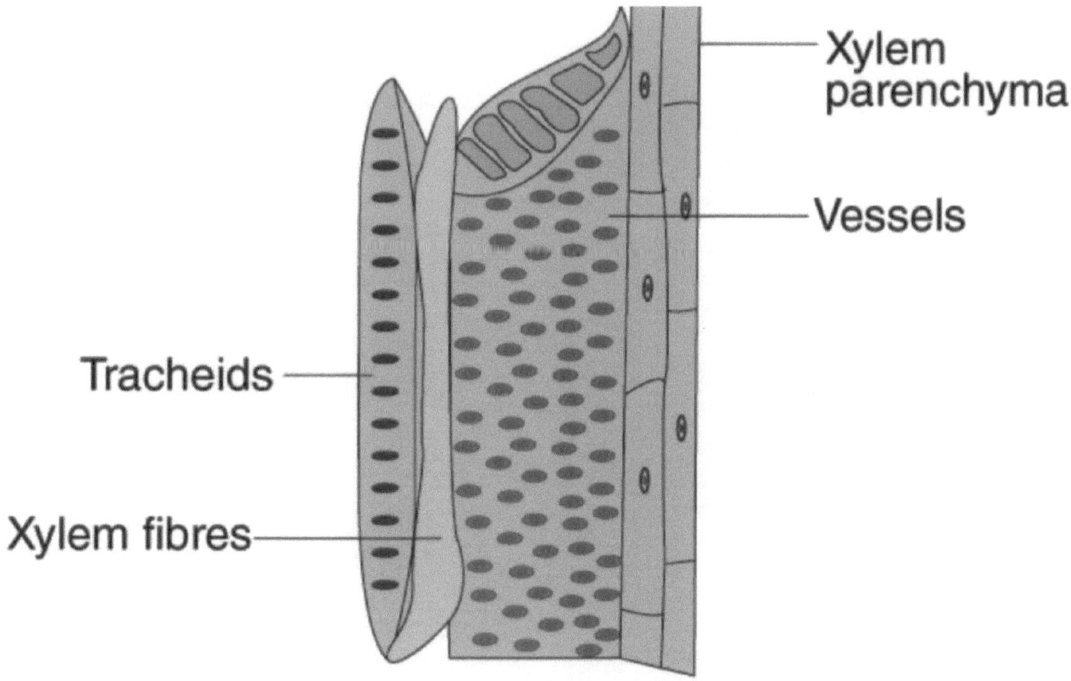

FIGURE 1.7 Xylem tissue.

Primary xylem is derived from the procambium, which arises from the apical meristem, whereas the secondary xylem is formed from the lateral vascular cambium. The secondary xylem forms the wood. When a continuous stream of water flows through tracheids, it passes through small, mostly rounded-to-elliptical areas called pits. Pits do not have secondary wall material and occur opposite one another. In contrast, vessel members are joined end to end to form completely open tubes or strips of wall material across the open ends. Vessels conduct water more efficiently than do the overlapping strands of tracheids. In addition to conducting cells, xylem includes fibers and parenchyma cells (ground tissue). The parenchyma cells, which are usually produced in horizontal rows (rays) by special ray initials of the vascular cambium, function in lateral conduction and food storage. A ray initial is a meristematic cell, which divides to produce another initial and a cell that differentiates into a ray cell. In cross sections of woody stems and roots, the rays can be seen radiating out from the center of the xylem like the spokes of a wheel. The presence of abundant fibers in some woody plants makes them very dense and heavy. As the arrangements of cells in the xylem are characteristic feature of the genera and species, it is possible to use wood characteristics in species identification.

Phloem: Phloem, which consists of two kinds of elongated cells (sieve cells and sieve-tube members), is responsible for distribution of products of photosynthesis from leaves to other organs (Figure 1.8). Both sieve cells and sieve-tube members are living but lack a nucleus at maturity. Sieve-tube members are more specialized than sieve cells. Both sieve cells and sieve-tube members have groups of pores called sieve areas that are more abundant on the overlapping ends of the cells and connect the protoplasts of adjoining sieve cells and sieve-tube members. Some sieve areas have larger pores that are called sieve plates. Sieve-tube members occur end to end and form longitudinal series called sieve tubes. The pores in all sieve areas are roughly of the same diameter, and each sieve-tube member is associated with an adjacent specialized parenchyma cell known as the companion cell. In angiosperms, a common initial cell divides asymmetrically to produce a sieve-tube member cell and its companion cell. Companion cells have nuclei and their numerous cytoplasmic connections

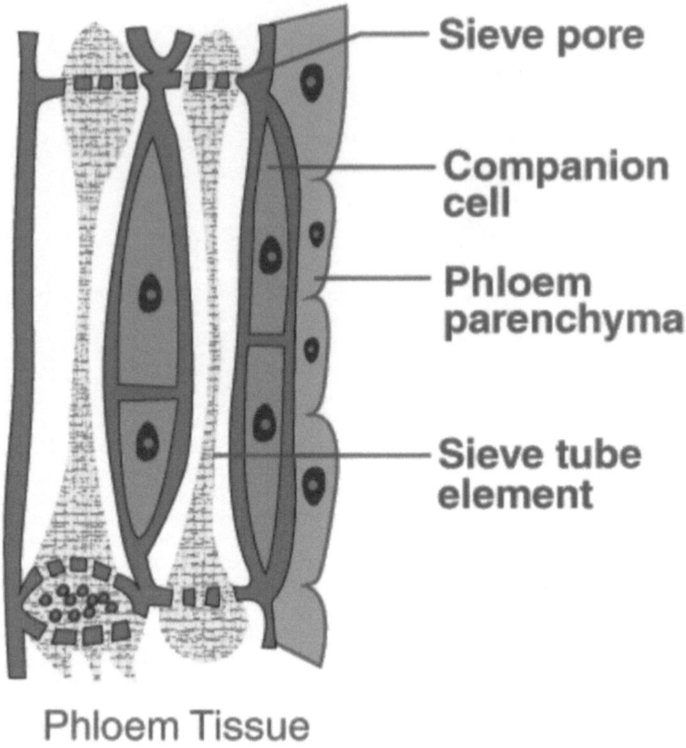

FIGURE 1.8 Phloem tissue.

between adjacent cells (plasmodesmata) to connect their cytoplasm with that of the associated sieve-tube members. Fibers and parenchyma cells are often abundant in phloem. Companion cells carry out metabolic functions that are needed to maintain the associated sieve-tube members.

1.4 PLANT BODY PARTS

External plant parts, such as roots, stems, leaves, flowers, fruits and seeds, are known as plant organs. Each organ contains an organized group of tissues that work together to perform a specific function. These organs can be divided into two groups: vegetative (roots, stems, shoots, nodes, buds and leaves) and sexual/reproductive (flower buds, flowers, fruit and seeds) organs. The vegetative organs are often used for asexual reproduction or vegetative propagation (e.g., cutting propagation, tissue culture or micro-propagation, budding, grafting). The individual plant organs are described in this section.

1.4.1 Structure and Functions of Roots

Following seed germination, the distal portion of the embryo (radicle) forms the root. The developmental progression of a root in carrot is shown in Figure 1.9. The roots lack nodes and never bear leaves or flowers directly. Roots absorb nutrients and moisture, anchor the plant in the soil, support the stem and, in some cases, store food. Roots of some plants are used for asexual propagation (e.g., fig, raspberry, phlox, carrot). Root system has a significant influence on plant size and vigor, adaptation to soil types and response to cultural practices. It also allows plants to cope up with water stress.

FIGURE 1.9 Development of roots in carrot.

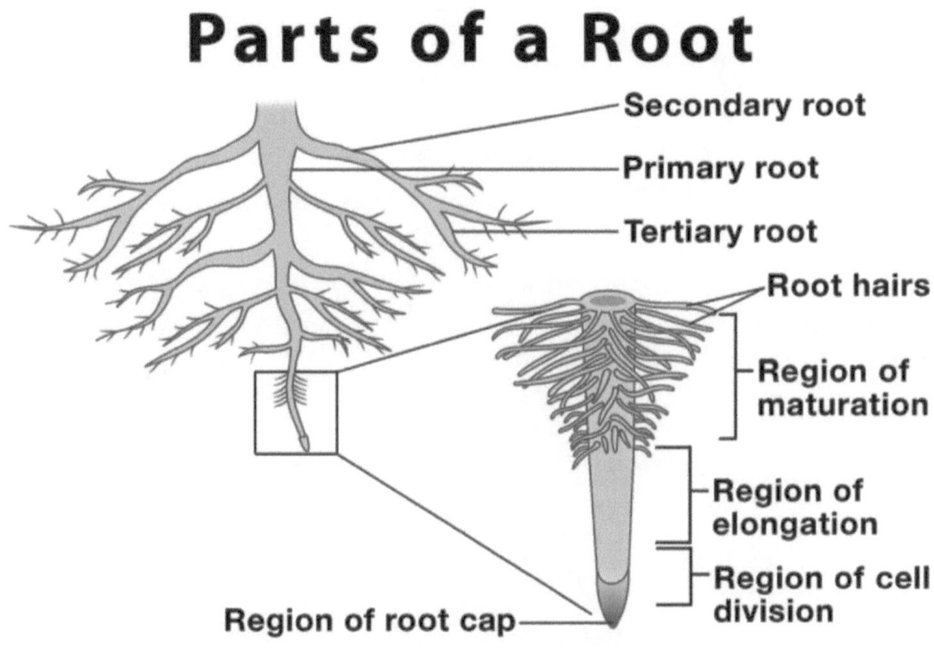

FIGURE 1.10 Parts of roots.

Roots have the simplest pattern of organization and development among the vegetative organs of a vascular plant. The developing roots have four distinct zones (Figure 1.10).

- **Root cap**. The daughter cells arising from the division of apical initials toward the tip of the root form the root cap cells.
- **Zone of cell division**. Also known as zone of meristematic growth, it is located just behind the root cap and is an area of active cell division and growth.
- **Zone of elongation.** This is the area just above the zone of cell division where cells increase in size through elongation or expansion due to the absorption of water and nutrients. As they grow, they push the root through the soil.
- **Zone of maturation**. This zone is above the zone of elongation and directly beneath the stem (crown at soil surface). In this zone, cells differentiate into specialized tissues such as epidermis, cortex or vascular tissue.

Root cap: The root cap consists of two types of cells, the inner column-like cells called columnella and the outer root cap cells that are continuously replenished by the root apical meristem (Figure 1.11). The root cap protects the delicate tissues behind it as roots grow through abrasive soil particles. Golgi bodies in the outer root cap cells release a slimy substance into the surrounding soil, forming a mucilaginous lubricant that helps roots to penetrate through soil particles. The root cap cells, whose average life span is less than a week, are constantly replenished from inside, besides the

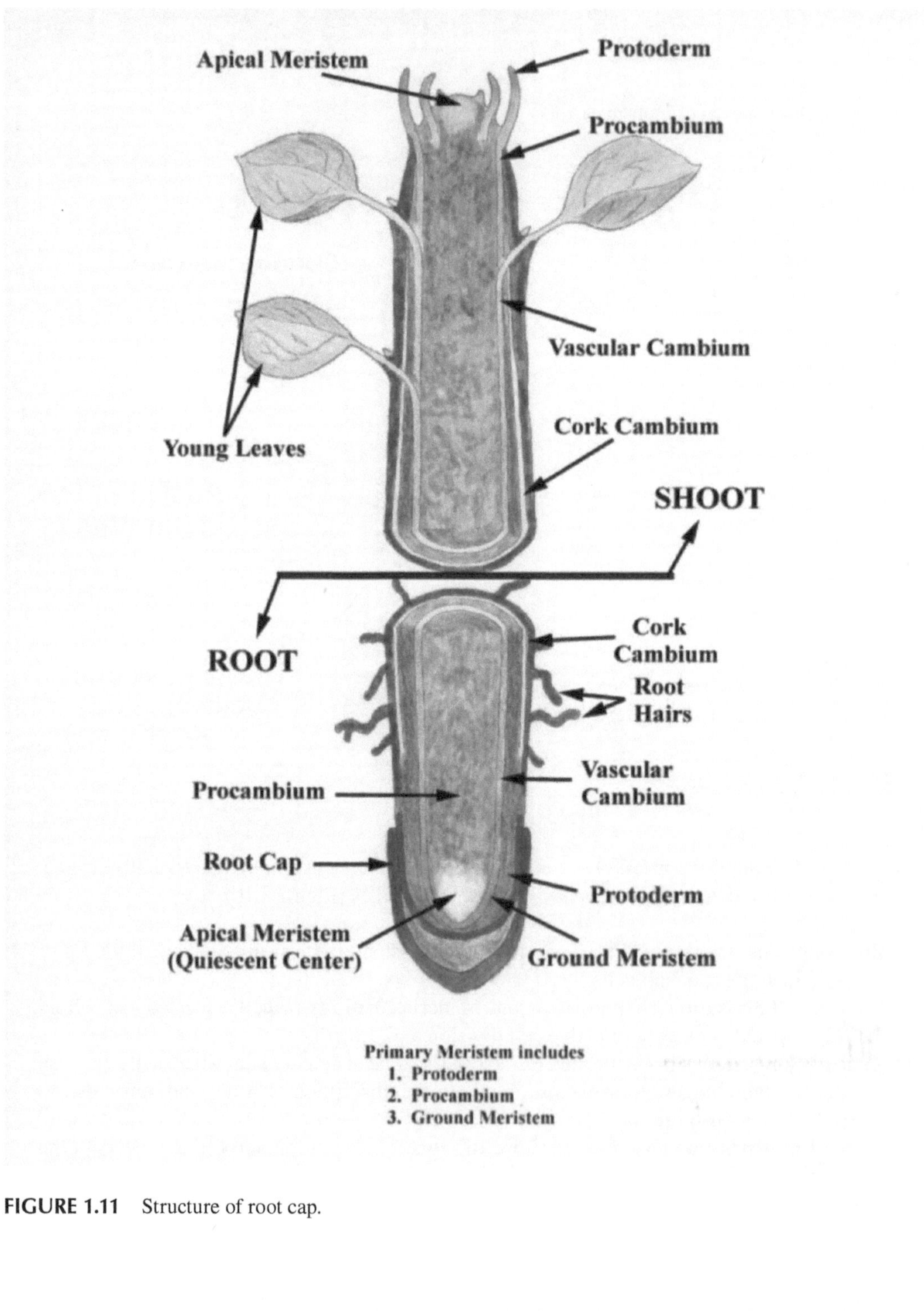

FIGURE 1.11 Structure of root cap.

damaged or accidentally removed outer root cap cells are continuously replaced by new ones. The slimy substance secreted by root cap cells also promotes the growth of beneficial microorganisms, such as nitrogen-fixing bacteria in leguminous roots.

The columnella cells have endoplasmic reticulum in the periphery and the nucleus at either the middle or the top of the cell, but they are devoid of large vacuoles. The columnella cells on the side facing the pull of gravity contain amyloplasts (plastids with starch grains). Columnella cells are highly specialized cells that sense the gravitational signals (gravitropism), although the exact nature of the gravitational response of columnella cells is not fully understood. Some reports suggest that calcium ions in the amyloplasts control the distribution of auxin within the cell. More recent reports suggest that the electrical signal that moves from the columnella cell to the cells in the zone of elongation is responsible for perception of gravity.

Zone of cell division: The zone of cell division is located close to the root tip and protected by root cap (Figure 1.10). The apical meristem is an inverted concave dome located in the center of the root tip close to the area protected by the root cap (Figure 1.12). The cells toward the edges of the dome divide once or twice a day. Most of the cells in this area are cuboidal in shape with a large centrally located nuclei and small vacuole. A group of cells in the central area of the dome divides relatively infrequently to form the quiescent center (QC). The daughter cells of apical meristem subdivide into three primary tissue systems, such as protoderm, procambium and ground meristem. Although the patterning of these cells begins in this zone, the anatomical and morphological expressions of patterning are fully evident only in the zone of maturation. The asymmetric cell division produces two cylinders of cells: the outer cell layer becomes the ground tissue and the inner cell layer forms the endodermis. The ground tissue performs the storage function, whereas the endodermis controls the intercellular flow of water and solutes into the vascular tissues of the root. Cells

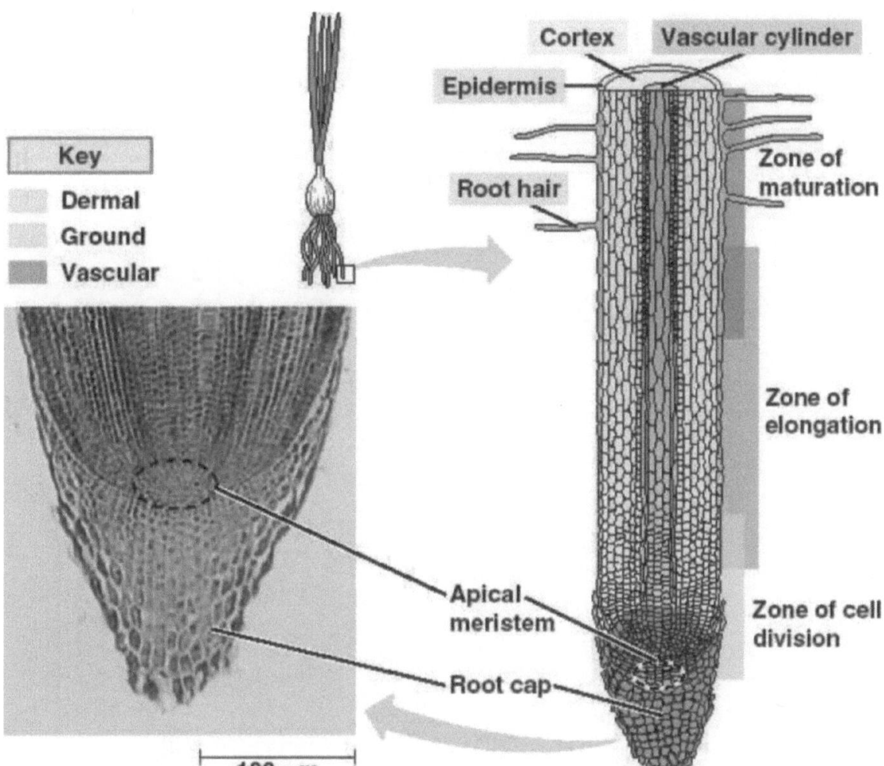

FIGURE 1.12 Zone of cell division in root.

in this region develop according to their position. Meristematic cells (also known as stem cells) are capable of unlimited self-renewal to generate differentiated descendants. In roots, individual stem cells are controlled by direct interaction with cells of the quiescent center that lies in the immediate neighborhood. A robust regulatory network keeps the balance between pluripotent stem cells and differentiating descendants.

Zone of elongation: As the name suggests, the cells in this zone are originally produced by the primary meristems and grow several times longer than wide, although their width also increases slightly (Figure 1.13). Most of the expansion in cell in this region occurs due to the expansion of vacuoles, and in a fully expanded cell, vacuoles occupy nearly 90% of the total volume of the cell. No further increase in cell size occurs above this zone.

Zone of maturation: The elongated cells in the zone of elongation are differentiated and become fully developed into specific cell types in the zone of maturation. The cells at the root surface grow into epidermal cells (Figure 1.14). Most of the epidermal cells may develop protrusions, structures or outgrowths (known as root hairs) that are not separated by a cross wall from the main part of the cell. These root hairs greatly increase the surface area, thereby significantly increasing the absorptive capacity of the root. The root hairs from the older part of the zone of maturation are shed, while new ones are continuously added near the zone of elongation. Symbiotic bacteria that fix atmospheric nitrogen enter the plant via root hairs.

Several layers of parenchyma cells are produced by the ground meristem and lined up just below the epidermis to form the cortex. The cortex is a food storage tissue. The inner boundary of the cortex differentiates into a single-cell layered cylinder of endodermis (Figure 1.15), whose

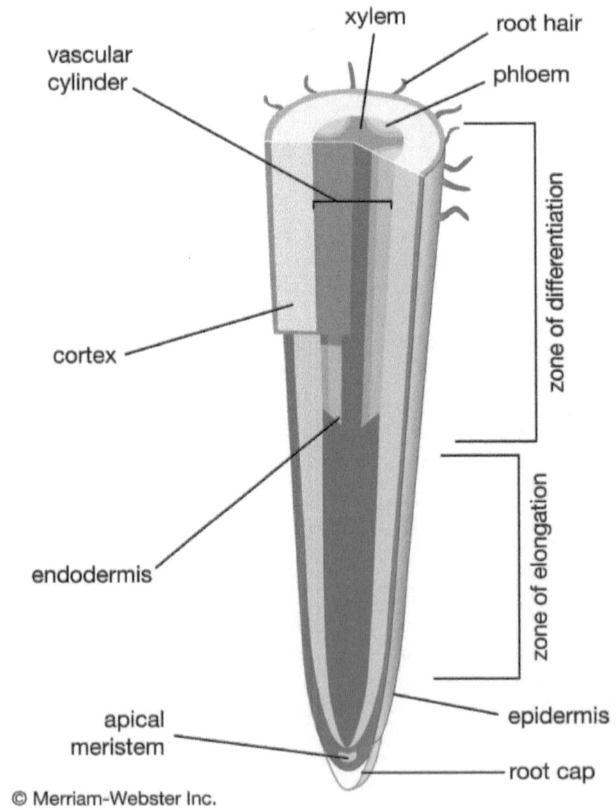

FIGURE 1.13 Zone of elongation.

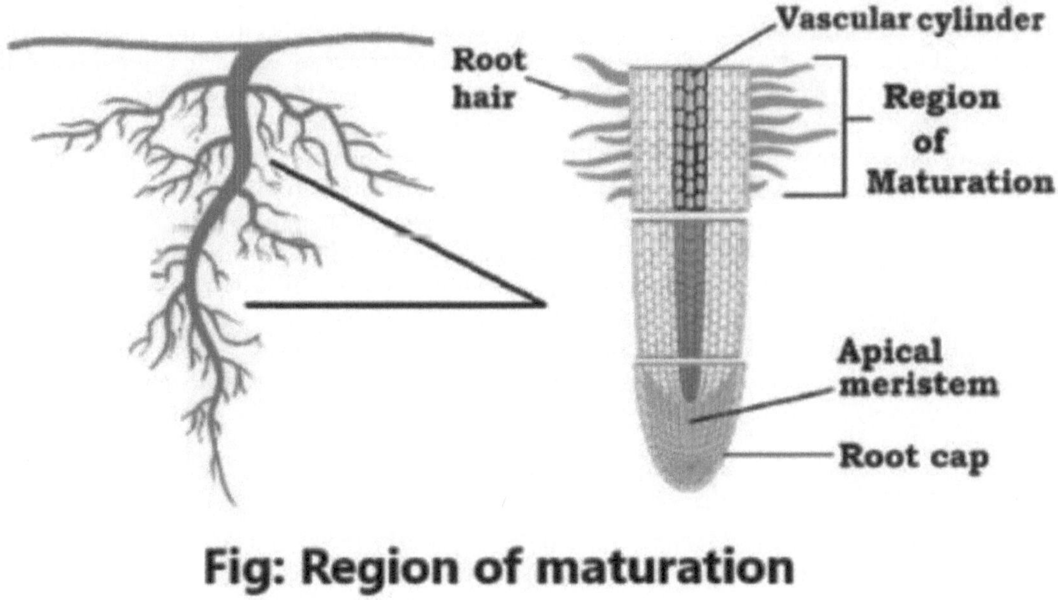

FIGURE 1.14 Zone of maturation in roots.

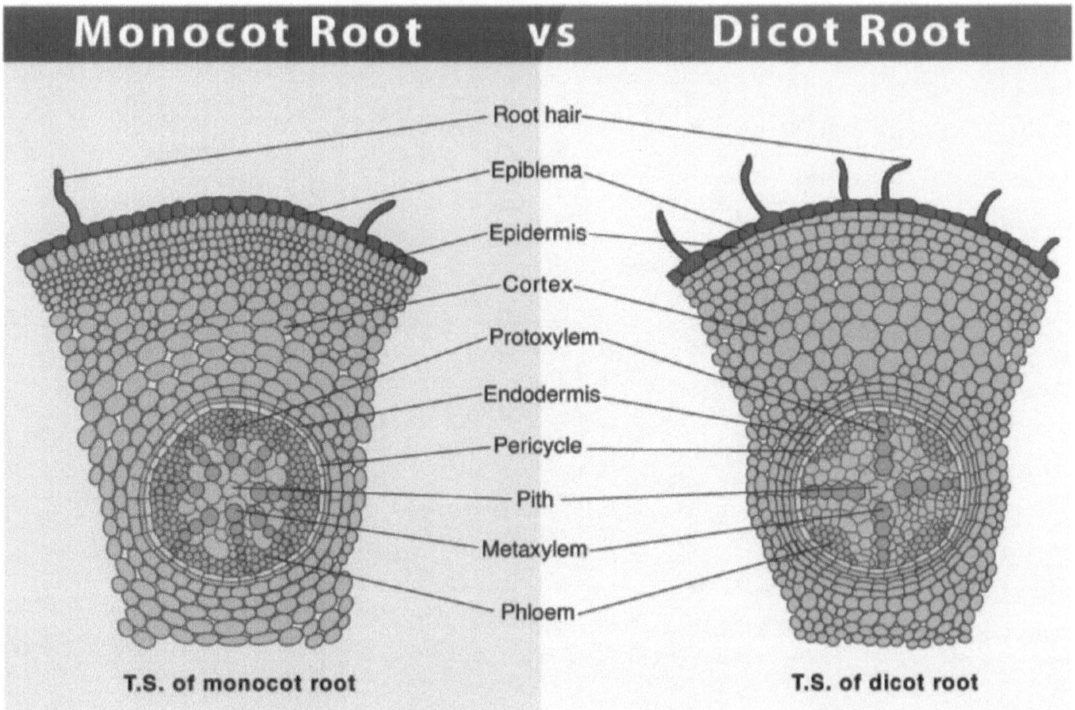

FIGURE 1.15 Monocot -vs.- dicot root.

primary walls are impregnated with a fatty substance called suberin. The suberin produced in bands called the Casparian strips surrounds each adjacent endodermal cell wall to prevent lateral transport of water between cells. The tissues inside the endodermis are called stele. Pericycle cells located immediately adjacent and interior to the endodermis can divide even after they mature to form lateral (branch) roots or become a part of the vascular cambium as in dicots. The water-conducting cells of the primary xylem are differentiated as a solid core in the center of young dicot roots. In monocot (and in a few dicot) roots, the primary xylem is in discrete vascular bundles arranged in a ring, which surrounds parenchyma cells, called pith, at the very center of the root. Primary phloem is differentiated in discrete groups of cells between the arms of the xylem in both dicot and monocot roots. In dicots and other plants with secondary growth, part of the pericycle and the parenchyma cells between the phloem patches and the xylem arms form the root vascular cambium, which starts producing secondary xylem to the inside and secondary phloem to the outside. The primary phloem, cortex and epidermis become crushed and are sloughed off as more secondary tissues are added.

1.4.2 Types of Roots

Roots are grouped into primary and lateral roots. The primary roots originate at the lower end of embryo and forms the taproot that continues to elongate downward. Taproot has limited secondary branching and becomes the central feature of the root system. Most trees and shrubs have taproots. The lateral or secondary roots are the side roots arising from the main root (Figure 1.16).

Fibrous root system is formed when the primary root ceases to grow and numerous lateral roots with similar diameter are produced. These lateral roots branch repeatedly to form the network of feeding roots. While monocots (e.g., grasses, onion, garlic, bulbs) produce a fibrous root system,

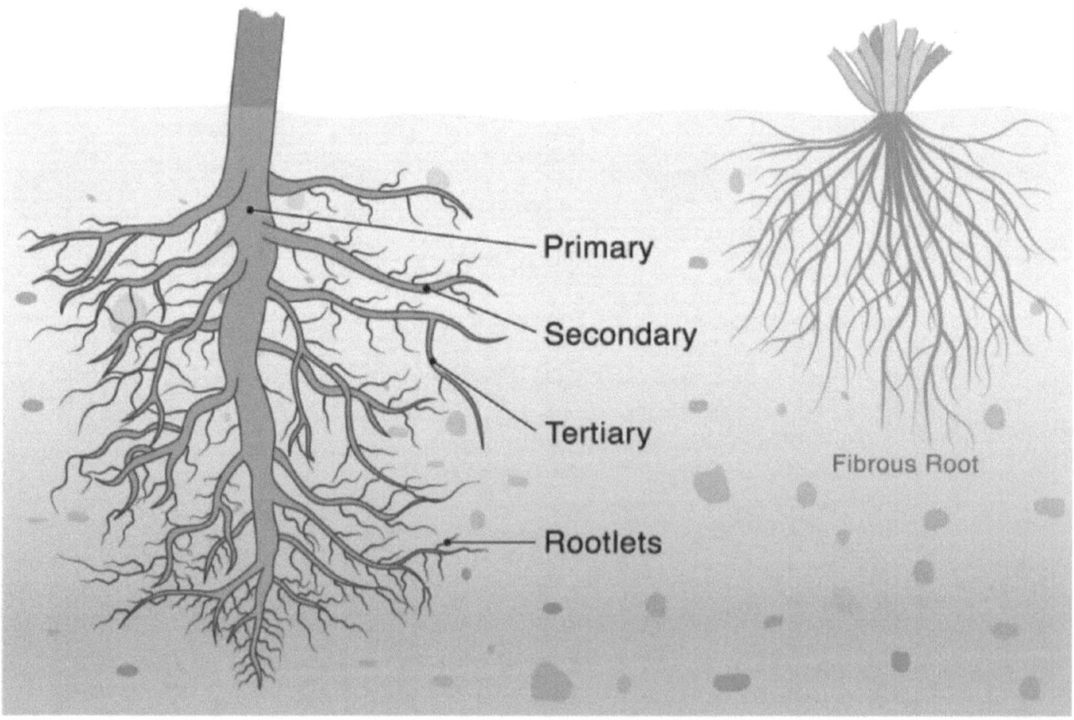

FIGURE 1.16 Types of roots, taproot (left); fibrous root (right).

severing of taproot of dicots encourages the development of fibrous roots. Fibrous roots are also produced at the base of stem cuttings during vegetative propagation (Figure 1.16).

1.4.3 Root Growth

The absorption of water and nutrients by the radicle of the seedling helps in the early growth of roots immediately after seed germination. Once the seedlings are established, the quantity and distribution of roots strongly influence seedling's ability to absorb moisture and nutrients. Both internal (species and varieties, nature of the root, age of the root, metabolic and microbial activities, hormonal levels in roots etc.) and environmental (soil temperature, soil moisture, oxygen levels in the soil, soil texture and soil structure, soil fertility, soil health etc.) factors influence root growth in plants.

1.4.3.1 Modified Roots

In some plants, the roots are modified to perform certain specialized functions, besides their normal functions (providing anchorage and absorption of water and nutrients), while some of the most common root modifications are briefly described in this section.

1.4.3.2 Aerial Roots

Plants such as banyan tree (*Ficus benghalensis*) and epiphytic orchid (e.g., *Dendrobium*, *Cattleya*) have roots that grow in the air (Figure 1.17). These aerial roots consist of several cell layer thick

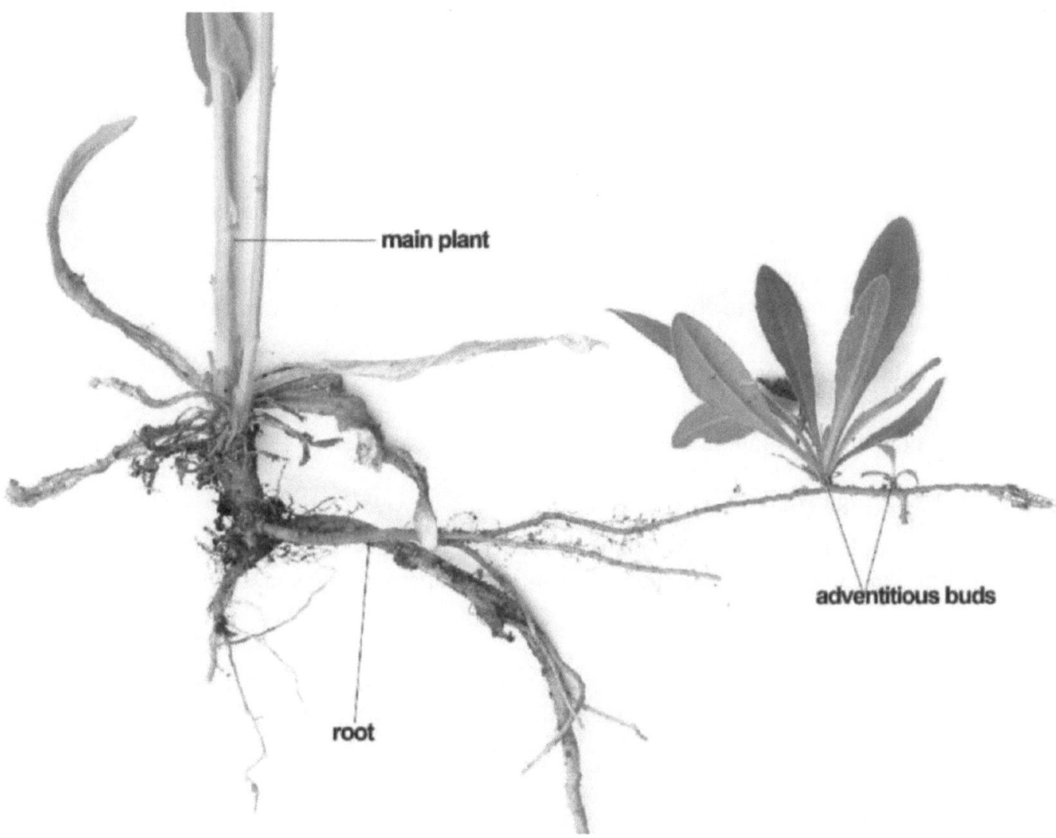

FIGURE 1.17A Aerial roots.

FIGURE 1.17B Aerial roots.

epidermis to reduce water loss. Aerial roots in some plants (e.g., vanilla plant) are green and are capable of photosynthesis, while in others such as mangrove and some tropical trees, aerial roots (prop roots, stilt roots or buttress roots) grow near the stem or trunk above the soil surface and go down to the ground to provide extra protection against wind and avoid dislodging. Climbing plants produce roots from their stems, which help them to cling to tree trunks of host plants or other supports (brick wall).

1.4.3.3 Pneumatophores

Mangrove plants produce pneumatophores or breathing roots. These are spongy outgrowths of underwater roots that facilitate in the supply of atmospheric oxygen to roots.

1.4.3.4 Contractile Roots

Roots of bulbous plants (e.g., *Allium*) contract (reduced) to nearly one-third of their original length as they spiral like a corkscrew to pull the plant a little deeper into the soil.

1.4.3.5 Parasitic Roots

Certain parasitic plants that lack chlorophyll (e.g., mistletoe or dodder) produce peglike roots called haustoria to penetrate and establish contact with the conducting tissues of the host plant to draw food and water from them.

1.4.3.6 Food Storage Roots

Several vegetables have enlarged or swollen roots that act as a storage organ. Best examples of such roots are carrots, beets, turnips, radish and sweet potato.

1.4.3.7 Water-Storage Roots

Some members of the *cucurbitaceae* family that grow in arid areas produce water-storage roots (e.g., *Cucurbita foetidissima*).

1.4.3.8 Buttress Roots

Certain species of fig and other tropical trees produce huge buttress plank roots toward the base of the trunk that provide extra support and stability. These shallow roots remain near the surface and vary in size from few centimeters to cover several meters.

1.4.4 Stem Structure and Types

Stems contain buds and leaves and serve as channels for carrying water and minerals from roots to leaves as well as transport photosynthetic products from leaves to other parts of the plants. Stems orient the leaves to allow them to absorb maximum sunlight.

1.4.4.1 Stem Structure

The stem is an axis from which leaves, buds and branches grow. The pattern of arrangement of leaves on the stem, called the phyllotaxy, could be either spiral around the stem or in pairs opposite one another; they may occur in whorls (circles) of three or more leaves, although spiral arrangement is the most common. The point of attachment of the leaf on the stem is called a node, and the portion of the stem between two consecutive nodes is called the internode. A leaf normally has a flattened surface called the blade (or the lamina) and a petiole (stalk) that attaches the leaf to the stem. The leaf without the petiole is called sessile. The angle between a petiole and the stem is the axil, and the bud in the axil is known as the axillary bud. Axillary buds develop into either branches or meristems that will produce flowers. Young stems are normally green with photosynthetic ability, and in herbaceous plants, stems commonly contain stomata and may have various types of trichomes (hairs). Stems in herbaceous plants die back after flowering and seed maturation. In contrast, woody stems in shrubs and trees persist over several years. The bud at the tip of the stem is called the terminal or the apical bud. These buds usually extend the length of the stem during the growing season. Terminal buds in most woody plants are covered with scales during winter to protect them from freezing temperatures. These scales drop off when normal weather returns leaving tiny bud scars. A pair of appendages called stipules (part of the leaf) develops at the base of some leaves. These stipules fall off leaving stipule scars. In deciduous plants, leaves drop in the fall and new leaves appear in the spring. The leaves after drop off leave scars on the stem.

The apical meristem produces primary tissues that are responsible for increasing the length of the stem. It gives rise to three types of meristems, the protoderm, which produces the epidermis, the ground meristem, which produces parenchyma cells, and the procambium, which produces cylinders of primary xylem and phloem. While the parenchyma cells in the center of the stem constitute the pith, those away from the center constitute the cortex (Figure 1.18). These spaces in the main cylinder of conducting tissues are called gaps. In dicots, a vascular cambium develops between the primary xylem and primary phloem, whereas in monocots, stems do not develop vascular cambiums and, consequently, do not produce secondary growth.

The vascular cambium cells divide to produce secondary xylem inside and phloem to the outside. In woody trees, the production of xylem is extensive. Rings in the wood of a tree reveal annual patterns of growth. In woody dicots, cork cambium that develops in the outer cortex produces cork cells to the outside and parenchyma-like phelloderm cells to the inside. The cork cambium, cork and phelloderm are collectively called the periderm. The outer bark is formed by cork tissues that are impregnated with suberin. The unsuberized cells produced by cork cells that permit gas exchange are called lenticels.

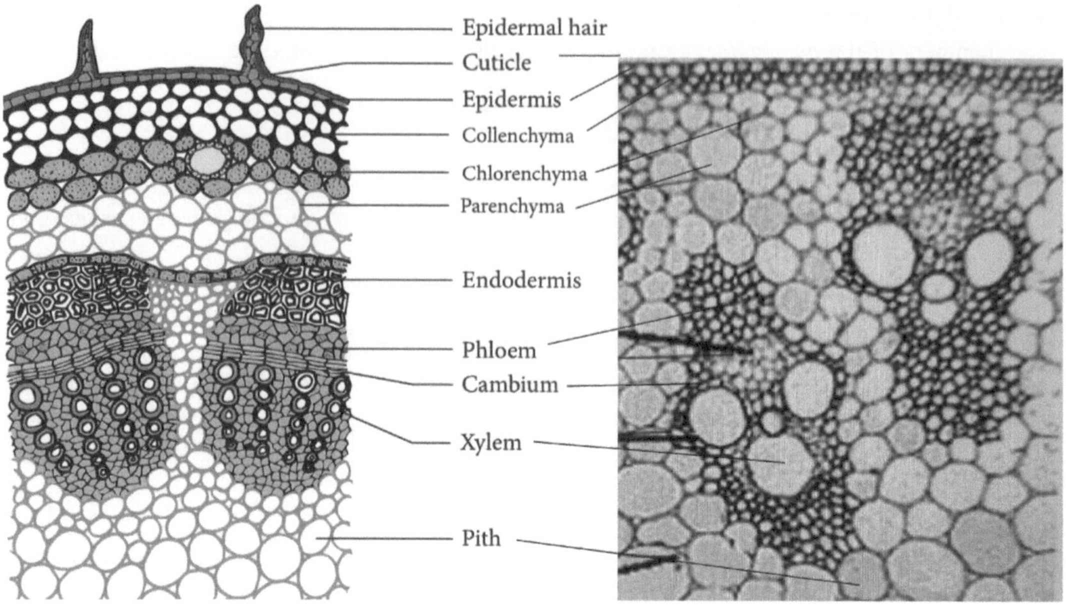

FIGURE 1.18 Transverse section of dicot (left) and monocot (right) stems.

TABLE 1.3
Comparison of Cross Section (CS) of Dicot and Monocot Stems

Features	Dicot stem Example: bean, tomato, cucumber	Monocot stem Example: banana, grass, onion
1. Epidermal trichomes	Present, multicellular	Absent
2. Hypodermis	Collenchymatous	Sclerenchymatous
3. Ground tissue	Differentiate into cortex and pith	Undifferentiated
4. Endodermoid layer and pericycle	Present	Absent
5. Stele	Eustele (xylem, cambium and phloem arranged in a ring of collateral bundles)	Atactostele (vascular tissue exist in scattered bundles)
6. Vascular bundles	Many open, wedge shaped	Numerous, closed, oval shaped
7. Lysigenous cavity	Absent	Present
8. Bundle cap	Present	Absent
9. Bundle sheath	Absent	Present
10. Secondary growth	Occurs later	Does not occur

The similarities and differences between cross sections (CSs) of monocot and dicot stems are presented in Table 1.3.

1.4.4.2 Modified Stems

The stems in certain plants are modified to perform special functions. These modified stems have nodes, internodes, leaves and buds, although some of them may be completely altered, modified or reduced as a means of adaptation to prevailing conditions. The most common modified stems are described in the following.

Bulb: The swollen underground stems of onion, garlic, lily etc. are large buds with adventitious roots at the base. These bulbs consist of fleshy leaves attached to a small, very compressed stem called the disc. In bulbs like onion, the fleshy leaves inside are surrounded by papery, scale-like leaves outside.

Corm: Corm is an underground stem without fleshy leaves. Instead, it consists of a stem with a few papery, nonfunctional leaves outside and adventitious roots below. Crocus, gladiolus and freesia are the best examples of corm.

Rhizome: Rhizomes are horizontal stems that grow underground, often close to the soil surface. While small scale-like leaves with an axillary bud are present at nodes, normal photosynthetic leaves occur at the tip. Adventitious roots are produced on the lower side of the node throughout the length of the rhizome. Perennial grasses (e.g., *Cyperus*), turmeric and ginger produce rhizomes.

Runner and Stolon: Runners and stolons are stems with long internodes usually growing on the surface of the ground. Runners grow along the surface and are horizontal. The stolons are similar to runners but have longer internodes and are arched that grow aboveground. Strawberry and grasses (*Cynodon*, *Paspalum*, etc.) produce runners and stolons, respectively.

Tubers: The underground swollen stems as in potato that act as a storage organ are called tubers. The "eyes" on tuber are axillary buds formed in the axils of scale-like leaves. The scale-like leaves drop off leaving a leaf scar. The examples of a tuber are potato, yam, cassava, artichoke, arrowroot, water chestnut and dahlia.

Tendrils: In several climbing plants (e.g., grapes, English ivy and *Ficus pumila*), stems are modified into tendrils that help them in climbing. Some tendrils, such as those in peas and pumpkins, are modified leaves or leaflets.

Cladophylls: The flattened, photosynthetic stems in cacti are called cladophylls. These structures resemble leaves, but the real leaves in these plants are reduced to spines.

1.4.5 Leaf Structure

Leaf, the seat of photosynthesis in plants, emerges as primordium and grow mostly by cell enlargement, although limited cell division does occur. Because the growth stops when they mature, leaves are considered determinate plant structures. The arrangement, form, size and internal structure play a pivotal role in the functions they perform. The leaves with one small vein are called the microphyll and is normally associated with the phylum, *Lycophyta*. However, in majority of plants, leaves have several veins and are called megaphylls. In most dicot plants, leaves have flattened blades formed by a shift from radial symmetry to dorsal-ventral (top-bottom) symmetry. The slender stalk that attaches the leaf to stem is called petiole. The place where the leaf is attached to the stem is called the node. In addition to petiole, a pair of stipules may be present at the base of the petiole. The stipules may look more like leaves or modified into spines or glands. They vary considerably in size from microscopic to almost half the size of the leaf blade. Leaves in grasses and other monocot plants usually lack a petiole but instead have sheath that covers the stem toward the base. Veins represent xylem and phloem and are distributed throughout the leaf blade. The main veins run parallel in most monocot leaves, whereas they form a reticulate network in dicots. Leaf blades may be oval to deeply lobed to having separate leaflets. In simple leaves, the blades are undivided and may have teeth, indentations or lobes of various sizes. In contrast, the leaf blade is divided into leaflets in compound leaves. The leaf is called pinnately compound if the leaflets are arranged in pairs along a common axis (rachis). However, in palmately compound leaves, leaflets radiate out from a common point at the blade end of the petiole. Both simple and compound leaves may be arranged alternately, in opposite pairs or in whorls.

The entire surface of a blade has a transparent epidermis that is devoid of chloroplasts but may be covered by waxy layer of cuticle of variable thickness or have different types of glands and trichomes (hairs). The lower epidermis (and occasionally the upper epidermis) of most leaves contains numerous stomata. These stomata are bordered by guard cells. The tissue between the upper and lower epidermis (mesophyll) is interspersed with veins (vascular bundles). In most dicot leaves, there are two distinct types of mesophylls, palisade mesophyll containing one to two rows

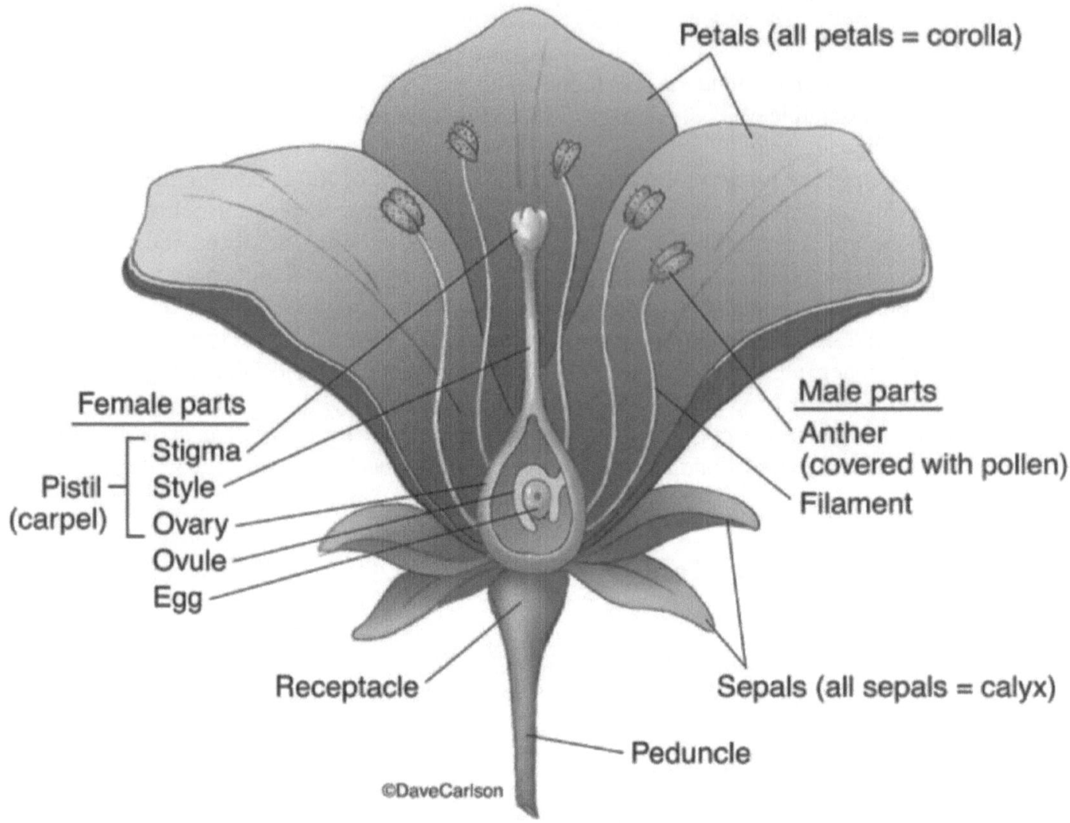

FIGURE 1.19 Transverse section of flower.

of tightly packed chlorenchyma cells (parenchyma with chloroplasts) close to the upper epidermis and the spongy mesophyll cells between the palisade mesophyll and the lower epidermis. These cells have many air pockets. The interconnected intercellular spaces along with the stomata are involved in gas exchange and the passage of water vapor from the leaves to the atmosphere. Since the mesophyll of monocot leaves is not differentiated into palisade and spongy layers, there is no distinction between the upper and lower epidermis. The structure of dicot and monocot leaf is depicted in Figure 1.19.

1.4.5.1 Modified Leaves

Leaves have evolved some remarkable adaptations to cope with the prevailing conditions in their immediate surroundings.

Floral leaves (bracts): Plants like poinsettia and *Bougainvillea* produce modified leaves, called bracts that are brightly colored. The true flower, which are tiny and inconspicuous are enclosed by these bracts.

Spines: The leaves of many cacti and berries are modified into spines. These spines are different from thorns (modified stems) or prickles (simply outgrowths of epidermis or the cortex).

Reproductive leaves: Plants like *Kalanchoë* and fern (*Asplenium rhizophyllum*) produce tiny plantlets along their leaf margins or tips of fronds (Figure 1.20). Each plantlet after they are separated from the leaf is capable of growing into independent full-sized plant.

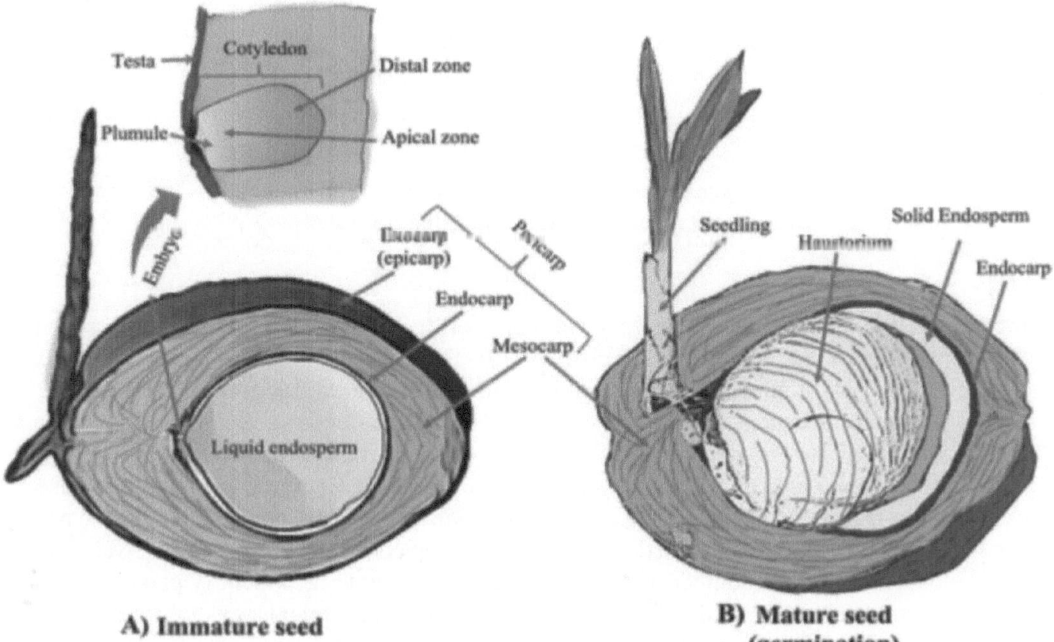

FIGURE 1.20 Transverse section of mango fruits and cross section of coconut.

Window leaves: Several plant genera growing in arid regions produce succulent, cone-shaped leaves with transparent tips. These leaves have thick epidermis and cuticle. They allow light to the hollow interiors, which is utilized for photosynthesis beneath the surface of the ground.

Shade leaves: Leaves produced under shade in some plants tend to be larger and thinner than those produced under direct light.

Insectivorous leaves: The leaves in plants that grow in acid swamps are modified to trap insects. Pitcher plants (*Sarracenia, Darlingtonia, Nepenthes*) have cone-shaped leaves with stiff, downward-pointing hairs at the rim. An insect falling into these leaves finds it very difficult to escape and eventually get trapped (Forterre et al., 2005). Other plants, such as sundews (*Drosera*), have glands that secrete sticky mucilage to trap and digest insects that fall on these leaves. The venus flytrap (*Dionaea muscipula*) produces leaves that look hinged at the midrib. When a moving insect stimulates tiny hairs on the leaf blade, the two halves of the leaf snap shut trapping the insect and digestive enzymes break down the soft parts of the trapped insect.

1.4.6 Buds

Bud is an underdeveloped shoot, which, when provided with right condition, is capable of producing leaves, flowers and fruits. In many plants, buds remain dormant under extreme temperature conditions and require exposure to critical temperature before they can sprout and produce new growth. Leaf buds are composed of a short stem with embryonic leaves and often are less plump and more pointed than flower buds. A flower bud, on the other hand, consists of a short stem with embryonic flower parts.

Buds located at the stem apex are called as terminal buds, while those found in the leaf axils are known as lateral buds. In some cases, each axil may have more than one bud. Adventitious buds arise at sites other than the terminal or axillary position. Adventitious buds allow stem, leaf and root cuttings to develop into new plants. Buds in some plants are enlarged to form the edible portion (e.g., Asparagus, artichoke).

1.4.7 Flower

Flower is the reproductive organ of the plant involved in sexual reproduction. The fragrance and nectar present in the flower attract insects and other pollinators.

1.4.7.1 Structure

Flowers contain a stamen (male flower part) and/or pistil (female flower part), plus accessory parts such as sepals, petals and nectar glands (Figure 1.20). The stamen consists of a pollen sac (*anther*) and a long supporting filament. The filament holds the anther in proper position, making the pollen available for dispersement by wind, insects or birds. The pistil consisting of a stigma, style and ovary is normally located in the center of flower. The ovary contains eggs inside ovules. The fertilized egg develops into a seed. Sepals are small, green, leaf-like structures located at the base of a flower. They protect the flower bud. Collectively, the sepals are called a calyx. Petals are the brightly colored portions of a flower. Like nectar glands, petals may contain perfume. Collectively, the petals are called a corolla. The number of petals on a flower is used to help identify plant families and genera. Flowers of dicots typically have four or five sepals and/or petals, or multiples thereof. In monocots, these floral parts typically come in three or multiples of three.

A complete flower has a stamen, pistil, petals and sepals, whereas the absence of any one of these parts makes the flower incomplete. In contrast, a perfect flower contains both the reproductive organs, i.e., functional stamens and pistils, although it may not contain petals and sepals. On the other hand, an imperfect flower has separate male and female flowers at different locations either on the same plant (monoecious) or on different plants (dioecious). In the case of dioecious plants, both male and female plants must be planted in close proximity for pollination to occur.

1.4.8 Types of Inflorescences

Plants produce either one (solitary flower) or a cluster of flowers (inflorescence) on each stem (each flower in the inflorescence is called a floret). Inflorescences are of two types: (i) racemes (florets start to open from the base and progress toward the tip) and (ii) cymes (the floret at the top opens first and progresses downward along the stem).

1.4.9 Fruits

Fruits comprise fertilized ovules (seeds) plus the ovary wall (Figure 1.21). Fruit may be fleshy (apple, pear, mango, banana, mangrove) or dry and hard (nuts). In some fruits, the seeds are enclosed within the ovary (e.g., mango, apples, peaches, oranges, squash and cucumbers), while in others, seeds present outside of fruit tissue (e.g., cashew nut and strawberries). The seed contains the genes from both the male and female parent, whereas other tissues in the fruit are developed from maternal tissues only.

1.4.10 Types of Fruit

Fruits can be simple (developing from a single ovary), aggregate (developing from a single flower but with several ovaries) or multiple (derived from a tight cluster of separate, independent flowers borne on a single structure). Simple fruits include fleshy fruits (mango, cherries and peaches [drupe]; pears and apples [pome] and tomatoes; black berry, strawberry [berries]) and dry fruits with either papery (almond, pistachio, beans, peas, etc.) or leathery hard wall (cashew nut, walnut).

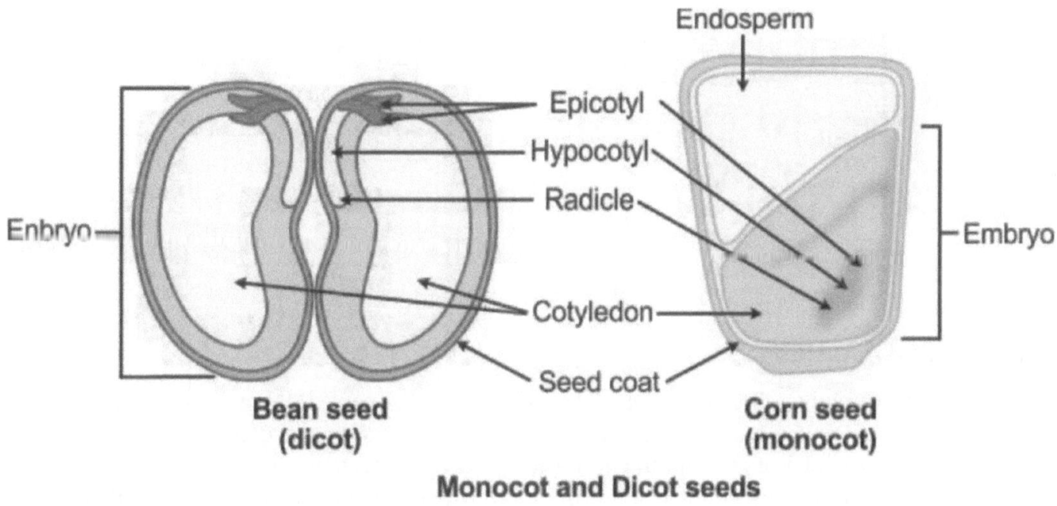

FIGURE 1.21 Transverse section of dicot and monocot seeds.

1.4.11 Seeds

A seed is embryonic plant containing all of the genetic information needed to develop into a complete plant. It can be divided into three parts: the embryo (a miniature plant in a dormant state of development), the endosperm and cotyledon (food storage tissue and the seed coat or testa [an outer hard covering that protects the seed from disease and insects as well as prevents water from entering the seed]). Various processes involved in the seed (development, dormancy and germination) are covered in other chapters.

REFERENCES

Forterre Y, Skotheim J, Dumais J and Mahadevan L. 2005. How the Venus flytrap snaps. *Nature* **433**: 421–425.
Graham L E, Graham J M and Wilcox L W. 2003. *Plant Biology*. Prentice Hall, Upper Saddle River, NJ.
Hirasuna D. 1985. *Vegetables*. Chronicle Books, San Francisco.
Mauseth J D. 1988. *Plant Anatomy*. Benjamin Cummings Publishing, Menlo Park, CA.
Mauseth J D. 2003. *Botany: An Introduction to Plant Biology*. 3rd edition. Jones and Bartlett, Sudbury, MA.
Raven P H and Johnson G B. 2001. *Biology*. 6th edition. The McGraw Hill Global Education Holding Ltd., New York, USA.

2 Embryogenesis and Early Development of Horticultural Plants

2.1 INTRODUCTION

In the majority of plants, after penetrating the degenerated synergid, the pollen tube releases the two sperm into the embryo sac, where one fuses with the egg to form a zygote and the other fuses with the two polar nuclei of the central cell forming the endosperm nucleus. This process is called the double fertilization because the true fertilization, which is the fusion of a sperm with an egg, is accompanied by another fusion process that of a sperm with the polar nuclei, which also resembles fertilization (Figure 2.1). Double fertilization is unique to angiosperms. The zygote is a diploid (i.e., having two sets of chromosomes), whereas the endosperm nucleus is a triploid (i.e., having three sets of chromosomes). The endosperm nucleus divides mitotically to form the endosperm of the seed, which stores food to be utilized by the developing embryo and subsequently by the germinating seed.

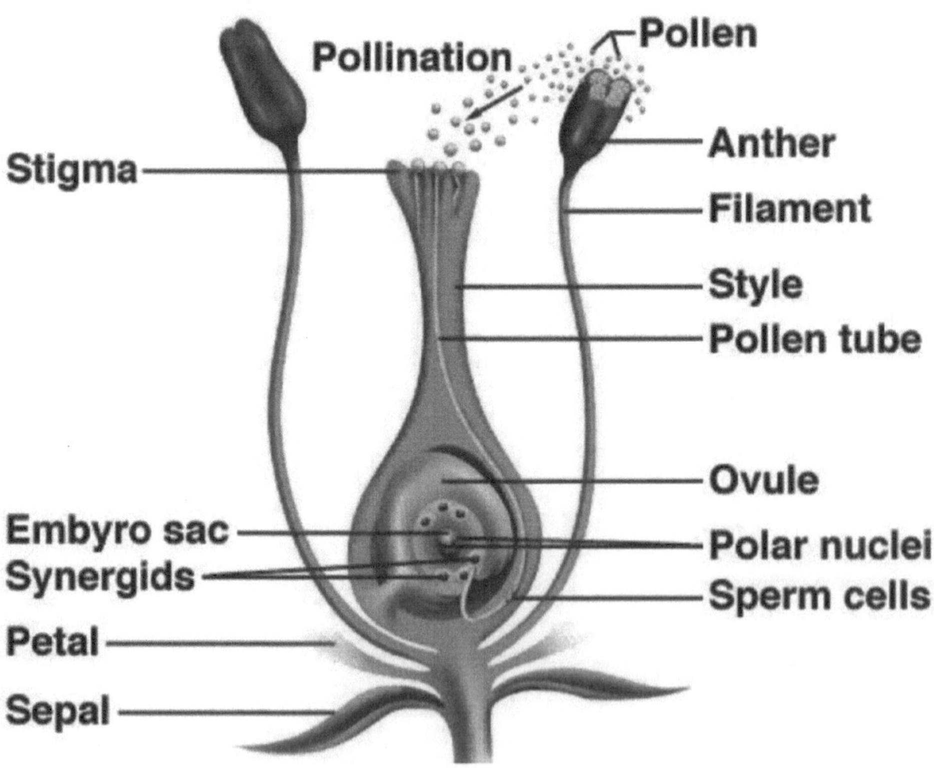

FIGURE 2.1 Double fertilization in higher plants.

DOI: 10.1201/9781003373889-3

2.2 EMBRYOMORPHOGENESIS

Embryogenesis begins with the formation of the unicellular zygote in the ovule. The zygote goes through species-specific cell division and histodifferentiation programs to form a morphologically complete embryo with shoot and root meristems, which, after germination, produce new plant organs (an embryonic stem or hypocotyl) and one or more cotyledons (embryonic leaves). The process of initiation and development of an embryo from the zygote is termed as zygotic embryogenesis (Figure 2.2a). As Goldberg and coworkers stated, the main purposes of embryogenesis in higher plants are to specify meristems and the shoot–root plant body pattern, differentiate the primary plant tissue types, generate a specialized storage organ essential for seed germination and seedling development and enable the sporophyte to lie dormant until conditions are favorable for postembryonic development (Goldberg et al., 1994).

Several plants have the ability to form embryos (somatic embryos) from cells other than the zygote, a trait unique to the plants and is known as "totipotency." There are some other plants, in which seeds are produced asexually from maternal tissues (ovule) without undergoing meiosis and double fertilization. This phenomenon is known as "apomixes." Offsprings produced apomictically are genetically identical to the maternal plant. Similarly, vegetative tissues can be induced to undergo somatic embryogenesis to form embryos *in vitro* (outside the living systems or under controlled environments), and the resulting embryos will have the same ploidy level and genetic composition as the donor plant (or the explants; Figure 2.2b). Alternatively, either male or female gametes or their associated accessory cells can be induced to form embryos ("gametophytic embryogenesis"). As these cells are derived post-meiotically, the resulting embryos represent the haploid progeny of the parent plant. Because of the large number of male gametes produced by a single anther compared to a single female gametophyte or ovule, haploid embryo induction from the male gametophyte is more common than that from the female gametophyte (ovule). Many different terms are used to describe gametophytic embryogenesis, including androgenesis, microspore embryogenesis and pollen embryogenesis. These refer to the *in vitro* culture of the immature male gametophyte, regardless of the developmental stage of the cells that form embryos. The haploid embryos from microspore embryogenesis can be used to produce mature plants, although these plants will be sterile. Normal ploidy (diploidy) can be restored through chromosome doubling, which could occur either spontaneously in culture or induced by the application of chromosome doubling agents such as colchicine. The resultant progeny will be able to reproduce normally.

The embryogenesis establishes the basic plant body (axis formation) and major organ systems of the plant as it prepares the embryo for dormancy and/or subsequent germination. For this purpose, both zygotic and somatic embryos go through distinct developmental stages involving an orderly sequence of cell division, enlargement and differentiation (Figure 2.2a, b). The embryonic development involves three overlapping stages: a period of morphogenesis, histogenesis and maturation phase. This is followed by the acquisition of desiccation tolerance and the loss of water coinciding with the onset of developmental arrest. The morphogenesis and histogenesis phases involve a high rate of cell division and differentiation leading to the establishment of basic body plan with major organs and tissue systems. This is followed by the maturation phase, which is characterized by cell expansion and the establishment of accessible food reserves. The fully formed embryos develop tolerance to desiccation stress by losing moisture and entering into a state of dormancy by arresting further development. The basic body plan of the "sporophyte" is established during embryogenesis; however, this plan is repeated and elaborated after dormancy is broken. The three stages of embryogenesis are described in the following.

ESTABLISHMENT OF THE PLANT BODY

The egg cell is highly polarized at the time of fertilization (Figure 2.3). The large vacuole moves toward the micropylar end (opening where the sperm tube enters the egg cell), and the nucleus and

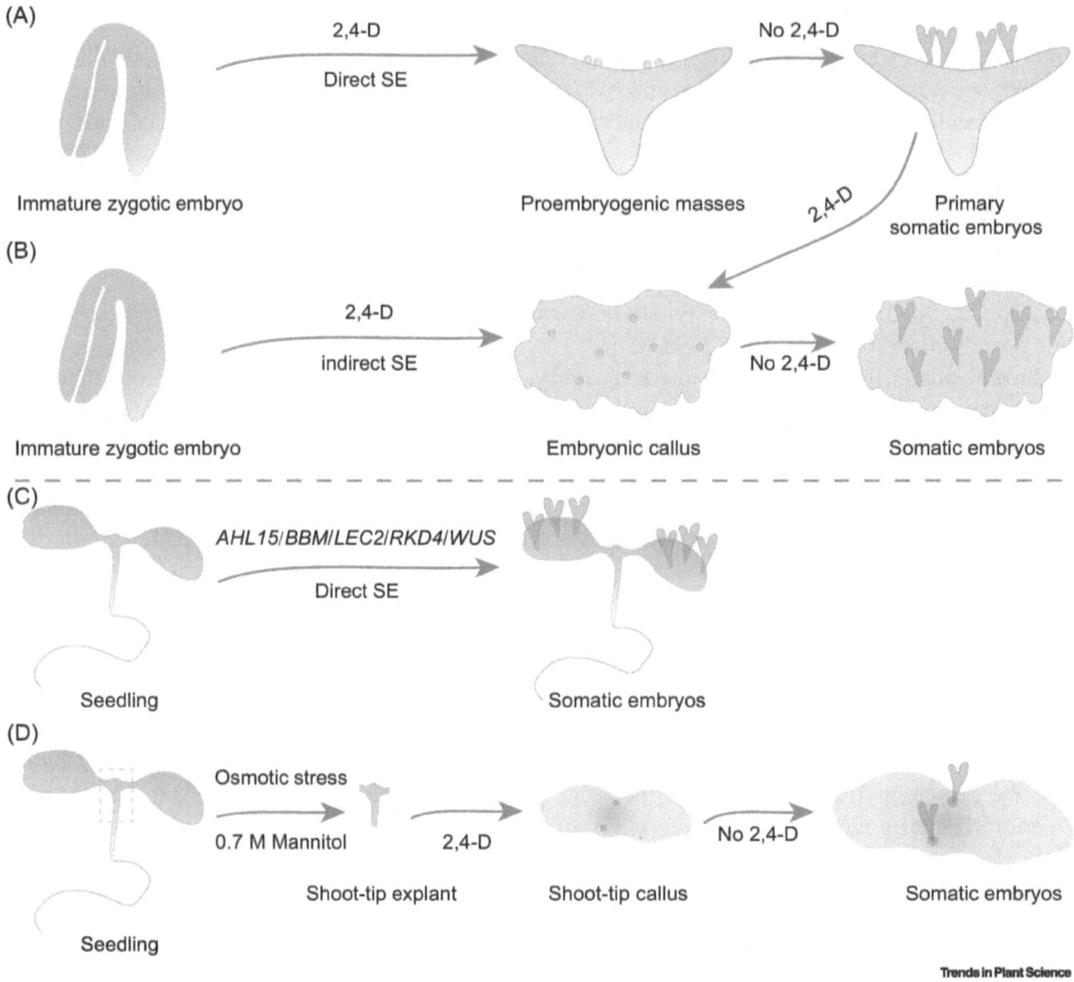

FIGURE 2.2 Zygotic embryogenesis and Somatic embryogenesis.

most of the cytoplasm contents position toward the chalazal end. Even after fertilization, this polarity is maintained in the zygote. The first cell division, which is asymmetric, occurs perpendicular to the axis of polarity. The apical cell receives most of the cytoplasm, while the basal cell is left with vacuole. The basal cell forms the suspensor and divides transversely to form a longitudinal file of cells and push the embryo proper into the nutrient-rich endosperm. Only the distal-most cell of the suspensor will contribute to the embryo proper.

The apical cell divides longitudinally, and the resulting two daughter cells divide again longitudinally to form a quadrant. These four cells then divide transversely to form an octant (eight cells). The transverse walls of these four cells divide the embryo in half along what is known as the "O" line. Then, all cells divide periclinally, and the first histologically distinct tissue, the protoderm, is formed. This stage is called the "globular" or "dermatogen" stage. The protodermal cells divide predominantly anticlinally to form a distinct cell lineage. The inner cells divide both anticlinally and periclinally to produce cell mass. At this stage, the lipid transfer protein gene begins to express in the protodermal cells. The distinct patterns of cell division and differential gene expression reflect the start of histological differentiation.

The two groups of cells divide periclinally leading to formation of bulges that emerge as cotyledon lobes. This represents a shift from radial to bilateral symmetry and marks the transition to

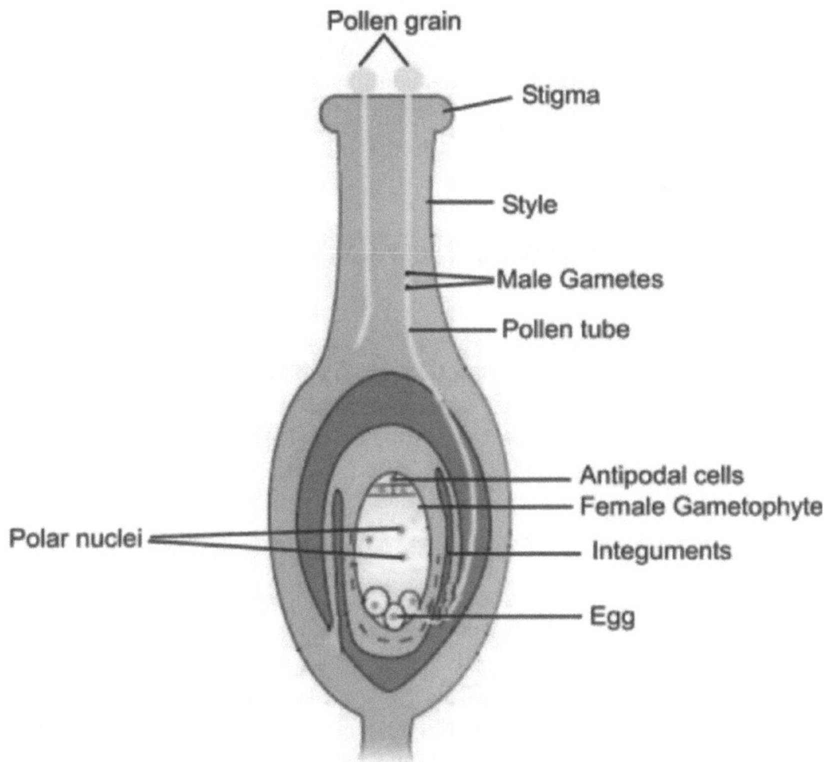

FIGURE 2.3 Embryosac in higher plants.

"heart stage." This also delineates the main embryonic organ systems (cotyledons and shoot axis). The root apical meristem (RAM) begins to take shape at this stage. The apical-most suspensor cell called the hypophysis becomes incorporated into the embryo proper. Both the hypophysis and apical cell derivatives contribute to the formation of the RAM. At this stage, procambium initials are evident. In the early heart stage, the differentiation of major tissue systems will be initiated. Continued cell division, growth and differentiation mark the beginning of the late-heart stage and beginning of the "torpedo" stage. Differentiation of the shoot apical meristem (SAM) is less obvious than the RAM. As the cotyledon lobes begin to grow, a group of cells between them divides infrequently, and the genes for storage proteins begin to express in the cotyledons and subtending tissues. The group of quiescent cells does not express these genes.

AXIS FORMATION

Plant tissues contain an inherently fixed polarity. Stem segments retain their original apical-basal polarity as is evident from the fact that when a stem segment is grafted in reverse orientation to a plant, roots will form at the original basal end.

2.3 REGULATION OF EMBRYOMORPHOGENESIS

Polar transport of auxins is needed for the establishment of bilateral symmetry. This was proven by treating the somatic embryos with the polar transport inhibitor of auxin, which led to failure of treated embryos to progress to the next stage of morphogenesis (globular embryos continue spherical growth without axis elongation or cotyledon formation). Abscisic acid (ABA) has been shown to control embryogenesis and maturation. ABA-deficient mutants show vivipary (i.e., the seeds germinate directly instead of entering into a period of dormancy following embryogenesis). Furthermore, germination of dormant seeds can be inhibited by the exogenous application of ABA. Gibberellic acid (GA) acts antagonistically to ABA to promote germination. GA-deficient mutants were shown to germinate poorly, and exogenous application of GA can overcome dormancy and promote vivipary.

2.4 CHEMICAL ENVIRONMENT OF THE ENDOSPERM AND EMBRYOGENESIS

Embryos in the cotyledon stage placed on a medium containing inorganic salts, vitamins and sucrose are able to grow into normal seedlings without exhibiting dormancy. However, those in the globular-heart (early-stage embryos) do not develop further in the above medium. A high osmoticum with 10% sucrose or mannitol allowed early-stage embryos to complete the development. The effect of osmoticum was shown to be independent of ABA effect.

2.5 GENETIC CONTROL OF DEVELOPMENTAL PROGRAM

The embryonic development in plant is under the influence of evolutionary genetic program. Considerable progress has been made in recent years to understand molecular-genetic mechanisms that regulate the embryonic development at the early stages of ontogenesis. The findings of these investigations have clearly proven the role of different classes of phytohormones in regulating the expression of key genes that play an important role in controlling plant embryogenesis (Tsygankova, 2015; Sun et al., 2010; Belmontea et al., 2012). Leafy cotyledons (*lec*) genes are a recessive mutation that causes the cotyledons to acquire the characteristics of leaves; they have trichomes (hairs), leaf-like vasculature and lack storage proteins typical of cotyledons (Lotan et al., 1998). The mutant also displays vivipary and lack desiccation tolerance. The phenotype suggests that *LEC* normally regulates embryonic development, and in mutants, embryonic development is disrupted allowing the default program of seedling growth to proceed. It also suggests homology between cotyledons and leaves. This relationship would not otherwise be obvious because leaves and cotyledons have very different ontogenies. Leaves are produced by apical meristems whereas cotyledons arise prior to the meristem formation during embryogenesis. Ectopic expression of the *LEC* gene in adult vegetative tissues induces embryo formation (Lotan et al., 1998).

2.5.1 Role of Suspensor in Embryogenesis

The suspensor is involved in the transport of nutrients to the embryo proper. In some species, the suspensor cells form haustoria or cell-wall invaginations aids in the transport of nutrients. Being metabolically active, they also provide growth regulators to the embryo proper. Under certain situations, especially when the influence of embryo proper is removed, suspensor proliferates to produce a mass of cells and even develops into another embryo.

2.6 EMBRYO MATURATION (DEVELOPMENTAL ARREST)

Cell division, for most part, ceases at the beginning of the maturation phase of the embryo development. Then on, the increase in the size of cotyledon and axis is essentially caused by cell expansion and accumulation of storage compounds (Goldberg et al., 1994; Mansfield et al., 1991; Mansfield and Briarty, 1991). Ground meristem cells within embryo and cotyledons become highly specialized and accumulate large amounts of storage proteins and oils. Storage proteins accumulate in the vacuole or as membrane-bound vesicles called protein bodies. Storage proteins represent an important source of amino acids, nitrogen and carbon for the germinating seedlings. The storage proteins are encoded by several multigene families with up to 55 different genes coding for a given storage protein. Protein synthesis is controlled at the transcriptional level by few regulatory genes each controlling a particular class of storage proteins.

The regulation of starch and lipid are produced by complex enzymatic pathways and their accumulation is not clearly understood. Each class of compound is a mixture of molecules with different chain lengths, chain branching characteristics, levels of saturation and other chemical modifications. Thus, the synthesis of these compounds is much less straight forward than storage proteins.

2.7 ACQUISITION OF DESICCATION TOLERANCE

The moisture content in a fully mature seed is about 5%. At such low-moisture levels, solutes tend to crystallize and precipitate, and with no aqueous phase, the membrane becomes unstable and leaky. These conditions are lethal to most plant tissues and membranes. Therefore, embryos develop programs that allow them to survive under such stressful conditions. At the late stage of maturation, a group of proteins called dehydrins are expressed. These proteins induce drought stress tolerance in postembryonic tissues. They also play an important role in ion sequestration and in forming a protective layer to stabilize membranes.

Morphogenesis and maturation appear to be controlled by independent developmental programs. Both hormonal and genetic controls are involved in the integration of these programs. ABA induces expression of genes involved in maturation and desiccation tolerance. An ABA-independent genetic program is also necessary to confer ABA sensitivity to the embryo, and mutants show ABA-insensitive vivipary. The *LEC* gene is known to coordinate these two different programs.

Major stages and different developmental/metabolic events of embryogenesis in flowering plants are shown in Table 2.1.

TABLE 2.1
Major Stages and Developmental Events of Embryogenesis in Flowering Plants

Sr. No.	Major stages of embryogenesis	Accompanying developmental events
1.	Zygote (postfertilization embryo)	• Terminal and basal cell differentiation. • Formation of suspensor and embryo proper.
2.	Globular-heart transition	• Differentiation of major tissue-type primordial. • Establishment of radial (tissue-type) axis. • Embryo proper becomes bilaterally symmetrical. • Visible appearance of shoot–root (apical-basal) axis. • Initiation of cotyledon and axis. • Hypocotyl-radicle development. • Differentiation of root meristem.
3.	Organ expansion and maturation	• Enlargement of cotyledons and axis by cell division and expansion. • Differentiation of shoot meristem. • Formation of lipid and protein bodies. • Accumulation of storage proteins and lipids. • Vacuolization of cotyledon and axis cells. • Cessation of RNA and protein synthesis. • Dehydration. • Prevention of precocious germination. • Onset of dormancy.

2.8 APOMIXIS

In the past, the term apomixis has been used for any form of asexual reproduction in plants, including vegetative propagation, but it is now used to describe asexual reproductive processes that occur in the ovule (Asker and Jerling, 1992; Koltunow, 1993). The apomictic embryo is produced from cells in the maternal ovule tissues rather than from the fusion of male and female gametes (Richards, 2003). Apomictic processes mimic a number of sexual reproduction events and produce fertile seeds containing embryos that have genetic constitution identical to that of the maternal parent. Major differences between apomictic embryo formation and somatic embryogenesis are given in Table 2.2.

Apomixis is known to occur in nearly 300 plant species from *Gramineae, Compositae, Rosaceae* and *Rutaceae* families. Because apomictically produced embryos are genetically identical to the female parent plant, this can be an inexpensive way to multiply a given genotype and commercially produce seeds of high-yielding food crops (Hanna and Bashaw, 1987).

2.9 SOMATIC EMBRYOGENESIS

Somatic plant cells maintain the "competence" to dedifferentiate and form an embryonic stem cell capable of producing an embryo *in vitro*. This process of producing somatic embryo from stem cells is referred as somatic embryogenesis (Verdeil et al., 2007; Ikeuchi et al., 2015; Ji et al., 2011). This process was first demonstrated in carrot (*Daucus carota*) cell cultures (Steward et al., 1958). Since then, somatic embryogenesis has successfully been employed in various plant species to produce somatic embryos (Tautorus et al., 1991; Jain et al., 2000). Several researchers have shown that both somatic embryos and zygotic embryos pass through the same developmental stages, such as globular, torpedo and cotyledonary stages in dicot; globular, scutellar and coleoptilar stages for monocots; or globular, early cotyledonary and late cotyledonary embryo stages in conifer (Mordhorst et al., 1997; Dodeman et al., 1997; Quiroz-Figueroa et al., 2006). Somatic embryogenesis plays a significant role in mass propagation, germplasm conservation and genetic improvement in woody plants (Chiancone and Germanà, 2013; Ozudogru and Lambardi, 2016).

Somatic embryogenesis (SE) can be induced by both direct and indirect pathways (Yang and Zhang, 2010). In the direct somatic embryogenesis pathway, somatic embryos are induced directly from the explant without any intermediate callus stage. In contrast, indirect somatic embryogenesis involves an intermediate callus stage before forming somatic embryos (Montalbán et al., 2012; Corredoira et al., 2013, 2015). Unlike in primary embryogenesis, new somatic embryos from secondary somatic embryogenesis are induced through existing somatic embryos (Raemakers et al., 1995). In woody plants, secondary somatic embryogenesis helps in maintaining the embryogenic

TABLE 2.2
Differences between Apomictic Embryo and Somatic Embryogenesis

Developmental event	Apomictic embryo	Somatic embryo
Location of embryo developments.	Occurs within the differentiated structure – maternal ovule tissues.	Derived from vegetative tissues called explant.
Type of cell involved in embryo formation.	Formed directly from a cell located in, or close to, a gametophytic structure, without the intervening callus phase.	Callus phase, which is often necessary.
Pattern of embryo formation.	It often indistinguishable from that which occurs in the nearest sexual relative.	Different from zygotic embryo formation pattern.

competence of cultures for several years (Martínez et al., 2015). SE technology is important for woody plants with a long-life cycle and that are difficult to propagate by conventional methods (Isah, 2016; Germanà and Lambardi, 2016). According to Giri et al. (2004), somatic embryogenesis offers several advantages over organogenesis, including the feasibility of single-cell origin, clonal propagation of desirable genotypes, genetic transformation and the possibility of producing synthetic seeds on a large scale for use in planting in place of zygotic seeds.

2.9.1 SE Pathways in Horticultural Plants

2.9.1.1 Direct SE Pathway

In the direct SE, explants are induced to directly produce somatic embryos. For example, the immature embryos and cotyledons of peach (Yan and Zhou, 2002), the young embryos of cherry (March et al., 1993), the leaves of apple (Da et al., 1996), the embryos of sugar beet (Zhang et al., 2008), the young leaves of Bartlett pear (Sun et al., 2003), the flowers of chrysanthemum (Mandal and Datta, 2005), the leaves of carnation (Liu et al., 1997; Yantcheva et al., 1998), the scale leaves of lily (Yantcheva et al., 1998), the tender leaf of *Kalanchoe* (Liu et al., 1997) and the leaves of orchid (Ma and Liu, 2003) could all induce somatic embryos from explants to form new plantlets.

2.9.1.2 Indirect SE Pathway

In the indirect SE, the explants first dedifferentiate into callus, and then, the resulting callus differentiates to form the somatic embryos. For example, the embryos of mango (Gow et al., 2009), the male inflorescences of banana (Rivera-Domínguez et al., 2004), the leaves of grape (Grapin et al., 1996), shoots of date palm, garlic bulb (Robache, 1993), the cucumber cotyledons (Fereol et al., 2002) and the young stems of poinsettia (Elmeer and Hennerty, 2008) and gypsophila (Chen et al., 1999) all can be induced to form somatic embryos via production of callus.

Additionally, some plants, i.e., chrysanthemum (Zamorano-Mendoz and Mejia-Munoz, 1994), cucumber (Jiang and Zhang, 2007), orchid (Jiang et al., 2008) and walnut (Ashok et al., 2003) have the ability to produce somatic embryos by both direct and indirect pathways.

2.9.2 Factors Impacting SE in Horticultural Plants

2.9.2.1 Genotype

The genotype plays a key role in the induction of SE. The extent and frequency of SE have been shown to vary, even within the same genus. Again, not all species within a genus can be induced to produce somatic embryos. Ji et al. (2011) reported that Yesanhan cucumber is the most suitable genotype for SE. Similarly, the ability of cherry cultivars to induce SE under identical cultural conditions and age ranged from 41.43% to only 14.88% (Ashok et al., 2003).

2.9.2.2 Physiological Status of Explants

The physiological status of explants has profound bearing on SE. In general, it is easier to induce SE in tissues with the highest level of metabolism and lowest level of differentiation (Bian et al., 2007) as is evident in callus induction in different poinsettia tissues, where young stems producing SE more than young inflorescences, which, in turn, is faster than that in young leaf (Werbrouck et al., 2000). In Multi-Blue climatis, shoot apexes have the greatest ability to induce SE among young leaves, stem tips and young stems (Niu et al., 2006). Similarly, in Freesia, only somatic embryos appeared at the original morphological lower end when inflorescence was used as explants, whereas morphological upper end did not produce embryos (Liu et al., 1992). In Anthurium, Xin et al. (2006) indicated that the leaf sections were better explants and expressed greater embryogenic potential than the other parts. Therefore, selection of a suitable explant is key to the success of SE.

2.9.2.3 Culture Medium

Horticultural plants have different requirement for the culture media depending on the genotypes, explants and species. Wang et al. (1998) indicated that the improved basal medium (Gamborg or B_5) was better than Murashige and Skoog (MS) or woody plant medium (WPM) in inducing SE in mango. Under the same conditions, the frequency of SE of young walnut embryos was higher in MS media than in the Driver and Kuniyuki Walnut (DKW) medium; nonetheless, the frequency of the SE of walnut cotyledons was higher in DKW media than in MS medium, indicating that the response to cultural medium depends on the species.

2.9.2.4 Plant Growth Regulators

Plant growth regulators (PGRs) are another important component of the culture medium that are impacting the success of SE. Proper balance of auxin and cytokinin (CTK) is necessary for successful somatic embryogenesis (SE). 2,4-Dichlorophenoxyacetic acid (2,4-D) is the most widely used auxin for SE of papaya (Dewald et al., 1989), grape (Grapin et al., 1996), peach (Yan and Zhou, 2002), American chestnut (Fitch, 1993), mango (Gow et al., 2009) and lily (Yantcheva et al., 1998). While not all plants can be induced to form somatic embryos in the medium containing 2,4-D, other auxins such as naphthalene acetic acid (NAA) and indole-3-acetic acid (IAA) can also be used to induce SE in citrus (Cui and Dai, 2000) and gladiolus (Zhang et al., 2005). In fact, 2,4-D even negatively impacted SE in cherry (March et al., 1993) and begonia (Stefaniak, 1994). Depending on the developmental stage of somatic embryos, 2,4-D could either promote the production of embryo callus or inhibit the embryoid differentiation stage. Picloram induces SE in banana rather than 2,4-D (Castillo and Smith, 1997).

A proper ratio of cytokinin (CK) and auxin is essential for inducing SE in horticultural plants. For example, while the higher ratio of auxin to CK would promote the SE of American chestnut (Fitch, 1993), begonia (Stefaniak, 1994) and cyclamen (Wei et al., 2007), the higher ratio of CK to auxin would promote SE in apple. Although gibberellic acid (GA) and abscisic acid (ABA) are not normally needed for SE induction, GA has been shown to promote SE in apple, pear and cherry and inhibit SE citrus (Chen and Lai, 2001). ABA has promoting or inhibitory effects depending on the type of explant. For example, increased concentrations of ABA produce high-quality somatic embryos in cherry (March et al., 1993).

2.9.2.5 Nitrogen, Carbon and Other Additives in the Medium

Both the type and concentrations of nitrogen, carbon and amino acids play a major role in SE. In general, high amounts of NH_4NO_3 in the medium induces SE. Ji et al. (2019 and Tulecke and Tulecke (1985) have shown that sucrose at 10–30 $g{\cdot}kg^{-1}$ promoted SE in garlic whereas growth of somatic embryos was inhibited at 60 $g{\cdot}kg^{-1}$ or higher sucrose level. Furthermore, inclusion of glucose in the medium promoted indirect SE, and the use of lactose in the culture medium promoted direct SE in cotyledons of melon (Ji et al., 2019). SE efficiency in citrus was increased by 6- to 12-folds by the addition of lactose and galactose to the culture medium (Yang et al., 2003).

Inclusion of hydrolyzed casein (CH), malt wort (MW) and coconut wort (CW) in the cultural media promoted SE in several horticultural species. For example, CH induced SE in grape, peach, sweet cherry, walnut (Chen and Lai, 2001), garlic (Robacher, 1993) and celery (Bian et al., 2007). Similarly, inclusion of CH promoted the formation of mature embryoids in lychee and high-quality SE in poinsettia (Nakagawa and Saijyo, 2002). CW, which contains many components, promoted SE in several horticultural plants.

2.9.3 Applications of SE

Somatic embryogenesis is useful for clonal propagation of genetically similar plant materials and elimination of pathogenic viruses. It also provides source tissues for genetic transformation experiments. It is possible to produce whole plants from single cells.

2.10 SYNTHETIC SEEDS

Synthetic seed, a concept first introduced in 1977 by Murashige, is an encapsulated single somatic embryo inside a matrix covering. Until now, synthetic seeds were used in some fruit and forest trees (Prewein and Wilhelm, 2003; Malabadi and Staden, 2005; Singh et al., 2007; Rai et al., 2009). The advanced encapsulation technology can effectively be used to protect the somatic embryo from mechanical damage, predation and unfavorable environmental conditions. It is also possible to supply nutrients for the developing embryo through encapsulation agents. Encapsulated synthetic seeds are easy to handle, store, transport and sow in the same manner as the zygotic seed (Rai et al., 2009). Sharma et al. (2013) employed the hydrated and desiccated encapsulation technology for large-scale production of synthetic seeds. Several coating agents, such as sodium alginate, potassium alginate, sodium pectate and carrageenan with sodium alginate, have been used to encapsulate synthetic seeds (Malabadi and Staden, 2005; Pintos et al., 2008; Gantait et al., 2015). Carbon sources, plant growth regulators and antimicrobial agents are also added to the hydrogel to facilitate growth, enhance seedling vigor and increase field survival rates of the encapsulated propagules (Bapat and Mhatre, 2005; Rai et al., 2008). Cryopreservation is used for long-term conservation of woody plant somatic embryos (Lambardi et al., 2008).

REFERENCES

Ashok K H G, Murthy H N and Pack K Y. 2003. Embryogenesis and plant regeneration from anther culture of *Cucumis sativus* L. *Scientia Horticulturae* **98**: 213–222.

Asker S E and Jerling L. 1992. *Apomixis in Plants*. CRC Press, Boca Raton.

Bapat V A and Mhatre M. 2005. Bioencapsulation of somatic embryos in woody plants. pp. 539–552. *Protocol for Somatic Embryogenesis in Woody Plants* (Eds.) Jain S M and Gupta P. Springer, Dordrecht.

Belmontea M F, Kirkbridea R C, Stonea S L, Pelletiera J M and Buib A Q. 2012. Comprehensive developmental profiles of gene activity in regions and subregions of the *Arabidopsis* seed. *Proceedings of National Academic Science, USA*: 1–10.

Bian F H, Qu F N, Zheng C X, You C R and Gong X Q. 2007. Recent advances in *Cyclamen persicum* Mill. somatic embryogenesis. *Northern Horticulture* **8**: 70–72.

Castillo B and Smith M A L. 1997. Direct somatic embryogenesis from *Begonia gracilis* explants. *Plant Cell Reports* **16**: 385–388.

Chen L P, Wang B L and Chen M F. 1999. Studies on somatic embryogenesis of *Euphorbia pulcherrima in vitro* culture. *Plant Physiology Communications* **35**: 463–465.

Chen Y J and Lai Z X. 2001. Researches and utilization of somatic embryogenesis in fruits and trees. *Journal of Fujian Agricultural University* **30**: 420–426.

Chiancone B and Germanà M A. 2013. Micropropagation of *Citrus* spp. by organogenesis and somatic embryogenesis. pp 99–118. *Protocols for Micropropagation of Selected Economically-Important Horticultural Plants* (Eds.) Lambardi M, Ozudogru E A and Jain S M. Springer Science + Business Media, NewYork.

Corredoira E, Ballester A, Ibarra M and Vieitez M. 2015. Induction of somatic embryogenesis in explants of shoot cultures established from adult *Eucalyptus globulus* and *E. saligna E. maidenii* trees. *Tree Physiology* **35**: 678–690.

Corredoira E, Valladares S, Martínez M T, Vieitez A M and SanJosé M C. 2013. Somatic embryogenesis in *Alnus glutinosa* (L.) Gaertn. *Trees* **27**: 1597–1608.

Cui K R and Dai R L. 2000. *Molecular Biology of Plant Somatic Embryogenesis*. pp 48–54. Science Press, Beijing, China.

Da K D, Zhang S, Li Y Z and Qi Z J. 1996. Direct somatic embryogenesis from *in vitro* leaves of apple. *Acta Horticulturae Sinica* **23**: 241–245.

Dewald S G, Litz R E and Moore G A. 1989. Optimizing somatic embryo production in mango. *Journal of American Society for Horticultural Science* **114**: 712–776.

Dodeman V L, Ducreux G and Kreis M. 1997. Zygotic embryogenesis versus somatic embryogenesis. *Journal of Experimental Botany* **48**: 1493–1509.

Elmeer K M S and Hennerty M J. 2008. Observations on the combined effects of light, NAA and 2,4-D on somatic embryogenesis of cucumber (*Cucumis sativus*) hybrids. *Plant Cell, Tissue and Organ Culture* **95**: 381–384.

Fereol L, Chovelon V and Causse S. 2002. Evidence of a somatic embryogenesis process for plant regeneration in garlic (*Allium sativum* L.). *Plant Cell Reporter* **21**: 197–203.

Fitch M M M. 1993. High frequency somatic embryogenesis and plant regeneration from papaya hypocotyl callus. *Plant Cell, Tissue and Organ Culture* **32**: 205–212.

Gantait S, Kundu S, Ali N and Sahu N C. 2015. Synthetic seed production of medicinal plants: a review on influence of explants, encapsulation agent and matrix. *Acta Physiologiae Plantarum* **37**: 1–12.

Germanà M A and Lambardi M. 2016. *In-vitro Embryogenesis in Higher Plants*. Springer Science + Business Media, New York.

Giri C, Shyamkumar B and Anjaneyulu C. 2004. Progress in tissue culture, genetic transformation and applications of biotechnology to trees: an overview. *Trees* **18**: 115–135.

Goldberg R B, de Paiva G and Ya R. 1994. Plant embryogenesis: from zygote to seed. *Science* **266**: 605–614.

Gow W P, Che J T and Chang W C. 2009. Effects of genotype, light regime, explant position and orientation on direct somatic embryogenesis from leaf explants of *Alaenopsis* Orchid. *Acta Physiologiae Plantarum* **31**: 263–269.

Grapin A, Schwendiman J and Teisson C. 1996. Somatic embryogenesis in plantain banana. *In Vitro Cellular and Developmental Biology – Plant* **32**: 66–71.

Hanna W W and Bashaw E C. 1987. Apomixis: its identification and use in plant breeding. *Crop Science* **27**: 1136–1139.

Ikeuchi M, Iwase A, Rymen B, Harashima H, Shibata M and Ohnuma M. 2015. PRC2 represses dedifferentiation of mature somatic cells in *Arabidopsis*. *Nature Plants* **1**: 1–7.

Isah T. 2016. Induction of somatic embryogenesis in woody plants. *Acta Physiologiae Plantarum* **38**: 1–22.

Jain S M, Gupta P K and Newton R J. 2000. *Somatic Embryogenesis in Woody Plants*. Springer Science + Business Media, Dordrecht.

Ji A, Geng X, Zhang Y, Yang H and Wu G. 2011. Advances in somatic embryogenesis research of horticultural plants. *American Journal Plant Science* **2**: 727–732.

Ji L, Mathioni S M, Johnson S, Tucker D, Bewick A J, Do Kim K and Schmitz R J. 2019. Genome-wide reinforcement of DNA methylation occurs during somatic embryogenesis in soybean. *The Plant Cell* **31**(10): 2315–2331.

Jiang X W, Chen F J, Lu M, Cai M and Zhang Q X. 2008. Direct somatic embryogenesis in groundcover chrysanthemum. *Journal of Beijing Forestry University* **30**: 65–70.

Jiang X W and Zhang Q X. 2007. Studies on transgenic acceptor system of groundcover *Chrysanthemum* via indirect somatic embryogenesis. *Forest Research* **20**: 328–333.

Koltunow A M. 1993. Apomixis: embryo sacs and embryos formed without meiosis or fertilization in ovules. *The Plant Cell* **5**: 1425–1437.

Lambardi M, Ozudogru E A and Benelli C. 2008. Cryopreservation of embryogenic cultures. pp. 177–210. *Plant Cryopreservation: A Practical Guide* (Ed.) Reed B. Springer, New York.

Liu S L, Han B W and Chen H Z. 1992. Somatic embryogenesis and cytological observation from petiole of walnut (*Juglans regia* L.). *Journal of China Agricultural University* **18**: 29–32.

Liu X M, Zhou P H, Qu S C, Lu X Y and Luo Z M. 1997. *In Vitro* induction of indefinite bulbs and somatic embryos from scale leaves of tetraploid 'Longya' Lily. *Acta Horticulturae Sinica* **24**: 353.

Lotan T, Ohto M, Yee K M, West M A, Lo R, Kwong R W, Yamagishi K, Fischer R L, Goldberg R B and Harada J J. 1998. *Arabidopsis* LEAFY COTYLEDON1 is sufficient to induce embryo development in vegetative cells. *Cell* **93**: 1195–1205.

Ma G H and Liu N. 2003. Direct somatic embryogenesis and shoot formation from cultured young leaf of *Kalanchoe blossfeldiana*. *Plant Physiology Communications* **39**: 625.

Malabadi R B and Staden J V. 2005. Storability and germination of sodium alginate encapsulated somatic embryos derived from the vegetative shoot apices of mature *Pinus patula* trees. *Plant Cell, Tissue and Organ Culture* **82**: 259–265.

Mandal A K A and Datta S K. 2005. Direct somatic embryogenesis and plant regeneration from ray florets of chrysanthemum. *Biologia Plantarum* **49**: 29–33.

Mansfield S G and Briarty L G. 1991. Early embryogenesis in *Arabidopsis thaliana*. II. The developing embryo. *Canadian Journal of Botany* **69**: 461–476.

Mansfield S G, Briarty L G and Erni S. 1991. Early embryogenesis in *Arabidopsis thaliana* I: the mature embryo sac. *Canadian Journal of Botany* **69**: 447–460.

March G, Grenier E, Miannay N, Sulmont G, David H and David A. 1993. Potential of somatic embryogenesis in *Prunus avium* immature zygotic embryos. *Plant Cell, Tissue and Organ Culture* **34**: 209–215.

Martínez M T, Vieitez A M and Corredoira E. 2015. Improved secondary embryo production in *Quercus alba* and *Q. rubra* by activated charcoal, silver thiosulphate and sucrose: influence of embryogenic explant used for subculture. *Plant Cell, Tissue, and Organ Culture* **121**: 531–546.

Montalbán I, DeDiego N and Moncaleán P. 2012. Enhancing initiation and proliferation in radiate pine *(Pinus radiate D. Don)* somatic embryogenesis through seed family screening, zygotic embryos tagging and media adjustments. *Acta Physiologiae Plantarum* **34**: 451–460.

Mordhorst A P, Toonen M A J, de Vries S C and Meinke D. 1997. Plant embryogenesis. *Critical Review in Plant Science* **16**: 535–576.

Murashige T. 1977. Plant cell and organ cultures as horticultural practices. *Acta Horticulturae* **78**: 17–30.

Nakagawa H and Saijyo T. 2002. Effect sugar and ABA on somatic embryo formation with cotyledon culture of melon. *Foreign Crop Breeding* **21**: 75.

Niu S S, Song S W, Yan F and Miao H X. 2006. Somatic embryogenesis and plantlet regeneration in *Citrullus lanatus* cv. Zhengkang No.4. *Journal of Fruit Science* **23**: 406–410.

Ozudogru E A and Lambardi M. 2016. Cryotechniques for the long-term conservation of embryogenic cultures from woody plants. pp. 537–550. *In-vitro Embryogenesis in Higher Plants* (Eds.) Germanà M A and Lambardi M. Springer Science + Business Media, New York.

Pintos B, Bueno M A, Cuenca B and Manzanera J A. 2008. Synthetic seed production from encapsulated somatic embryos of cork oak (*Quercus suber* L.) and automated growth monitoring. *Plant Cell, Tissue and Oran Culture* **95**: 217–225.

Prewein C and Wilhelm E. 2003. Plant regeneration from encapsulated somatic embryos of pedunculate oak (*Quercus robur* L.). *In-vitro Cellular and Developmental Biology – Plant* **39**: 613–617.

Quiroz-Figueroa F R, Rojas-Herrera R, Galaz-Avalos R M and Loyola-Vargas M. 2006. Embryo production through somatic embryogenesis can be used to study cell differentiation in plants. *Plant Cell, Tissue and Organ Culture* **86**: 285–301.

Raemakers C, Jacobsen E and Visser R. 1995. Secondary somatic embryogenesis and applications in plant breeding. *Euphytica* **81**: 93–107.

Rai M K, Asthana P, Singh S K, Jaiswal V S K and Jaiswal U. 2009. The encapsulation technology in fruit plants: a review. *Biotechnology Advances* **27**: 671–679.

Rai M K, Jaiswal V S and Jaiswal U. 2008. Effect of ABA and sucrose on germination of encapsulated somatic embryos of guava (*Psidium guajava* L.). *Scientia Horticulturae* **117**: 302–305.

Richards A J. 2003. Apomixes in flowering plants: an overview. *Philosophical Transactions of the Royal Society of London. Series B* **358**: 1085–1093.

Rivera-Domínguez M, Manzanilla-Ramírez M A, Robles-González M and Gómez-Lim M A. 2004. Induction of somatic embryogenesis and plant regeneration of 'Ataulfo' mango (*Mangifera indica*). *Plant Cell, Tissue and Organ Culture* **79**: 101–104.

Robacher C. 1993. Somatic Embryogenesis and plant regeneration from Muscadine grape leaf explant. *HortScience* **28**: 53–55.

Sharma S, Shahzad A and daSilva J A T. 2013. Synseed technology: a complete synthesis. *Biotechnology Advances* **31**: 186–207.

Singh B, Sharma S, Rani G, Virk G S, Zaidi A A and Nagpal A. 2007. In vitro response of encapsulated and non-encapsulated somatic embryos of Kinnow mandarin (*Citrus nobilis* Lour, *C. deliciosa* Tenora). *Plant Biotechnology Reports* **1**: 101–107.

Stefaniak B. 1994. Somatic embryogenesis and plant regeneration of Gladiolus (Gladiolus hort.). *Plant Cell Reports* **13**: 386–389.

Steward F, Mapes M O and Mears K. 1958. Growth and organized development of cultured cells II. Organization in cultures grown from freely suspended cells. *American Journal of Botany* **45**: 705–708.

Sun Q R, Liu Q Z and Zhao R H. 2003. Somatic embryo genesis from *in vitro* leaves of pear. *Acta Horticulturae Sinica* **30**: 85–86.

Sun X, Shantharaj D, Kang X and Ni M. 2010. Transcriptional and hormonal signaling control of *Arabidopsis* seed development. *Current Opinion Plant Biology* **13**: 611–620.

Tautorus T, Fowke L and Dunstan D. 1991. Somatic embryogenesis in conifers. *Canadian Journal Botany* **69**: 1873–1899.

Tsygankova V A. 2015. Genetic control and hormonal regulation of plant embryogenesis. *International Journal Medical and Biotechnology Genetics* **3**: 9–20.

Tulecke W and Tulecke G. 1985. Somatic embryogenesis and plant regeneration from cotyledons of Walnut, *Juglans regia* L. *Plant Science* **40**: 57–63.

Verdeil J L, Alemanno L, Niemenak N and Tranbarger T J. 2007. Pluripotent versus totipotent plant stem cells: dependence versus autonomy? *Trends in Plant Science* **12**: 245–252.

Wang L, Bao X M, Huang B Q and Hao S. 1998. Somatic embryogenic potential determined by the morphological polarity of the explant in tissue cultures of *Freesia refracta. Acta Botanica Sinica* **40**: 138–143.

Wei X X, Liu Y and Gao R F. 2007. Cyanide-resistant respiration and its role of *Paeonia lactiflora* cut flowers during storage time. *Journal of Beijing Forestry University* **29**: 206–210.

Werbrouck S, Eeekhaut T and Debe T. 2000. Induction and conversion of somatic embryogenesis on the anther filament of *Spathiphyllum* Schott. *Acta Horticultuare* **520**: 263–269.

Xin W J, Xu B, Wang G D, Guo W M, Wen F D and Jin J P. 2006. Somatic embryogenesis and plant regeneration of *Anthurium andraeanum. Acta Horticulturae Sinica* **33**: 1281–1286.

Yan G H and Zhou Y. 2002. Plant regeneration from excised immature embryos of peach (*Prunus persica* L.). *Acta Horticulturae Sinica* **29**: 480–482.

Yang A F, Zhu Y M and Hou A J. 2003. Several factors affecting somatic embryos derived from cotyledons of cucumber (*Cucumis sativus*). *Plant Physiology Communications* **39**: 206–208.

Yang X and Zhang X. 2010. Regulation of somatic embryogenesis in higher plants. *Critical Reviews in Plant Science* **29**: 36–57.

Yantcheva A, Vlahova M and Antanassov A. 1998. Direct somatic embryogenesis and plant regeneration of carnation (*Dianthus caryophyllus* L.). *Plant Cell Report* **18**: 143–153.

Zamorano-Mendoza J J and Mejia-Munoz J M. 1994. *In Vitro* propagation of gypsophila (*Gypsophila paniculata* L.) cv. Perfecta. *Revista-Chapingo Horticultural Services* **1**: 67–71.

Zhang C L, Chen D F, Kubalakova M, Zhang J, Scott N W, Elliott M C and Slater A. 2008. Efficient somatic embryogenesis in sugar beet (*Beta vulgaris* L.) breeding lines. *Plant Cell, Tissue and Organ Culture* **93**: 209–221.

Zhang T, Cao Z Y and Wang X Y. 2005. Induction of somatic embryogenesis and plant regeneration from cotyledon and hypocotyls explants of *Eruca sativa* Mill. *In Vitro Cellular and Developmental Biology – Plant* **41**: 655–657.

3 Protoplast Structure, Cell Cycle and Cell Elongation

3.1 HISTORICAL PERSPECTIVE

Every living organism is composed of basic units called cells. While algae and fungi consist of a single cell (unicellular), angiosperms care made of many cells (multicellular). In 1665, Robert Hooke described and proposed the name "cell" to mean room or cubicles. In the mid-19th century, Matthias Schleiden and Theodor Schwann recognized the cell to be the basic unit of life. They also suggested the "cell theory," which states that all organisms are composed of cells, and all cells rose from preexisting cells.

3.2 STRUCTURE OF THE PLANT CELL

Plant cell consists of a outer wall (cell wall) enclosing the inner protoplast, which contains the nucleus and various organelles dispersed in the cytoplasm. The structure of a eukaryotic (cell containing distinct, membrane-bound nucleus) plant cell is illustrated in Figure 3.1. The cell wall and plasmodesmata set apart plant cells from rest of the organisms. It should be recognized that not all functional plant cells are living, although many cells live only for a short time. The cell walls remain intact and provide strength and rigidity to the plant body. For instance, the dead cells of xylem tissue act like pipes for conducting water between the roots and the aboveground organs and continue to provide support to the plant body for the entire life span of the organism. The components of a eukaryote plant cell are described in the following.

3.2.1 Cell Wall

The cell wall is the outmost layer of the plant cell encircling the protoplast. It may comprise one or two layers, an outermost primary wall made mostly of cellulose, which is formed early in the life of the cell, and a highly lignified secondary wall located inside the primary wall. The cell wall is synthesized by the cytoplasm and deposited outside of the plasma membrane, the outer boundary of the protoplast. The wall and plasma membrane, together, regulate the passage of various materials into or through the cell, thus making it possible for the cell to maintain an identity different from its surroundings. The wall is not simply an inert structure providing strength and giving structure to plant cells but is the site of several active metabolic processes including absorption and secretion of substances, detection of bacterial and fungal pathogens and even regulating the growth and development.

During the formation of the primary wall, adjacent cellulose molecules link together forming bundles called microfibrils. The microfibrils, in turn, twist together like a rope, producing strong cords called macrofibrils. Hemicelluloses and pectins are deposited among the fibril network and chemically bind the whole fibril network together. The entire process of synthesis and assembly of fibrils is coordinated by the plasma membrane. On the other hand, the lignin in the secondary wall gives strength and protection to the cell against attack by pathogens. Other substances that are deposited in walls of tissues serve a specific function. For example, cell wall in the outer layer of leaves often contains cutin or waxes, both of which effectively reduce the loss of water from leaves. Similarly, the walls of cork cells in tree bark contain suberin, another type of protective material. Only the cells involved in the transport and protective function have secondary walls.

DOI: 10.1201/9781003373889-4

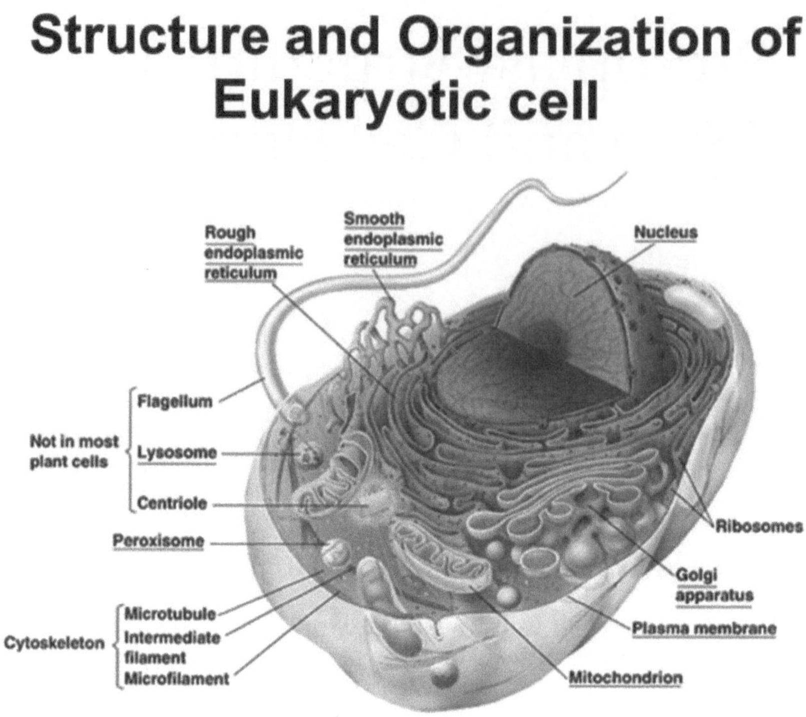

FIGURE 3.1 Structure of eukaryotic cell.

Cell walls have minute pores called pits, which help in the transfer of materials across the cells. The cytoplasmic strands, called plasmodesmata, protrude through these pits to the protoplast of the adjacent cell to allow the movement of materials across the cells. A sticky layer made of pectins, called the middle lamella, glues walls of neighboring cells together.

3.2.2 Cytoplasm

The cytoplasm is jelly-like fluid that fills a cell and houses the organelles and cell components. It contains up to 90% water with the remaining made of proteins, other organic molecules, ions and enzymes. The part of the cytoplasm that is confined to the boundaries of the cell without the organelles is called cytosol. Cytoplasm also includes cytoskeleton made of proteinaceous microtubules and microfilaments. The main functions of cytoplasm in the cell are as follows:

- It is the site of many vital biochemical reactions.
- It is the site where cell expansion and growth occur.
- It provides a medium for the organelles to remain suspended and facilitates their movement within the cell.
- The cytoskeleton in the cytoplasm provides shape to the cell and facilitates their movement.
- The enzymes in the cytoplasm metabolize macromolecules into small components, so that it can be easily made available for the other cellular organelles, such as mitochondria.
- The cytoplasm allows the transport of genetic material and products of cellular respiration.
- The cytoplasm acts as a buffer and protects the genetic material and cellular organelles from damage caused due to movement and collision with other cells.

- The cytoplasmic inclusions, which are non-soluble molecules floating in the cytoplasm, store fats and sugars that are readily available for cellular respiration.
- The cytoplasm and the proteins prevent the grouping of organelles in one place due to gravity that would impede their function.

3.2.3 Plasma Membrane

The outermost layer of the protoplast lining the cell wall is the plasma membrane. It is composed of phospholipids and proteins (Figure 3.2). The basic structure of all biological membranes is the same – comprising two lipid layers within which proteins are embedded. Some of the protein molecules extend through the lipid layer with their ends projecting out on opposite sides. Others remain on or partially embedded in one or the other surface. When carbohydrate molecules are attached to the membrane proteins and lipids, they are called glycoproteins and glycolipids, respectively. The carbohydrates are usually short chains of about five to seven monosaccharides.

Most membranes are selectively (differentially) permeable, permitting or preventing materials from leaving or entering the cell. When toxic substances destroy the selective permeability and the membrane becomes freely permeable, materials leak unchecked in and out of the cell, eventually leading to the death of the cell. The main features of membrane are as follows:

(i) It is the site of a number of biochemical processes of life.
(ii) It is in constant flux, changing both their physical and chemical structure and composition as long as the cell is alive.
(iii) It partitions the cell into compartments in which metabolic reactions take place independently of processes occurring in the nearby compartments, thus permitting a variety of reactions to occur simultaneously within a single cell.
(iv) Plasma membrane translates biochemical signals that control cell growth and differentiation and regulates the formation of the cell wall.
(v) The lipid layer in the plasma membrane provides flexibility to the membrane, whereas protein maintains the chemical climate in the cell.
(vi) It protects the integrity of the interior of the cell by allowing only the selected materials to enter the cell.

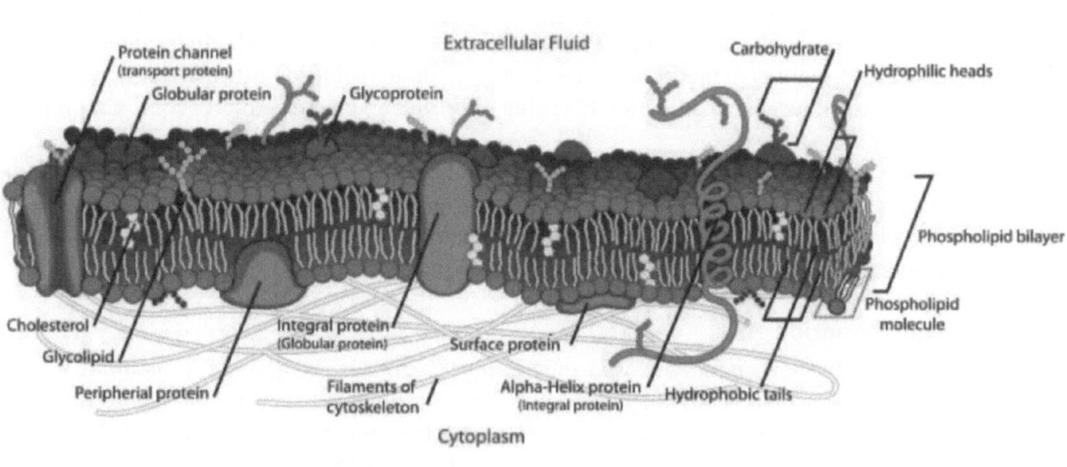

FIGURE 3.2 Structure of plasma membrane.

3.2.3.1 Movement of Materials In and Out of the Cell

There is a constant exchange of materials between cells and their environment by diffusion. Diffusion is the spontaneous movement of particles or molecules from areas of higher concentration to areas of their lower concentration. However, the semipermeable membranes, such as the plasma membrane, act as selective barriers to this free movement of molecules. They permit the diffusion of certain molecules while preventing the passage of other molecules. Several molecules are too large to diffuse through membranes. The water molecules move freely through membranes by the process called osmosis. The water molecules move from the area of higher concentration to the area of lower concentration of solutes. Therefore, if the cell is placed in distilled water, then the water will enter the cell through membrane. On the other hand, if the cells are placed in a highly concentrated solution (called hypertonic solution) for any length of time, the cell will lose water and protoplast shrinks away from the cell wall. This process is called plasmolysis, and the affected cells are referred to as plasmolyzed cells.

In a wilted plant, most of the cells in leaves are plasmolyzed. In contrast, if the cell is placed in pure water or a weak (hypotonic) solution, water enters the cell through membrane until the solute concentration inside and outside becomes nearly the same. At this stage, the vacuoles in the cell will be fully extended and push the cytoplasm up against the cell wall. Such cells are called turgid cells, and the phenomenon is called turgidity. This happens when plants are provided with sufficient moisture. The crispiness and crunch of fresh celery are due to its turgid cells. There will be no net movement of water if the cells are placed in isotonic solution (same concentration) and such cells will not become turgid. Thus, both diffusion and osmosis occur along the concentration gradient, i.e., from higher to lower concentrations. However, movement of solutes against the concentration gradient can happen in living cells. Such movements require energy and, hence, are called active transport. Proteins present in the membrane are involved in transporting substances against the concentration gradient across the membrane.

3.2.4 Organelles

A wide variety of organelles, most of which are membrane bound, are distributed within the cytoplasm of the cell. The membranes of these organelles are similar in structure and function to the plasma membrane. These organelles are described in this section.

3.2.4.1 Plastids

Plastid is a collective name for chloroplasts, leucoplasts and chromoplasts. The chloroplasts present in green leaves resemble discs with double membranes. Chloroplasts contain several pigments, such as chlorophylls (predominant), carotenes and xanthophylls within their internal membranes. The chlorophyll is most concentrated in membranous stacks called grana which are interconnected and embedded in a protein-rich medium called stroma. Although chloroplasts are easily seen with a light microscope, the internal organization is visible only with an electron microscope. Chloroplasts are involved in photosynthesis.

Two other organelles, namely, leucoplasts and chromoplasts are also present in some plants. Leucoplasts are colorless organelles that store starch (amyloplast in potato tubers is a type of leucoplast). Chromoplasts, on the other hand, contain orange, red or yellow pigments that are present in abundant quantities in colored plant parts (petals and fruits).

3.2.4.2 Mitochondrion

Mitochondrion, another organelle enclosed by a double membrane, is the site of cellular respiration in eukaryotic cells. The size, shape and number of mitochondria vary among cells, but all mitochondria have a smooth outer membrane and an infolded inner membrane called cristae. The compartment enclosed by cristae (matrix) contains enzymes that are needed for cellular respiration. The cristae are the sites of ATP formation.

3.2.4.3 Nucleus

The nucleus is the seat of hereditary information. The nucleus is surrounded by a double membrane with small openings called nuclear pores, which lead to the cytoplasm. At places, the nuclear membrane is connected to the endoplasmic reticulum (ER). The nucleus contains granular chromatin, consisting of DNA (the hereditary material), RNA and proteins. The nucleus also contains one or two dark staining nucleoli that is not membrane bound and is nearly spherical. Nucleoli are involved in the formation of ribosomes.

3.2.4.4 Vacuole

Most mature plant cells are characterized by a large central vacuole that is separated from the rest of the cytoplasm by its own membrane. In some cells, the vacuole takes up almost 90% of the cell volume, pushing the cytoplasm into a thin layer against the plasma membrane. The vacuole contains the cell sap, a watery solution of sugars, salts, amino acids, proteins and crystals. The cell sap is often acidic, and some of the substances in the vacuole constitute waste products. The concentrations of these materials in the vacuole may become so high that they precipitate out as crystals and could even be toxic to other organelles in the cell.

Vacuoles may also contain certain water-soluble pigments like anthocyanin in beetroot, grapes, onion and red cabbage. Unlike the pigments of the chloroplasts and chromoplasts, the pigments present in the vacuoles are distributed uniformly in the cell sap and released readily when the wall is ruptured.

3.2.4.5 Endoplasmic Reticulum (ER)

All these structures are involved in the synthesis, packaging and transport of various materials within the cell. The ER represents a network of membranous channels in the cytoplasm and is covered with minute bodies called the ribosomes that contain RNA and protein. Ribosomes are involved in protein synthesis. The ERs that are attached to ribosomes are known as rough ERs and are actively involved in protein synthesis. Portions of the ER without ribosomes (smooth ER) are involved in the synthesis of lipids and transport and packaging of proteins.

3.2.4.6 Golgi Apparatus

Golgi apparatus is a stack of flattened hollow sacs with bloated edges. Small vesicles are pinched off the edges of these sacs. The Golgi apparatus functions in the storage, modification and packaging of proteins that are produced by the ER. Golgi sacs form complex biological molecules from proteins transported from ER. Often, carbohydrates are added to proteins to form glycoproteins. The vesicles that are pinched off contain products that will be secreted from the cell. Some of the polysaccharides other than cellulose found in the cell wall are also secreted by these Golgi vesicles.

3.2.4.7 Microbodies

These are small, spherical organelles involved in various enzymatic reactions. Plant cells contain two types of microbodies: peroxisomes, which are found in leaves and play a limited role in photosynthesis under certain conditions, and glyoxysomes, which are involved in the conversion of stored fats into sugars in some seeds.

3.3 CELL CYCLE

The cell passes through orderly set of events between their formation and division into two daughter cells. Since the cloning of the first plant gene in the early 1990s (Feiler and Jacobs, 1990; Ferreira et al., 1991; Hirt et al., 1991), the plant cell cycle has been intensely investigated to understand the relationship between cell cycle control and development, on the one hand, and between hormones

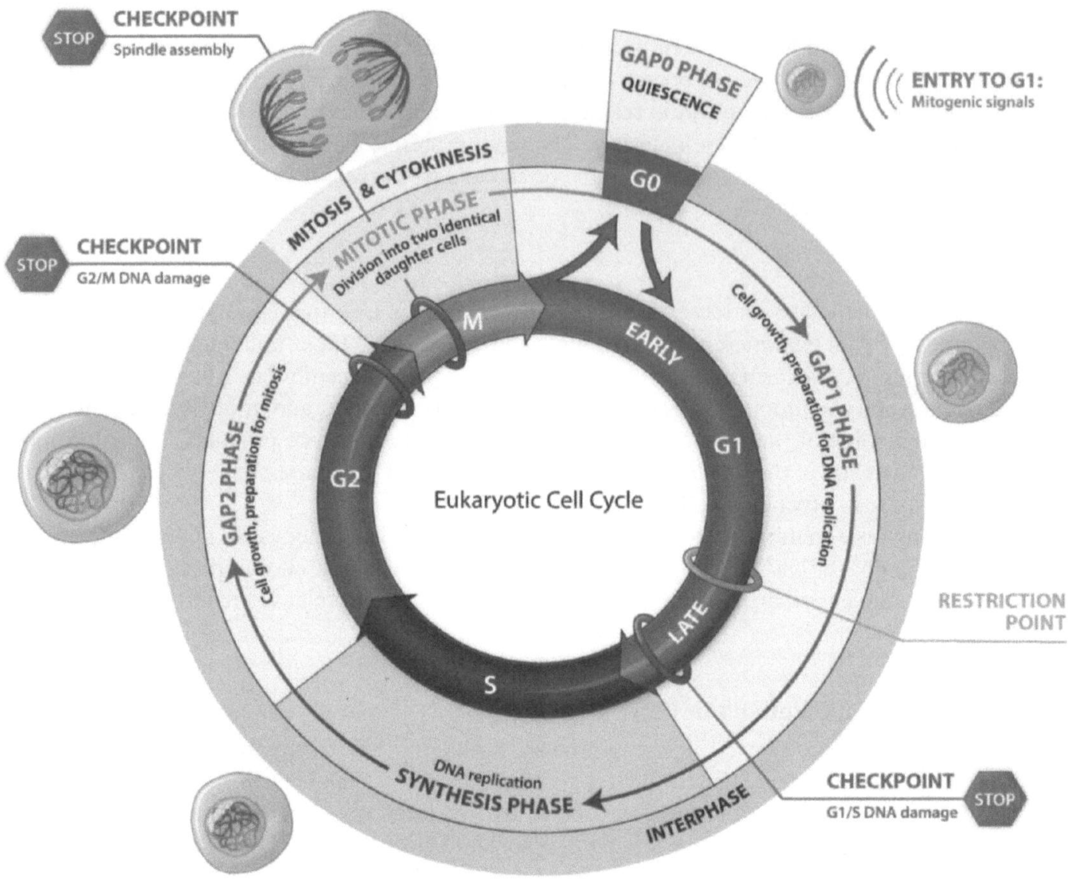

FIGURE 3.3 Cell cycle in eucaryotic cell.

and environmental signals, on the other hand. These events repeat itself in a cyclic manner. The cell cycle consists of four distinct phases: Gap 1 phase (G_1), Synthesis phase (S), Gap 2 phase (G_2) and Mitosis (M) phase with cytokinesis (C) (Figure 3.3). The G_1, S and G_2 phases (the first three phases) together constitute the interphase. Therefore, the cycle of duplication and division is also called the cell cycle. The total length of cell cycle, relative lengths of individual stage and the timing of cell cycle events are determined by the type of cell and prevailing conditions. The length of cell cycle could be as short as 8 h and as long as 100 days or more. There have been excellent comprehensive reviews in the recent literature about the mitotic cell cycle (de Veylder et al., 2007; Gutierrez, 2009), endocycle control (de Veylder et al., 2011) and the meiotic cell cycle (Wijnker and Schnittger, 2013).

Different stages of cell cycle are described in this section.

3.3.1 Interphase

The first stage of the cell cycle is known as interphase. During the interphase, the cell continues to grow by accumulating nutrients and prepares itself for division by replicating DNA. Interphase is the longest and the most active phase in the cell cycle.

3.3.1.1 G_1 Phase (Gap or Growth Phase)

G_1 stage separates the end of mitosis and the start of the Synthesis (S) phase. This stage is shortest in actively growing tissues, whereas the slow-growing cells remain in this phase for relatively longer duration. Therefore, the length of the G1 stage determines the length of the cell cycle, for most part. After cytokinesis, the resulting daughter cells are rather small and have limited energy. Therefore, cell grows in size, and their cytoplasm undergoes rapid growth during the G_1 phase. The daughter cells attain the size of the mother cell; the chromosomes become thread-like and are dispersed in the cytoplasm. There is no change in DNA content with each chromosome containing only a single molecule of DNA (unduplicated or unreplicated chromosome). The cell in this phase prepares the enzymatic machinery and increases its metabolic rate. The cell in this phase also undergoes heightened biochemical and cellular synthesis and produces various cell organelles. The nucleolus produces rRNA, mRNA and ribosomes. In this phase, cells synthesize structural and functional (enzymes) proteins, amino acids, nucleotides and ATP. During G_1 phase, a cell may follow one of the three options:

- Cells reach the restriction point (R point). After a short lag, the cycle continues and cells divide.
- Cells permanently stop dividing and enter G_0 or quiescent stage.
- Cell cycle is arrested at a specific point of G_1 phase, and cells are withdrawn from the cell cycle. When conditions change and growth is resumed, the cell reenters the G_1 phase.

3.3.1.2 Synthetic Phase (S)

This stage is sandwiched between G1 and G_2 phases. This phase is marked by the synthesis of new DNA with each chromosome containing two molecules of DNA. The DNA content of the nucleus is doubled, and more proteins are synthesized in this phase. Also, during this stage, every double-helical DNA molecule is duplicated, and two strands of DNA that are exactly identical are produced. The two DNA strands attach together by specific proteins at a short sequence of DNA to form a centromere. Each chromosome becomes two chromatids. Once DNA duplication is complete, histone proteins are synthesized to cover the DNA strand (chromatin synthesis). Each chromosome is changed from a single stranded to a two-stranded form.

3.3.1.3 G_2 Phase

The cell undergoes a second growth and energy acquisition stage called the G_2 phase. In this phase, cells continue to synthesize RNA and protein, but DNA synthesis stops. The mitotic spindle proteins are formed with the multiplication of chloroplasts and mitochondria and production of materials needed for mitosis. The mitotic spindle is a structure that is involved in the movement of chromosomes during mitosis. The chromatin begins to condense into the relatively compact structures called chromosomes. These chromosomes remain attached through their centromeres. Preparations for chromosomal segregation are initiated in this phase although actual segregation does not happen in this phase. The duration of the S phase, the G_2 phase and mitosis is more or less constant for most cell types.

3.3.2 Mitosis Phase

This stage is the shortest phase in the cell cycle. Mitosis is a repetitive process involving the condensation and separation of the replicated chromosomes. Continuity of life depends on cell division. Unicellular plants reproduce asexually by binary fission in which two identical cells similar to the parent cell are produced from each cell. In multicellular plants, however, life starts with a single fertilized cell (zygote) that divides and redivides to form various organs that are highly specialized for performing different functions. Plant cell divides by one of the following three ways:

- **Amitosis** or direct cell division (Robert Remak – 1855)
- **Mitosis** or indirect cell division (Flemming – 1882)
- **Meiosis** or reduction division (Farmer and Moore – 1905)

3.3.2.1 Amitosis

In this type of cell division, a simple cleavage of the nucleus appears first without the change in its structure and is then followed by the division of the cytoplasm. It is not the usual mode of division, occurring mainly in highly specialized cells that are not capable of long multiplication, those with transitory structures, and those that are in early stages of degeneration. This type of cell division is found mainly unicellular: prokaryotic plants, e.g., cynobacteria, bacteria, yeast, internodal cells of *Chara zeylanica* and *Chara contraria* and endosperm cells of developing seed.

Stages of amitosis: The nucleus elongates and a constriction appears in the center, resulting in a dumbbell shaped structure. The constriction deepens and the nucleus gets divided into two. There is no spindle formation or formation of chromosomes. The nuclear division is followed by the division of cytoplasm, and the cell gets divided into two daughter cells. The two daughter nuclei are of similar shape and size. Thus, amitosis is known as a quantitative division.

3.3.3 Mitotic Cell Division

Mitosis is the most common method of nuclear division and usually occurs in vegetative (or somatic) cells. Therefore, it is also known as somatic division. In mitosis, the mother cell produces two genetically identical daughter cells, which resemble each other and also the parent cell both qualitatively and quantitatively. The separation of sister chromatids into two new cells with exactly the same number of chromosomes and half the amount of nuclear DNA is known as mitosis. In mitotic division, not only the chromosomes are replicated but all necessary cytoplasm constituents and organelles are divided between the two daughter cells. Mitosis is observed in all types of cells (haploid, diploid or polyploid cells).

Mitosis division is usually restricted to the diploid cells, although, in some lower plants and insects, haploid cells also divide by mitosis. Mitosis results in the production of diploid daughter cells with identical genetic complement. Since growth of multicellular organisms essentially occurs due to mitosis, it is necessary for the cell to divide to restore the nuclear and cytoplasmic contents of the cell. Mitotic divisions in apical and lateral cambiums ensure continuous growth of plants. The production of new cells through mitosis also enables plants to repair wounds and regenerate lost cells. Cell division leads to the production of new, genetically identical individuals known as clones and forms the basis for asexual or vegetative reproduction. Continuous division of lateral meristems increases radial growth.

3.3.4 Stages of Mitosis

Prophase: In this phase, the nucleus condenses and thickens, coiling up into chromosomes (Figure 3.4a). Each chromosome comprises two identical chromatids, representing the condensed duplicated strands of chromatin. The chromatids are joined at the center by centromere. The formation

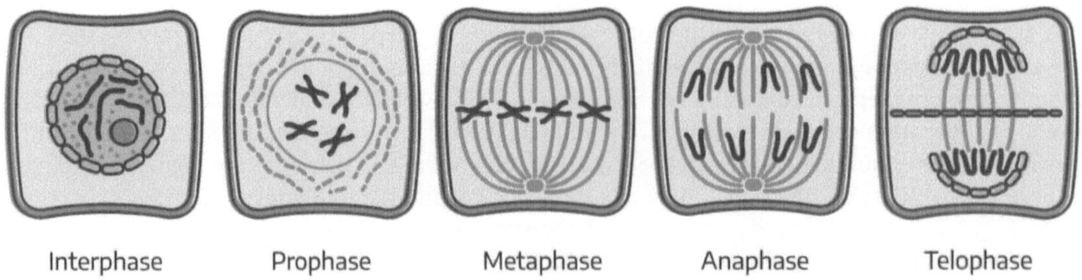

FIGURE 3.4 Early (left) and late (right) prophase phase of mitosis of higher plant cell.

of chromosomes is completed by the end of prophase. In addition, the nuclear membrane and the nucleoli disperse into the cytoplasm, setting the chromosomes free in the cytoplasm.

Metaphase: The chromosomes are arranged across the center of the cell during metaphase (Figure 3.4b). The formation of spindle is clearly evident in this stage, and the microtubules in spindle fibers extend from each end of the cell to the kinetochore of each chromatid. Kinetochores formed during late prophase attach each chromatid to the spindle. Other spindle fibers stretch from each pole to the equator.

Anaphase: In this phase, chromatids of each chromosome are pulled by the spindle fibers to opposite ends of the cell and separate, thus dividing the genetic material into two identical sets, each with the same number of single stranded chromosomes. At the end of anaphase, the spindle is less evident (Figure 3.4c).

Telophase: The chromatin appears as the chromatids at each end of the cell and begins to unwind and lengthen. The nuclear membrane reappears around the chromatin forming two distinct nuclei. Nucleoli become visible within each nucleus (Figure 3.4c).

3.3.4.1 Cytokinesis

The division of the cytoplasm (cytokinesis) separates the two identical daughter nuclei into two cells (Figure 3.4d). Cytokinesis, which begins at the later part of the anaphase, is completed in the telophase. The vesicles, microtubules and portions of ER in the phragmoplast accumulate across the center and coalesce to form the cell plate. The cell plate finally becomes the cell wall separating the newly formed daughter cells.

3.3.5 Meiosis

Meiosis is the reduction division in which the chromosome number is reduced by half in the daughter cells. It takes place during the gametogenesis in plants. The key features of meiosis are as follows:

- Meiosis involves two sequential cycles (nuclear and cell division) known as meiosis I and meiosis II but the DNA replication takes place only once.
- Meiosis I is initiated after the parental chromosomes have replicated to produce identical sister chromatids in the S phase.
- Meiosis involves pairing of homologous chromosomes and recombination between them.
- Four haploid cells are formed at the end of meiosis II.

Meiosis helps in the conservation of specific number of chromosomes in sexually reproducing organisms. It also increases the genetic variability in the population of organisms. Meiotic events can be grouped under the following stages.

3.3.6 Meiosis I

Prophase I: This phase is very complex involving five steps, such as leptotene, zygotene, pachytene, diplotene and diakinesis. The first step (leptotene) involves the compaction of chromosomes becoming more obvious. In the second step (zygotene), the chromosomes start pairing together through the process of synapsis to produce homologous chromosomes. This is accompanied by the formation of complex structure called synaptonemal complex. The paired synapsed homologous chromosomes are called a bivalent or a tetrad. This is followed by the pachytene stage, during which recombination nodules appear, facilitating crossing over or the exchange of genetic material between non-sister chromatids of the homologous chromosomes. Crossing over is mediated by recombinase enzyme and leads to the recombination of genetic material on the two homologous chromosomes. This process is completed by the end of pachytene. The chromosomes are linked at the sites of crossing over. Then, the dissolution of the synaptonemal complex takes place, which marks the beginning of the diplotene stage. In this stage, the recombined homologous chromosomes of the bivalents separate from each other except at the sites of crossovers forming X-shaped structures, called chiasmata. The final stage of meiotic prophase I (diakinesis) is reflected in the terminalization of chiasmata. The other events in the stage include completion of chromosomes condensation, assembling of meiotic spindle, disappearance of nucleolus and breaking down of nuclear apparatus. Diakinesis marks the transition to metaphase.

Metaphase I: This phase is represented by the alignment of bivalent chromosomes at the equatorial plate and attachment of microtubules from the opposite poles of the spindle to the pair of homologous chromosomes.

Anaphase I: The homologous chromosomes separate, while sister chromatids remain associated at their centromeres in this stage.

Telophase I: In this phase, the nuclear membrane and nucleolus reappear. This is followed by cytokinesis to form a diad of cells.

The short stage between the two meiotic divisions is called interkinesis, which makes way to prophase II.

3.3.7 Meiosis II

Meiosis II is initiated immediately after cytokinesis and resembles a normal mitosis (Figure 3.5). (Figure 3.6)

Prophase II: The nuclear membrane disappears and the chromosomes again become compact by the end of prophase II.

Metaphase II: In this stage, the chromosomes align at the equator, and the microtubules from opposite poles of the spindle get attached to the kinetochores of sister chromatids.

Anaphase II: This phase involves the simultaneous splitting of the centromere of each chromosome allowing them to move toward opposite poles of the cell.

Telophase II: Telophase II marks the end of meiosis II. In this stage, the two groups of chromosomes once again are enclosed by a nuclear envelope. This is followed by cytokinesis, resulting in the formation of four haploid daughter cells (tetrads).

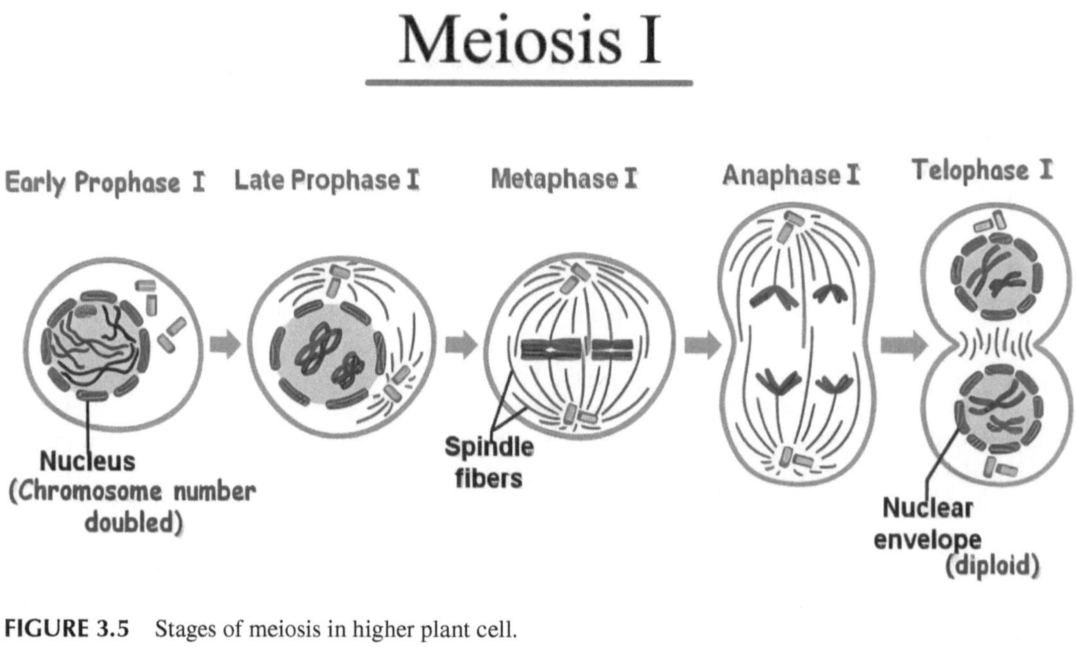

FIGURE 3.5 Stages of meiosis in higher plant cell.

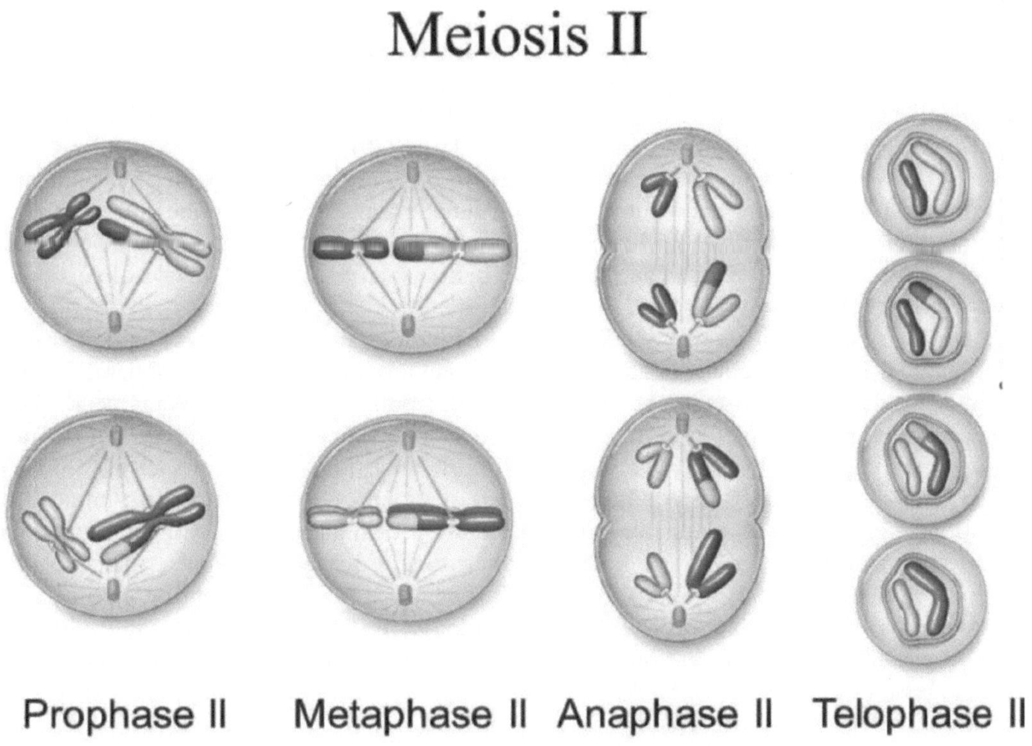

FIGURE 3.6 Stages of meiosis I.

3.4 CELL ELONGATION AND GROWTH

The newly formed daughter cells enlarge in volume by as much as 100,000 or 1,000,000 and in surface area by up to 10,000 times of their original size (McCann et al., 2001). Most of this expansion is accomplished through the increased size of the vacuole. The cell-wall expansion arises from two processes: the establishment of the plane of division in the cell, thus determining the positions of new walls and, then, directing the deposition of new cell-wall material at specific sites of the cell surface (McCann et al., 2001). The primary cell wall formed in the cell plate during cell division is capable of expanding to a great extent. In many plants, mature cells have both a primary and a distinct secondary cell wall.

The expansion of cell wall among the neighboring cells needs to be coordinated properly. Although the size, stature and form of various organs is genetically controlled, the environmental factors also play an important role in regulating the size of individual cells. The environmental signals, such as light, water stress and nutrient status regulate the pattern of cell growth and the ultimately determine the cell size.

The developmental pathways that are responsible for elongation of various cell types involve several growth factors and signal transduction mechanisms. Thus, the cells must be able to integrate a highly complex network of intracellular and extracellular signals to start growth, expansion and, then, stop the growth in a coordinated fashion. Plant cell growth is a cellular process involving the "loosening" of the cell wall, the outward push of the protoplast driven by osmotic forces, and the deposition and intercalation of new wall materials to maintain thickness of the cell wall as the existing wall expands. While the orientation of cellulose microfibrils on the inner surface of the wall establishes the eventual direction of cell expansion, the cytoskeleton essentially determines cell orientation. The direction of cell growth may be influenced by cortical microtubules, whereas the rate of growth is regulated by wall loosening and wall deposition. The cessation of growth is a least understood process, although it plays a key role. The ultimate cell size is known to be associated with chromosome number, chromosome doubling, endoreduplication or polyteny (McCann et al., 2001). At the end of growth, various crosslinking mechanisms in the wall are activated to lock the cell into its final size and shape (McCann et al., 2001). Several models have been proposed to explain the physiological mechanisms of cell expansion. Additionally, new molecular and genetic tools that determine cell identity and biochemical and physiological mechanisms of expansion and differentiation are being used to unravel the process of cell expansion. A detailed discussion of various pathways of cell elongation is available (McCann et al., 2001).

Cell expansion, which is accompanied by a change in microtubule organization in many cases, is under hormonal control. The major hormones involved are auxin, gibberellins (GA), brassinosteroids (BRs) and ethylene (Le et al., 2005; Polko et al., 2012; Wang et al., 2012). However, it is not completely clear whether hormone-induced microtubule reorientation is always accompanied by reorientation of cellulose microfibrils (McCann et al., 2001). At least in short-term, usually in a matter of couple of hours, hormone-treatment-induced cell elongation might not involve concomitant reorientation of microtubules and cellulose microfibrils (Shibaoka, 1993, 1994). For instance, microtubule-disrupting agents, such as colchicine and cellulose synthesis inhibitor (2,6-dichlorobenzonitrile), do not inhibit auxin-induced cell elongation in maize coleoptiles (Hogetsu et al., 1974b; Edelmann et al., 1989). Similarly, GA-induced microtubule reorientation involves different and complex mechanisms (Shibaoka, 1972; Hogetsu et al., 1974a; Shibaoka and Hogetsu, 1977). BRs act through the PM receptor-like kinase (BRI1) to activate members of the BRASSINAZOLE-RESISTANT (BZR) family of transcription factors (Sun et al., 2010). BR-deficient/insensitive mutants are known to affect hypocotyl cell elongation. BR might control cell expansion by changing the mechanical properties of cell walls and may also be affecting microtubule organization (Kauschmann et al., 1996). The BR signaling regulates the reorientation of microtubules from longitudinal to transverse orientation (Wang et al., 2012).

REFERENCES

De Veylder L, Beeckman T and Inzé D. 2007. The ins and outs of the plant cell cycle. *Nature Reviews of Molecular Cell Biology* **8**: 655–665.

De Veylder L, Larki J C and Schnittger A. 2011. Molecular control and function of endoreplication in development and physiology. *Trends in Plant Science* **16**: 624–634.

Edelmann H B, Bergfeld R and Schopfer P. 1989. Role of cell wall biogenesis in the initiation of auxin-mediated growth in coleoptiles of *Zea mays* L. *Planta* **179**: 486–494.

Feiler H S and Jacobs T W. 1990. Cell division in higher plants: a cdc2 gene, its 34-kDa product, and histone H1 kinase activity in pea. *Proceedings of the National Academy of Science, USA* **87**: 5397–5401.

Ferreira P C, Hemerly A S, Villarroel R, Van Montagu M and Inzé D. 1991. The *Arabidopsis* functional homolog of the p34cdc2 protein kinase. *The Plant Cell* **3**: 531–540.

Gutierrez C. 2009. The *Arabidopsis* cell division cycle. *Arabidopsis Book* **7**: e0120.

Hirt H, Pay A, Gyorgyey J, Badó L, Mémeth K, Bogri L, Schweyen R J, Heberle-Bors E and Dudits D. 1991. Complementation of a yeast cel1 cycle mutant by an alfalfa cDNA encoding a protein kinase homologous to p34'd'Z. *Proceedings of National Academy of Sciences, United States of America* **88**: 1636–1640.

Hogetsu T S, Shibaoka H and Shimokoriyama M. 1974a. Involvement of cellulose synthesis in actions of gibberellin and kinetin on cell expansion: gibberellin – coumarin and kinetin – coumarin interactions on stem elongation. *Plant Cell Physiology* **15**: 265–272.

Hogetsu T S, Shibaoka H and Shimokoriyama M. 1974b. Involvement of cellulose synthesis in actions of gibberellin and kinetin on cell expansion: 2,6-dichlorobenzonitrile as a new cellulose-synthesis inhibitor. *Plant Cell Physiology* **15**: 389–393.

Kauschmann A, Jessop A, Koncz C, Szekeres M, Willmitzer L and Altmann T. 1996. Genetic evidence for an essential role of brassinosteroids in plant development. *Plant Journal* **9**: 701–713.

Le J, Vandenbussche F, De Cnodder T, van Der Straeten D and Verbelen J P. 2005. Cell elongation and microtubule behavior in the *Arabidopsis* hypocotyl: responses to ethylene and auxin. *Journal of Plant Growth Regulation* **24**: 166–178.

McCann M C, Stacey N J, Dahiya P, Milioni D, Sado P E and Roberts K. 2001. Zinnia: everybody needs good neighbors *Plant Physiology* **127**: 1380–1382.

Polko J K, van Zanten M, van Rooij J A, Maree A F M, Voesenek L A C J, Peeters J M and Pierik R. 2012. Ethylene-induced differential petiole growth in *Arabidopsis thaliana* involves local microtubule reorientation and cell expansion. *New Phytologist* **193**: 339–348.

Shibaoka H. 1972. Gibberellin – colchicine interaction in elongation of azuki bean epicotyl sections. *Plant Cell Physiology* **13**: 461–469.

Shibaoka H. 1993. Regulation by gibberellins of the orientation of cortical microtubules in plant cells. *Australian Journal of Plant Physiology* **20**: 461–470.

Shibaoka H. 1994. Plant hormone-induced changes in the orientation of cortical microtubules: alterations in the cross-linking between microtubules and the plasma-membrane. *Annual Review of Plant Physiology and Plant Molecular Biology* **45**: 527–544.

Shibaoka H and Hogetsu T. 1977. Effects of ethyl N-phenylcarbamate on wall microtubules and on gibberellin-controlled and kinetin-controlled cell expansion. *The Botanical Magazine Tokyo* **90**: 317–321.

Sun Y, Fan X Y, Cao D M, He K, Tang W, Zhu I-Y, He J-X, Bai M, Zhu S, Oh E, Patil S, Kim T-W, Ji H, Wong W H, Rhee S Y and Wang Z-Y. 2010. Integration of brassinosteroid signal transduction with transcription network for plant growth regulation in Arabidopsis. *Developmental Cell* **19**: 765–777.

Wang X L, Zhang J, Yuan M, Ehrhardt D W, Wang Z Y and Mao T L. 2012. *Arabidopsis* microtubule destabilizing protein40 is involved in brassinosteroid regulation of hypocotyl elongation. *Plant Cell* **24**: 4012–4025.

Wijnker E and Schnittger A. 2013. Control of the meiotic cell division program in plants. *Plant Reproduction* **26**: 143–158.

4 Meristems, Differentiation and Dedifferentiation

4.1 INTRODUCTION

The term meristem, first used by Carl Wilhelm von Nägeli in 1958, refers to the unique structures made up of pluripotent stem cells, a transitory population of indeterminate cells and determinate organ primordia formed at the periphery (Scofield and Murray, 2006; Doerner, 2003). The term meristem is derived from a Greek word "*merizein*" meaning to divide. Meristems are classified based on their position (apical meristems, lateral meristems, leaf meristems, intercalary meristems or floral meristems), origin (promeristem, primary meristems or secondary meristems) or plane of division (mass meristem, plate meristem or rib meristems). The apical (shoot and root) meristems are formed during embryogenesis but are not activated until after germination (Jürgens, 2001). The shoot apical meristems (SAM) in dicots appear in the center between the two cotyledons, whereas they arise laterally at the base of a single cotyledon in monocots (Machida et al., 2013). During the embryogenesis, SAM produces two (dicots) to five (monocot) embryonic leaves. Once the seeds germinate, their meristems become active to resume the production of aboveground organs (stem, leaf, floral parts) from SAM and belowground organs (root cap, root proper) from root apical meristem (RAM) (Barton, 2010). As the plant passes through several developmental phases, such as juvenile vegetative, adult vegetative, flowering and reproduction stages, shoot meristems change their identity (Machida et al., 2013; Doerner, 2003). For instance, in *Arabidopsis*, which is the model system used to study the functioning of meristems, SAM represents leaves during the initial vegetative growth, leaves and axillary meristems during the transition to flowering, and floral meristems and bracts in the inflorescence meristem during reproductive growth (Doerner, 2003). However, root meristems do not change their identity during the development.

Although SAM and RAM are known to be the two most important meristems that provide cells for postembryonic growth and development, other types of meristems (lateral meristem, intercalary meristems, floral meristems and leaf meristems) are formed *de novo* during the life cycle and contribute for the growth of specific organs. The growth and development of plants depend on the balance between the cell proliferation and differentiation in these meristems. By producing new organs almost continuously throughout their life cycle, plants are able to replace the body parts that have been lost due to disease, environmental stress and predators.

Most of what we know about SAM and RAM are the findings of the studies carried out in model systems, such as *Arabidopsis*, snapdragon, tomato, petunia, maize and rice (Barton, 2010). While the same principles apply to all species for most part, it is often possible to identify genes with similar roles in meristem development (Barton, 2010). Because of complete characterization of gene sequence and the availability several mutant lines, the progress in gaining insight into the functioning of apical meristem is in very advanced stage in *Arabidopsis* (The Arabidopsis Genome Initiative, 2001; Alonso et al., 2003; Daxinger et al., 2007).

4.1.1 Model System

Arabidopsis (family: *Brassicaceae*) has a small fully sequenced genome, short life cycle, high transformational efficiency and powerful reverse and forward genetics (Chen et al., 2004; Meyerowitz and Pruitt, 1985; Arabidopsis Genome Initiative, 2000; Somerville and Koornneef, 2002; Figure 4.1). Genome duplication and polyploidization are common in *Arabidopsis* and related species. Hybridization

DOI: 10.1201/9781003373889-5

FIGURE 4.1 Model system of *Arabidopsis thaliana*.

occurs frequently in *Arabidopsis*, giving rise to allopolyploids, and hence, *Arabidopsis* species are either autopolyploid or allopolyploid. They provide a powerful system for understanding evolutionary mechanisms or processes associated with genome duplication and polyploidization. While *Arabidopsis* autopolyploids and allopolyploids occur in nature, they can also be readily be formed in the laboratory. These autopolyploids and allopolyploids can be used for comparing the changes in gene expression and genome structure among relatively "young" and "established" or "ancient" polyploids. Additionally, *Arabidopsis* offers an exceptional system for understanding regulatory mechanisms of gene and genome duplication.

4.2 STRUCTURE AND ORGANIZATION OF MERISTEMS

The meristematic cells are immature, small cells with small vacuole, dense cytoplasm and differentiated rudimentary plastids. They have thin cell wall with a proportionately large nucleus. Meristematic cells are closely packed without any intercellular spaces. Division of a meristematic cell produces two cells; one remains meristematic, while the other differentiates into tissues and various organs of the plant body. In this way, the population of meristem cells is frequently maintained. Molecular evidence suggested that stem cells in animals and meristem cells in plants might share some common molecular mechanisms. However, unlike most stem cells in animals, cells produced by plant meristems have the ability to differentiate as any cell type. Thus, some prefer to think of meristems as populations of embryonic cells because the cells produced go through the entire process of organogenesis, pattern formation, histogenesis and differentiation. Vegetative meristems in plants are indeterminate meaning they can theoretically grow indefinitely. These meristems are also highly regulative and repetitive, often described as reiterative or meristic. Thus, primary shoot growth involves the repetitive addition of stem segments and associated leaves to the end of the shoot. After a phase of vegetative growth during which only stem tissues and leaves are formed, the SAM and each axillary meristem turn into a flower meristem (FM) that bear flowers. FMs have the same organization as the SAMs but generate only flowers, which are composed of sepals, petals, stamens and carpels in four whorls from the outside to the inside of the flower, and have a transient activity, meaning that their activity stops after enough cells have been produced for their recruitment into most important organs (Milksche and Brown, 1965; Steeves and Sussex, 1989). Therefore, they are called the terminating structures. In contrast, the RAM cells give rise to root tissues. Although the SAM and RAM have different structures, they utilize same cell-to-cell communication to maintain proper meristematic activities and meristem organization and to coordinate new organ formation.

4.2.1 Shoot Apical Meristem

The SAM is a collection of cells that continuously divide to provide cells for new shoot organs (Bowman and Eshed, 2000; Haecker and Laux, 2001; Fletcher, 2002; Figure 4.2). During the embryogenesis, SAM is formed in three stages: the first phase determines the axes at the start of the embryogenesis, the second phase generates and senses the regional information to locate the SAM inside of the embryo, and in the final stage, actual formation takes place (Azizi et al., 2015). The SAM continuously produces cells that become incorporated into stem tissue, lateral organs (leaves and flowers) and axillary meristems. The aboveground part of the plant can be divided into iterative modules, called phytomers, each containing a leaf (or leaves) attached at a node, an axillary meristem at the base of the leaves and a subtending internode. After a phase of vegetative growth during which stem tissues and leaves are formed, the SAM and each axillary meristem turn into an inflorescence meristem that bear flowers.

SAM in most angiosperms can be described by the tunica-corpus theory of meristem organization. Tunica is the outer covering of small cells that forms epidermis and outer cortex, whereas corpus is the inner mass of the cells forming the inner core. Schmidt in 1924 proposed the tunica-corpus theory. The tunica layers are characterized by having anticlinal (perpendicular to the surface) cell divisions, whereas the corpus layers have periclinal (parallel to the surface) cell

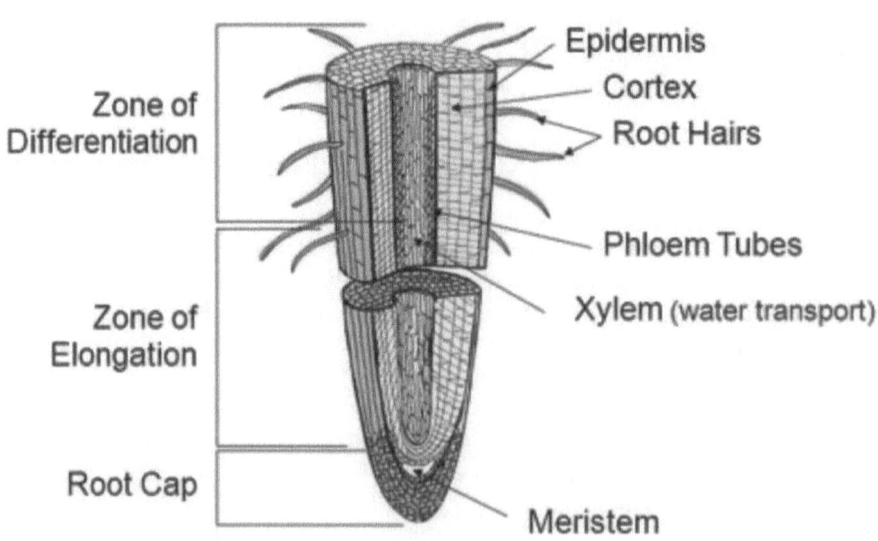

FIGURE 4.2 Shoot apical meristem.

divisions. The cytohistological analyses of SAM in several angiosperm species revealed that it is organized into cell layers (Steeves and Sussex, 1989). The surface region (tunica) consists of one to five cell layers in which cells divide in a single plane. Tunica in most dicots consists of two layers (L1 and L2), whereas the monocots usually have one layer. The underlying corpus (L3) is an arrangement of cells that divide in all planes. These layers generally maintain distinct cell lineages, each generating specific derivatives. In *Arabidopsis*, the L1 contributes to the epidermis of shoots, leaves and flowers, the L2 produces the ground tissues and germ cells, and the L3 contributes to the vascular tissues of the stem and the most internal tissues of leaves and flowers (Carles and Fletcher, 2003). Szymkowiak and coworkers have shown that the development of each clonal layer is flexible and can adapt to cell proliferation changes occurring in other layers (Szymkowiak and Sussex, 1996). The SAM cell layers comprise three meristematic zones having distinct functions. The peripheral zone (PZ) and the rib zone (RZ) contain cells that will become incorporated into lateral organs and the stem core, respectively. The central zone (CZ), which is characterized by a lower mitotic activity (Carles and Fletcher, 2003), constitutes the self-renewing stem cell reservoir that is the source of cells for the PZ and RZ. According to Carles and Fletcher (2003), during primordia initiation, sets of cells in the PZ and RZ become specified as the founder cells of young plant tissues that are replaced by dividing pluripotent cells provided by the CZ. In this way, the SAM sustains itself as a stable structure by homeostasis, in spite of the constant flow of cells passing through it (Betsuyaku et al., 2011; Uchida et al., 2011, 2012, 2013). Shifts in gene expression occur within the cells as their positions within the meristem change to sustain the modular growth of the shoot and maintain SAM homeostasis. Thus, SAM cells determine their appropriate gene expression states based on signal exchanges with other cells in the meristem rather than through a strict lineage-based predetermination of cell fate (Szymkowiak and Sussex, 1996).

In *Arabidopsis*, tunica-corpus structure is not evident until the torpedo stage of embryogenesis, but the apical histological zonation is visible before cotyledon initiation in some species (Kaplan and Cooke, 1997; Barton and Poethig, 1993). Therefore, either the SAM is formed by the apical portion of the globular embryo, or alternatively, it is not formed until the tunica-corpus structure is evident at the late-heart or early-torpedo stage of embryogenesis. Using gene expression patterns as histological markers to analyze the development of the apical portion of the *Arabidopsis* embryo between the globular and the torpedo stages, Long and Barton (1998) and Aida et al. (1999) have concluded that the complex gene expression patterns (histology) of the SAM develop gradually during embryogenesis, both independent and interdependent relationships exist among genes directing SAM establishment and maintenance. The apical portion of the globular embryo is divided into domains demarcated by gene expression patterns with distinct developmental fates.

WUSCHEL (*WUS*) is the first gene expressed in SAM. This is limited to a small group of cells underneath the outer three layers (L3) in the mature SAM, but it is first expressed in the apical subepidermal cells at the 16-cell stage of embryogenesis (Mayer et al., 1998). The expression pattern of *WUS* becomes limited to deeper regions during the formation of the SAM. This suggests that cell–cell interactions dictate the boundaries of its expression domain. Long and Barton (1998) and Aida et al. (1999) suggested that genes, SHOOT *MERISTEMLESS* (*STM*), *CUP-SHAPED COTYLEDON2* (*CUC2*) and *AINTEGUMENTA* (*ANT*), are first expressed in the late globular embryo stage. The expression patterns of these genes divide the apical portion of the embryo into three regions, namely, a central region that give rise to the SAM (*STM* and *CUC2*), a peripheral region, which is further subdivided to form regions that will produce cotyledons (*ANT*) and regions where growth will be suppressed, which form the boundaries between the cotyledons (*STM, CUC2* and *ANT*). *CUC2* expression becomes limited to the boundary regions between the cotyledons and the SAM in the heart stage. This restriction is influenced by the *STM* activity (Aida et al., 1999). *CUC2* is shown to have a role in the separation of organs from the meristem and from each other, in a manner analogous to that of *NO APICAL MERISTEM* in petunia (Aida et al., 1997; Souer et al.,

1996). By early heart stage, after cotyledon primordia have formed, *UNUSUAL FLORAL ORGANS* (*UFO*) and *CLAVATA1* (*CLV1*) are activated in the L2 and L3 of the presumptive SAM (Long and Barton, 1998). *CLV1* acts with *CLV3* and *WUS* to maintain the integrity of the central zone. Both initial *UFO* expression and maintenance of *CLV1* expression requires *STM* activity. This means that *STM* initiates a developmental program required to establish or maintain several components of the SAM (Barton and Poethi, 1993). Therefore, it is proposed that the apical region of the globular embryo is progressively subdivided during development. The establishment of the functional regions of the SAM is a gradual and dynamic process occurring during embryonic pattern formation. The earliest acting genes (*WUS* and *STM*) are required for establishment or maintenance of stem cell fate or, alternatively, repression of differentiation. The genes such as *CLV1* are involved in regulating the size of the central zone (Clark et al., 1993, 1995, 1997; Stone et al., 1998). A range of other mutations with abnormal meristem size has been described. For instance, cell proliferation within the meristem in *wuschel* appears to be reduced, whereas *fasciata* and *mgoun* (*mgo*), like *clv*, show meristem overgrowth (Leyser and Furner, 1992; Laufs et al., 1998b). One of the proposed functions for the *CLV1* and *CLV3* genes is the limitation of cell division rates in the CZ (Clark et al., 1993, 1995, 1997; Laux and Schoof, 1997). Meyerowitz (1997) has proposed a direct role in the coordination of cell division for a range of genes. However, there are no in-depth studies describing the spatial and temporal distribution of cell division and expansion in any mutant.

In spite of continuous production of lateral organs, the number and their dividing derivatives do not change, a remarkable quality referred as homeostasis (Meyerowitz, 1997; Irish, 1991; Poethig, 1989; Heidstra and Sabatini, 2014). This suggests that cells within the SAM have to continually decide to divide, differentiate or remain as they are (Irish, 1991; Poethig, 1989), failure to do so would lead to either an accumulation of cells within the SAM, or SAM will lose its ability to maintain. The accumulation of cells in the central zone is caused by a reduction in the rate of cells being recruited into the peripheral zone (Clark et al., 1993, 1995; Laufs et al., 1998a, 1998b). Studies have suggested *WUS* and *CLV1/CLV* genes, which have opposite effects on the stem cell population of the central zone, either promote or restrict cell division rates, respectively, within the central zone. Based on genetic interactions, the expression patterns and nature of the encoded gene products of *CLV1*, *CLV3* and *WUS* scientists have developed a model of their action (Clark et al., 1997; Fletcher et al., 1999; Laux et al., 1997). *CLV1*, whose mRNA is present mainly in the L3 of the central zone, has been shown to encode a leucine-rich repeat (LRR) receptor kinase (Clark et al., 1997; Fletcher et al., 1999). *CLV3*, whose mRNA is primarily restricted to the L1 and L2 of the central zone, encodes a small, putatively secreted protein (Fletcher et al., 1999). Because *CLV1* and *CLV3* have been shown to act in a common pathway (Clark et al., 1995), they might act as a receptor–ligand pair in a signal transduction cascade that restricts cell division rates in the central zone (Fletcher et al., 1999). *CLV1* and *CLV3* genes act to regulate central zone size. It has been suggested that *WUS*-expressing cells act as an organizing center, conferring stem cell identity to overlying neighboring cells in a similar manner of the quiescent center in the root meristem (Mayer et al., 1998; Barlow, 1974; van den Berg et al., 1997). The *CLV1/CLV3* signaling pathway may potentially act to negatively regulate the activity of *WUS* directly. Thus, it is possible that *WUS* promotes stem cell fate non-cell autonomously among cells of the central zone and that the *CLV1/CLV3* signaling pathway dampens this promotion by restricting cell division within the peripheral zone. One of the hypotheses states that the lateral organ primordia communicate their formation to the SAM, resulting in a replenishment of the peripheral zone from the cells derived from the central zone.

The effects of signals emanating from mature leaves on the fate of the apical meristem have already been established. However, based on the localized exogenous application of the cell-wall-loosening protein EXPANSIN to the organ anlage of live tomato apices, it has been shown that such effects also occur during primordia initiation (Fleming et al., 1997; Reinhardt et al., 1998; Green, 1999). Complexities of the relationships between SAMs and lateral organs are further exposed by the analysis of the maize ortholog of *PHAN – ROUGH SHEATH2* (*RS2*) (Timmermans et al., 1999; Tsiantis et al., 1999). Leaves of *rs2* mutants appear similar to gain-of-function alleles of the SAM-specific *KNOTTED* class I genes (Schneeberger et al., 1998; Freeling and Hake, 1985; Vollbrecht et al., 1991). Several genes of that group were shown to be misregulated in either *phan* or *rs2* mutants (Timmermans et al., 1999; Tsiantis et al., 1999), leading to the concept that *PHAN* and *RS2* might have different functions in *Antirrhinum* and *Zea* leaves, respectively (Timmermans et al., 1999; Tsiantis et al., 1999; Schneeberger et al., 1998). Specifically, it was suggested that *RS2* could be involved in establishing the proximal–distal axis rather than the abaxial–adaxial axis in developing leaves (Tsiantis et al., 1999). However, the development of these two axes might be linked (Waites and Hudson, 1995; Waites et al., 1998; Siegfried et al., 1999). Thus, analysis of orthologous genes in other species might be required to explain this issue.

4.2.2 Root Apical Meristem

Compared to SAMs, the root apical meristem (RAM) or root apex received lesser attention because they are underground structures and considered to be responsible only for anchoring the plants. However, a number of researchers attempted to characterize vascular plants based on the organization of root apical meristems (Heimsch, 2008). Recently, researchers have tried to understand the basis of root structure and development using the mutants of *Arabidopsis*.

Primary root growth occurs due to root apical meristem (RAM) located above the protective root cap. Unlike the SAM, the RAM produces cells in two directions. It produces tissues of the root cap, which covers the distal tip of roots. The root cap protects the root tip as it grows through the soil. Cells are continuously sloughed off from the outer surface of the root cap. The RAM also produces cells inside that contribute to the root proper, but unlike the SAM, the RAM produces no lateral appendages.

The root can be divided into three regions: (i) zone of cell division, (ii) zone of elongation and (iii) zone of maturation. The zone of cell division contains the apical meristem. Most of the root growth occurs in the elongation zone, where cells expand. The maturation zone produces the root hairs and lateral roots. The protoderm forms the epidermis, the procambium becomes vascular tissues and the ground meristem becomes the cortex. The center of the root apical meristem called the quiescent zone contains a region where relatively fewer cell divisions occur. Although cell division is slow in this region, the quiescent zone plays an important role in establishing tissue organization in the root meristem.

Cell Signaling and RAM Organization: Laser ablation studies in *Arabidopsis* indicated that the position determines the cell fate (identity). For example, if a cortical initial is ablated, a pericycle cell will invade the void, assume the pattern of cell division characteristic of a cortical initial and give rise to cortical and endodermal derivatives. Thus, the cell changes the fate from pericycle to cortical initial when its position is altered. It has also been demonstrated that the differentiated cells proximal to the RAM influence the fate of cells in the RAM. The more mature, differentiated cells direct the pattern of division and differentiation of the initial cells (van den Berg et al., 1995). The quiescent center regulates the differentiation of neighboring cells. If one or two QC cells were ablated, the columnella initials that were no longer in direct contact with a QC cell cease dividing and undergo differentiation as a columnella cell. Similarly, cortical initial cells that were no longer in direct contact with a QC cell behave as cortical daughter cells and undergo asymmetric division to generate cortical and endodermal cell files. Therefore, direct contact with the QC inhibits the differentiation of initial cells in the RAM (van den Berg et al., 1997). The cells in the quiescent center position the stem cell niche and prevent the differentiation of the surrounding stem cells or initials (van den Berg et al., 1997). Polar auxin transport is required to position and restrict stem cells along the proximal-distal primary axis. Elevated auxin concentration leads to an expansion of the stem cell population (Sabatini et al., 1999). The genes responsible for patterning the meristem and maintaining stem cells have remained obscure; however, the function of *SCARECROW* (*SCR*) in stem cell maintenance and positioning of the stem cell niche has been more closely characterized (Sabatini et al., 2003). *SCR* and related genes have a joint role in the establishment and maintenance of root and shoot meristems (Stuurman et al., 2002; Sabatini et al., 2003; Benfey et al., 1993).

Cytokinin and RAM Function: Cytokinin (CK) inhibits root growth, and reduced CK level promote the expression of CKX genes (Werner et al., 2001, 2003). An increase in root meristem size was observed in the ipt3, ipt5, ipt7 triple mutant and the ahk3 receptor mutant in which CK biosynthesis or signaling is reduced (Dello Ioio et al., 2007). Based on detailed characterization of the enlarged root meristems, the same researchers showed that the differentiation rate of meristematic cells are decreased in mutants which leads to the accumulation of meristematic cells in the root meristem (Dello Ioio et al., 2007). Thus, in roots, CKs control the rate of meristematic cell differentiation. These authors have further indicated that this function is mediated in a spatial manner by CKs (Dello Ioio et al., 2007).

Auxin and RAM Organization: Disruption of localized accumulation by auxin transport inhibitors, exogenous auxin application or mutations in genes involved with auxin response disrupts the organization of cell types in the root apex, as visualized by promoter trap cell markers. Thus, proper localization and perception of auxin are important for organizing the RAM (Sabatini et al., 1999).

4.2.3 Floral Meristems

Under proper conditions, a vegetative SAM becomes broader (less dome shaped) and transit to a floral meristem. The floral appendages are initiated in whorls at the periphery of the meristem in certain order starting with the outer sepals, petals, stamens and innermost pistil.

4.2.4 Lateral Meristems

Lateral meristems are responsible for increase in diameter (or axial growth) of the stem or root. Lateral meristems include two basic types, the vascular cambium and the cork cambium (phellogen). The meristematic cells in the cambium are called fusiform initials. The fusiform initials are longer and wide and are wedge shaped at both ends. They are either storied or non-storied cambiums. The cells are shorter in storied cambium, while they are more elongated in non-storied cambium and have greater overlapping of cells. These cells undergo periclinal divisions to create cells to both sides within the cambial zone. Xylem cells are produced to one side and phloem cells to the other side of the cambium. One set of division produces xylem and another set gives rise to phloem.

The cambium tissue has another type of initials called the ray initials, which are formed either as cutout portion of fusiform initial or the entire fusiform initial may form transverse section. These initials are smaller and less elongated than fusiform initials and produce ray cells. The ray cells form files of living cells within secondary vascular systems and serve the storage function.

In young stems, the xylem and phloem tissues in vascular bundles appeared to be separated. The cambium within the vascular bundle is called fascicular cambium, whereas the one between vascular bundle is called the interfascicular cambium. With age, the vascular system and cambium form a complete ring within the stem to form a secondary vascular system.

In addition to vascular cambiums, higher plants contain cork cambiums (phellogens) that are responsible for axial growth (axillary growth). The cork cambium produces periderm cells, which replaces epidermis and protects the bark tissues in woody stem. The periderm is made of three layers, the outer phellem (cork layer), the middle phellogen (cork cambium) and the inner phelloderm (parenchyma cells).

4.2.5 Leaf Meristems

Leaves of higher plants also contain apical and marginal meristems. While the apical meristems produce longitudinal growth, the marginal meristems present along the edge of the leaves are responsible for differential development of the leaf. The two meristems, together, are responsible for the size and various shapes of leaves. It is currently thought that the leaf primordium produces topological meristematic regions along the leaf margin termed blastozones. Segmentation and differential development within the blastozone lead to the observed variations in leaf shape. Leaf formation in higher plants follows three basic stages, initiation of a leaf primordium on either side of the apical dome, followed by primary morphogenesis where meristematic blastozones are formed along the leaf margin and, finally, expansion and differentiation of cells into specific tissue types. The lateral growth of a leaf occurs with a relatively uniform thickness. This type of growth where parallel tissue layers divide anticlinally is termed a plate meristem.

4.2.6 Intercalary Meristem

The meristem present at the nodes of monocots shoots is called intercalary meristems. Monocots produce numerous compact nodes behind the apical meristem that later elongate using intercalary meristems. The growth and cell division that occur away from where the cells were initially formed in the original apical meristem are called the intercalary growth. In the tall monocots like palms and bamboo, the extension growth from nodal intercalary meristems can be quite dramatic. In these plants, the apical meristem produces numerous nodes early in development near the plant base. Stem extension growth proceeds with cells at each node dividing and elongating from the intercalary meristem. Some monocot leaves also produce an intercalary meristem in the form of a narrow band in the collar region near the leaf base. These meristems account for much of the leaf blade expansion growth. They can resume growth if a portion of the immature leaf is removed (mowing turf grass or lawn).

4.3 DIFFERENTIATION

Young cells are multifunctional, but during plant development, their fate depends on their developmental context. This is accomplished through acquisition of one specific function and unique shape, morphological, biochemical and physiological properties. This process, called "differentiation," creates orderly structures and complex compounds. The common examples of cell differentiation in horticultural plants include changes in color of leaves in autumn, production of wax on the skins, production of phenols, vitamins, aromatic substances and tannin in the tree's bark, ripening of fruits, the growth of shoot ends in a terminal bud, and onset of dormancy in seeds and vegetative buds.

Growth and differentiation can occur simultaneously or very close to each other. For example, production of axillary buds above each leaf appears as multifunctional buds in spring. These cells differentiate into either flower-bud cells or leaf-bud cells, depending on conditions, and again in late summer, amino acids and carbohydrates are stored in these buds as reserve for growth in the following spring. Growers can accelerate differentiation by shoot pruning and/or subjecting the plants to different photoperiods and vernalization treatments.

Differentiation processes in the carrot are reflected in the production of finely divided leaves and the production of scents and pigments, which give the carrot its characteristic taste and aroma. Therefore, a well-differentiated carrot has more intense color and characteristic aroma than rapidly growing, less-differentiated carrots.

4.3.1 Balance Between Growth and Differentiation

Proper balance between cell growth and differentiation is necessary to secure optimum yield and product quality in horticultural crops. Growers try to strike a balance between growth and differentiation by following modern horticultural practices.

4.4 TISSUE TYPES

During periods of growth, the cells of apical meristems divide and continually add more cells to the tips of a seedling's body to increase the length. Primary growth in plants is brought about by the apical meristems and plant tissues arising from primary growth are called primary tissues.

The elongation of the root and stem forms produces the primary plant body, which is made up of primary tissues. The apical meristem gives rise to three types of embryonic tissue systems called primary meristems. These are protoderm, which forms the epidermis, the procambium, which produces primary vascular tissues (primary xylem and primary phloem), and the ground meristem, which differentiates further into ground tissue and composed of parenchyma cells. Many herbaceous plants exhibit only primary growth, but others exhibit secondary growth arising from lateral meristems. Tissues formed by lateral meristems are known as secondary tissues and are collectively called the secondary plant body.

4.5 DEDIFFERENTIATION AND REDIFFERENTIATION

Plants show another interesting phenomenon. The living differentiated cells lose their capacity to divide. However, under certain conditions, they can resume their capacity to divide (totipotency), and their cells start to divide again. This phenomenon is termed as dedifferentiation. The formation of interfascicular and cork cambiums from fully differentiated parenchyma cells is an example of this phenomenon. While doing so, such meristems are able to divide and produce cells but lose their capacity to perform specific functions. These cells can differentiate again through the process of redifferentiation.

REFERENCES

Aida M, Ishida T, Fujisawal H and Tasaka M. 1997. Genes involved in organ separation in *Arabidopsis*: an analysis of the *cup-shaped cotyledon* mutant. *Plant Cell* **9**: 841–857.

Aida M, Ishida T and Tasaka M. 1999. Shoot apical meristem and cotyledon formation during *Arabidopsis* embryogenesis: interaction among the *CUP-SHAPED COTYLEDON* and *SHOOT MERISTEMLESS* genes. *Development* **126**: 1563–1570.

Alonso J M, Stepanova A N, Leisse T J, Kim C J, Chen H, Shinn P, Stevenson D K, Zimmerman J, Barajas P, Cheuk R, Gadrinab C, Heller C, Jeske A, Koesema E, Meyers C C, Parker H, Prednis L, Ansari Y, Cho N, Deen H, Geralt M, Hazari N, Hom E, Karnes M, Mulholland C, Ndubaku R, Schmidt I, Guzman P, Aguilar-Henonin L, Schmid M, Weigel D, Carter D E, Marchand T, Risseeuw E, Brogden D, Zeko A, Crosby W L, Berry C C and Ecker J R. 2003. Genome-wide insertional mutagenesis of *Arabidopsis thaliana*. *Science* **301**: 653–657.

Arabidopsis Genome Initiative. 2000. Analysis of the genome sequence of the flowering plant *Arabidopsis thaliana*. *Nature* **6814**: 796–815.

The Arabidopsis Genome Initiative. 2001. Analysis of the genome sequence of the flowering plant Arabidopsis thaliana. *Nature* **408**: 796–815.

Azizi P, Rafii M A, Maziah M, Abdullah S N A, Hanafi M M, Latif M A, Rashid A A and Sahebi M. 2015. Understanding the shoot apical meristem regulation: a study of the phytohormones, auxin and cytokinin, in rice. *Mechanisms of Development* **135**: 1–15.

Barlow P. 1974. Regeneration of the cap of primary roots of *Zea mays*. *New Phytologist* **73**: 937–954.

Barton M K. 2010. Twenty years on: the inner workings of the shoot apical meristem, a developmental dynamo. *Developmental Biology* **341**: 95–113.

Barton M K and Poethig R S. 1993. Formation of the shoot apical meristem in *Arabidopsis thaliana*: an analysis of development in the wild type and in the shoot meristemless mutant. *Development* **119**: 823–831.

Benfey P N, Linstead P J, Roberts K, Schiefelbein J W, Hauser M T and Aeschbacher R A. 1993. Root development in *Arabidopsis*: four mutants with dramatically altered root morphogenesis. *Development* **119**: 57–70.

Betsuyaku S, Takahashi F, Kinoshita A, Miwa H, Shinozaki K and Fukuda H. 2011. Mitogen-activated protein kinase regulated by the CLAVATA receptors contributes to shoot apical meristem homeostasis. *Plant and Cell Physiology* **52**: 14–29.

Bowman J L and Eshed Y. 2000. Formation and maintenance of the shoot apical meristem. *Trends in Plant Science Review* **5**: 110–115.

Carles C C and Fletcher J C. 2003. Shoot apical meristem maintenance: the art of a dynamic balance. *Trends in Plant Science* **8**: 394–401.

Chen F, Ro D K, Petri J, Gershenzon J, Bohlmann J, Pichersky E and Tholl D. 2004. Characterization of a root-specific Arabidopsis terpene synthase responsible for the formation of the volatile monoterpene 1, 8-cineole. *Plant Physiology* **135**: 1956–1966.

Clark S E, Running M P and Meyerowitz E M. 1993. CLAVATA1, a regulator of meristem and flower development in Arabidopsis. *Development* **119**: 397–418.

Clark S E, Running M P and Meyerowitz E M. 1995. CLAVATA3 is a specific regulator of shoot and floral meristem development affecting the same processes as CLAVATA1. *Development* **121**: 2057–2067.

Clark S E, Running M P and Meyerowitz E M. 1997. The CLAVATA1 gene encodes a putative receptor kinase that controls shoot and floral meristem size in Arabidopsis. *Cell* **89**: 575–585.

Daxinger L, Hunter H, Sheikh M, Jauvion V, Gasciolli V, Vaucheret H, Matzke M and Furner I. 2007. Unexpected silencing effects from T-DNA tags in Arabidopsis. *Trends Plant Science* **13**: 359–367.

Dello Ioio R, Scagila Linhares F, Scacci E, Casamitjana-Martinez E and Sabatini S. 2007. Cytokinins determine Arabidopsis root-meristem size by controlling cell differentiation. *Current Biology* **17**: 678–682.

Doerner P. 2003. Plant meristems: a merry-go-round of signals. *Current Biology* **13**: R368–R374.

Fleming A J, McQueen-Mason S, Mandel T and Kuhlemeier C. 1997. Induction of leaf primordia by the cell wall protein expansin. *Science* **276**: 1415–1418.

Fletcher J C. 2002. Coordination of cell proliferation and cell fate decisions in the angiosperm shoot apical meristem. *Bioessays* **24**: 27–37.

Fletcher J C, Brand U, Running M P, Simon R and Meyerowitz E M. 1999. Signaling of cell fate decisions by *CLAVATA3* in *Arabidopsis* shoot meristems. *Science* **283**: 1911–1914.

Freeling M and Hake S. 1985. Developmental genetics of mutants that specify knotted leaves in maize. *Genetics* **111**: 617–634.

Green P B. 1999. Expression of pattern in plants: combining molecular and calculus-based biophysical paradigms. *American Journal of Botany* **86**: 1059–1076.

Haecker A and Laux T. 2001. Cell-cell signaling in the shoot meristem. *Current Opinion in Plant Biology* **4**: 441–446.

Heidstra R and Sabatini S. 2014. Plant and animal stem cells: similar yet different. *National Review of Molecular Cell Biology* **15**: 301–312.

Heimsch C and Seago Jr J L. 2008. Organization of the root apical meristems in angiosperms. *American Journal of Botany* **95**: 1–21.

Irish V F. 1991. Cell lineage in plant development. *Current Opinion in Genetics and Development* **1**: 169–173.

Jürgens G. 2001. Apical – basal pattern formation in Arabidopsis embryogenesis. *The EMBO Journal* **20**: 3609–3616.

Kaplan D R and Cooke T J. 1997. Fundamental concepts in the embryogenesis of dicotyledons: a morphological interpretation of embryo mutants. *Plant Cell* **9**: 1903–1919.

Laufs P, Dockx J, Kronenberger J and Traas J. 1998b. *MGOUN1* and *MGOUN2*: two genes required for primordium initiation at the shoot apical and floral meristems in *Arabidopsis thaliana. Development* **125**: 1253–1260.

Laufs P, Grandjean O, Jonak C, Kiêu K and Traas J. 1998a. Cellular parameters of the shoot apical meristem in Arabidopsis. *Plant Cell* **10**: 1375–1389.

Laux T, Mayer K F, Berger J and Jürgens G. 1997. The WUSCHEL gene is required for shoot and floral meristem integrity in Arabidopsis. *Development* **122**: 87–96.

Laux T and Schoof H. 1997. Maintaining the shoot meristem – the role of CLAVATA1. *Trends in Plant Science* **2**: 325–327.

Leyser H M O and Furner I J. 1992. Characterisation of three shoot apical meristem mutants of *Arabidopsis thaliana. Development* **116**: 397–403.

Long J A and Barton D K. 1998. The development of apical embryonic pattern in *Arabidopsis*. *Development* **125**: 3027–3035.

Machida Y, Fukaki H and Araki T. 2013. Plant meristems and organogenesis: the new era of plant developmental research. *Plant and Cell Physiology* **54**: 294–301.

Mayer F F X, Schoof H, Haecker A, Lenhard M, Jürgens G and Laux T. 1998. Role of *WUSCHEL* in regulating stem cell fate in the *Arabidopsis* shoot meristem. *Cell* **95**: 805–815.

Meyerowitz E M. 1997. Genetic control of cell division patterns in developing plants. *Cell* **88**: 297–308.

Meyerowitz E M and Pruitt R E. 1985. Arabidopsis thaliana and plant molecular genetics. *Science* **229**: 1214–1218.

Milksche J P and Brown J A M. 1965. Development of vegetative and floral meristems of *Arabidopsis thaliana. American Journal Botany* **52**: 533–553.

Poethig S. 1989. Genetic mosaics and cell lineage analysis in plants. *Trends in Genetics* **5**: 272–277.

Reinhardt D, Wittwer F, Mandel T and Kuhlemeir C. 1998. Localized upregulation of a new expansin gene predicts the site of leaf formation in the tomato meristem. *Plant Cell* **10**: 1427–1437.

Sabatini S, Beis D, Wolkenfelt H, Murfett J, Guilfoyle T, Malamy J, Benfey P, Leyser O, Bechtold N and Weisbeek P. 1999. An auxin-dependent distal organizer of pattern and polarity in the Arabidopsis root. *Cell* **99**: 463–472.

Sabatini S, Heidstra R, Wildwater M and Scheres B. 2003. *SCARECROW* is involved in positioning the stem cell niche in the *Arabidopsis* root meristem. *Genes and Development* **17**: 354–358.

Schneeberger R, Tsiantis M, Freeling M and Langdale J A. 1998. The *rough sheath2* gene negatively regulates homeobox gene expression during maize leaf development. *Development* **125**: 2857–2865.

Scofield S and Murray J A. 2006. The evolving concept of the meristem. *Plant Molecular Biology* **60**: 5–7.

Siegfried K R, Eshed Y, Baum J F, Otsuga D, Drews G N and Bowman J L. 1999. Members of the YABBY gene family specify abaxial cell fate in *Arabidopsis. Development* **126**: 4117–4128.

Somerville C and Koornneef M. 2002. A fortunate choice: the history of Arabidopsis as a model plant. *Nature Reviews Genetics* **3**: 883–889.

Souer E, van Houwelingen A, Kloos D, Mol J and Koes R. 1996. The *No Apical Meristem* gene of petunia is required for pattern formation in embryos and flowers and is expressed at meristem and primordia boundaries. *Cell* **85**: 159–170.

Steeves T A and Sussex I M. 1989. *Patterns in Plant Development*. Cambridge University Press, Cambridge.

Stone J M, Trotochaud A E, Walker J C and Clark S E. 1998. Control of meristem development by CLAVATA1 receptor kinase and kinase-associated phosphatase interactions. *Plant Physiology* **117**: 1217–1225.

Stuurman J, Jäggi F and Kuhlemeier C. 2002. Shoot meristem maintenance is controlled by a GRAS-gene mediated signal from differentiating cells. *Genes & Development* **16**: 2213–2218.

Szymkowiak E J and Sussex I M. 1996. What chimeras can tell us about plant development. *Annual Review of Plant Physiology and Plant Molecular Biology* **47**: 351–376.

Timmermans M C P, Hudson A, Becraft P W and Nelson T. 1999. ROUGH SHEATH2: a myb protein that represses knox homeobox genes in maize lateral organ primordia. *Science* **284**: 151–153.

Tsiantis M, Schneeberger R, Golz J F, Freeling M and Langdale J A. **1999**. The maize *rough sheath2* gene and leaf development programs in monocot and dicot plants. *Science* **284**: 154–156.

Uchida N, Igari K, Bogenschutz N L, Torii K U and Tasaka M. 2011. Arabidopsis ERECTA-family receptor kinases mediate morphological alterations stimulated by activation of NB-LRR-type UNI proteins. *Plant and Cell Physiology* **52**: 804–814.

Uchida N, Lee J S, Horst R J, Lai H H, Kajita R and Kakimoto T. 2012. Regulation of inflorescence architecture by intertissue layer ligand – receptor communication between endodermis and phloem. *Proceedings National Academy of Sciences of USA* **109**: 6337–6342.

Uchida N, Shimada M and Tasaka M. 2013. ERECTA-family receptor kinases regulate stem cell homeostasis via buffering its cytokinin responsiveness in the shoot apical meristem. *Plant and Cell Physiology* **54**: 343–351.

van den Berg C, Willemsen V, Hage W, Weisbeek P and Scheres B. 1995. Cell fate in the Arabidopsis root meristem determined by directional signaling. *Nature* **378**: 62–65.

van den Berg C, Willemsen V, Hendricks G P, Weisbeek P and Scheres B. 1997. Short-range control of cell differentiation in the Arabidopsis root meristem. *Nature* **390**: 287–289.

Vollbrecht E, Veit B, Sinha N and Hake S. 1991. The developmental gene *Knotted-1* is a member of a maize homeobox gene gamily. *Nature* **350**: 241–243.

Waites R and Hudson A. 1995. *phantastica*: a gene required for dorsoventrality of leaves in in Antirrhinum majus. *Development* **121**: 2143–2154.

Waites R, Selvadurai H R N, Oliver I R and Hudsom A. 1998. The *phantastica* gene encodes a MYB transcription factor involved in growth and dorsoventrality of lateral organs in *Antirrhinum. Cell* **93**: 779–789.

Werner T, Motyka V, Laucou V, Smets R, Van Onckle H and Schmüllin T. 2003. Cytokinin-deficient transgenic Arabidopsis plants show multiple developmental alterations indicating opposite functions of cytokinins in the regulation of shoot and root meristem activity. *Plant Cell* **15**: 2532–2550.

Werner T, Motyka V, Strnad M and Schmulling T. 2001. Regulation of plant growth by cytokinin. *Proceedings National Academy of Sciences of USA* **98**: 10487–10492.

5 Vascular Cambium, Xylem and Phloem

5.1 ORIGIN OF VASCULAR CAMBIUM

Meristems allow plants to continue their growth and to produce new structures throughout their lifetime (Evert, 2006a). Plant stem cells, the "meristematic cells" or the "initials," are the cell reservoirs that can both produce new cells for the generation of new tissues and organs and self-maintain indeterminately via asymmetric cell division (Kucukoglu, 2015). The asymmetric division of an initial cell generates two daughter cells; one remains as an undifferentiated initial, whereas the other one (or the "derivative of an initial") is destined to proceed with differentiation pathways (Evert, 2006a). These derivatives can directly differentiate or may divide several rounds before they differentiate into different tissues (Evert, 2006a). Plant stem cells are "pluripotent" in nature, so they can produce most types of cells in the plant body (Verdeil et al., 2007). In fact, all plant cells are "developmentally totipotent" because they are competent to dedifferentiate and generate new cells and even a new plant under certain circumstances (Evert, 2006b). The study of vascular cambium is fundamental to gaining insight into secondary growth in woody plants and advancing our understanding of the influence of internal and external factors on the quality of wood (Jurra et al., 2005).

During embryogenesis, zygote cells divide to generate three primary meristematic regions (protoderm, ground meristem and procambium) as well as the root and shoot apical meristems (ten Hove et al., 2015). Procambium, whose developmental pattern is established in the embryo itself, contains the stem cells to produce conductive tissues of the primary plant body (Beck, 2010). The primary vascular system in stems and roots as well as in tissues associated with them is called "stele." In dicots, the procambium within the vascular bundle is referred to as "fascicular cambium," and the region between vascular bundles is called the "interfascicular region." The periclinal divisions (in parallel to the surface plane of an organ) of the procambium cells generate primary xylem and primary phloem leading to the formation of collateral bundles. In these collateral bundles, the primary xylem is formed toward the pith, and primary phloem is formed toward the surface of the plant (Kucukoglu, 2015). Unlike in the stems, the primary vascular system in roots is arranged in a bisymmetric pattern in which central primary xylem is enclosed by primary phloem with procambium cells lying between xylem and phloem (Beck, 2010).

The outermost layer of the primary meristem forms the epidermis, and central ground tissues (pith and cortex) lie beneath the epidermis, called the pith and cortex. These tissues merge with the procambium to form the innermost layer of the primary meristem. While the primary meristem produces longitudinal growth (upwards), the secondary meristem generates lateral growth (increase in girth of the organ) outwards. The vascular cambium lets plants to expand in girth and ultimately forms the protective layers of fibers which are necessary for growth and development. While the vascular cambium produces secondary xylem and phloem, the cork cambium produces cork cells, which promote the growth of bark and provide protection from the elements and environmental damage.

As the plants grow larger, a layer of undifferentiated procambium cells within the vascular bundles remains active even when the primary growth is completed (Evert, 2006c, 2006d; Beck, 2010). These cells get dedifferentiated into interfascicular cambium cells and regain their ability to divide. The fascicular cambium and the interfascicular cambium together constitute the "vascular cambium" and create a complete cylinder of meristematic cells within the stem (Evert, 2006c, 2006d; Beck, 2010).

DOI: 10.1201/9781003373889-6

In roots, vascular cambium arises from the activity of the procambium and the pericycle cells (Bishop et al., 2013). Vascular cambium contains the vascular stem cells to produce the secondary vascular tissues like secondary phloem (Evert, 2006c, 2006d; Beck, 2010). The cork cambium forms a multilayered protective structure called the "periderm" on the surface of the shoots throughout the secondary growth by producing the phellem (cork) and phelloderm (cortex) (Evert, 2006c; Beck, 2010).

5.2 OCCURRENCE AND DISTRIBUTION OF VASCULAR CAMBIUM

Vascular cambium is present in angiosperms, gymnosperms and extinct group of pteridophytes. In stems, it can be fascicular (FC) or interfascicular (IC) in position (Gokhale and Kavathekar, 2012). The fascicular cambium originates from procambium within a vascular bundle (called fascicle). It divides periclinally to produce secondary xylem inside and secondary phloem outside. Interfascicular cambium arises between the vascular bundles (Gokhale and Kavathekar, 2012).

5.3 STRUCTURE OF CAMBIAL CELLS

Unlike other meristematic cells, cambial cells do not have internal cellular organization, have less dense cytoplasm and are highly vacuolated (Gokhale and Kavathekar, 2012). The radial walls of primary xylem and phloem are thicker than tangential walls. Cells of procambium have gabled ends and do not differentiate into long fusiform and isodiametric ray cells. Mature cambial cells are morphologically heterogeneous and can be distinguished into two fusiform and ray initial cells (RCCs). Fusiform initial cells are highly elongated, prismatic in the center and wedge shaped toward the ends (Gokhale and Kavathekar, 2012; Evert, 2006d; Beck, 2010). The derivatives of fusiform initials always divide periclinally and show characteristic growth patterns involving progressive elongation reaching a plateau leading eventually to death. However, in some species, plateau is not reached even after hundreds of years. Storied cambia have shorter fusiform initials than non-storied cambia. Fusiform initials are the only cell types that divide along their longitudinal axis, thereby defying Errera's law of cell division which states that "the cell plate, at the time of its formation, adopts the geometry that a soap film would take under the same conditions" (Errera, 1886; Wright and Smith, 2008).

The length of the fusiform initials varies depending on the species and the developmental age of the vascular cambium with older fusiform initials being longer than the newly formed fusiform initials (Evert, 2006d; Beck, 2010). The products of fusiform initials are tracheids, vessel members, xylem fibers and xylem parenchyma. Together, these fusiform products form the secondary xylem. They also produce secondary phloem components, such as phloem parenchyma, phloem fibers, sieve cells, aluminous cells (Strasbuger's cells) and sieve-tube members along with companion cells (Evert, 2006d; Beck, 2010).

The nearly isodiametric RCCs produce the ray parenchyma cells extending from the secondary phloem to the secondary xylem inside the trunks of stems or roots and are responsible for the translocation of water, nutrients and photosynthetic assimilates, apart from serving as storage tissues (Beck, 2010). The periclinal or the additive divisions in the vascular cambium increase the number of cells within the secondary xylem and the secondary phloem, although they occur in a greater frequency on the xylem side than on the phloem side (Evert, 2006d; Beck, 2010). In active cambial cells, cell division occurs at a much higher rate than cell differentiation leading to formation of a cambial zone. This zone comprises a single layer of true cambial cells (initials), their immediate derivatives and undifferentiated or partially differentiated xylem and phloem cells (Plomion et al., 2001). The cambial zone is not affected by continued differentiation of the cambial derivatives into xylem and phloem. The cambial zone is wide in active cambium, whereas it is only single layer thick in the dormant state. Cells of proper cambial zone are mitotically active but their mitotic rate decreases as one moves toward the periphery of the zone (Evert, 2006d). Fusiform and ray initial cells are inter-convertible (Gokhale and Kavathekar, 2012). Formation of ray initial from fusiform initial may be brought about by lateral partitioning, direct partitioning and septation. Reverse pattern of transformation of ray initial to fusiform initial is although present, but uncommon.

The vascular cambium cells undergo significant changes during the dormant period. Their walls thicken; large central vacuoles get divided to form smaller vacuoles, cytoplasmic streaming stops and the level of storage products (lipid) drops, but starch grains accumulate in the cell (Evert, 2006d). With the return of favorable weather in early spring, the cambial cells expand and regain normal features and resume normal functions (Evert, 2006d; Beck, 2010).

5.3.1 Cambium Types

Depending upon the number and arrangement of fusiform initials in tangential view, cambium can be either stratified (fusiform initials are arranged in horizontal rows in such a way that their ends are approximately at the same level) or non-stratified (fusiform initials partially overlap and their ends do not lie at the same level).

5.3.2 Seasonal Activity

Being highly sensitive to seasonal changes, plants prepare themselves to face these challenges well in advance. This is evident especially in temperate trees, in which the cell size is reduced and the thickness of the vessels is increased and foliages are dropped in preparation to face the cold temperatures during winter. These trees regain the normal size of the vessel with the onset of spring to cope with increased demands from emerging new foliage for higher conductance of water and nutrients. Increasing scientific evidence has shown the role of hormones in cambial cell activities and dynamics of xylogenesis and the mechanism of vessel size control along the stem.

5.4 SECONDARY GROWTH IN STEMS AND ROOTS

The main shoot, root, lateral branches and sometimes leaves of gymnosperms, woody plants and herbaceous dicots exhibit secondary growth. The fascicular and interfascicular cambium together produces secondary xylem centripetally and secondary phloem centrifugally. The ring formation continues in different cycles of secondary growth; each cylinder has their axial and radial systems in turn derived from fusiform and ray initials, respectively. Formation of secondary xylem and secondary phloem often puts a lot of pressure on the primary xylem and phloem. In a way, secondary xylem and phloem intrude the space between primary xylem and phloem and lead to significant alteration in the internal structure of a stem. The pith shrinks in size as its space is almost taken over by the secondary xylem, which also takes over the primary xylem structurally and functionally. Likewise, the primary phloem is pushed toward outside and is eventually crushed due to increasing internal pressure; it also becomes redundant in function. Sometimes in dicots, primary phloem develops into fibers.

Unlike dicots, most monocots do not show typical secondary growth, but some monocots, like palms, develop a thick stem by higher rate of cell division and enlargement by parenchyma cells of the ground tissue. Such secondary growth is called diffuse secondary growth. *Agave*, *Aloe*, *Yucca* and *Dracaena* produce secondary vascular bundles and parenchyma toward inside and parenchyma outside. This is mediated by a specialized cambium, arising in parenchymatous tissue outside the vascular bundle and contributes to secondary growth.

Secondary growth in roots is mediated by the activity of vascular cambium. The cambial layer occurs as strips of undifferentiated cells lying between primary xylem and primary phloem. These strips may be two in number in diarch roots and three in number in triarch roots. As secondary growth progresses, some cells of pericycle located outside the xylem ridges become active and start functioning as cambium. This leads to formation of a complete cambium encircling the xylem core.

Such arrangement gives diarch root an oval structure and triarch root a triangular look. Cambium situated on the inner face of phloem becomes active first and produces xylem and phloem cells by periclinal divisions. Anticlinal divisions lead to an increase in the circumference.

5.5 HORMONAL CONTROL OF PROCAMBIUM INITIATION

Recent studies reveal that several factors regulate the initiation and development procambium and cambium cells. These factors have been summarized in Table 5.1.

5.6 STRUCTURE AND FUNCTION OF XYLEM

5.6.1 Xylem Water Transport

By allowing the plants to continuously replace old water-conducting tissues with new ones, secondary xylems enhance the lifespan of plants (Myburg and Sederoff, 2001). Additionally, continuous production of xylem tissues is associated with the increased efficiency of water transport and the increased mechanical support in plants. Although the exact mechanism of water transport in plants is still not fully understood, research has strongly suggested that water moves from the region of less negative water potential (from roots) to the area of more negative water potential (to leaves and eventually to the air surrounding the leaves). Xylem elements or tracheary connects these two areas of steep water potential gradients by forming a continuous column of water between roots and leaves. With diffusion of water vapor from leaves into the air, more negative water potential is created in

TABLE 5.1
Role of Hormones in Functioning of Vascular Cambium

Sr. No.	Hormones	Role in vascular cambium activity	References
1.	Cytokinin	Induces cell division in cambium and differentiation in tracheary elements	Nieminen et al. (2008), Matsumoto-Kitano et al. (2008)
2.	Auxin	Induces cell division, maintains cell identity, induces growth of xylem initials Enhance xylem vessel density	Bhalerao and Bennett (2003), Carlsbecker and Helariutta (2005), Moyle et al. (2002), Sieburth and Deyholos (2006), Uggla et al. (1996, 2001)
3.	Brassinosteroids	Promotes cell division, induces programmed cell death, secondary wall deposition in tracheary elements.	Cano-Delgado et al. (2004), Fukuda (1997), Iwasaki and Shibaoka (1991)
4.	Gibberellin	Along with auxin induces cell proliferation and elongation of xylem fibers.	Turner et al. (2007), Yamamoto et al. (1997, 2001), Bjo¨ rklund et al. (2007), Dayan et al. (2010), Eriksson et al. (2000), Israelsson et al. (2005)
5.	ABA	Inhibits cambial growth	Digby and Wareing (1966), Eriksson et al. (2000), Ragni et al. (2011)
6.	Ethylene	Promotes cell division and expansion	Andersson-Gunnera's et al. (2003), Du and Yamamoto (2003), Little and Pharis (1995), Love et al. (2009), Agusti et al. (2011)
7.	Thermospermine	Suppresses xylem differentiation, promotes stem elongation	Love et al. (2009)
8.	TE differentiation inhibition factor (TDIF)	Maintain vascular meristem, inhibits differentiation of procambial cells and enhances their proliferation.	Yashimoto et al. (2012), Kakehi et al. (2008), Hirakawa et al. (2010)

the leaf cells, causing water to gradually move upward in xylem elements by cohesive force by conductance. The conductance of water molecules in xylem elements is affected by the weight of the water column, adhesion of water to the walls of xylem elements, and adhesion of the water to soil particles (Myburg and Sederoff, 2001). The conductance of water by xylem elements follows the Hagen-Poiseuill's law which states that a slight increase in diameter of the element will significantly increase the conductance. The secondary wall thickening of vessel elements and tracheids prevent the inward collapse caused by negative tension of water columns. In xylem elements with smaller diameter, hydrogen bonding of water molecules to the wall (adhesion force) serves to reinforce and strengthen the water column. However, under severe tension, the water column may break (cavitate), and an "air bubble" (known as embolism) may be created in the element. In vessel, it can spread from element to element through the perforations that link consecutive vessel elements and make the whole vessel dysfunctional for water transport. Embolism is more serious in large-diameter elements than in small-diameter elements.

Longer tracheids reduce the number of pit membranes that a water molecule has to cross on its way to the leaves. Under non-limiting soil moisture conditions, most of the water is transported by large-diameter vessels; however, under water stress conditions, water moves through small-diameter vessels and tracheids.

5.6.2 Structural Support

The cell walls of xylem consist of cellulose microfibrils that provide resistance against internal forces (turgor) as well as externally applied forces (tension). Highly lignified xylem tissues provide additional strength. Fibers in xylem tissues provide most of the mechanical support in dicots. The elastic nature of fibers helps them in retaining their original length even after subjecting to large tension forces. Vessels and tracheids with secondary thickened walls also contribute to structural support in xylem. In conifer wood, all the structural support is provided by tracheids.

Wood contains crystalline cellulose embedded in a matrix of hemicellulose and lignin, a highly cross-linked phenolic polymer. The cellulose fibrils in the secondary wall are deposited in layers, each layer with a different microfibril angle. The xylem cells in wood are arranged in cylinders parallel to the long axis of the stem. Additionally, the annual growth rings form concentric layers of wood tissue with different wall and lumen dimensions making them more complex and stronger than reinforced concrete.

5.6.3 Biosynthesis and Differentiation of Xylem

The process by which procambial and cambial initials differentiate into mature xylem cells is called xylogenesis. This process consists four stages: (i) cell division and enlargement followed by (ii) cell-wall thickening, (iii) lignification and (iv) programmed cell death. Cell-wall biosynthesis is an integral part of xylem formation. Tracheary elements are derived from either the procambium (primary xylem) or vascular cambium (secondary xylem). The differentiation of cambial initials into xylem elements is thought to be initiated by plant hormones. The immature xylem cells have dense protoplasm, small vacuoles and thin primary walls. Soon after cell division, these cells undergo cell elongation and an increase in the size of the vacuole and nuclei. Most xylem cell walls undergo secondary thickening. The deposition of secondary walls begins sometime before tracheary elements and fibers reach their full size. The cellulose, lignin, hemicellulose and protein components of the secondary wall are synthesized and deposited cooperatively during secondary wall thickening. The onset of secondary wall thickening is associated with the formation of arrays of microtubules under those regions of the plasma membrane where active secondary wall deposition will take place. Microtubules may play a role in defining the pattern of secondary walls by guiding dictyosome-derived vesicles with cell-wall material to the sites of deposition on the cell membrane. Cellulose microfibrils are produced at the membrane surface of the cell by complex rosette structures, which consist of several different proteins. The movement of these rosette complexes in the plasma membrane may also be directed by microtubules. The cell walls and intercellular regions of xylem cells are lignified. Following secondary thickening of the xylem cell walls, lignin is deposited between the newly formed tracheary elements and within their walls. The areas between the cells, called the middle lamella, and the primary walls are rapidly lignified, followed by a more gradual lignification of the secondary walls. Lignin is a very complex, cross-linked, three-dimensional polymer of aromatic phenolic monomers, called cinnamyl alcohols. The lignin monomers are delivered to the cell wall via Golgi and endoplasmic reticulum-derived vesicles and polymerized into lignin by wall-bound enzymes. The aromatic nature of the lignin monomers makes lignin hydrophobic. Lignin, therefore, provides a hydrophobic inner surface to the cell wall that facilitates water transport. The three-dimensional nature of the lignin polymer provides rigidity and compressive strength to the cell wall, while the chemical stability of lignin provides protection against pathogens.

At the completion of secondary wall deposition and lignification, tracheary elements undergo autolysis, an example of programmed cell death in higher plants. Soon after the initiation of secondary thickening, hydrolytic enzymes (DNAases, RNAases and proteases) start accumulating in the vacuole. The autolytic process is initiated when the tonoplast ruptures, causing the hydrolytic enzymes to spill out into the cytoplasm. This leads to the complete degradation of the cell contents and partial digestion of the unprotected regions of the primary wall. Only regions covered by lignified secondary wall material are protected from degradation. The end walls of differentiating vessel elements are degraded at the perforation sites to allow direct cell-to-cell movement of water and nutrients. Pit membranes are often partially degraded to leave mats of cellulose fibrils. This enhances the movement of water through pit-pairs, which is the only way water can enter and leave tracheids.

5.6.4 Genetic Manipulation of Xylem Formation

The content and composition of xylem cell walls affect the commercial value of various biological materials, such as wood and plant fibers, fodder crops, cereals, fruits and vegetables. Xylem properties are specified by a large number of genes and proteins. The composition and structure of xylem cell walls are determined by the coordinated expression of a large number of genes and proteins during xylogenesis. Variation in the developmental program and levels of expression of individual genes determine the variation in cell-wall architecture within and between different species. Although the general composition and structure of xylem cell wall is known, very little is known of the organization and biosynthesis of cell-wall components. Xylem cell walls contain hundreds of proteins and enzymes involved in the formation of the primary cell wall, which provide the framework for the synthesis of the secondary wall. Biosynthesis of the secondary wall involves precisely in regulating the formation of cellulose microfibrils, assembly of hemicellulose – cellulose complexes and polymerization of a network of the phenolic polymer lignin. Only a few genes have been studied sufficiently to establish their specific roles. Arabidopsis, Zinnia, tobacco and maize have been used to identify specific genes and proteins involved in cell-wall formation. Recently, high-throughput automated techniques are being used to identify the expressed genes in higher plants and to learn their function. This approach, called genomics, is expected to rapidly advance our knowledge of these genes and proteins forming the primary and secondary cell walls of xylem.

5.7 STRUCTURE AND FUNCTION OF PHLOEM

Phloem loading drives assimilates from green leaves to other organs for utilization in growth and storage. It is linked with symplasmic and apoplastic pre- and postphloem transport processes. Phloem transport depends on a finely balanced equilibrium between photosynthesis in source leaves and growth-related assimilate consumption in sink organs (Schulz and Thompson, 2009). Disturbance of this regulation points at the possibility of feedback mechanism and signaling along

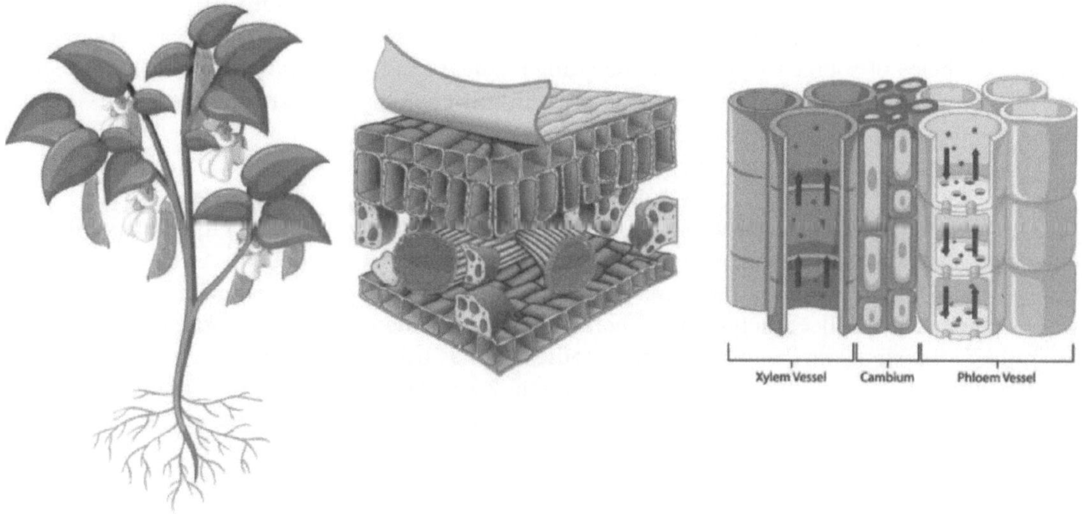

FIGURE 5.1 Phloem tissue.

the phloem pathway. Furthermore, major physiological events such as growth and flowering are influenced by signaling molecules transported through the phloem.

5.7.1 Phloem Development

Phloem comprises phloem fibers, phloem parenchyma, sieve elements and their accompanying companion cells. Sieve elements are elongated structures arranged end to end in files referred to as sieve tubes. Sieve elements are interconnected through sieve pores. These pores collectively form sieve plates (Figure 5.1). The ability of sieve tubes to transport materials is dependent on a developmentally programmed degeneration of the sieve element cell contents to produce an open membrane-bound tube. These sieve tubes provide a longitudinal network which conducts phloem sap. Sieve elements are associated with one or more companion cells, forming a sieve element–companion cell complex that plays an important role in the transport. In mature sieve element–companion cell complex, relatively open sieve elements contrast with adjacent companion cells containing dense, ribosome-rich cytoplasm with a prominent nucleus and abundant mitochondria and rough endoplasmic reticulum. High densities of extensively branched plasmodesmata in contiguous walls of sieve elements and companion cells account for intense intercellular coupling in sieve element–companion cell complex. Thus, companion cells perform the metabolic functions required for maintenance of sieve elements. A detailed account of phloem development has been described by several authors (Schulz and Thompson, 2009; Heo et al., 2014).

5.8 PATHWAYS OF TRANSLOCATION

Current theories explaining the mechanisms of assimilate translocation are based on the pressure-flow hypothesis first proposed by Ernst Münch in 1927. Water crossing a semipermeable membrane into a region of high solute concentration and out of a connected region of low solute concentration creates a turgor pressure gradient in the pathway that passively transports solutes until equilibrium is achieved (Schulz and Thompson, 2009). In terms of phloem transport, the osmotic differential is maintained as assimilates are "loaded" into sieve elements in source tissues and "unloaded" from sieve elements in sink tissues. The uptake of sugar leads to an osmotic pressure difference across the plasma membrane that stimulates the influx of water and the development of a high turgor pressure in sieve elements of source tissues. The unloading of sugars and water from sieve elements lowers the turgor pressure in the sink tissues. The pressure differential between source and sink drives sucrose and water transport in the phloem.

There are three kinds of phloem, the collection phloem, the transport phloem and the release phloem. The smaller veins in leaves represent the collection phloem in higher plants. The maintenance of a short distance between photosynthetically active mesophyll cells and assimilate-exporting sieve element ensures proper loading of photosynthetic products. As compared with transport and release phloem, the companion cells in collection phloem are much larger than the sieve elements. The collection phloem is well equipped with membrane transporters that allow for assimilate loading against a concentration gradient. It has been shown that the transport phloem has the ability to actively take up assimilates (Schulz and Thompson, 2009). Although the basic properties of phloem remain the same, different types of phloem are adapted to serve different functions.

5.8.1 Loading of Assimilates

In contrast to xylem transport, the "mass flow" mode of phloem transport requires an intact plasma membrane. The sugars are moved into the phloem sieve elements at the source end of the phloem (area where the sugars are synthesized – leaf). The water potential of the cell decreases due to the loading of sugars leading to water flow from surrounding areas by osmosis, which causes increase in pressure that forces movement sugar, water and amino acids to move toward the sink tissue (where sugars are utilized). The transported sugars are off-loaded at the sink tissue. This increases the water potential of the sink tissue and causes water to flow out of the phloem by osmosis. These sieve elements will have to remain alive to keep plasma membrane intact in order to control the movement of sugars into and out of the sieve element (Figure 5.2). Three distinct pathways are used by higher plants for phloem loading (Schulz and Thompson, 2009). These loading pathways are explained in this section (Figure 5.2).

5.8.1.1 Apoplastic Loading of Sucrose

Aplastic loading involves the movement of sugars from apoplast (extracellular cell-wall space) across the plasma membrane into the cell. Plasma membrane invaginations (lyrinth) in the companion cells are used for loading sugars in the leaf. The wall labyrinth increases the surface area of the plasma membrane. Companion cells having this wall labyrinth are termed transfer-type companion cells (Schulz and Thompson, 2009). Plasma membrane transporter proteins are required for apoplastic loading. Apoplastic loading is energized by plasma membrane-bound adenosine triphosphatases (ATPases) that hydrolyze ATP to pump protons from the cytoplasm into the apoplast. The sucrose/proton cotransporter has a 1:1 stoichiometry, transporting one proton with one sucrose molecule from the apoplast into the cytoplasm. In intact leaves, differences in solute potential and indirect measurements indicate a two- to threefold higher concentration of sucrose in the sieve element–companion cell complexes compared with mesophyll cells. However, considering the lower concentration of sugars in the apoplast, the capacity of sucrose transporters must be higher and the real accumulation factor should be in the range of 3 to 30.

5.8.1.2 Symplasmic Loading of Sugars

Plants translocating raffinose or higher polymers of sucrose–galactosyl sugars (stachyose, verbascose) appear to load assimilates symplasmically into the collection phloem of mature leaves (Schulz and Thomspon, 2009). An exceptionally high number of plasmodesmata linking the sieve

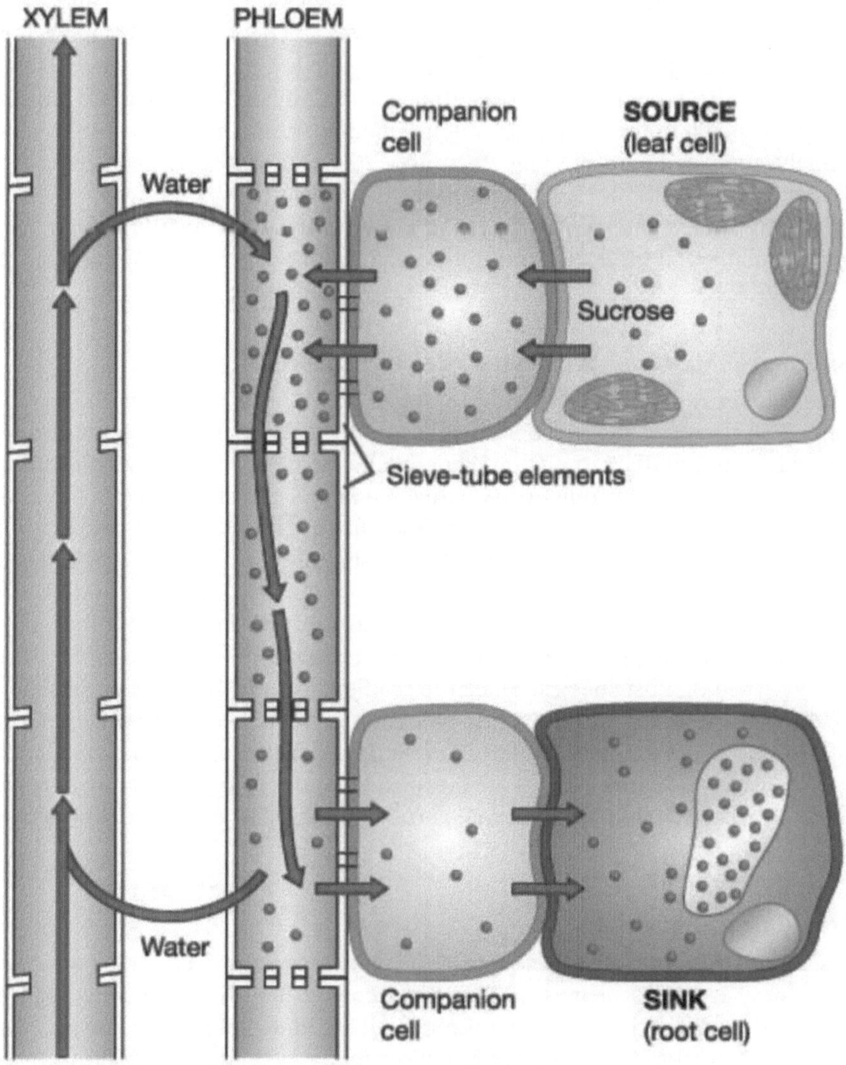

FIGURE 5.2 Loading of sugar in phloem tissue.

element with the rest of the leaf symplasm is a prerequisite for symplasmic loading. These plants have structurally distinctive intermediary-type companion cells with many small vacuoles and numerous plasmodesmata linking them to bundle sheath cells. The plasmodesmata are predominantly branched on the companion cell side where the orifices appear very small. In developing leaves, plasmodesmatal branching occurs concomitantly with the sink–source transition. Initially, symplasmic loading via plasmodesmata appears to contradict mechanistically the concept that sugars must accumulate to higher concentrations within the sieve elements than in the surrounding cells for osmotically driven translocation. The "polymer trap hypothesis," illustrated in *Coleus blumei* and members of the cucurbitaceae family, proposes a solution for the problem of loading sugars through plasmodesmata. According to this hypothesis, sucrose diffuses symplasmically through the prephloem pathway into the intermediary cells. Enzymes involved in raffinose and stachyose synthesis from sucrose and galactose have been localized to the intermediary cells. Synthesis of sucrose–galactosyl polymers (raffinose and stachyose) in intermediary cells result in sugar molecules that are too large to diffuse back through the narrow orifices of the branched plasmodesmata into the prephloem pathway. Indeed, the highest carbohydrate concentration in mature leaves was shown in the sieve element–intermediary cell complexes of microdissected tissue. Interestingly, symplasmic loading members appear to be less flexible than apolastic loaders in adapting to high light conditions. Loading capacity is determined by the minor vein density and plasmodesmal connectivity in symplasmic loaders, which are anatomically limiting as compared to increasing the membrane surface of transfer-type companion cells of apoplastic loaders (Schulz and Thompson, 2009).

5.8.1.3 Alternative Modes of Symplasmic Phloem Loading

Although the phloem in a number of woody plants has numerous plasmodesmal connections between sieve elements and the mesophyll tissue, they neither translocate sucrose–galactosyl polymers nor have intermediary-type companion cells. In conifers, the symplasmic pathway is continuous from mesophyll via transfusion parenchyma and Strasburger cells flanking the sieve cells of the central vascular bundle in needles. Specifically, dome-shaped wall portions provide numerous branched plasmodesmata between the Strasburger cells. These regions are covered by extended ER complexes. However, the physiology of phloem loading in conifers still remains unclear. In Salix, it is postulated that photoassimilates are transferred from mesophyll into the sieve elements by diffusion along the concentration gradient, rather than against the concentration gradient observed in most species. Since the gradient extends from the assimilate-producing mesophyll cells to the sink tissues, the term "phloem loading" does not imply uphill transport in this loading mode. The only membrane step, the exchange of triose phosphate against orthophosphate across the chloroplast envelope, could be considered the "motor" of phloem transport, possibly modulated by the interchange of sucrose across the tonoplast along the pre- and postphloem transport pathway.

5.8.1.4 Transport of Assimilates

The ultrastructure of the conducting elements of the phloem is optimized for a low-resistance pathway. However, sieve elements of most angiosperms contain highly polymerized aggregations of phloem proteins (P proteins) that, if present throughout the lumen of a sieve element, would inhibit the flow of sugars. P proteins are most often found accumulating at the sieve plates. Observations of intact sieve elements using confocal laser scanning microscopy support observations that P proteins are involved in wound responses, either to maintain turgor pressure within damaged sieve elements or to protect the valuable contents of sieve elements from leakage. The term "sieve element occlusion proteins" has been proposed to describe proteins with this common function.

Two major structural proteins, the phloem filament protein and phloem lectin, appear to be the principal components of highly polymerized aggregates and filaments in sieve elements of these species. Crosslinking of evenly dispersed polymerized filament protein with the carbohydrate-binding lectin has been hypothesized as a mechanism to anchor the protein polymers to membrane glycoconjugates, allowing the assimilate stream to remain undisturbed. In cucurbits, the unpolymerized sieve element occlusion proteins are freely mobile in the assimilate stream, and the mobile phloem lectin also appears to function as a ribonucleic acid (RNA)-binding protein. The equilibrium between polymerized and unpolymerized forms might be critical for alternative physiological functions of these enigmatic proteins within the phloem. Forisomes are protein aggregates composed of multiple protein subunits that undergo a calcium-dependent conformational change from the aggregated crystalloid (contracted) to an amorphous (expanded) form, which can rapidly respond to calcium fluxes within sieve elements to reversibly block the translocation stream. Sieve element occlusions occur in response to plasma membrane leakage and rapid changes in pressure that characterize sieve element damage. In addition to the sieve element occlusion proteins, a large number of soluble proteins are known to be transported in the assimilate stream.

5.8.2 Phloem-Feeding Insects

Aphids and other phloem-feeding insects have evolved elaborate strategies to withdraw nitrogenous assimilates from the sieve tubes for their nutritional requirements. These insects insert specialized mouthparts (stylets) between plant mesophyll cells and penetrate sieve elements to feed on phloem sap. Calcium binding salivary proteins injected into sieve elements prior to feeding prevent sieve element occlusion and result in copious excretion of "sugary" honeydew from the aphids. Aphids feeding from sieve elements alter normal source-to-sink relations and carbohydrate partitioning in host plants. These insects also transmit pathogenic viruses. Aphids and other phloem-feeding insects are important experimental tools for translocation studies.

5.8.2.1 Unloading of Assimilates

Phloem unloading refers to release of assimilates from phloem elements. It is followed by post-phloem transport in sink tissues. The mechanisms involved in the removal of assimilates from the phloem are still unclear. In most cases, unloading of assimilates from the phloem is a passive process according to the assimilate gradient, which is steepest toward an active sink tissue. Root and shoots apices with high mitotic activity and extension growth are the strongest utilization sinks. In contracts, tubers, fruits and seeds that accumulate sucrose, lipids and/or starch are the storage sinks. Phloem unloading in utilization and storage sinks is primarily a symplasmic process where assimilates are removed from sieve elements into the postphloem pathway via plasmodesmata. Plasmodesmal length or relative diameter could regulate the rate of unloading. Apoplastic unloading occurs in specific situations where sieve element–companion cell complexes are symplasmically isolated, or the number of plasmodesmata is insufficient to allow symplastic transport.

Studies of phloem unloading in potato tubers indicate alternation of symplasmic and apoplastic pathways during subsequent stages of development (Schulz and Thompson, 2009). In fruits, sucrose is symplasmically unloaded from sieve elements into phloem parenchyma, and an apoplastic step often occurs during postphloem transport. Physical separation between cells in the postphloem pathway of developing seeds results in an apoplastic unloading.

5.8.3 Phloem-Sap Signaling

Schulz and Thompson (2009) reported the involvement of proteins with regulatory functions, such as deoxyribonucleic acid (DNA)-binding transcription factors, RNA-binding, mRNA translation, ubiquitin-mediated proteolysis and macromolecular and vesicular trafficking. Furthermore, small, noncoding RNAs, such as short interfering and micro RNAs, have been shown to be involved in posttranscriptional gene regulation. Noninvasive tracing techniques and the appearance of the macromolecule in a graft partner are commonly used methods for validating the mobility of macromolecules in the phloem (Schulz and Thompson, 2009). Further investigation to elucidate the mechanism of phloem-sap signaling is in progress.

REFERENCES

Agusti J, Herold S, Schmarz M, Sanchez P, Ljung K, Dun E A, Brewer P B, Beveridge C A, Sieberer T, Sehr E M and Greb T. 2011. Strigolactone signaling is required for auxin-dependent stimulation of secondary growth in plants. *Proceedings of National Academy of Sciences United States of America* **108**: 20242–20247.

Andersson-Gunnerås S, Hellgrent J M and Björklund S. 2003. Asymmetric expression of a poplar ACC oxidase *controls ethylene production during gravitational induction of tension wood. Plant Journal* **34**: 339–349.

Beck C B. 2010. *An Introduction to Plant Structure and Development*. Cambridge University Press, Cambridge.

Bhalerao R P and Bennett M J. 2003. The case for morphogens in plants. *Nature Cell Biology* **5**: 939–943.

Bishop A, El-Showk S and Helariutta Y. 2013. Vascular development in Arabidopsis roots. pp. 7-1–7-8. *Plant Roots*. CRC Press, Boca Raton, FL.

Björklund S, Antti H, Uddestrand I, Moritz T and Sundberg B. 2007. Cross-talk between gibberellin and auxin in development of Populus wood: gibberellin stimulates polar auxin transport and has a common transcriptome with auxin. *Plant Journal* **52**: 499–511.

Cano-Delgado A, Yin Y H, Yu C, et al. 2004. BRL1 and BRL3 are novel brassinosteroid receptors that function in vascular differentiation in Arabidopsis. *Development* **131**: 5341–5351.

Carlsbecker A and Helariutta Y. 2005. Phloem and xylem specification: pieces of the puzzle emerge. *Current Opinion in Plant Biology* **8**: 512–517.

Dayan J, Schwarzkopf M, Avni A, et al. 2010. Enhancing plant growth and fiber production by silencing GA 2-oxidase. *Plant Biotechnology Journal* **8**: 425–435.

Digby J and Wareing P F. 1966. The effect of applied growth hormones on cambial division and the differentiation of the cambial derivates. *Annals of Botany* **30**: 539–548.

Du S and Yamamoto F. 2003. Ethylene evolution changes in the stems of Metasequoia glyptostroboides and *Aesculus turbinata* seedlings in relation to gravity-induced reaction wood formation. *Trees Structure and Functions* **17**: 522–528.

Eriksson M E, Israelsson M, Olsson O, et al. 2000. Increased gibberellin biosynthesis in transgenic trees promotes growth, biomass production and xylem fiber length. *Nature Biotechnology* **18**: 784–788.

Errera L. 1886. Sur une condition fondamentale d'équilibre des cellules vivantes [On a fundamental condition of equilibrium for living cells]. *C R Hebd Seances Academy of Sciences* **103**: 822–824 (in French).

Evert R F. 2006a. Meristems and differentiation. pp. 103–131. *Esau's Plant Anatomy – Meristems, Cells, and Tissues of the Plant Body: Their Structure, Function, and Development*. 3rd edition. John Wiley & Sons, Inc., NJ.

Evert R F. 2006b. Phloem: Secondary phloem and variations in its structure. pp. 407–425. *Esau's Plant Anatomy – Meristems, Cells, and Tissues of the Plant Body: Their Structure, Function, and Development*. 3rd edition. John Wiley & Sons, Inc., Hoboken, NJ, United States.

Evert R F. 2006c. Structure and development of the plant body – An overview. pp. 1–13. *Esau's Plant Anatomy – Meristems, Cells, and Tissues of the Plant Body: Their Structure, Function, and Development*. Chapter 1. 3rd edition. John Wiley & Sons, Inc., Hoboken, NJ, United States.

Evert R F. 2006d. Vascular cambium. pp. 323–355. *Esau's Plant Anatomy – Meristems, Cells, and Tissues of the Plant Body: Their Structure, Function, and Development*. Chapter 12. 3rd edition. John Wiley & Sons, Inc., Hoboken, NJ, United States.

Fukuda H. 1997. Tracheary element differentiation. *Plant Cell* **9**: 1147–1156.

Gokhale S P and Kavathekar A K. 2012. *Vascular Cambium*. Lesson Prepared for NME ICT. https://nmeict.ac.in/.

Heo J, Roszak P, Furuta K M and Helariutta Y. 2014. Phloem development – Current knowledge and future perspective. *American Journal of Botany* **101**: 1393–1402.

Hirakawa Y, Kondo Y and Fukuda H. 2010. TDIF peptide signaling regulates vascular stem cell proliferation via the WOX4 homeobox gene in Arabidopsis. *Plant Cell* **22**: 2618–2629.

Israelsson M, Sundberg B and Moritz T. 2005. Tissue-specific localization of gibberellins and expression of gibberellin-biosynthetic and signaling genes in wood-forming tissues in aspen. *Plant Journal* **44**: 494–504.

Iwasaki T and Shibaoka H. 1991. Brassinosteroids act as regulators of tracheary-element differentiation in isolated Zinnia mesophyll cells. *Plant Cell Physiology* **32**: 1007–1014.

Jurra J, Wloch W, Kojs P, Wilczek A and Szendera W. 2005. Current trends in the structural investigations of the vascular cambium. *Bulletin of Botanical Gardens* **14**: 43–47.

Kakehi J, Kuwashiro Y, Niitsu M and Takahashi T. 2008. Thermospermine is required for stem elongation in *Arabidopsis thaliana*. *Plant Cell Physiology* **49**: 1342–1349.

Kucukoglu M. 2015. *Molecular Regulation of Vascular Cambial Identity and Activity*. Doctoral Thesis. Swedish University of Agricultural. Science, Umea, Sweden.

Little C H A and Pharis R P. 1995. Hormonal control of radial and longitudinal growth in the tree stem. pp 281–319. *Plant Stems: Physiology and Functional Morphology* (Ed.) Gardner B L. Academic Press, San Diego.

Love J, Björklund S, Vahalab J, et al. 2009. Ethylene is an endogenous stimulator of cell division in the cambial meristem of Populus. *Proceedings of National Academy of Sciences United States of America* **106**: 5984–5989.

Matsumoto-Kitano M, Kusumoto T, Tarkowski P, et al. 2008. Cytokinins are central regulators of cambial activity. *Proceedings of National Academy of Sciences United States of America* **105**: 20027–20031.

Moyle R, Schrader J, Stenberg A, et al. 2002. Environmental and auxin regulation of wood formation involves members of the Aux/IAA gene family in hybrid Aspen. *Plant Journal* **31**: 675–685.

Myburg A A and Sederoff R R. 2001. Xylem structure and function. *Encyclopedia of Life Sciences Nature Publishing Group*. Available at www.els.net.

Nieminen K, Immanen J, Laxell M, et al. 2008. Cytokinin signaling regulates cambial development in poplar. *Proceedings of National Academy of Sciences United States of America* **105**: 20032–20037.

Plomion C, Leprovost G and Sokes A. 2001. Wood formation in trees. *Plant Physi*ology **127**: 1513–1523.

Ragni L, Nieminen K, Pacheco-Villalobos D, et al. 2011. Mobile gibberellin directly stimulates Arabidopsis hypocotyl xylem expansion. *Plant Cell* **23**: 1322–1336.

Schulz A and Thompson G A. 2009. Phloem structure and function. *Encyclopedia of Life Sciences*. John Wiley and Sons, Ltd., Chichester.

Sieburth L E and Deyholos M K. 2006. Vascular development: the long and winding road. *Current Opinion in Plant Biology* **9**: 48–54.

ten Hove C A, Lu K J and Weijers D. 2015. Building a plant: cell fate specification in the early *Arabidopsis* embryo. *Development* **142**: 420–430.

Turner S, Gallois P and Brown D. 2007. Tracheary element differentiation. *Annual Review of Plant Biology* **58**: 407–433.

Uggla C, Magel E, Moritz T, et al. 2001. Function and dynamics of auxin and carbohydrates during earlywood/latewood transitions in Scots pine. *Plant Physiology* **125**: 2029–2039.

Uggla C, Moritz T, Sandberg G, et al. 1996. Auxin as a positional signal in pattern formation in plants. *Proceedings of National Academy of Sciences United States of America* **93**: 9282–9286.

Verdeil J L, Alemanno L, Niemenak N and Tranbarger T J. 2007. Pluripotent versus totipotent plant stem cells: dependence versus autonomy? *Trends in Plant Science* **12**: 245–252.

Wright A and Smith L. 2008. Division plane orientation in plant cells. pp. 33–57. *Cell Division Control in Plants* (Eds.) Verma D and Hong Z. Springer, Berlin.

Yamamoto R, Demura T and Fukuda H. 1997. Brassinosteroids induce entry into the final stage of tracheary element differentiation in cultured Zinnia cells. *Plant Cell Physiology* **38**: 980–983.

Yamamoto R, Fujioka S, Demura T, et al. 2001. Brassinosteroid levels increase drastically prior to morphogenesis of tracheary elements. *Plant Physiology* **125**: 556–563.

Yashimoto K, Noutoshi Y, Hayashi K, Shirasu K, Takahashi T and Motose H. 2012. A chemical biology approach reveals an opposite action between thermospermine and auxin in xylem development in Arabidopsis thaliana. *Plant Cell Physiology* **53**: 635–645.

6 Reproduction in Horticultural Plants

6.1 INTRODUCTION

Most flowering plants go through a dominant diploid sporophyte (the multicellular stage of the plant or alga that alternates with a multicellular haploid gametophyte phase and produces asexual spores) and a transient haploid gametophyte phase in their life cycle. They adopt two distinct strategies to reproduce themselves; sexual reproduction consisting of alternating haploid and diploid generations and asexual reproduction involving vegetative parts, such as bulb, rhizomes, tubers and corms, are used to multiply plants. In the sexual reproduction, the embryo sac represents the haploid female gametophyte, whereas the pollen constitutes the haploid male gametophyte. The haploid egg and sperm fuse to form diploid zygote, from which new sporophyte develops. The offsprings from sexual reproduction are genetically diverse, whereas those from asexual reproduction are genetically identical to the mother plant. In contrast, the offsprings can also be produced without fertilization (Apomixis). Polyembryonic development of seeds in citrus is an example of this phenomenon.

It is important for a plant breeder and horticulturist to know how the plant reproduces naturally, predicting the field performance of the progeny, and alternative reproduction methods that can be used for developing new varieties (Fryxell, 1957). To produce hybrid lines/population in self-pollinated crops, the breeder has to remove anthers (male reproductive part) from the female parent and artificially pollinate with pollens from the desired male parent. This requires prior knowledge of floral biology (the time of anthesis, period of stigma receptivity and pollen viability) of both the parents.

6.2 FLOWER MORPHOLOGY

A typical flower consists of four parts, each arranged in whorls and performing a distinct function. The outermost whorl is the calyx consisting of sepals (Figure 6.1), which are normally green and enclose and protect the developing bud. Next to calix is the corolla, which consists of petals, whose function is to facilitate pollination by attracting pollinators (insects, birds and other animals). The calyx and corolla together form the perianth. The whorl next to the corolla is androecium (derived from the Greek word, *andros*, meaning "man," and *oikos*, "house"), which consists of male organs called stamens. Each stamen typically comprise a slender stalk or filament attached to the flower at its base and carrying on its free, upper end structure called an anther. Anther houses the pollen, which is the male gametophyte or microsporophyte. The pollen grains mature either in the anther or during the germination of the pollen grain or growth of the pollen tubes on the pistil. It will have two sperm cells and one vegetative nucleus. The sperm cells, with thin cell walls, are located within the cytoplasm of the vegetative cell. The transcription-active vegetative nucleus and its metabolism-active cytoplasm functions together allow pollen tube to germinate on stigma, to go along the style and to escort two sperm cells to the female gametophyte (or called embryo sac).

The innermost whorl of the flower is the gynoecium (derived from the Greek word, ṅ, meaning "woman," and *oikos*, "house"), which consists of female organ called carpels. Each carpel comprises a basal ovary containing the ovules, a slender column-shaped structure (style) that terminates with expanded stigma. The main function of stigma is to receive the pollen grains and support the growth of pollen tube. Additional whorls, such as the epicalyx, may be present in some flowers, such as Cotton.

Embryo sac formed within the ovule typically consists of one egg, two synergid, one central (two nuclei) and three antipodal cells (seven cells and eight nuclei; Figure 6.1). Similar to the pollen

DOI: 10.1201/9781003373889-7

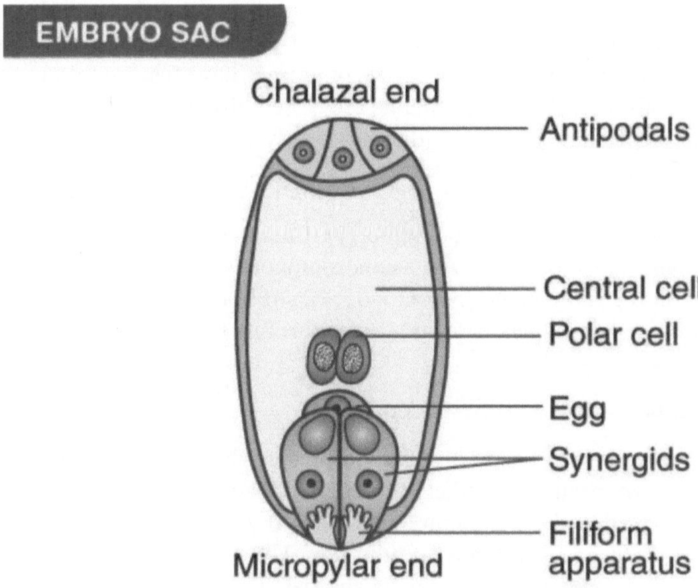

FIGURE 6.1 Embryo sac structure.

grain, the embryo sac can be considered as one gigantic cell with two nuclei (central nucleus). Within the cytoplasm of this cell (all with cell wall), six specialized cells are located. The egg cell and the synergid cells are located at the micropylar pole of the embryo sac, while the antipodal cells are near chalazal pole (Figure 6.1).

Flowers in several species lack one or more parts, or even some parts may be fused with other parts. Such flowers are called atypical flowers. For example, in tulips, sepals and petals look alike. Similarly, brightly colored leaves and bracts surrounding the flower in *Bougainvillea* can easily be confused with petals. Many species during evolution lost some parts of the flower or lost sexual function, allowing them to specialize in the other. The petals may be fused to make a tube, as in a petunia flower. Flowers may combine to form what is called an inflorescence as in Brassicas.

6.3 EVOLUTION OF THE FLOWER

Pollination in angiosperms does not involve direct contact between the pollen grain and the ovule. Mature pollens are transported from anthers to stigma of another flower by insects, birds or other animals. Pollen grains germinate on the stigma to produce pollen tube carrying the sperm nuclei to the embryo sac. Following double fertilization, development of endosperm begins. The seed matures within the ripening fruit; germination of the seed takes place away from the mother plant. Mutations either spontaneous or induced in any one of the parents can reduce the chances for natural reproduction. Similarly, if parents flower at different times or their morphology is altered, pollinators may not be able to carry out successful pollination. Thus, the floral morphology is much more complex and diverse than simple initiation and maturation of floral organs.

6.3.1 Hermaphrodity and Unisexuality

Androecium and gynoecium are the essential parts of a flower because they are directly involved in sexual reproduction. The flowers, in which one of the essential parts is missing, are called unisexual flowers, which are either pistillate/female flowers (when only gynoecium is present) or staminate/

male flowers (when only androecium is present). Different families have different types of flowers, legumes have bisexual flowers with petals modified as banner petal, wing petals and keel, and cucurbits usually have unisexual flowers but may sometimes have bisexual flowers.

Based on the type of flowers produced, plants are called bisexual/hermaphrodite if it bears only bisexual/hermaphrodite flowers, unisexual (dicliny) if they bear flowers containing only one sexual organ or subdioecy if plants bear both bisexual and unisexual flowers. The unisexual flowers are further subdivided as monoecious (cucumber, coconut) if both the male and female flowers occur on the same plant or dioecious (papaya, date, mulberry) if male and female flowers occur on different plants. Subdioecy plants are subcategorized as andromonoecious if both the male and hermaphrodite flowers present on the same plant; gynomonoecious if both female and hermaphrodite flowers are present; trimonoecious if female, male and hermaphrodite flowers are formed on the same plant.

6.3.2 Floral Specialization and Symmetry

Evolutionary trends have led to a wide diversity in present-day flowering plants. In some flowers, separate floral parts have grouped together, or fused to form a tube, while in others, some floral parts have been lost or reduced. In the more advanced angiosperms, the number of parts in each whorl has often been reduced from many to only few. The central axis in many flowers has been shortened, and the whorls are placed close to one another. These modifications often reflect the predominant pollination mechanisms. While much of the floral diversity is the result of natural selection related to pollination, breeding and artificial selection had major impact on flower morphology. Flowers could be either radically symmetrical (if the flower is cut anywhere through the center, it will result in two roughly equal halves) or bilaterally symmetrical (i.e., the flower can be cut in two equal halves only along a single plane). Flowers in snapdragons, peas, violets, mints and orchids are bilaterally symmetrical. Floral symmetry is regulated by genes like *cyclodia* in snapdragon. In the absence of this gene, flowers may have a different symmetry. For example, in the absence of *cyclodia* gene, the flowers become more radial.

6.3.3 Gamete Formation

Pollen grains are the male gametophytes or microgametophytes, whereas the embryo sac is the female gametophyte, or megagametophyte. In angiosperms, both pollen grains and the embryo sac are produced in separate, specialized structures. Reproductive success depends on uniting the gametes (egg and sperm) found in the embryo sacs and pollen grains of flowers. Plant sexual life cycles are characterized by an alternation of a diploid sporophyte generation and a haploid gametophyte generation.

Pollen grains form in the two pollen sacs located in the anther. Each pollen sac contains specialized chambers in which the microspore mother cells are enclosed. The microspore mother cells undergo meiosis to form four haploid microspores. Subsequently, mitotic divisions form four pollen grains. The generative cell inside each pollen grain will divide to produce two sperm cells. Pollen grain shapes are specific to flower species, and most pollen grains have a furrow from which the pollen tube emerges.

Eggs develop in the ovules of the flower. Within each ovule is a megaspore mother cell, which undergoes meiosis to produce four haploid megaspores. Only one of these megaspores survives, while remaining three are absorbed by the ovule. The one remaining megaspore undergoes repeated mitotic divisions to produce eight haploid nuclei that are enclosed within a seven-celled embryo sac. Within the embryo sac, the eight nuclei are arranged in precise positions, one nucleus located near the opening of the embryo sac in the egg cell, two located in a single cell in the middle of the embryo sac (polar nuclei), two nuclei are housed in synergids cell, and the remaining three nuclei reside in cells called the antipodals, located at the end of the sac, opposite to the egg cell. The pollen grains germinate on the stigma of the carpel and pollen tube grows toward the embryo sac.

6.4 POLLINATION

The first step in sexual reproduction is the union of gametes (egg and sperm) found in the embryo sacs and pollen grains of flowers. Pollination is the process by which pollen is transported to the stigma. Based on the source of the pollen that pollinates the plant, sexually reproducing plants can be grouped into self-pollinated, where the pollen grains from anther is transferred to the stigma of the same flower or another flower on the same plant or cross-pollinated if the pollen from a flower on one plant pollinates the stigma of a flower on another plant. Stebbins (1950) observed that there is a relationship between the length of life cycle of a plant and its reproduction mode. For example, in annual plants, which are short-lived, self-pollination is the key to reproductive assurance. On the other hand, perennials tend to outcross because they have more opportunities to genetically recombine in a life span spread over by many years (Duvick, 1966). The terms self-pollinating and cross-pollinating crops just mean that one method of pollination is more predominant than the other in that crop because some amount of outcrossing in self-pollinating crops and selfing in cross-pollinating crops do occurs. Table 6.1 presents the mode of pollination in some common horticultural crops.

6.4.1 Mechanisms Influencing the Mode of Pollination

6.4.1.1 Self-pollination

Bisexuality: Presence of male and female organs in the same flower is known as bisexuality. The presence of bisexual flowers is a must for self-pollination. All the self-pollinated plants are hermaphrodites.

Homogamy: Maturation of anthers and stigma of a flower at the same time, or the homogamy is a prerequisite for self-pollination.

Cleistogamy: When pollination and fertilization occur in unopened flower bud, it is known as cleistogamy. It ensures self-pollination and prevents cross-pollination. Cleistogamy has been reported in some varieties of wheat, barley, oats and several other grass species.

Chasmogamy: Opening of flowers only after the completion of pollination is known as chasmogamy. This also promotes self-pollination and is found in crops like wheat, barley, rice and oats.

Position of anthers: In some species, stigmas are surrounded by anthers in such a way that self-pollination is ensured. Such situation is found in tomato and brinjal. In some legumes, the stamens and stigma are enclosed by the petals to facilitate self-pollination. Examples are greengram, blackgram, soybean, chickpea and pea.

6.4.1.2 Cross-Pollination

Dicliny: It refers to unisexual flowers. This is of two types, viz. (i) monoecy and (ii) dioecy. When male and female flowers are separate but present in the same plants, it is known as monoecy. In some

TABLE 6.1
Pollination Types in Some Vegetables

Self-pollinated crops	Cross-pollinated crops
Common beans	Banana
Egg plant	Carrot
Lettuce	Cassava
Pea	Cucumber
Peach	Onion
Pepper	Potato
Tomato	Watermelon
	Sweet potato

crops, the male and female flowers are present in the same inflorescence such as in mango, castor and banana. In some cases, they are on separate inflorescence as in maize. Other examples are cucurbits, grapes, strawberry, cassava and rubber. When staminate and pistillate flowers are present on different plants, it is called dioecy. It includes papaya, date palm, spinach, hemp and asparagus.

Dichogamy: It refers to maturation of anthers and stigma of the same flower at different times. Dichogamy promotes cross-pollination even in the hermaphrodite species. Dichogamy is of two types, viz. (i) protogyny and (ii) protandry. When pistil matures before anthers, it is called protogyny such as in pearl millet. When anthers mature before pistil, it is known as protandry. It is found in maize, sugar beet and several other species.

Heterostyly: When styles and filaments in a flower are of different lengths, it is called heterostyly. It promotes cross-pollination.

Herkogamy: Hindrance to self-pollination due to some physical barriers such as presence of hyaline membrane around the anther is known as herkogamy. Such membrane does not allow the dehiscence of pollen and prevents self-pollination as in alfalfa.

Self-incompatibility: The inability of fertile pollens to fertilize the same flower is referred to as self-incompatibility. It prevents self-pollination and promotes cross-pollination. Self-incompatibility is found in several crop species like *Brassica, Radish, Nicotiana* and many grass species. It is of two types, sporophytic and gametophytic.

Genetically controlled pollination systems: The sperm cells are highly compacted and has very little cytoplasm. Large scale of EST analysis of the sperm cell-expressed genes indicated that sufficient mRNAs are present in the sperm cells (Engel et al., 2003). These authors reported that none of genes is really sperm-specific. The transcripts that are present in sperm cells are also available in unicellular or bicellular microspores, suggesting that the expression of these sperm-carried mRNA is expressed during microspore development. The first identified sperm cell-expressed gene is *GENERATIVE CELL SPECIFIC 1 (GCS1)* from lily (Xu et al., 1998, 1999). *GCS1* encodes a sperm-cell surface protein. *GCS1* mutant of *Arabidopsis* did not diminish pollen tube growth *in vitro* or in the pistil, but it reduced ovule targeting by two folds. Furthermore, the *gcs1* sperms transported to the ovules fail to initiate the fertilization, suggesting the possible role of GCS1 in the recognition or fusion between male and female gametes (von Besser et al., 2006). Genetic analysis of pollen development has revealed two loci *duo1* and *duo2* during fertilization. The *duo1* pollen fails to enter mitosis at G2-M phase, while *duo2* enters PMII but arrested at pro-metaphase (Durbarry et al., 2005). *DUO1* encodes a novel R2R3 MYB transcription factor. DUO1 proteins accumulate in the nuclei of sperm cells. (Rotman et al., 2005).

6.4.2 Pollinating Agents

6.4.2.1 Wind

Early seed plants were pollinated passively, by the action of the wind. Individual plants of any given species must grow relatively close (<100 m) to one another for such a system to operate efficiently. This short distance is significant compared with the long distances pollen is routinely carried by certain insects, birds and other animals. Many wind-pollinated plants have stamen- and carpel-containing flowers separated among individuals or on a single individual. If the pollen-producing and ovule-bearing flowers are separated, it is certain that pollen released to the wind will reach a flower other than the one that sheds it, a strategy that greatly promotes outcrossing. Some wind-pollinated plants flower in the spring, prior to the development of their leaves, thereby preventing interference of leaves with the wind-borne pollen.

6.4.2.2 Bees

The spreading of pollen from one plant to another by pollinators visiting flowers of specific species has played an important role in the evolutionary success of the group. As flowers become increasingly specialized, so do their relationships with particular groups of insects and other animals. Bees initially locate sources of food by odor and then orient themselves on the flower or group of flowers by its shape, color and texture. Flowers that bees visit are blue or yellow often with stripes or lines of dots that indicate the location of the nectaries. Some bees collect nectar. Pollen is used to provide food in cells where bee larvae complete their development. These bees produce several generations a year and must shift their attention to different kinds of flowers as the season progresses. To maintain large colonies, they also need to use more than one kind of flower as a food source at any given time. The highly specialized relationships exist between bees and flowers they visit. This has resulted in modifications, over time, in both the flowers and the bees. Some of the examples of these relationships are reflected in the synchronization of flower opening with the time of the day when bees appear; the mouthparts of the bees might get modified in relation to length of flowers, etc. These interactions provide both an efficient mechanism of pollination for the flowers and a constant source of food for the bees that "specialize" on them.

6.4.2.3 Insects Other Than Bees

A few groups of insects belonging to *Lepidoptera* family (butterflies and moths) also act as pollinators. For example, flowers such as phlox often have flat "landing platforms" on which butterflies perch. They also tend to have long, slender floral tubes filled with nectar that is accessible to the long, coiled proboscis of butterflies and moths. Flowers like jimsonweed, evening primrose and others visited regularly by moths are lightly colored (white, yellow) and heavily scented, thus making them visible at night. Ultraviolet light is highly visible to insects. Carotenoids, yellow or orange pigments frequently found in plants, are highly visible to bees and other insects.

6.4.2.4 Birds

Several plants are regularly visited and pollinated by birds like hummingbirds and sunbirds. Such plants produce large amounts of nectar to provide enough food to these birds. Red color flowers signal the presence of abundant nectar and attract the birds. Again, red fruits are commonly dispersed by birds.

6.4.2.5 Other Animals

Other animals including bats, small rodents and monkeys may also facilitate pollination and dispersal of seeds.

6.5 FERTILIZATION

In angiosperms, fertilization is a complex, in which two sperm cells combine with an egg cell in a unique process called double fertilization, which involves the fertilization of the egg, and the formation of a nutrient-rich endosperm. Once a pollen grain falls on the upper surface of the stigma, it adheres to it by the sticky, sugary substance that covers the stigma. The pollens then begin to grow and produce a tube that pierces the style. The pollen tube, nourished by the sugary substance, continues to grow until it reaches the ovule in the ovary. The generative cell within the pollen tube cell undergoes division to form two sperm cells. At the entry to the embryo sac, the tip of the pollen tube bursts and releases the two sperm cells. Simultaneously, the two nuclei that flank the egg cell disintegrate, and one of the sperm cells fertilizes the egg cell, forming a zygote. The other sperm cell fuses with the two polar nuclei located at the center of the embryo sac, to form the primary endosperm nucleus, which eventually develops into the endosperm. Once fertilization is complete, the embryo develops by repeated cell division. The protective tissues enclosing the embryo develop into seed. The seed, in turn, is enclosed in another structure called the fruit.

6.5.1 Homozygosity and Heterozygosity

The reproductive system largely determines the genetic structure of a plant. Asexual reproduction results in genetically identical progenies. They resemble their parents, although environmental conditions and a rare genetic mutation could produce genetic changes in asexually produced offsprings.

Plants reproduce sexually by either self-pollination or cross pollination. The mode of reproduction influences the genetic structure of the population. Self-pollination increases homozygosity due to accumulation of similar alleles resulting from selfing over several generations, whereas cross pollination increases heterozygosity due to frequent recombination and segregation. So the genetic structure of a self-pollinated population is heterogeneous with homozygous individuals and that of a cross-pollinated population is homogeneous with heterozygous individuals.

6.5.1.1 Inbreeding

Inbreeding is the natural mode of breeding in self-pollinated crops, and it produces desirable results by increasing homozygosity and uniformity of the plants. There is no adverse effect of inbreeding on the self-pollinated plants because due to continuous selfing over the generations, the recessive deleterious alleles are eliminated from the progenies. However, enforced inbreeding in naturally cross-pollinating species may lead to appearance of several lethal and subvital types in early generations of selfing. This may be followed by the separation of population into distinct lines that are increasingly uniform within and increasingly distinct from other lines over several generations of selfing. Many of these lines show general decrease in their vigor and fertility and are difficult to maintain even in the most favorable cultural conditions. In the end, even the lines that survive exhibit lower size and vigor.

6.5.1.2 Hybrid Vigor/Heterosis

Hybrid vigor or heterosis refers to increased vigor in the hybrids as compared to both of its parents. The basic mechanisms involved in the heterosis are dominance and overdominance. Dominance means that the dominant allele masks the effect of recessive allele. Overdominance means that the combination of genotypes from two different parents leads to supplementing the effect of each other; therefore, the effects lead to increased vigor (Chahal and Gosal, 2002).

6.6 SELF-INCOMPATIBILITY

Self-incompatibility (SI) is the mechanism utilized by flowering plants to prevent self-pollination (Silva and Goring, 2001). SI has been effectively used to avoid inbreeding and encourage outcrossing. Since the use of SI eliminates the laborious emasculation of the female parent in breeding, breeders use SI to achieve a more efficient F_1 hybrid production system. Because SI prevents self-fertilization and promotes outcrossing, F_1 hybrid seeds can be produced readily from two parental SI lines with the help of pollinators.

6.6.1 Mechanisms of Self-Incompatibility

Flowering plants have developed unique mechanisms for SI. Some plants produce unisexual flowers, which are either male/staminate or female/pistillate, thereby acting as natural barrier for self-fertilization (McCubbin and Kao, 2000). However, the vast majority of flowering plants have perfect/hermaphrodite/bisexual flowers, containing both male and female reproductive organs within close proximity on the same flower (Kao and McCubbin, 1996).

6.6.2 SI Types

Self-incompatibility is widespread in flowering plants to prevent inbreeding and promote outcrossing. The SI response is genetically controlled by one or more multi-allelic loci and relies on a series of complex cellular interactions between the self-incompatible pollen and pistil. The following two types of SI are noticed in flowering plants:

Heteromorphic SI: In this system, the morphological differences in reproductive parts are visible in flowers to coincide with crossibility. The features such as varying style length, filament length, pollen size and exine sculpturing in the two parents promote SI. The presence or absence of other SI mechanisms will not affect cross-pollination in heteromorphic SI. There are three subcategories of this type of SI. In the heteromorphic distyle (e.g., primula) type, the flowers have style at two different heights (thrum flower with short style and long anther and pin flower with short anthers and long style). While the homozygous thrum (SS) with homozygous pin (ss) flowers (SS × ss) are cross compatible, partnering of heterozygous thrum with homozygous or heterozygous thrum (SS × SS; Ss × Ss) or homozygous pin (ss × ss) are cross incompatible. However, pairing of homozygous pin and heterozygous thrum types (ss × Ss) and heterozygous thrum and homozygous pun (Ss X ss) is also cross compatible.

In heteromorphic "Tristyle" type (e.g., Lythrum or loosestrife), SI is due to a short style with mid- or long stamens, mid style with short or long stamens, or long style with short and mid-stamens. Crossing of homomorphic flowers results in incompatible mating, whereas crossing of heteromorphic flowers results in compatible mating.

Homomorphic SI: Mating types in homomorphic SI cannot be recognized by morphological features. The SI in this type of flowers is controlled by "S" alleles; fertilization occurs only when different parental "S" alleles are crossed. This type of SI is subdivided into sporophytic self-incompatibility (SSI) and gametophytic self-incompatibility (GSI). The rejection of pollen in SSI is controlled by "S" loci which is dominant. This type of self-incompatibility is controlled by diploid genotypes of the sporophyte (pollen). Pollen will not germinate on the stigma of the flower that contains either of the two alleles in the pollen. SSI is less rare compared to GSI.

The rejection or acceptance of pollen in GSI is controlled by haploid genotype of pollen itself. S_1 pollen can germinate on pistil of S_1S_2, but due to common S_1 allele, the pollen tube does not grow. Similarly, in S_2, the pollen tube growth seizes. There will be normal growth of pollen tube and fertilization when S_3 pollen comes in contact with pistil of S_1S_2. This is also called as partial compatibility; however, S_3 and S_4 pollens can pierce into the S_1S_2 style and produce normal fertilization. This is called as complete compatibility. The mating systems in a single gene control can be explained as follows.

Pollen rejection and acceptance of pollen involve complex interactions, such as pollen–stigma interaction seen in SSI, pollen–style interaction seen in GSI, pollen tube–ovule interaction seen in GSI, and pollen tube–ovule interaction. In pollen tube–ovule interaction, which is very rare (seen in some cases of cocoa), the pollen tube reaches ovule and effect fertilization but incompatible combination will lead to degeneration of the embryo at the early-stage development without any seed set.

S_1S_2 (pistil) × S_1S_2 (pollen): 0% compatibility
S_1S_2 × S_2S_3: 50% compatibility
S_1S_2 × S_3S_4: 100% compatibility

6.6.3 Application of SI in Plant Breeding

SI can be used for producing F_1 hybrid by avoiding contamination by self-pollinated seeds (Franklin-Tong, 2008). The main problem of utilizing SI is the establishment of two pure line SI parents. Breeders choose the SI lines from a large number of seed stocks and use physiological (treatment with CO_2 gas, irradiation and pistil grafting) or genetic techniques to break SI to produce pure lines (Franklin-Tong, 2008; Nettancourt, 2001). In the SSI system, receptor protein S-receptor kinase (SRK) covers the stigmatic surface just prior to anthesis and acts as a barrier for penetration by germinating pollen grains (Silva and Goring, 2001). Seed sets of pure line parents can be obtained if the pollen is applied after bud opening and before the SRK protein barrier is formed (Franklin-Tong, 2008). This technique is called the bud pollination. However, the efficiency of production of a large number of parental seeds of F1-hybrid by bud pollination is low (Franklin-Tong, 2008). Furthermore, none of the major field crop plants use SSI system.

6.6.4 Breakdown of Self-Incompatibility

The physiological modifications, being temporary, cannot be transmitted from one generation to the next (Nettancourt, 2001). On the other hand, genetic modifications may or may not be permanent and result in various effects (Nettancourt, 2001). The hybrid self-fertility resulting from epigenetic changes in the expression of the S-locus genes has been demonstrated by Nasrallah et al. (2007) to be effective in overcoming SI. Epigenetic mechanism is also reversible.

6.6.4.1 Physiological Breakdown of Self-Incompatibility

Because SI phenotype of the pollen and pistil is fully determined in the mature flower, selfing can be induced by using immature material in which the S phenotype is not yet expressed (bud pollination) or by using old flowers and aged pollen, in which SI effect is becoming weaker (Mable, 2008). Several studies also point out that heat treatment at 32°C to 60°C can lead to breakdown of SI in the pistil during the first two days following pollination in lily, tomato, radish and *Brassica* (Nettancourt, 2001). Treatment with 3–5% CO_2 gas can also break SI (Franklin-Tong, 2008), although the timing of the CO_2 gas treatment is critical.

The "mentor effect" is another method to break SI. In this method, pollination with mixtures of compatible (mentor) and incompatible pollen is used to overcome SI (Nettancourt, 2001). The mentor effect can be enhanced by using inactivated or dead pollen (Nettancourt, 2001). The mentor effect has been applied successfully in *Citrus, Crocus, Lotus, Paspalum* and *Theobroma* (Nettancourt, 2001). However, the mentor effects are not always reproducible (Nettancourt, 2001).

Other methods of physiological breakdown of SI include irradiation, hormones and pistil grafting (Nettancourt, 2001).

6.6.4.2 Genetic Breakdown of Self-Incompatibility

Genetic breakdown of SI occurs in nature and can also be induced by breeders artificially. Tetraploid relatives (or induced tetraploids) derived from GSI diploid progenitors usually display a self-compatibility (Horandl, 2010). Breakdown of SI occurs in pollen, whereas the pistil maintains the function of identifying incompatible pollen and then rejects it. Self-fertilization results from the loss of the incompatible phenotype in the diploid pollen produced by the tetraploid plant (Nettancourt, 2001). For example, pollen from SC *Petunia axillaris* tetraploid ($S_1S_1S_2S_2$) can grow in both SI progenitor (S_1S_2) diploid and SC tetraploid ($S_1S_1S_2S_2$) pistils (Nettancourt, 2001). Remarkably, SI breakdown functions only in diploid heteroallelic pollen (S1S2), which is called competitive interaction (Nettancourt, 2001). On the other hand, modern biotechnologies including site-directed mutagenesis, the use of antisense DNA, and exploitation of the silencing effects of certain duplications or the swapping of S-locus DNA sequences have been demonstrated to be effective in making irreversible switches from SI to self-compatibility (Horandl, 2010).

6.7 MALE STERILITY

Hybrid seed production requires a female plant, which has no viable male gametes of its own by emasculating the female line or parent. Another simple way to establish a female line for hybrid seed production is to identify or create a line that is unable to produce viable pollen. This male sterile line is, therefore, unable to self-pollinate and seed formation is dependent upon pollen from the male line. In hermaphrodite flower, pollens are nonfunctional or inactive or sterile while female gametes function normally. It is the inability of plant to produce or release functional pollen as a result of the failure of formation or the development of functional stamens, microspores or gametes. This prevents autogamy and permits cross pollination, thereby promoting heterozygosity. Sterility could be due to nuclear genes (genetic male sterility), cytoplasmic (cytoplasmic male sterility) gene or both (genetic – cytoplasmic sterility). In hybrid seed production, the male sterile female parental line is crossed with a male fertility restorer line to get F hybrid.

6.7.1 Interspecific Hybridization

Interspecific hybridization involves crossing distantly related or sometimes unrelated species. In some cases, unrelated plants can be easily crossed but the hybrid formed is sterile. In other cases, the zygote resulting from such crosses get aborted at embryonic stages. In the worst situations, there are some morphological reproductive barriers which do not even allow the crossing and fertilization to occur easily. Various strategies such as bridge crossing, somatic hybridization, induced polyploidy, embryo rescue and use of plant growth regulators can be employed to overcome these problems and accomplish successful interspecific hybridization in widely related species. When two species cannot be intercrossed to produce a fertile hybrid, a third species which can be easily crossed with both the parental types can be used to produce a bridge cross. R. C. Buckner and his colleagues have used this method to cross Italian ryegrass (*Lolium multiflorum*) with tall fescue (*Festuca arundinacea*) using meadow fescue (*Fescue pratensis*) as the bridging species (Acquaah, 2007).

6.7.2 Somatic Hybridization

Somatic hybridization is a technique that can bypass the mating barriers between widely related species. The individual somatic cell protoplasts from both the species are fused using polyethylene glycol (PEG) or electric fusion, and a somatic cell hybrid is formed. These somatic hybrids are later regenerated into a full hybrid plant. *Solanum brevidens* and *S. phureja* have been used in somatic hybridization technique to introduce disease resistance for Potato leaf roll virus and Potato virus X into the *S. tuberosum* species. The nuclear genome can be rendered inactive by treating with Indolacetate to get the desired hybrids (Chahal and Gosal, 2002). Embryo rescue can be employed in situations when there are risks of embryo abortion or loss of viability of the hybrid seeds. In studies involving the production of hybrids between *Cicer arietinum* and *C. bijugum* (wild variety), it was demonstrated that the hybrid produced from the cross did not yield viable seeds. However, if the embryos are cultured on the suitable media they can be regenerated into plantlets and then transferred to soil (Clarke et al., 2006). Plant growth regulators such as auxins and gibberellins can be used to save the weak hybrid plants from the crosses that do not survive under natural conditions. Induced polyploidy is helpful in partially or completely overcoming the sterility in wide intercross hybrids (Fehr and Hadley, 1980).

6.7.3 Sex Modification

Sex expression in dioecious and monoecious plants can be modified to a varying extent by environmental conditions, such as mineral nutrition, photoperiod, temperature and hormones. Among these, hormones have been found to be most effective agents for sex modification and their role in regulation of sex expression in flowering plants has been well documented. The morphological differences in various sex types and their specific metabolic characteristics result from the presence of specific patterns of proteins, enzymes and other molecules. Modification of sex expression in cucurbits has been achieved both by changing the environmental conditions and by exogenous application of growth regulators (Girek et al., 2013). Auxin (NAA), ethylene (Etherl, Ethephon), Cytokinins (BA) and Brassinosteriods treatments increase the femaleness. In contrast, gibberellic acid (GA), $AgNO_3$, abscisic acid (ABA), thioporpinic acid, pthalimide and paclobutrazol induce maleness.

In cucumber, $AgNO_3$ sprayed when the first true leaf is fully expanded was found to be a potent inhibitor of ethylene action leading to femaleness. Similarly, gibberellic acid spray caused excessive elongation, weakening of plants and an increased number of malformed male flowers with less pollen. In gynoecious cucumber, $AgNO_3$ and gibberellic acid increased the number of male nodes.

6.7.4 Environmental Modification of Sex

Environment has the greatest influence on the sex expression. Male sex expression in several plant species has been shown to be favored by high temperatures, whereas low temperature induces female flowers. This was demonstrated by increasing the growing temperature for cucumber from 19°C to 23°C wherein the node at which the first female flower appeared and number of flowers increased. Similarly, the male sterile mutant of tomato produced only male sterile flowers when grown at high temperature (above 30°C) and normal flowers under lower temperatures. In Brussels sprouts, low temperature affected the development of androecium.

6.8 COMMERCIAL UTILIZATION OF MALE STERILITY IN HORTICULTURAL CROPS

According to Kumar and Singh (2004), nearly 19 male-sterile-based hybrids have been released in tomato (*Lycopersicon esculentum*), 17 of them are based on functional sterility controlled by gene positional sterility (ps-2). The ps-2 mutant has been used on commercial scale in Czech Republic, Moldova and Bulgaria (Atanassova, 1999). The MS-12 (ms-10ms-10) line is being utilized to produce seeds of CH-1 and CH-3 hybrids of hot pepper in India. In Hungary, male sterile lines possessing ms-3 gene are being utilized to produce hybrid seeds (Kumar et al., 2000). In South Korea (Shifriss, 1997), China (Boaxi et al., 2000) and India, CMS lines are being utilized to develop hybrids of hot pepper. However, nonavailability of Rf genes in most of the sweet pepper genotypes is still a handicap in developing CMS-based commercial sweet pepper hybrids. Although GMS-based experimental crosses have been developed in cole vegetables (*Brassica* spp.), it has not been commercially utilized mainly because of the difficulty in multiplication of male sterile seeds and availability of self-incompatibility and CMS systems.

6.9 ASEXUAL REPRODUCTION

Two types of asexual propagation methods are commonly used. Some plants naturally reproduce vegetatively through tubers (e.g., potato), rhizomes (e.g., ginger), bulbs (e.g., onion), etc. Others are artificially propagated vegetatively using various methods like cutting, grafting, layering and budding. It should be noted that most of the species that reproduce vegetatively can also reproduced sexually (Burdon, 1980).

Apomixis is the process by which plants reproduce asexually through seed (Nogler, 1984). Nogler (1984) divided apomixis according to the origin and development of the maternal embryos into three groups, apospory (embryo sacs are formed from the nucellar cells), diplospory (megaspore develops from the reproductive tissue without normal meiosis), and adventitious embryony (embryo develops directly from a somatic cell of the megagametophyte). Apomixis is said to be facultative when some progeny from a plant may result from a normal meiosis and/or normal fertilization in addition to the apomictic progeny. Apomixis is obligate when all the progeny are maternal. Because the hybrids formed by apomixis may be sterile and they can only be asexually propagated, apomictic plants are uniform and free from virus. Apomixis is less common than vegetative reproduction, although it has been reported from at least 30 families of flowering plants (Grant, 1981).

REFERENCES

Acquaah G. 2007. *Principles of Plant Genetics and Breeding*. Blackwell Publications, Malden, MA.

Atanassova B. 1999. Functional male sterility (*ps-2*) in tomato (*Lycopersicon esculentum*) and its application in breeding and hybrid seed production. *Euphytica* **107**: 13–21.

Boaxi Z, Sanwen H, Guimei Y and Jiazhen G. 2000. Two RAPD markers linked to a major fertility restorer gene in pepper. *Euphytica* **113**: 155–169.

Burdon J and Harper J. 1980. Relative growth rates of individual members of a plant population. *Journal of Ecology* **68**: 953–957.

Chahal G S and Gosal S S. 2002. *Principles and Procedures of Plant Breeding*. Alpha Science International Ltd., Pangbourne.

Clarke H J, Wilson J G, Kuo I, Lulsdorf M M, Mallikarjuna N, Kuo J and Siddique K H M. 2006. Embryo rescue and plant regeneration *in vitro* of selfed chickpea, *Cicer arietinum* L. and its wild annual relatives. *Plant Cell Tissue and Organ Culture* **85**: 197–204.

Durbarry A, Vizir I and Twell D. 2005. Male germ line development in Arabidopsis. *duo* pollen mutants reveal gametophytic regulators of generative cell cycle progression. *Plant Physiology* **137**: 297–307.

Duvick D N. 1966. Influence of morphology and sterility on breeding methodology. pp. 85–139. *Plant Breeding*. University Press, Ames, IA.

Engel M L, Chaboud A, Dumas C and McCormick S. 2003. Sperm cells of *Zea mays* have a complex complement of mRNAs. *Plant Journal* **34**: 697–707.

Fehr W R and Hadley H H. 1980. *Hybridization of Crop Plants*. American Society of Agronomy. Crop Science Society of America, Madison, 765pp.

Franklin-Tong V E E. 2008. *Self-Incompatibility in Flowering Plants Evolution Diversity and Mechanisms*. Springer-Verlag, Berlin and Heidelberg.

Fryxell P A. 1957. Mode of reproduction of higher plants. *Botanical Review* **23**: 135–233.

Girek Z, Prodanovic S, Zdravkovic J, Zivanovic T, Ugrinovic M and Zdravkovic M. 2013. The effect of growth regulators on sex expression in melon (*Cucumis melo* L.). *Crop Breeding and Applied Biotechnology* **13**: 165–171.

Grant V. 1981. *Plant Speciation*. 2nd edition. Columbia University Press, New York, xii, 563p.

Horandl E. 2010. The evolution of self-fertility in apomictic plants. *Sex and Plant Reproduction* **23**: 73–86.

Kao T H and McCubbin A G. 1996. How flowering plants discriminate between self and non-self-pollen to prevent inbreeding. *Proceedings of National Academy of Sciences, United States of America* **93**: 12059–12065.

Kumar S, Banerjee M K and Kalloo G. 2000. Male sterility: mechanisms and current status on identification, characterization and utilization in vegetables. *Vegetable Science* **27**: 1–24.

Kumar S and Singh P K. 2004. Mechanism for hybrid development in vegetables. *Journal of New Seeds* **6**(4): 300–407.

Mable B K. 2008. Genetic causes and consequences of the breakdown of self-incompatibility: case studies in the Brassicaceae. *Genetics Research* **90**: 47–60.

McCubbin A G and Kao T. 2000. Molecular recognition and response in pollen and pistil interactions. *Annual Review of Cell Development and Biology* **16**: 333–364.

Nasrallah J B, Liu P, Sherman-Broyles S, Schmidt R and Nasrallah M E. 2007. Epigenetic mechanisms for breakdown of self-incompatibility in interspecific hybrids. *Genetics* **175**: 1965–1973.

Nettancourt D D. 2001. *Incompatibility and Incongruity in Wild and Cultivated Plants*. 2nd edition. Springer, Berlin, Heidelberg and New York.

Nogler G A. 1984. Gametophytic apomixis. pp 475–518. *Embryology of Angiosperms*. Springer-Verlag, Berlin.

Rotman N, Durbarry A, Wardle A, Yang W C, Chaboud A, Faure J E and Twell D. 2005. A novel class of MYB factors controls sperm-cell formation in plants. *Current Biology* **15:** 244–248.

Shifriss C. 1997. Male sterility in pepper (*Capsicum annuum* L). *Euphytica* **93**: 83–88.

Silva N F and Goring D R. 2001. Mechanisms of self-incompatibility in flowering plants. *Cell Molecular Life Science* **58**: 1988–2007.

Stebbins G L Jr. 1950. *Variation and Evolution in Plants*. Columbia University Press, New York.

von Besser K, Frank A C, Johnson M A and Preuss D. 2006. Arabidopsis *HAP2* (*GCS1*) is a sperm-specific gene required for pollen tube guidance and fertilization. *Development* **133**: 4761–4769.

Xu H, Swoboda I, Bhalla P L, Sijber A M, Zhao C, Ong E K, Hoeijmakers J H and Singh M B. 1998. Plant homologue of human excision repair gene *ERCC1* points to conservation of DNA repair mechanisms. *Plant Journal* **13**: 823–829.

Xu H, Swoboda I, Bhalla P L and Singh M B. 1999. Male gametic cell-specific gene expression in flowering plants. *Proceedings of National Academy of Sciences United States of America* **96**: 2554–2558.

Part II

Genetic Aspects of Plant Growth and Development

Until recently, plant developmental biology investigations resulted in rather descriptive answers or solution to highly complex phenomena. Although a wealth of information was obtained by a judicious choice of model plants, it was clear that the molecular and genetic techniques have a much higher potential to improve and rapidly enlarge our understanding of plant developmental programs. Although *in vitro* recombinant techniques were available for several decades, it is recently that transgenic plants were developed and used to unravel the developmental signals. Such a long delay in the application of molecular techniques in plants compared to animal research was partly because the plant morphologists and physiologists were bit slow in introducing the molecular biology approaches.

Now, we are living in a tremendously exciting time, perhaps the one that in the future will be replete with technological and conceptual breakthroughs. Application of novel analytical tools and techniques has led to significant biological discoveries and hypotheses that are conceptually transformative. This synergy is particularly evident in the study of plant genome evolution, in which massively parallel sequencing approaches have revealed genomic diversity in very much detail, which has led to many insights into genome function and evolution. This section highlights the progress made in understanding plant genomes, with special focus on horticultural crops and on recent key insights.

DOI: 10.1201/9781003373889-8

7 Plant Genome and Regulation of Gene Expression

7.1 INTRODUCTION

Several approaches such as hybridization breeding, mutation breeding and transgenic breeding have been used to improve yield and quality in horticultural crops. Using these technologies, new varieties with novel traits have been evolved in a number of horticultural crops. In the past, individuals with desirable traits, such as larger fruits, higher yield, and better-tasting and nutritious fruits were selected and multiplied. At the beginning of the century, attempt of artificial pollination was initiated to develop individuals with desirable traits. This marked the beginning of hybridization breeding, a characteristic of modern agriculture and horticulture. Since then, there has been a phenomenal increase in interest and the rate of crop improvement. By intentional hybridization and selection, breeders have developed several fruit and vegetable crop varieties in strawberry, apple, mango, sweet orange, tomato, cucumber, sweet pepper, squash and many other vegetables and fruits. Another application of hybridization breeding is the generation of seedless watermelon by crossing diploid and tetraploid parents.

With the advancement in the understanding of plant genetics, breeders began focusing on specific genes to improve individual traits. This led to the development of diseases resistant varieties. During the process of crop evolution, spontaneous variations with new traits occurred naturally and these have been preserved for use in breeding of new varieties. This approach has at least three shortcomings, which include low frequency of natural mutation, their noninheritable or reversible nature, and possibility of appearance of chimera. These can be overcome by exposing seeds, plants parts, tissues, cuttings, pollens to mutagens. The progress in conventional breeding of fruit trees has been constrained by their extended juvenile phase, seasonality in growth and reproduction, large size and a marketable product that cannot be assessed until plants attain physiological maturity (Luby and Shaw, 2001; Rikkerink et al., 2007). A number of biotic and abiotic stresses that affect growth, fruiting and fruit quality also complicate genetic improvement in these crops.

During the early stages, breeders selected phenotypes with merit without knowing the genotype. The advent of molecular genetics led to the development of modern biotechnological breeding. By knowing the genetic control and inheritance of desirable traits, it has become possible to precisely manipulate the gene encoding to create novel genotypes through the use of DNA recombinant or the transgenic technology. With this technology, it is possible to transfer the cloned gene regardless of the source or recipient of the genes. The transgenic technology integrates the desired foreign genes into the host plant genome using the Agrobacterium-mediated methods, particle bombardment methods or protoplast transformation methods. Although the Agrobacterium-mediated method is the easiest and most convenient method, most horticulture crops are not susceptible to Agrobacterium and therefore, cannot be transformed with this method. The bombardment of protoplast method can be used to overcome the limitations of the Agrobacterium-mediated method. Even these methods also have certain shortcomings (e.g., the bombardment method requires specific facilities and the protoplast method is dependent on handling skills). Protoplast transformation and regeneration of a whole plant from a single cell is an alternative strategy for many horticulture crops. Through transgenic approaches, breeders can introduce genes encoding for new traits into plants, such as resistance to certain pests or diseases. The primary advantage is its precision and the improvement of a trait without altering the genetic constitution of an elite genotype. This is particularly useful for many highly heterozygous perennial horticultural crops. Slow softening (decaying) FlavrSavr tomato and ring spot disease

DOI: 10.1201/9781003373889-9

resistant papaya are the best examples of transgenic lines developed in the 1990s (Bruening and Lyons, 2000; Clark et al., 2004; Azad et al., 2014; Xiong et al., 2015). Several transgenic lines of horticultural plant have been developed using this technology. Although transgenic approach has been very successful, this technology still faces several challenges (environmental and food safety concerns and several of economically important plants are recalcitrant to genetic transformation). Recently developed highly efficient, cost-effective genomics-based approaches have opened up new avenues in fruit breeding (Luo et al., 2007; Xiong et al., 2015; Chia and Ware, 2011; Meneses and Orellana, 2013; Myles, 2013; van Nocker and Gardiner, 2014; He et al., 2014; Varshney et al., 2014; Iwata et al., 2016).

7.2 GENOME EDITING

In the past decade or so, genome editing or genome editing with engineered nucleases (GEEN) technology, a type of genetic engineering, has become available. These technologies rely on engineered endonucleases (EEN) that cleave DNA in a sequence-specific manner. Through recognition of the specific DNA sequence, these nucleases can efficiently and precisely cleave the targeted genes. The double strand breaks (DSBs) of DNA consequently result in cellular DNA repair mechanisms, including homology-directed repair (HDR) and error-prone non-homologous end joining (NHEJ) breaks, leading to gene modification at the target sites (Pal, 2012; Iwata et al., 2016; Sonah et al., 2011). Genomic resources like whole genome sequences, expressed sequence tags (ESTs), genomic survey sequences (GSSs) and high-throughput genome sequences (HTGs) have been used for sequencing and annotation, mapping and closing of genes or quantitative-trait loci (QTL) and marker-assisted selection (MAS) in important horticultural crops (Sonah et al., 2011). Genome maps also allow breeders to select for characters in young seedlings without the need for laborious, time-consuming field evaluation for traits that are expressed only in mature plants, such as abiotic stress tolerance, flower quality or fruit quality. Unlike cereal crops, the progress in genomic studies in horticultural plants is rather limited, although with the recent availability of next-generation sequencing (NGS) technologies, such as FLZ-454, Illumina, SOLiD and Helicose, the progress in the development and use of genomic resources of horticultural crops is expected to grow rapidly.

7.3 PLANT GENOMES

Hans Winkler coined the word "genome" by combining gene and chromosome (Lederberg and McCray, 2001). Lederberg and McCray (2001) defined the genome as the total DNA amount or gene complement of the haploid chromosome set. It is expressed as megabases (1 mb = 1,000,000 bases). Plant genomes are best described in terms of genome size, gene content, extent of repetitive sequences and polyploidy/duplication events. Although plants also possess mitochondrial and chloroplast genomes, nuclear genome is the largest and highly complex. The genome size of some representative plants is presented in Table 7.1.

Thomas Roderick first used the term "genomics" in 1986 to describe a discipline of genetics that employs recombinant DNA, DNA-sequencing methods and bioinformatics to sequence, assemble and analyze the function and structure of genomes (the complete set of DNA within a single cell of an organism) (McKusick, 1997). It includes the determination of the entire DNA sequence of organisms and fine-scale genetic mapping as well as studies of intragenomic phenomena such as heterosis, epistasis, pleiotropy and other interactions between loci and alleles within the genome. There are three types of genomics, namely, structural genomics (dealing with characterization of structure of genome), functional genomics (involving elucidation of gene functions and its interactions based on information generated from genome sequencing projects) and comparative genomics (dealing with comparison of genomic features of different organisms). Genomic studies are required to know the sequence of the whole genome, locate specific genes on the chromosome, locate promoters and functional motifs, and discover and utilize genetic variations, genome manipulation and evolutionary studies (Pal, 2012).

TABLE 7.1
Genome Size of Selected Plants

Plants	Genome size (Mb)	Chromosome number (1n)
Onion	15,290	8
Garden pea	3,947	9
Asparagus	1,308	15
Tomato	907	12
Sugar beet	758	11
Apple	743	17
Bean	637	9
Brinjal	1,100	12
Potato	840	12
Grape	483	19
Cantaloupe	454	12
Brassica oleracea	600	9
Papaya	370	9
Capsicum annum	3,000	12
Citrus sinensis	380	9
Cucumber	370	7
Carrot	470	9

Source: Compiled from Campos-de Quiroz, 2002; Sonah et al. (2011).

7.4 SEQUENCING OF HORTICULTURAL CROP GENOMES

Several projects on genome sequencing of horticultural crops were undertaken in recent years. The most important ones include tomato, potato, papaya and grapes (Sonah et al., 2011; Pal, 2012; Sharma et al., 2017). The conventional DNA sequencing (di-deoxynucleotide sequencing or Sanger's method of DNA sequencing) provides large enough read length with quality sequence but it is a laborious and labor-intensive process. With the advent of next-generation sequencing (NGS) technologies, such as FLX454 (Roche), Solexa (Illumina) and SOLiD (Applied Biosystems), there has been appreciable progress in DNA sequencing (Cheung et al., 2006). Genomic research has excellent potential to revolutionize the molecular biology research in horticultural crops. The whole genome sequencing of *Arabidopsis thaliana* laid the foundation of plant genomics research (Arabidopsis Genome Initiative – AGI, 2000). Since then, enormous progress has been made in decoding complete genome sequence of poplar, grapevine (Jaillon et al., 2007), transgenic papaya (Ming et al., 2008), cucumber (Huang et al., 2009) and several other crops. The draft sequence of cassava, potato, tomato, shepherd's purse and peach are now available in the database (Sonah et al., 2011). Among the solanaceous crops, tomato and potato are used as model system for genomic studies. The international solanaceae genome project (ISGP) was set up in 2003 to sequence tomato (Mueller et al., 2009). Similarly, Potato Genome Sequencing Consortium was set up for sequencing of potato genome. There was other multinational genome sequencing projects in brassica, cassava, grapevine, papaya and others (Sonah et al., 2011). With the establishment of genome and transcriptome sequencing projects for several horticultural crops, the huge wealth of sequence information has been generated (Azim et al., 2014).

7.5 MOLECULAR MARKERS

DNA-based molecular markers are no longer looked as a finger printing or forensic tools but rather as versatile tools to develop marker-based gene tags, map-based cloning of genes, variability studies, phylogenetic analysis and marker-assisted selection of desirable genotypes. Ever since their development, they are constantly being modified to enhance their utility and bring about automation in genome analysis. The DNA markers offer several advantages over traditional phenotypic markers, as they provide data that can be analyzed objectively. The discovery of polymerase chain reaction (PCR) proved to be a unique process that led to a new class of DNA profiling markers. Molecular markers include biochemical constituents (e.g., secondary metabolites in plants) and macromolecules (e.g., proteins and deoxyribonucleic acids).

7.5.1 DNA-Based Molecular Markers

DNA finger printing/profiling is a versatile tool for investigating various aspects of plant genomes, such as characterization of genetic variability, genome fingerprinting, genome mapping, gene localization, analysis of genome evolution, population genetics, taxonomy, plant breeding and diagnostics. The properties of an ideal DNA marker include highly polymorphic nature, codominant inheritance (determination of homozygous and heterozygous states of diploid organisms), frequent occurrence in genome, selective neutral behavior (the DNA sequences of any organism are neutral to environmental conditions or management practices), easy availability, high reproducibility and easy exchange of data between laboratories. Although it is extremely difficult to find a molecular marker that would meet all the above criteria, depending on the type of study to be undertaken, a marker system can be identified that would fulfill at least a few of the above characteristics.

Molecular markers are classified as hybridization-based markers and PCR-based markers. In the former, DNA profiles are visualized by hybridizing the restriction enzyme-digested DNA, to a labeled probe, which is a DNA fragment of known origin or sequence. PCR-based markers involve *in vitro* amplification of particular DNA sequences or loci with the help of specifically or arbitrarily chosen oligonucleotide sequences (primers) and a thermostable DNA polymerase enzyme. The amplified fragments are separated electrophoretically and banding patterns are detected by different methods such as staining and autoradiography. The primer sequences are chosen to allow base-specific binding to the template in reverse orientation. PCR is extremely sensitive and has opened up new possibilities in the field of molecular biology. The main types of markers are described here.

7.5.1.1 Restriction Fragment Length Polymorphism (RFLP)

RFLPs have their origin in the DNA rearrangements that occur due to evolutionary processes, point mutations within the restriction enzyme recognition site sequences, insertions or deletions within the fragments, and unequal crossing over. In RFLP analysis, restriction enzyme-digested genomic DNA is resolved by gel electrophoresis and then blotted on to a nitrocellulose membrane. Specific banding patterns are then visualized by hybridization with labeled probe. These probes are mostly species-specific single locus probes of about 0.5–3.0 kb in size, obtained from a cDNA library or a genomic library. The selection of appropriate source for RFLP probe varies, with the requirement of particular application under consideration. RFLPs, being codominant markers, can detect coupling phase of DNA molecules, as DNA fragments from all homologous chromosomes are detected. However, their utility has been hampered due to the large amount of DNA required for restriction digestion and blotting. The assay also is time-consuming and labor-intensive, and only one out of several markers may be polymorphic, which is highly inconvenient especially for crosses between closely related species. Their inability to detect single base changes restricts their use in detecting point mutations occurring within the regions at which they are detecting polymorphism.

7.5.2 Restriction Landmark Genomic Scanning (RLGS)

This method is a new finger printing technique based on the principle that restriction enzyme sites can be used as landmarks. It employs direct labeling of genomic DNA at the restriction site and two-dimensional (2D) electrophoresis to resolve and identify these landmarks. The technique has proven its utility in genome analysis of closely related cultivars and for obtaining polymorphic markers that can be cloned by spot target.

7.5.2.1 RFLP Markers Converted Into PCR-Based Markers Sequence-Tagged Sites (STS)

RFLP probes specifically linked to a desired trait can be converted into PCR-based STS markers to obtain specific amplicon. This approach is extremely useful for studying the relationship between various species.

7.5.2.2 Allele-Specific Associated Primers (ASAPs)

These markers tag specific alleles in the genome and are more or less similar to SCARs.

7.5.2.3 Expressed Sequence Tag Markers (EST)

ESTs are popularly used in full genome sequencing and for identifying active genes, thus helping in identification of diagnostic markers. Moreover, an EST that appears to be unique helps to isolate new genes.

7.5.2.4 Single Strand Conformation Polymorphism (SSCP)

This is a powerful and rapid technique for gene analysis particularly for detection of point mutations and typing of DNA polymorphism. SSCP can identify heterozygosity of DNA fragments of the same molecular weight and can even detect changes of a few nucleotide bases.

7.5.2.5 Multi-Locus Probes Repetitive DNA (LPRD)

The repetitive DNA regions play an important role in absorbing mutations in the genome. The markers belonging to this class are both hybridization-based and PCR-based.

7.5.2.6 Microsatellites and Minisatellites

The methodology has been derived from RFLP, and specific fragments are visualized by hybridization with a labeled micro- or minisatellite probe. Microsatellites and minisatellites form an ideal marker system creating complex banding patterns by simultaneously detecting multiple DNA loci. They are dominant fingerprinting markers and codominant STMS (sequence-tagged microsatellites) markers.

7.5.2.7 Minisatellite and Microsatellite Sequences Converted Into PCR-Based Markers Sequence-Tagged Microsatellite Site Markers (STMS)

This method includes DNA polymorphism using specific primers designed from the sequence data of a specific locus.

7.5.2.8 Direct Amplification of Minisatellite DNA Markers (DAMD-PCR)

DAMD-PCR clones being individual-specific DNA fingerprinting pattern have the potential as markers for species differentiation and cultivar identification.

7.5.2.9 Inter-Simple Sequence Repeat Markers (ISSR)

In this technique, primers based on microsatellites are utilized to amplify inter-SSR DNA sequences. These are mostly dominant markers, though occasionally a few of them exhibit codominance.

7.5.2.10 Other Repetitive DNA-Type Markers Transposable Elements

A large number of transposable repeat elements have been studied in plants; however, only a few have been exploited as molecular markers.

7.5.2.11 Randomly Amplified Polymorphic DNA Markers (RAPD)

This procedure detects nucleotide sequence polymorphisms in DNA using a single primer of arbitrary nucleotide sequence. RAPD assay has been used by several groups as an efficient tool for identification of markers linked to agronomically important traits that are introgressed during the development of near isogenic lines. Some variations in the RAPD technique include DNA amplification fingerprinting (DAF), which can be tailored by employing various modifications such as predigesting of template. This technique has been useful in genetic typing and mapping. The arbitrary primed polymerase chain reaction (AP-PCR) is a special type of RAPD, wherein discrete amplification patterns are generated by employing single primers of 10–50 bases in length in PCR of genomic DNA. Compared to DAF, AP-PCR is not very popular as it involves autoradiography.

7.5.2.12 Sequence-Characterized Amplified Regions for Amplification of Specific Band (SCAR)

In this technique, the RAPD marker termini are sequenced, and longer primers are designed for specific amplification of a particular locus. Compared to arbitrary primers, SCARs exhibit several advantages in mapping studies. SCARs also allow comparative mapping or homology studies among related species, thus making it an extremely adaptable concept in the near future.

7.5.2.13 Cleaved Amplified Polymorphic Sequences (CAPs)

These polymorphic patterns are generated by restriction enzyme digestion of PCR products. These markers are codominant in nature.

7.5.2.14 Randomly Amplified Microsatellite Polymorphisms (RAMPO)

In this method, genomic DNA is first amplified using arbitrary (RAPD) primers. The amplified products are then electrophoretically separated, and the dried gel is hybridized with microsatellite oligonucleotide probes. This technique has been successfully employed in the genetic fingerprinting of tomato and kiwi fruit.

7.5.2.15 Amplified Fragment Length Polymorphism (AFLP)

This technique is based on the detection of genomic restriction fragments by PCR amplification. This technique is extremely useful in detection of polymorphism between closely related genotypes. It involves three steps, namely, restriction of DNA using a rare cutting and a commonly cutting restriction enzyme simultaneously (such as *Mse*I and *Eco*RI) followed by ligation of oligonucleotide adapters of defined sequences including the respective restriction enzyme sites; selective amplifications of sets of restriction fragments, using specifically designed primers; and gel analysis of the amplified fragments. AFLPs are extremely useful tools for DNA fingerprinting and cloning and mapping of variety-specific genomic DNA sequences.

7.6 APPLICATIONS OF MOLECULAR MARKERS IN PLANT GENOME ANALYSIS AND BREEDING

The most popular markers are SSR, SNP and conserved ortholog set (COS). The ESTs have been used for the generation of SSR and SNP markers in several horticultural crops (Frary et al., 2005; Sargent et al., 2006; Ekué et al., 2009). SSR markers are relatively easy and cheap to develop and will remain a choice of markers for crop research community. However, ESTs derived markers have limited use in developing linkage maps and DNA fingerprints. Single nucleotide polymorphism (SNP) markers have been developed in many species including grape and apple (Gilchrist et al., 2006; Hyten et al., 2008; Salmaso et al., 2008; Vezzulli et al., 2008). SNP has been used successfully for the construction of genetic linkage maps and for aligning physical and genetic maps and as potential transferable markers across the Rosaceae family (Chagne et al., 2008; Folta and Gardiner,

2009). Ortholog specific EST-SSR markers have been used for genetic mapping and genotyping applications in Iris. SSR markers can also be used for studying synteny between distantly related species (Dirlewanger et al., 2004a; Howad et al., 2005; Celton et al., 2009).

Markers tightly linked to major genes responsible for the expression of disease/pest resistance, fruit/nut quality, self-incompatibility and so on are being used for marker-assisted selection to improve breeding efficiency in horticultural crops (Dirlewanger et al., 2004b; Zhu and Barritt, 2008; Iezzoni et al., 2010; Bliss, 2010). Iezzoni et al. (2010) proposed an eight-stage process to improve utilization of markers for prioritization of available marker-locus trait associations, demonstration of screening efficiency of markers, improving markers for specific loci and traits, validating the markers across crops, demonstrating marker utility within target populations of specific breeding programs, making strategic decisions on MAPS, performing cost and logistics analyses based on marker-assisted selection (MAS), and actually demonstrating utility of MAS for crop improvement. The development of ultra-dense genetic linkage (UDGL) map will serve as the basis for map-based gene cloning and marker-assisted breeding in horticultural crops. Mapping resources are now available for grape, apple, citrus and *Prunus* spp. (Liebhard et al., 2003; Celton et al., 2009; Chen et al., 2008; Olmstead et al., 2008).

7.6.1 Identification and Cloning of Candidate Genes

Molecular biologists use genomic resources to identify candidate genes in the genome sequences and clone them from the related species by PCR amplification and sequencing (Sonah et al., 2011). Several researchers have identified quantitative-trait loci (QTL) that govern most of the important traits in fruit and vegetable crops (Kenis et al., 2008; Olmstead et al., 2008; Zhang et al., 2010; Kubo et al., 2010; Hatakeyama et al., 2010). Using QTL, several authors have attempted to improve nutritional quality of fruits and vegetables. Cloning of genes is also used in nutritional genomics for antinutritional, nutraceutical and pharmaceutical applications (Davuluri et al., 2005; Zhang et al., 2009; Blas et al., 2010).

7.6.2 Genome Array Analysis

DNA microarray analysis is a powerful tool for global expression of genes in an organism; however, it needs well characterized cDNA or expressed sequence tags (EST). The cDNA arrays are used in potato (Rensink et al., 2005), citrus (Forment et al., 2005), gerbera (Laitinen et al., 2005), *Brassica oleracea* (Soeda et al., 2005) and strawberry (Aharoni et al., 2004). Oligonucleotide arrays are commercially available for many crop species including *Solanum lycopersicon* and *Vitis vinifera* (www.affymatrix.com).

7.6.3 Comparative Genomics

Comparative genomics is the study of relationship between structural and functional attributes of genomes across different species (Sonah et al., 2011). This approach has been utilized in *Fabaceae* (Zhu et al., 2005; Cronk et al., 2006), *Rosaceae* (Dirlewanger et al., 2004a, 2004b), *Solanaceae* (Mueller et al., 2009), *Asteraceae* (The Compositae Genome Project: www.compgenomics.ucdavis.edu) and Brassicaceae (Lysak and Lexer, 2006). Several studies have demonstrated the use of *Arabidopsis* genome for fine mapping of QTLs and cloning of genes in Brassica (Formanová et al., 2010; Zhao et al., 2010).

7.7 CHLOROPLAST AND MITOCHONDRIAL GENOMES

7.7.1 Chloroplast

Besides photosynthesis, chloroplasts and other plastid play vital roles in the synthesis of amino acids, nucleotides, fatty acids, phytohormones, vitamins and a number of metabolites that are involved in abiotic (heat, drought, salt, light, etc.) and biotic stress responses (pests and diseases) (Daniell et al., 2016a, 2016b; Bobik and Burch-Smith, 2015). The chloroplast genome is known to encode many key proteins that are involved in photosynthesis and other metabolic processes (Daniell et al., 2016b). Since the first chloroplast genome was sequenced in 1986 from tobacco (*Nicotiana tabacum*), lot of progress has been made with over 800 complete chloroplast genomes sequenced of which 300 are from trees (Shinozaki et al., 1986; Daniell et al., 2016a). The next-generation sequencing (NGS) methods are faster and cheaper compared to the older methods for sequencing chloroplast genomes (Moore et al., 2006). A third-generation sequencer (PacBio), which uses single molecule real-time (SMRT) sequencing, is now widely used for chloroplast genome sequencing (rom tobacco (*Nicotiana tabacum*) (Jackman et al., 2016; Ferrarini et al., 2013; Wu et al., 2014; Li et al., 2014; Redwan et al., 2015; Chen et al., 2015).

7.7.2 Chloroplast Genome

The chloroplast genomes comprise a single circular molecule with a quadripartite structure that includes two copies of an IR region that separate large and small single-copy (LSC and SSC) regions (Daniell et al., 2016b). The chloroplast genome includes 120–130 genes, primarily participating in photosynthesis, transcription and translation. Recent studies have identified considerable diversity within noncoding intergenic spacer regions, which often include important regulatory sequences (Daniell et al., 2016a). Chloroplast genome size varies among species, ranging from 107 kb (*Cathaya argyrophylla*) to 218 kb (*Pelargonium*). The genome size is independent of nuclear genome size (Daniell et al., 2016b, 2006).

Chloroplast genomes serve as a valuable tool for identifying plants that are likely to be closely related and, therefore, genetically compatible (Daniell et al., 2016b). Understanding the genetic relationships between cultivated crops and their wild relatives facilitates introduction of specific advantageous traits into cultivated crops.

7.7.3 Mitochondrial Genome

Both mitochondria and chloroplasts were once believed to be independent free-living organisms that were engulfed by larger cells (Morley and Neilson, 2017; Rogalski et al., 2015). Over time, mitochondria lost their DNA as its genes were incorporated into the nuclear genome of the hosts they inhabited. At some point, these changes caused mitochondria to lose their ability to survive on their own. Therefore, even today, mitochondria and chloroplasts still possess and maintain their own DNA (Morley and Neilson, 2017). This residual DNA encodes only some of the genes required for respiration and photosynthesis (Knoop, 2004; Schuster and Brennicke, 1994).

Plants have large mitochondrial genomes that can vary in size anywhere from 200 to 2,000 kbp (Gualberto et al., 2014; Kolesnikov and Gerasimov, 2012). Plant mitochondrial genomes also contain significant amounts of relatively short nuclear and chloroplast genomic sequences that appear to have been integrated at some point in time during evolution (Knoop, 2004; Schuster and Brennicke, 1994; Schultze and Kondorosi, 1998; Knoop and Brennicke, 2002; Gray et al., 1989). The genes encoded in mitochondria include rRNAs, tRNAs, ribosome synthesis proteins and oxidative phosphorylation proteins (Unseld et al., 1997; Burger et al., 2003). The fragmented nature of plant mitochondrial genomes also leads to frequent DNA recombination events. Unlike nuclei, a single mitochondrion possesses many copies of its own genome and the number of copies per organelle can vary between 50 and 500, depending on the tissue type and age of the plant.

7.8 GENE EXPRESSION IN PLANTS

Gene expression in higher plants is regulated quantitatively and qualitatively by specific DNA sequences that are known as gene promoters (Dutt et al., 2014). Promoters regulate gene expression through DNA recognition sequences, which interact with basic transcription initiation complexes and numerous transcription factors (Katagiri and Chua, 1992). Such DNA recognition sequences include a core promoter with upstream enhancer sequences located close to the structural portion of the gene (Komarnytsky and Borisjuk, 2003). Promoters are divided into two regions: a core promoter region and upstream regulatory regions (Singh, 1998). The core promoter comprises a 50–100 bp sequence adjacent to the transcription initiation site and flanking sequences (Wu et al., 2001). This region interacts with the general transcription machinery and ensures accurate initiation of transcription by RNA polymerase II (Novina and Roy, 1996; Juven-Gershon and Kadonaga, 2010). The core promoter consists of two key genetic elements: the TATA box (present in most promoters) and/or an initiator (Inr) element overlapping the transcription start site (Breathnach and Chambon, 1981; Smale and Baltimore, 1989). The initiator element binds trans-acting factors for the placement of the start site and can also mediate transcription initiation in some TATA-less promoters (Means and Farnham, 1990; Mukherjee et al., 1995; Seto and Shenk, 1991; Zenzie-Gregory et al., 1992). The cis-regulatory elements contained in upstream promoter regions of 1–2 kb or more serve as the binding sites for gene specific regulators (Wu et al., 2001). Several regulatory sequences present upstream of the 59 regions of plant genes include multiple cis-regulatory elements whose distribution and presence contribute to the expression pattern of that particular gene (Koltunow et al., 1990; Wen et al., 2014). This interaction between the cis-acting elements and the transcription factors is key to the regulation of gene expression (Yang et al., 2000). The presence of several cis-acting elements contributes to the complex expression profile of a particular gene and their differential combinatorial interactions with the transcription factors result in expression of the adjacent gene to be either constitutive, induced by external factors, tissue-specific or some combination of these (Mitsuhara et al., 1996).

7.8.1 Constitutive Gene Expression

Constitutive promoters consist of a core DNA sequence (core promoter) and other regulatory elements, such as enhancers, silencers and other DNA sequences (Dutt et al., 2014). These interact with DNA binding proteins (transcription factors) to drive transgene expression in various plant cells (Benfey et al., 1990). Constitutive promoters may provide ectopic gene expression in transgenic plants, not otherwise observed under normal conditions (Dutt et al., 2014). The constitutive promoter may generate different gene expression in monocotyledonous and dicotyledonous species making it important to identify candidate promoters for specific groups to ensure high transgene expression levels (Gupta et al., 2001). Most constitutive promoters used for production of transgenic plants derive their origin from viral sequence (Dutt et al., 2014). With the advent of genome sequencing and availability of public genomic databases have led to the identification of numerous plant-derived constitutive promoters, which are increasingly being used in plant transformation (Dutt et al., 2014). The CaMV 35S or simply 35S (cauliflower mosaic virus) promoter is by far the most widely studied and used promoter in plant transformation (Benfey et al., 1990; Birch, 1997; Porto et al., 2014). It consists of two domains A and B with several subdomains (Benfey et al., 1990). These subdomains have specific cis-elements that are involved in gene expression in specific tissues in both above and belowground plant parts. The exact gene expression patterns depend on the combinations of cis-elements of the 35S promoter thereby suggesting the need for interaction between cis-elements for expression at various plant growth and developmental stages (Benfey et al., 1990). Gene expression by the 35S promoter was found to be influenced by species, environmental factors and physiological state of plant development (Schnurr and Guerra, 2000; Pretova et al., 2001). A number of studies indicated that high GUS expression levels were observed in pollen of transgenic strawberry plants when a 35S promoter was used, but no expression was detected in transgenic tomato or petunia plants with similar construct configurations (de Mesa et al., 2004; Mascarenhas and Hamilton, 1992; de Jong et al., 1994; Fitch et al., 1990; Gonsalves et al., 2004; Scorza et al., 1994, 2001, 2013). The 35S promoter has been extensively used in horticultural crops for improving abiotic and biotic stress tolerance, for quality traits and for modification of plant architecture (Fitch et al., 1990; Gonsalves et al., 2004; Scorza et al., 1994, 2001, 2013).

7.8.2 Tissue-Specific Gene Expression

The use of constitutive promoters has certain limitations, especially when temporal and spatial gene expression patterns are required to achieve manipulation of specific plant organs or developmental stages or may not be even active in certain tissues (Kasuga et al., 1999; Hsieh et al., 2002). In such cases, tissue-specific promoters may be useful for driving transgenic expression without interfering with normal plant growth and development processes. A number of promoters involved in various stages of growth, maturity and ripening have been identified and are being used in genetic engineering to improve yield, quality and postharvest shelf life. Studies conducted by Karaaslan and Hrazdina (2010) in sour cherry revealed the presence of a TATA box and several CAAT boxes that are conserved among promoter sequences. In strawberry, Bustamante et al. (2009) showed the presence of several hormones, light and biotic stress-related regulatory regions in addition to the TATA box and several CAAT boxes. A number of promoter sequences that are involved in the expression of genes controlling fruit composition and sucrose accumulation during development and ripening have been determined in peach, water melon and banana (Rahim et al., 2014; Yamagata et al., 2002; Yin et al., 2009; Choudhury et al., 2008). An excellent review of gene expression in horticultural plants is provided by Dutt et al. (2014).

7.8.3 Promoters in Other Organs

The expression of genes involved in the production of seed storage proteins is highly regulated making them ideal candidates for improving seed-specific traits. Thomas (1993), and van der Geest and Hall (1996) attempted to identify proximal regions that confer seed specificity and distal regions that are responsible for modulating gene expression. The 2S albumin gene promoter from a number of horticultural species has been used to direct seed-specific gene expression (Li and Gray, 2005; Kroj et al., 2003). Legumin gene promoters have also been well studied in a number of plant species.

A typical flower consists of four organs that are formed in an ordered pattern (Endress, 2001). Each of these whorls comprises unique genes targeted to the specific organ and several homeotic genes that affect the fate of organ primordial (Coen et al., 1990). Some of the traits that can be engineered in the floral tissues include increased vase life, modifications in flower color, fragrance and male and female sterility, and so on (Yang et al., 2011; Tanaka et al., 2005).

Several root-specific promoters have also been evaluated in horticultural plants. For example, the SLREO gene isolated from tomato is highly expressed in roots, but had a very low level of expression in aerial plant organs (Jones et al., 2009). The RB7 protein from tobacco (Yamamoto et al., 1999) is another root-specific gene and has been used to drive the *Arabidopsis* thionin (Thi2.1) gene in tomato (Chan et al., 2005). Other promoters identified include a 2 kb promoter fragment of the MipB gene from *Mesembryanthemum crystallinum* expressed strongly in the tobacco root.

7.8.4 Inducible Promoters

Inducible promoters are induced by either physical (biotic and abiotic factors) or chemical agents. Some of these promoters have been isolated from horticultural crops or used for horticultural plant improvement (Dutt et al., 2014).

The aim of genomics and bioinformatics is to relate the available data for understanding the molecular mechanism involved in various developmental processes. Many economically important horticultural crops have very less or no genomic resources available which need to be taken on priority basis. The most important challenge in many of the fruit species is unavailability of well-defined molecular-genetic linkage maps. Many times, these species are difficult to study at genetic and molecular levels because of their perennial nature. However, the advances in genomics, along with powerful statistical tools can be used for developing markers linked to a specific trait. One of the challenges in understanding genome structure of woody plants is unavailability of ESTs or other genomic resources. Being complex genomes, it would also be difficult to sequence these species de novo using NGS technologies. Though one can generate enough sequences from any species, de novo genome assembly becomes a stumbling block. Therefore, it would be desirable to first determine the genome size of species and also develop sufficient DNA markers from the ESTs or other GSS.

REFERENCES

Aharoni A, Giri A P, Verstappen F W A, Bertea C M, Sevenier R and Sun Z. 2004. Gain and loss of fruit flavor compounds produced by wild and cultivated strawberry species. *Plant Cell* **16**: 3110–3331.

Arabidopsis Genome Initiative genomeanalysis@tgr.org genomeanalysis@gsf.de. 2000. Analysis of the genome sequence of the flowering plant Arabidopsis thaliana. *Nature* **408**(6814): 796–815.

Azad M A, Amin L and Sidik N M. 2014. Gene technology for papaya ring spot virus disease management. *The Scientific World Journal* **2014**: 768038.

Azim K M, Khan I A and Zhang Y. 2014. Characterization of mango (Mangifera indica L.) transcriptome and chloroplast genome. *Plant Molecular Biology* **85**: 193–208.

Benfey P N, Ren L and Chua N-H. 1990. Tissue-specific expression from CaMV 35S enhancer subdomains in early stages of plant development. *The EMBO Journal* **9**: 1677–1684.

Birch R G. 1997. Plant transformation: problems and strategies for practical application. *Annual Review of Plant Biology* **48**: 297–326.

Blas A L, Ming R, Liu Z, Veatch O J, Paull R E, Moore P H and Yu Q. 2010. Cloning of the papaya chromoplast-specific lycopene beta-cyclase, CpCYC-b, controlling fruit flesh color reveals conserved microsynteny and a recombination hot spot. *Plant Physiology* **152**: 2013–2022.

Bliss F A. 2010. Marker assisted breeding in horticulture crops. pp. 339–350. (Ed) Bassil N V and Martin R. *Proceedings of International Symposium on Molecular Markers in Horticulture. Acta Horticulturae* **859**.

Bobik K and Burch-Smith T M. 2015. Chloroplast signaling within, between and beyond cells. *Frontiers in Plant Science* **6**: 781.

Breathnach R and Chambon P. 1981. Organization and expression of eukaryotic split genes coding for proteins. *Annual Review of Biochemistry* **50**: 349–383.

Bruening G and Lyons J. 2000. The case of the FLAVR SAVR tomato. *California Agriculture* **54**: 6–7.

Burger G, Gray M W and Lang B F. 2003. Mitochondrial genomes: anything goes. *Trends in Genetics* **19**: 709–716.

Bustamante C A, Civello P M and Martínez G A. 2009. Cloning of the promoter region of bxylosidase (FaXyl1) gene and effect of plant growth regulators on the expression of FaXyl1 in strawberry fruit. *Plant Science* **177**: 49–56.

Campos-de Quiroz H. 2002. Plant genomics: an overview. *Biological Research* **35**: 385–399.

Celton J M, Tustin D S, Chagné D and Gardiner S E. 2009. Construction of a dense genetic linkage map for apple rootstocks using SSRs developed from Malus ESTs and Pyrus genomic sequences. *Tree Genetics and Genomes* **5**: 93–107.

Chagne D, Gasic K, Crowhurst R N, Han Y, Bassett H C and Bowatte D R. 2008. Development of a set of SNPmarkers present in expressed genes of the apple. *Genomics* **92**: 353–358.

Chan Y-L, Prasad V, Sanjaya, Chen K H, Liu P C, Chan T and Cheng C-P. 2005. Transgenic tomato plants expressing an Arabidopsis thionin (Thi2.1) driven C P by fruit inactive promoter battle against phytopathogenic attack. *Planta* **221:** 386–393.

Chen C, Bowman K D, Choi Y A, Dang P M, Rao M N, Huang S, Soneji J R, McCollum T G and Gmitter F G. 2008. EST-SSR genetic maps for Citrus sinensis and Poncirus trifoliate. *Tree Genetics and Genomes* **4**: 1–10.

Chen X, Li Q, Li Y, Qian J and Han J. 2015. Chloroplast genome of *Aconitum barbatum* var. puberulum (*Ranunculaceae*) derived from CCS reads using the PacBio RS platform. *Frontiers in Plant Science* **6**: 42.

Cheung F, Haas B J, Goldberg S M, May G D, Xiao Y and Town C D. 2006. Sequencing *Medicago truncatula* expressed sequenced tags using 454 Life Sciences technology. *BMC Genomics* **7**: 272.

Chia J M and Ware D. 2011. Sequencing for the cream of the crop. *Nature Bio*technology **29**: 138–139.

Choudhury S R, Roy S, Das R and Sengupta D N. 2008. Differential transcriptional regulation of banana sucrose phosphate synthase gene in response to ethylene, auxin, wounding, low temperature and different photoperiods during fruit ripening and functional analysis of banana SPS gene promoter. *Planta* **229**: 207–223.

Clark D, Klee H and Dandekar A. 2004. Despite benefits, commercialization of transgenic horticultural crops lags. *California Agriculture* **58**: 89–98.

Coen E S, Romero J M, Doyle S, Elliott R, Murphy G and Carpenter R. 1990. Floricaula: a homeotic gene required for flower development in *Antirrhinum majus*. *Cell* **63**: 1311–1322.

Cronk Q, Ojeda I and Pennington R T. 2006. Legume comparative genomics: progress in phylogenetics and phylogenomics. *Current Opinion in Plant Biology* **9**: 99–103.

Daniell H, Chan H-T and Pasoreck E K. 2016a. Vaccination through chloroplast genetics: affordable protein drugs for the prevention and treatment of inherited or infectious human diseases. *Annual Review of Genetics* **2016**. In Press.

Daniell H, Lee S B, Grevich J, Saski C, Quesada-Vargas T and Guda C. 2006. Complete chloroplast genome sequences of *Solanum bulbocastanum*, *Solanum lycopersicum* and comparative analyses with other *Solanaceae* genomes. *Theoretical and Applied Genetics* **112**: 1503–1518.

Daniell H, Lin C, Yu M and Chang W J. 2016b. Chloroplast genomes: diversity, evolution, and applications in genetic engineering. *Genome Biology* **17**: 134.

Davuluri G R, van Tuinen A, Fraser P D, Manfredonia A, Newman R, Burgess D and Brummell D A. 2005. Fruit-specific RNAi-mediated suppression of DET1 enhances carotenoid and flavonoid content in tomatoes. *Nature Biotechnology* **23**: 890–895.

de Jong J, Mertens M M and Rademaker W. 1994. Stable expression of the GUS reporter gene in chrysanthemum depends on binary plasmid T-DNA. *Plant Cell Reporter* **14**: 59–64.

de Mesa M C, Santiago-Doménech N, Pliego-Alfaro F, Quesada M and Mercado J. 2004. The CaMV 35S promoter is highly active on floral organs and pollen of transgenic strawberry plants. *Plant Cell Reporter* **23**: 32–38.

Dirlewanger E, Cosson P, Howad W, Capdeville G, Bosselut N and Claverie M. 2004a. Microsatellite genetic linkage maps of myrobalan plum and an almond-peach hybrid – location of root-knot nematode resistance genes. *Theoretical and Applied Genetics* **109**: 827–838.

Dirlewanger E, Graziano E, Joobeur T, Garriga-Calderé F, Cosson P, Howad W and Arús P. 2004b. Comparative mapping and marker-assisted selection in *Rosaceae* fruit crops. *Proceedings of National Academy of Science, USA* **101**: 9891–9896.

Dutt M, Dhekney S A, Soriano L, Kandel R and Grosser J D. 2014. Temporal and Spatial control of gene expression in horticultural plants – A review. *Horticulture Research* **1**: 14047. https://doi.org/10.1038/hortres.2014.47.

Ekué M R M, Gailing O and Finkeldey R. 2009. Transferability of simple sequence repeat (SSR)markers developed in *Litchi chinensis* to *Blighia sapida* (Sapindaceae). *Plant Molecular Biology Report* **27**: 570–574.

Endress P K. 2001. Origins of flower morphology. *Journal of Experimental Zoology* **291**: 105–115.

Ferrarini M, Moretto M, Ward J A, Šurbanovski N, Stevanović V and Giongo L. 2013. An evaluation of the PacBio RS platform for sequencing and de novo assembly of a chloroplast genome. *BMC Genomics* **14**: 670.

Fitch M M, Manshardt R M, Gonsalves D, Slightom J L and Sanford J C. 1990. Stable transformation of papaya via microprojectile bombardment. *Plant Cell Reporter* **9**: 189–194.

Folta K M and Gardiner S E. 2009. *Genetics and Genomics of Rosaceae*. p. 600. (Ed.) Jorgensen R. Springer, New York.

Formanová N, Stollar R, Geddy R, Mahé L, Laforest M, Landry B S and Brown G G. 2010. High resolution mapping of the *Brassica napus* Rfp restorer locus using *Arabidopsis* derived molecular markers. *Theoretical and Applied Genetics* **120**: 843–851.

Forment J, Gadea J, Huerta L, Abizanda L, Agusti J and Alamar S. 2005. Development of a citrus genome-wide EST collection and cDNA microarray as resources for genomic studies. *Plant Molecular Biology* **57**: 375–391.

Frary A, Xu Y, Liu J, Mitchell S, Tedeshi E and Tanksley S. 2005. Development of a set of PCR-based anchor markers encompassing the tomato genome and evaluation of their usefulness for genetics and breeding experiments. *Theoretical and Applied Genetics* **111**: 291–312.

Gilchrist E J, Haughn G W, Ying C C, Otto S P, Zhuang J and Cheung D. 2006. Use of Ecotilling as an efficient SNP discovery tool to survey genetic variation in wild populations of *Populus trichocarpa*. *Molecular Ecology* **15**: 1367–1378.

Gonsalves D, Gonsalves C and Ferreira S. 2004. Transgenic virus resistant papaya: From hope to reality for controlling papaya ring spot virus in Hawaii. *APSnet Feature*, July, Ferreira. DOI:10.1094/APSnetFeature-2004-0704.

Gray M V, Cedergren R, Abel Y and Sankoff D. 1989. On the evolutionary origin of the plant mitochondrion and its genome. *Proceedings of National Academy of Science, USA* **86**: 2267–2271.

Gualberto J M, Mileshina D, Wallet C, Niazi A K, Weber-Lotfi F and Dietrich A. 2014. The plant mitochondrial genome: dynamics and maintenance. *Biochimie* **100**: 107–120.

Gupta P, Raghuvanshi S and Tyagi A K. 2001. Assessment of the efficiency of various gene promoters via biolistics in leaf and regenerating seed callus of millets, *Eleusine coracana* and *Echinochloa crus-galli*. *Plant Biotechnology* **18**: 275–282.

Hatakeyama K, Horisaki A, Niikura S, Narusaka Y, Abe H and Yoshiaki H. 2010. Mapping of quantitative trait loci for high level of self-incompatibility in *Brassica rapa* L. *Genome* **53**: 257–265.

He J, Zhao X, Laroche A, Lu Z X, Liu H and Li Z. 2014. Genotypingby-sequencing (GBS), an ultimate marker-assisted selection (MAS) tool to accelerate plant breeding. *Frontiers in Plant Science* **5**: 484.

Howad W, Yamamoto T, Dirlewanger E, Testolin R, Cosson P and Cipriani G. 2005. Mapping with a few plants: using selective mapping for microsatellite saturation of the Prunus reference map. *Genetics* **171**: 1305–1309.

Hsieh T H, Lee J T and Yang P T. 2002. Heterology expression of the Arabidopsis C-repeat/dehydration response element binding factor 1 gene confers elevated tolerance to chilling and oxidative stresses in transgenic tomato. *Plant Physiology* **129**: 1086–1094.

Huang S, Li R, Zhang Z, Li L, Gu X and Fan W. 2009. The genome of the cucumber, *Cucumis sativus* L. *Nature Genetics* **41**: 1275–1281.

Hyten D L, Song Q, Choi I Y, Yoon M S, Specht J E and Matukumalli L K. 2008. High-throughput genotyping with the Golden Gate assay in the complex genome of soybean. *Theoretical and Applied Genetics* **16**: 945–952.

Iezzoni A, Weebadde C, Luby J, Yue C, van de Weg E and Fazio G. 2010. RosBREED: enabling marker-assisted breeding in Rosaceae. pp. 385–394. (Eds) Bassil N V and Martin R. *Proceedings of International Symposium on Molecular Markers in Horticulture. Acta Horticulturae* **859**.

Iwata H, Minamikawa M F, Kajiya-Kanegae H, Ishimori M and Hayashi T. 2016. Genome-assisted breeding in fruit crops. *Breeding Science* **66**: 100–115.

Jackman S D, Warren R L, Gibb E A, Vandervalk B P, Mohamadi H and Chu J. 2016. Organellar genomes of white spruce (Picea glauca): assembly and annotation. *Genome Biology and Evolution* **8**: 29–41.

Jaillon O, Aury J M, Noel B, Policriti A, Clepet C and Casagrande A. 2007. The grapevine genome sequence suggests ancestral hexaploidization in major angiosperm phyla. *Nature* **449**: 463–467.

Jones M O, Kamarainen-Karppinen T, Koskimaki J J, et al. 2009. The promoter from SlREO, a highly-expressed, root-specific Solanum lycopersicum gene, directs expression to cortex of mature roots. *Functional Plant Biology* **35**: 1224–1233.

Juven-Gershon T and Kadonaga J T. 2010. Regulation of gene expression via the core promoter and the basal transcriptional machinery. *Developmental Biology* **339**: 225–229.

Karaaslan M and Hrazdina G. 2010. Characterization of an expansin gene and its ripening specific promoter fragments from sour cherry (*Prunus cerasus* L.) cultivars. *Acta Physiologiae Plantarum* **32**: 1073–1084.

Kasuga M, Liu Q, Miura S, Yamaguchi-Shinozaki K and Shinozaki K. 1999. Improving plant drought, salt, and freezing tolerance by gene transfer of a single stress-inducible transcription factor. *Nature Biotechnology* **17**: 287–291.

Katagiri F and Chua N H. 1992. Plant transcription factors: present knowledge and future challenges. *Trends in Genetics TIG* **8**: 22–27.

Kenis K, Keulemans J and Davey W M. 2008. Identification and stability of QTLs for fruit quality traits in apple. *Tree Genetics and Genomes* **4**: 647–661.

Knoop V. 2004. The mitochondrial DNA of land plants: peculiarities in phylogenetic perspective. *Current Genetics* **46**: 123–139.

Knoop V and Brennicke A. 2002. Molecular biology of the plant mitochondrion. *CRC Critical Review of Plant Science* **21**: 111–126.

Kolesnikov A A and Gerasimov E S. 2012. Diversity of mitochondrial genome organization. *Biochemistry-Moscow* **77**: 1424–1435.

Koltunow A M, Truettner J, Cox K H, Wallroth M and Goldberg R B. 1990. Different temporal and spatial gene expression patterns occur during anther development. *Plant Cell* **2**: 1201–1224.

Komarnytsky S and Borisjuk N. 2003. Functional analysis of promoter elements in plants. pp. 113–141. *Genetic Engineering* (Ed.) Setlow J. Volume 25. Springer, Berlin.

Kroj T, Savino G, Valon C, Giraudat J and Parcy F. 2003. Regulation of storage protein gene expression in *Arabidopsis. Development* **130**: 6065–6073.

Kubo N, Saito M, Tsukazaki H, Kondo T, Matsumoto S and Hirai M. 2010. Detection of quantitative trait loci controlling morphological traits in *Brassica rapa* L. *Breeding Science* **60**: 164–171.

Laitinen R A, Immanen J, Auvinen P, Rudd S, Alatalo E and Paulin L. 2005. Analysis of the floral transcriptome uncovers new regulators of organ determination and gene families related to flower organ differentiation in *Gerbera hybrida* (Asteraceae). *Genome Research* **15**: 475–486.

Lederberg J and McCray A T. 2001. Ome sweet omics – a genealogical treasury of words. Genealogical treasury of words. *Scientist* **15**: 8.

Li Q, Li Y, Song J, Xu H, Xu J and Zhu Y. 2014. High-accuracy de novo assembly and SNP detection of chloroplast genomes using a SMRT circular consensus sequencing strategy. *New Phytologist* **204**: 1041–1049.

Li Z T and Gray D. 2005. Isolation by improved thermal asymmetric interlaced PCR and characterization of a seed-specific 2S albumin gene and its promoter from grape (*Vitis vinifera* L.). *Genome* **48**: 312–320.

Liebhard R, Koller B, Gianfranceschi L and Gessler C. 2003. Creating a saturated reference map for the apple (Malus×domestica Borkh.) genome. *Theoretical and Applied Genetics* **106**: 1497–1508.

Luby J J and Shaw D V. 2001. Does marker-assisted selection make dollars and sense in a fruit breeding program? *HortScience* **36**: 872–879.

Luo K M, Duan H, Zhao D G, Zheng X L, Deng W and Chen Y Q. 2007. GM-gene-deletor: fused loxP-FRT recognition sequences dramatically improve the efficiency of FLP or CRE recombinase on transgene excision from pollen and seed of tobacco plants. *Plant Biotechnology Journal* **5**: 263–274.

Lysak M and Lexer C. 2006. The era of comparative evolutionary genomics in *Brassicaceae. Plant Systematics and Evolution* **259**: 175–198.

Mascarenhas J P and Hamilton D A. 1992. Artifacts in the localization of GUS activity in anthers of petunia transformed with a CaMV 35S-GUS construct. *Plant Journal* **2**: 405–408.

McKusick V A. 1997. Genomics: structural and functional studies of genomes. *Genomics* **45**: 244–249.

Means A L and Farnham P J. 1990. Transcription imitation from the dihydrofolate reductase promoter is positioned by HIPI binding at the initiation site, MOI. *Cell Biology* **10**: 653–661.

Meneses C and Orellana A. 2013. Using genomics to improve fruit quality. *Biological Research* **46**: 347–352.

Ming R, Hou S, Feng Y, Yu Q, Dionne-Laporte A and Saw J H. 2008. The draft genome of the transgenic tropical fruit tree papaya (*Carica papaya* Linnaeus). *Nature* **452**: 991–996.

Mitsuhara I, Ugaki M and Hirochika H. 1996. Efficient promoter cassettes for enhanced expression of foreign genes in dicotyledonous and monocotyledonous plants. *Plant Cell Physiology* **37**: 49–59.

Moore M J, Dhingra A, Soltis P S, Shaw R, Farmerie W G, Folta K M and Soltis D E. 2006. Rapid and accurate pyrosequencing of angiosperm plastid genomes. *BMC Plant Biology* **6**: 17.

Morley S A and Neilson B L. 2017. Plant mitochondrial DNA. *Frontiers in Bioscience, Landmark* **22**: 1023–1032.

Mueller L A, Lankhorst R K, Tanksley S D, Giovannoni J J, White R and Vrebalov J. 2009. A snapshot of the emerging tomato genome sequence. *Plant Genome* **2**: 78–92.

Mukherjee B, Burma S and Hasnain S E. 1995. The 30-kDa protein binding to the 'initiator' of the baculovirus polyhedrin promoter also binds specifically to the coding strand. *Journal of Biological Chemistry* **270**: 4405–4411.

Myles S. 2013. Improving fruit and wine: what does genomics have to offer? *Trends in Genetics* **29**: 190–196.

Novina C D and Roy A L. 1996. Core promoters and transcriptional control. *Trends in Genetics* **12**: 351–355.

Olmstead J W, Sebolt A M, Cabrera A, Sooriyapathirana S S, Hammar S and Iriarte G. 2008. Construction of an intra-specific sweet cherry (*Prunus avium* L.) genetic linkage map and synteny analysis with the Prunus reference map. *Tree Genetics and Genomes* **4**: 897–910.

Pal S. 2012. Genome assisted breeding of fruit crops. *Biotechnology Articles* **1**: 1–9.

Porto M S, Pinheiro M P and Batista V G. 2014. Plant promoters: an approach of structure and function. *Molecular Biotechnology* **56**: 38–49.

Pretová A, Obert B and Wetzstein H. 2001. Leaf developmental stage and tissue location affect the detection of b-glucuronidase in transgenic tobacco plants. *Biotechnology Letters* **23**: 555–558.

Rahim M A, Busatto N and Trainotti L. 2014. Regulation of anthocyanin biosynthesis in peach fruits. *Planta* **240**: 913–929.

Redwan R, Saidin A and Kumar S. 2015. Complete chloroplast genome sequence of MD-2 pineapple and its comparative analysis among nine other plants from the subclass Commelinidae. *BMC Plant Biology* **15**: 196.

Rensink W A, Lee Y, Liu J, Iobst S, Ouyang S and Buell C R. 2005. Comparative analyses of six solanaceous transcriptomes reveal a high degree of sequence conservation and species-specific transcripts. *BMC Genomics* **14**: 124.

Rikkerink E H A, Oraguzie N C and Gardiner S E. 2007. Prospects of association mapping in perennial horticultural crops. pp. 249–269. *Association Mapping in Plants* (Eds.) Oraguzie N C, Rikkerink E H A, Gardiner S E and Silva H N. Springer, New York.

Rogalski M, Vieira L N, Fraga H P and Guerra M P. 2015. Plastid genomics in horticultural species: importance and applications for plant population genetics, evolution, and biotechnology. *Frontiers in Plant Science* **6**: 586.

Salmaso M, Malacarne G, Troggio M, Faes G, Stefanini M and Grando M S. 2008. A grapevine (Vitis vinifera L.) genetic map integrating the position of 139 expressed genes. *Theoretical and Applied Genetics* **116**: 1129–1143.

Sargent D J, Clarke J, Simpson D W, Tobutt K R, Arus P and Monfort A. 2006. An enhanced microsatellite map of diploid *Fragaria*. *Theoretical and Applied Genetics* **112**: 1349–1359.

Schnurr J and Guerra D. 2000. The CaMV-35S promoter is sensitive to shortened photoperiod in transgenic tobacco. *Plant Cell Reporter* **19**: 279–282.

Schultze M and Kondorosi A. 1998. Regulation of symbiotic root nodule development. *Annual Review of Genetics* **32**: 33–57.

Schuster W and Brennicke A. 1994. The plant mitochondrial genome: physical structure, information content, RNA editing, and gene migration to the nucleus. *Annual Review of Plant Physiology and Plant Molecular Biology* **45**: 61–78.

Scorza R, Callahan A and Dardick C. 2013. Genetic engineering of Plum pox virus resistance: 'HoneySweet' plum-from concept to product. *Plant Cell Tissue and Organ Culture* **115**: 1–12.

Scorza R, Callahan A, Levy L, Damsteegt V, Webb K and Ravelonandro M. 2001. Posttranscriptional gene silencing in plum pox virus resistant transgenic European plum containing the plum pox potyvirus coat protein gene. *Transgenic Research* **10**: 201–209.

Scorza R, Ravelonandro M and Callahan A M. 1994. Transgenic plums (*Prunus domestica* L.) express the plum pox virus coat protein gene. *Plant Cell Reporter* **14**: 18–22.

Seto E, Shi Y and Shenk T. 1991. YY1 is an initiator sequence-binding protein that directs and activates transcription in vitro. *Nature* **354**: 241–245.

Sharma S, Kaur R and Singh A. 2017. Recent advances in CRISPR/Cas mediated genome editing for crop improvement. *Plant Biotechnology Reports* **11**: 193–207.

Shinozaki K, Ohme M, Tanaka M, Wakasugi T, Hayashida N and Matsubayashi T. 1986. The complete nucleotide sequence of the tobacco chloroplast genome: its gene organization and expression. *EMBO Journal* **5**: 2043–2049.

Singh K B. 1998. Transcriptional regulation in plants: the importance of combinatorial control. *Plant Physiology* **118**: 1111–1120.

Smale S T and Baltimore D. 1989. The 'initiator' as a transcription control element. *Cell* **57**: 103–113.

Soeda Y, Konings M C J M, Vorst O, van Houwelingen A M M, Stoopen G M and Maliepaard C A. 2005. Gene expression programs during *Brassica oleracea* seed maturation, osmopriming, and germination are indicators of progression of the germination process and the stress tolerance level. *Plant Physiology* **137**: 354–368.

Sonah H, Deshmukh R K, Singh V P, Gupta D. K, Singh N K and Sharma T. 2011. Genomic resources in horticultural crops: status, utility and challenges. *Biotechnology Advances* **29**: 199–209.

Tanaka Y, Katsumoto Y, Brugliera F and Mason J. 2005. Genetic engineering in floriculture. *Plant Cell, Tissue, and Organ Culture* **80**: 1–24.

Thomas T L. 1993. Gene expression during plant embryogenesis and germination: an overview. *Plant Cell* **5**: 1401–1410.

Unseld M, Marienfeld M P, Brandt P and Brennicke A. 1997. The mitochondrial genome of *Arabidopsis thaliana* contains 57 genes in 366,924 nucleotides. *Nature Genetics* **15**: 57–61.

van der Geest A H and Hall T C. 1996. A 68 bp element of the beta-phaseolin promoter functions as a seed-specific enhancer. *Plant Molecular Biology* **32**: 579–588.

van Nocker S and Gardiner S E. 2014. Breeding better cultivars, faster: applications of new technologies for the rapid deployment of superior horticultural tree crops. *Horticulture Research* **1**: 14022.

Varshney R K, Terauchi R and McCouch S R. 2014. Harvesting the promising fruits of genomics: applying genome sequencing technologies to crop breeding. *PLoS Biology* **12**: e1001883.

Vezzulli S, Troggio M, Coppola G, Jermakow A, Cartwright D and Stefanini M. 2008. A functional integrated map for cultivated grapevine (*Vitis vinifera* L.) from three pedigrees, based on 283 SSR and 501 SNP markers. *Theoretical and Applied Genetics* **117**: 499–511.

Wen Z, Yang Y Z and Zhang J J. 2014. Highly interactive nature of flower-specific enhancers and promoters, and its potential impact on tissue-specific expression and engineering of multiple genes or agronomic traits. *Plant Biotechnology Journal*, in press.

Wu K, Malik K and Tian L. 2001. Enhancers and core promoter elements are essential for the activity of a cryptic gene activation sequence from tobacco, tCUP. *Molecular Genetics and Genomics* **265**: 763–770.

Wu Z, Gui S, Quan Z, Pan L, Wang S and Ke W. 2014. A precise chloroplast genome of Nelumbo nucifera (Nelumbonaceae) evaluated with Sanger, Illumina MiSeq, and PacBio RS II sequencing platforms: insight into the plastid evolution of basal eudicots. *BMC Plant Biology* **14**: 289.

Xiong J-S, Ding J and Li Y. 2015. Genome-editing technologies and their potential application in horticultural crop breeding. *Horticulture Research* **2**: 1509.

Yamagata H, Yonesu K, Hirata A and Aizono Y. 2002. TGTCACA motif is a novel cis-regulatory enhancer element involved in fruit-specific expression of the cucumisin gene. *Journal of Biological Chemistry* **277**: 11582–11590.

Yamamoto S, Suzuki K and Shinshi H. 1999. Elicitor responsive, ethylene-independent activation of GCC box-mediated transcription that is regulated by both protein phosphorylation and dephosphorylation in cultured tobacco cells. *Plant Journal* **20**: 571–579.

Yang Y, Li R and Qi M. 2000. In vivo analysis of plant promoters and transcription factors by agroin filtration of tobacco leaves. *Plant Journal* **22**: 543–551.

Yang Y, Singer S and Liu Z. 2011. Petunia AGAMOUS enhancer-derived chimeric promoters specify a carpel, stamen, and petal-specific expression pattern sufficient for engineering male and female sterility in tobacco. *Plant Molecular Biology Reports* **29**: 162–170.

Yin T, Wu H, Zhang S, Liu J, Lu H, Zhang L, Xu Y and Daming Chen. 2009. Two negative cis-regulatory regions involved in fruit-specific promoter activity from watermelon (*Citrullus vulgaris* S.). *Journal of Experimental Botany* **60**: 169–185.

Zenzie-Gregory B, O'Shea-Greenfield A and Smale S T. 1992. Similar mechanisms for transcription initiation mediated through a TATA box or an initiator element. *Journal of Biological Chemistry* **267**: 2823–2830.

Zhang G, Sebolt A M, Sooriyapathirana S S, Wang D, Bink M C A M and Olmstead J W. 2010. Fruit size QTL analysis of an F1 population derived from a cross between a domesticated sweet cherry cultivar and a wild forest sweet cherry. *Tree Genetics and Genomes* **6**: 25–36.

Zhang J, Tao N, Xu Q, Zhou W, Cao H, Xu J and Deng X. 2009. Functional characterization of Citrus PSY gene in Hongkong kumquat (*Fortunella hindsii* Swingle). *Plant Cell Reporter* **28**: 1737–1746.

Zhao J, Kulkarni V, Liu N, Del Carpio D P, Bucher J and Bonnema G. 2010. BrFLC2 (FLOWERING LOCUS C) as a candidate gene for a vernalization response QTL in *Brassica rapa*. *Journal of Experimental Botany* **61**: 1817–1825.

Zhu H, Choi H K, Cook D R and Shoemaker R C. 2005. Bridging model and crop legumes through comparative genomics. *Plant Physiology* **137**: 1189–1196.

Zhu Y and Barritt B. 2008. Md-ACS1 and Md-ACO1 genotyping of apple (Malus×domestica Borkh.) breeding parents and suitability for marker-assisted selection. *Tree Genetics and Genomes* **4**: 555–562.

Part III

Quantitative Aspects of Plant Growth and Development

Plant growth is a multi-phase process, operating from the cellular through the whole-plant to plant-community level. The choice of the level at which growth is measured depends heavily on the reason for measuring. Research on the mechanistic understanding of growth, based on measured phenotypic traits, their correlation or causal relation, and their variability in response to the atmospheric and belowground environment continues to deliver data for quantitative approaches. The latter assist in breeding and crop improvement in line with next-generation sequencing and the development of mapping populations and diversity panels. Plant growth modeling has become a key research tool in agriculture, horticulture, forestry and environmental sciences. Due to growth of computer resources and the sharing of experiences between biologists, computer scientists and information specialists, the development of plant growth models has progressed enormously during the last two decades. The use of an interdisciplinary approach is necessary to advance research in plant growth modeling and simulation further but is one of the most challenging aspects to be encountered so far.

This part presents the current knowledge of growth measurements and crop modeling to help the students to appreciate its importance and the potential it offers in various areas of plant science.

DOI: 10.1201/9781003373889-10

8 Growth Stages, Growth Analysis and Growth Modeling

8.1 INTRODUCTION

As mentioned in Chapter 2, the life of the multicellular plant starts with a single-celled zygote (fertilized egg) and passes through a programmed developmental process to emerge as an embryo. The embryo then passes through a series of growth stages, beginning with germination, the development of leaves, a stem elongation stage, adult vegetative phase, flowering and fruiting stage ultimately leading to senescence and death. Although these phenological stages are continuous and gradual, it is possible to recognize the stage at which an individual plant or crop is at on a particular date. It is also well known that plants show seasonal patterns in their development. For example, plants show rapid meristematic growth in the spring and a virtual cessation of growth in the summer (Leather, 2010). These seasonal patterns in growth are reflected in the chemical composition in the plant's tissues and water content of their foliage. Many physiological processes play a key role during these growth stages.

Growth in the context of individual plant means an irreversible change with time (Hunt, 2003). In cells, such changes are simply an irreversible increase in number or volume, whereas it reflects an increase in both number and size or volume of cells in tissues and organs. At the whole-plant level, it means an irreversible increase in form, length, girth, fresh and dry (bacteria, fungi, etc.). It is obvious that several parameters can be used to measure growth depending upon the situation and the needs of the observer. However, whatever parameter is used to measure, basically, growth reflects an irreversible increase in volume or size (Hunt, 2003). While the growth determines the external form of individual organ of a plant, development refers to the sum of all subtle changes that a cell, tissue, organ or organism goes through in its life cycle. Development is most visibly manifested as changes in the form of an organ or organism, such as the transition from embryo to seedling, from a leaf primordium to a fully expanded leaf, or from the production of vegetative organs to the production of floral structures. Embryogenesis, vegetative development and reproductive development are the main stages of sporophytic development of higher plants.

Cell division and expansion determine the final size and shape of organ and whole plant (Sablowski and Carnier, 2014; Wuyts et al., 2015). In plants, the balance between these two processes requires coordinated regulation of macromolecular synthesis (cytoplasmic growth), turgor-driven cell-wall extension, mitotic cycle and endocycle (Sablowski and Carnier, 2014). Therefore, there is a need to view these processes within a spatial and temporal framework, at the tissue, organ and whole-plant level, on the one hand, and the influences of the plant's environment, on the other hand (Wuyts et al., 2015). Although there has been a considerable progress in our understanding of regulatory processes involved in growth at cellular level, the complexity of growth at the tissue, organ and whole-plant level is yet to be fully revealed (Wuyts et al., 2015). Measurement of growth at different levels helps in advancing of our understanding of growth as well as in defining the spatial and temporal resolution of sampling strategies in the model plant (*Arabidopsis thaliana*) and crop plants (Nelissen et al., 2014).

Plant growth is a multiple-level process from the cellular to the whole-plant and community levels. Proper understanding of growth across these levels is necessary (Wuyts et al., 2015). Any decision on the choice of a particular level at which the growth is measured will depend on the purpose for which the growth is being analyzed.

DOI: 10.1201/9781003373889-11

8.2 CLASSIFICATION OF GROWTH

8.2.1 Primary Growth

Increases in the length of the central axis of the plant caused by repeated mitotic divisions of meristematic cells in the root and shoot tips (apical meristem) are known as the primary growth.

8.2.2 Secondary Growth

The growth resulting from repeated division of lateral or secondary meristems is known as the secondary growth.

8.2.3 Unlimited Growth

The root and the shoot system of plants grow continuously throughout the life span of the plant. This growth is referred to as "unlimited" or "indeterminate" growth.

8.2.4 Limited Growth

The leaves, fruits and flowers stop growing after attaining certain size. This type of growth is called "limited" or "determinate" growth.

8.2.5 Vegetative Growth

The growth that produces roots, leaves, stems, branches and other vegetative plants of the plant is called "vegetative growth," and the phase is called the vegetative phase.

8.2.6 Reproductive Growth

The growth that gives rise to reproductive parts (flowers, fruits and seeds) is called the reproductive growth, and the stage of development is called the reproductive phase.

8.3 GROWTH ANALYSIS

Quantification procedures are used to describe, analyze and interpret the performance of plants under a set of natural, seminatural or controlled conditions. These quantification procedures are known as growth analysis (Hunt, 2003). Using simple parameters like weight (dry and fresh), area (leaf girth, leaf area index), volume (canopy) and content of either the plant or its components (organs like shoot, roots), the growth analysis provides holistic and integrative approach to investigate plant form, functions and processes. Price and Munns (2016) defined growth analysis as a conceptual framework for resolving the nature of genotype × environment interactions in plant growth and development. Under natural environments, growth and developmental cycles are completed within a fixed time frame dictated mostly by environmental conditions, such as light, temperature, soil moisture and nutrients. These factors often limit the expression of genetic potential and the adaptive mechanisms that plants develop to counter such constraints and sustain relative growth rate. These impacts can be revealed via growth analysis under contrasting conditions. Under managed environments, crop plants experience similar restrictions, but their economic yield is subject to genetic control. Crop scientists need to explain plant's growth and reproductive performance in quantitative terms. Sources of variation in productivity can then be resolved into those processes responsible for converting external resources into biomass and those responsible for partitioning biomass into usable sinks such as fruits, grains and storage organs.

In the beginning, plant growth analysis was mainly used by plant physiologists and agronomists to compare growth of different species or varieties within each species, but more recently, it is used by plant ecologists and molecular biologists also (Hunt, 2003). Plant growth is a fundamental biological process studied in a wide range of scientific fields involving different scales from physiology to community dynamics and ecosystem structure (McMahon and Bonner, 1983). There is, however, a basic distinction between individual plants and the population requiring different approaches to measure growth. The computational plant growth modeling put forward by Paine et al. (2012) and Zhang et al. (2012) provides a deeper insight and more accurate predictions for a wide range of plant physiological pathways responsive to biotic or abiotic stresses. They stated that computational plant growth model accurately and precisely captures the exact changes in plant growth rates by explaining how these pathways control photosynthesis or energy distribution.

8.4 MEASUREMENT OF GROWTH

Growth is the increase in dry mass, volume, length or area resulting from the division, expansion and differentiation of cells. Therefore, simple methods were first used to measure growth in the early 19th century (Blackman, 1919; West et al., 1920). These authors considered the absolute increase in mass over time (called the absolute growth rate or AGR) as a measure of plant growth. However, the absolute growth rate varies depending on the initial size of the plant. Therefore, the concept of relative growth rate (RGR) was introduced to overcome this problem (Hunt, 2003). RGR is the exponential increase in size relative to the size of the plant at the start of a given time interval. It is used to growth rates among species and individuals that differ widely in size. Measurements of leaf, stem and root mass, and leaf area (LA) provides a good insight into growth variation in a relatively simple way. These parameters are related to allocation (leaf-mass fraction, the fraction of plant biomass allocated to leaf), leaf morphology (specific leaf area or SLA) and physiology (the rate of increase in plant biomass per unit LA or rate of photosynthesis per unit LA; also known as net assimilation rate) (Poorter, 2002).

Growth analysis requires the destructive harvest of two or more groups of plant individuals. For accurate results, the test plants should be acclimated to the current growth conditions before they can be sampled. At least one initial and one final harvest need to be carried out and the actual number of plants to be sampled each time varies with the degree of variability in the population. Variability in size can be minimized by maintaining a larger population and selecting only a limited number of representative plants. Alternatively, plants can be grouped into different categories, with the number of plants per category equal to the number of harvests (Hunt, 2003). Destructive harvests provide enormous amount of information, but apart from being labor-intensive, it destroys at least one subset of the materials being studied. Alternatively, growth can be followed nondestructively by measuring an aspect of plant size at different times during the life span of the plant. By repeated measurements on the same individuals, a more accurate estimate of RGR can be obtained.

The RGR allows a fair comparison of the intrinsic growth physiology of different genotypes or species, irrespective of the initial size (Hunt and Cornelissen, 1997). Unfortunately, the classical RGR again is not the perfect measure of growth because plants or tissues become increasingly inefficient with age or as they grow larger. These processes lead to systematic decline in RGR with size (Evans, 1972; Hunt, 1982; Metcalf et al., 2003). Thus, even within species, differently sized individuals will have different RGRs. RGR with increasing size is accompanied by systematic changes in physiology, morphology and allocation of resources which is referred to as ontogenetic drift (Evans, 1972; Hunt, 1982). Therefore, adequate care must be exercised to overcome the size dependency of the variables while comparing species or varieties (Evans, 1972; Hunt, 1982; Hunt and Lloyd, 1987; Coleman et al., 1994). Coleman et al. (1994) clearly demonstrated the need to account for plant size (ontogenetic drift) when comparing experimental treatments that influence growth. While some of the published studies have compared the species based on common size, others did not (Poorter and Remkes, 1990). Despite these problems, RGR is considered as the best measure for analyzing growth, provided the size issue is adequately addressed.

The RGR calculated based on net assimilation rate (NAR) does not always explain inconsistencies observed in crop yields because any comparison of growth based on LAI has certain inherent inconsistencies (Gunasena and Harris, 1969, 1971; Ifenkwe, 1975). This discrepancy mainly arises because all leaves, irrespective of their age, arrangement and angle of exposure, are considered equal while calculating the LAI. Hence, NAR can be considered as only one of several criteria for measuring growth. Monteith (1977) proposed a new approach in which growth was considered to be regulated by the amount of energy intercepted by the leaf surface and the efficiency of its use. The amount of assimilate respired during the whole season was taken as a constant proportion of assimilated dry matter (Winzeler et al., 1976). In potato, the NAR was determined by the amount of intercepted radiation, especially in the early stages of development (Allen and Scot, 1980). Similar linear responses were reported in barley, wheat, maize and soya beans (Bisooe and Gallagher, 1977).

In leafy vegetables, increase in fresh weight over time is used to measure growth. However, fresh weight can fluctuate considerably in response to variations in available soil moisture and the physiological status of the plant. Cell number is another commonly used parameter to measure the growth in unicellular organisms like green alga, *Chlamydomonas*. In multicellular plants, however, use of cell number can be misleading because cells can divide without increasing in volume. For example, during the early stages of embryogenesis, the zygote divides to form progressively smaller cells with no net increase in the size of the embryo. The increase in volume begins only after it reaches the eight-cell stage.

One of the common measures of plant growth is the rate of volume change, which appears as the rate of change in length of a radially symmetrical organ (root, stem internode, single plant and coleoptile) or the rate of change in area of leaves (Silk and Erickson, 1979). Such expansion is uniform neither in magnitude nor in direction over the plant surface and is frequently confined to the apical regions of roots, stems and branches only. It also changes with developmental stage of growing organ and prevailing environmental conditions. This is another reason for the lack of correlation between cell number and growth. Since the developing plant cells do not migrate relative to neighboring cells and plant tissue is deformed only by the expansions and contractions of component cells, the observed growth at a given point is the sum of the volume changes that occur between the reference origin and the point. In view of this, they suggested the "kinematic" approach to describe the growth patterns of organs mathematically in terms of the expansion patterns of the component cells. These quantitative and localized descriptions of growth provide a useful tool to develop models for estimating growth rate.

The "classical" and "functional" approaches are the two widely used methods to analyze the growth of individual plants. In the "classical" approach, events occurring less frequently, destructive in nature and relatively large (e.g., harvest) are used to analyze growth (Hunt, 2003). This includes total yield, biomass, number and weight of roots, etc. On the other hand, the "functional" approach uses more repetitive or frequent events (plant height at different times, number of leaves, NAR, etc.) to analyze the growth. The two approaches are not mutually exclusive, although scientists are forced to choose one of the methods depending on purpose for which the growth analysis is being conducted.

As per the classical approach, the RGR is estimated as follows (Hunt, 1982):

$$W_2 = W_1 e^{r(t_2 - t_1)} \tag{1}$$

where W_1 and W_2 are plant dry weights at times t_1 and t_2, and r is RGR of individual plant. By using the natural logarithm-transformed dry weights, the formula can be rearranged as follows:

$$r = \frac{\ln(W_2) - \ln(W_1)}{t_2 - t_1} \tag{2}$$

Destructive harvesting is required to determine plant dry weight. Although equation (2) is widely used growth analysis approach, this formula does not always show how RGR should be calculated, especially when samples of individual plants are used. One way is to substitute W_1 and W_2 with sample means $_1$ and $_2$. The formula for estimating RGR (first estimator) will be

$$\hat{r}_1 = \frac{\ln\left(\overline{W}_2\right) - \ln\left(\overline{W}_1\right)}{t_2 - t_1} \tag{3}$$

Alternatively, RGR can be calculated from the mean natural logarithm-transformed plant weights as follows:

$$\hat{r}_2 = \frac{\overline{\ln\left(W_2\right)} - \overline{\ln\left(W_1\right)}}{t_2 - t_1} \tag{4}$$

where $\overline{\ln(W)_t}$ is the mean of the ln-transformed plant weights at time t. This is referred as estimator 2. In the first estimator, plant weights are averaged before ln-transforming, whereas in the second estimator, the weights are ln-transformed before averaging. These estimators almost always yield different values due to the fact that $\overline{\ln(W)_t}$ is not equal to $\ln(W)$ if there is variation in plant weight among individuals.

The "classical" approach, although appears to be straightforward, is often considered unsatisfactory in explaining the variations in RGR of individuals because the time interval of the growth could be severely masked by fluctuations in RGR between consecutive harvest intervals (Poorter, 1989; Causton and Venus, 1981). Additionally, a fixed relation between leaf area and plant weight has to be assumed in the calculation of the net assimilation rate (NAR), and it is difficult to statistically evaluate differences in RGR (Evans, 1972; Poorter and Lewis, 1986; Hunt et al., 2002).

The "functional" approach was introduced to overcome the shortcomings of "classic" approach (Vernon and Allison, 1963; Hughes and Freeman, 1967). This method uses two variables, dependent variable Y (ln-transformed weight of the plant) and independent variable X (time of harvest). The polynomial equation of the form $Y = b_0 + b_1X_1 + b_2X_2 + \ldots + b_nX_n$ is fitted through the growth data, and the RGR equation is obtained through differentiating polynomial equation. The advantages of functional approach include clearer perception of ontogenetic drift, avoidance of calculation of mean values of NAR, and integration of statistical analysis with growth analytical procedure. Poorter and Lewis (1986) demonstrated that the observed differences in RGR had only limited biological meaning.

In the functional approach, RGR can be calculated by grouping the plants into pairs of similarly sized individuals before the first harvest and harvesting one plant from each pair on the first harvest date. The second plant from each pair is harvested on the second harvest date and so on. RGR is then calculated for each pair, and the values averaged over all pairs. The value of RGR obtained by this method will be identical to that derived from equation four above. Alternatively, a curve can be fit to the ln-transformed plant weights through time series, and RGR at any particular time can be calculated as the slope of the curve.

8.5 MODELING OF RELATIVE GROWTH RATE

Modeling and simulation are used as tools to study plant growth and developmental processes. By expressing experimentally obtained knowledge in the form of a system of interacting mathematical equations, it allows horticulturists and agronomists to gain mechanistic understanding of highly complex behavior of living systems. It principally integrates the output of molecular regulatory networks with two cellular processes, namely, cell division and cell expansion that drive growth and development of the living organ. In turn, these cellular processes are controlled at the molecular level by hormonal signaling. Integration of modeling framework with regulatory modules of cell division, expansion and hormone signaling forms the basis of growth simulation model for a functional organ. While the regulation of the cell cycle and hormone transport within the organ or whole plant has been modeled extensively and many such modules have been integrated, the progress in the modeling of cell expansion is rather limited mainly because the current generation models are largely qualitative. Together with experimental methods to measure crucial model parameters, these modeling techniques could provide a basis to develop a "systems biology" approach to gain a fundamental insight into the relationship between gene function and whole organ behavior. The computational growth model enables more accurate predictions for a wide range of plant physiological conditions.

First developed at the beginning of the 20th century at the Sheffield University (Hunt, 1982), it came to be known as British School of plant growth analysis. Another group at Tharandt/Dresden Technical University in Germany developed a population model and a size class model based on relative growth rate concept to predict the growth of trees (Wenk et al., 1990; Wenk, 1994).

8.5.1 Modeling Individual Plant Growth

Pommerening and Muszta (2015) provided the steps in modeling individual plant growth, which include the selection of a suitable function of relative growth rate, identification of a primary plant size characteristic (e.g., tree volume), selection of secondary plant size characteristics (e.g., tree height and tree diameter) that are linked to the function of relative growth rate of the primary characteristic through allometric relations, and simultaneous estimation of the two to three model parameters of RGR function and the two allometric coefficients using nonlinear regression. Jones et al. (2009) described the procedure for calculation of complex types of nonlinear regression in R using the function optim. Tree volume can be replaced by weight or biomass. To illustrate this combined methodology Pommerening and Muszta (2015) used stem analysis data of four Sitka spruce (*Picea sitchensis* (Bong.) Carr.) trees from the same forest stand. A number of researchers believe that RGR is size dependent, meaning that individuals with a smaller initial size have a larger relative growth rate (Hunt and Parson, 1974; Turnbull et al., 2008; Rose et al., 2009; Rees et al., 2010). This can mask important relationships, making it difficult to conclude whether a tree grows slowly because it is large or because it is pursuing a slow growth strategy. To overcome this, Pommerening and Muszta (2015) suggested the use of size-standardized RGR. They used the relationship between RGR and growth multiplier to predict future growth from current size and size dominance.

8.5.2 Modeling of Plant Population

It is possible to model the whole-plant populations. In this case, the growth rates are calculated from population summary characteristics as suggested previously (Nicke, 1988; Wenk, 1994). The methodology is almost similar to the modeling of individual plant with the exception that the trees leaving the plant stand due to management or natural mortality or can be a combination of both. They used the additional loss function to collectively take care of death events as a result of disturbances. Pommerening and Muszta (2015) provided a step-wise procedure for modeling of RGR of in Douglas fir (*Pseudotsuga menziesii*) plantation. They have demonstrated that modeling of RGR is straight forward process, especially when size-standardized relative growth rates are used and growth functions have few model parameters.

8.5.3 High-Throughput Plant Growth Analysis

Plant growth models quantify either absolute growth rate (AGR) or relative growth rate (RGR), both of which are measured from biomass or plant area at successive time points. Then, plant growth is often fitted with linear or exponential equations such as a logistic model (one- or two-parameter version) (Hunt, 1982; Vanclay, 1994). This comprises initial exponential growth and a term that reduces RGR as the area increases, resulting in an asymptotic maximum area (Zeide, 1993). Although simple logistic models require relatively few observations, they do not often fit with observations well (Paine et al., 2012) and the traditional approaches to modeling growth using linear and exponential models to be inadequate due to their oversimplified assumptions (Paine et al., 2012). Several researchers have considered nonlinear models, including three- and four-parameter logistic (Gottschalk and Dunn, 2005; Pinheiro and Bates, 2000), power-law, (Gompertz, 1825; Heinen, 1999) and monomolecular model (Heinen, 1999; Richards, 1959) to provide sufficient flexibility to obtain the best fit between models and observations (Paine et al., 2012). Taking advantage of the current rapid development in imaging systems and computer vision algorithms, growth modeling based on high-throughput computational phenotyping techniques has been developed (Arvidsson et al., 2011; El-Lithy et al., 2004; Granier et al., 2006; Zhang et al., 2012). In this method, top-view images are captured periodically, and a growth curve is generated using the observed value of the plant area over time (Zhang et al., 2012). However, the observed value of plant area from a top view is remarkably affected by complex leaf overlap during growth (in addition to leaf twisting and curling, and circadian movement), resulting in inaccurate growth patterns. HPGA was used to estimate leaf overlap percentage to measure plant area more precisely (Tessmer et al., 2013). This approach has the following advantages.

- Unlike the existing approaches that simply count the number of valid pixels in an image, HPGA estimates plant areas by explicitly taking leaf overlaps into consideration (Sack et al., 2012; Tessmer et al., 2013). The leaf overlap problem is addressed with a four-step approach: plant center identification, leaf tip identification, leaf area estimation and plant area measurement.
- HPGA avoids the leaf segmentation problem to recognize all the leaves of a plant from a top-view image, which has been considered to be a challenging problem in the computer vision community due to high plant-to-plant variations (Wang et al., 2008).
- With high-throughput phenotyping technique, researchers are able to generate hundreds or even thousands of observations for every plant automatically. Feeding enough observations to a nonlinear model ensures the robustness and precision of plant growth modeling.
- In HPGA, functional data analysis is applied on growth curves for better interpretation of the plant growth scenarios.

8.6 GROWTH RATES AND GROWTH CURVES

In multicellular plants, many aspects of growth are associated with newly formed cells produced by meristematic tissues. The newly formed cells in apical meristems first enlarge slowly but later expand more rapidly as they are displaced into sub-apical regions. The resulting increase in cell volume can be as much as 100-fold, depending on the species and environmental conditions. The predictable and site-specific ways in which these derivatives expand are closely linked to the final size and shape of the primary plant body. In essence, the total growth of the plant can be thought of as the sum of the local patterns of cell division and expansion. In such cases, the final axis length can be related to the number of cells produced. Changes in phenotype during the growth period can be modeled by growth curves, such as generalized logistic or Gompertz growth curves (Karadavut et al., 2008). If we plot the increase in cell number (growth rate) against time, a typical S-shaped curve is obtained. This is called growth curve or sigmoid growth curve.

(i) Lag Phase – This is the initial phase of growth when the rate of growth is very slow.
(ii) Log Phase – It shows rapid growth and is maximum during the entire life span.
(iii) Stationary Phase – Here, the rate of growth starts decreasing and finally it stops.

The total time period during which the fastest growth occurs is called "grand period of growth." The shape of the curve can change depending on the nature and number of parameters used, function used to fit the curve (linear, exponential or logistic), species and environmental conditions. It is important to select a proper curve and its parameter in order to evaluate the growth response. Recent studies have combined kinematic analysis, flow cytometry and microarray analysis to characterize cell cycle regulation during the growth process of leaves and roots.

REFERENCES

Allen E J and Scott R K. 1980. An analysis of growth of the potato crop. *Journal of Agricultural Science Cambridge* **94**: 583–606.

Arvidsson S, Perez-Rodriguez P and Mueller-Roeber B. 2011. A growth phenotyping pipeline for *Arabidopsis thaliana* integrating image analysis and rosette area modeling for robust quantification of genotype effects. *New Phytologist* **191**: 895–907.

Bisooe P V and Gallagher J N. 1977. Weather, dry-matter production and yield. *Environmental Effects on Crop Physiology* (Eds.) Landsberg J J and Cutting C V. Academic Press, London.

Blackman V H. 1919. The compound interest law and plant growth. *Annals of Botany* **33**: 353–360.

Causton D R and Venus J C. 1981. *The Biometry of Plant Growth*. Edward Arnold, London.

Coleman J S, McConnaughay K D M and Ackerly D D. 1994. Interpreting phenotypic variation in plants. *Trends in Ecology and Evolution* **9**: 187–191.

El-Lithy M E, Clerk E J, Ruys G J, Koornneef M and Vreugdenhil D. 2004. Quantitative trait locus analysis of growth-related traits in a new arabidopsis recombinant inbred population. *Plant Physiology* **135**: 444–458.

Evans G C. 1972. *The Quantitative Analysis of Plant Growth*. Blackwell Scientific Publications, Oxford.

Gompertz B. 1825. On the nature of the function expressive of the law of human mortality, and on a new mode of determining the value of life contingencies. *Philosophical Transactions of the Royal Society of London* **115**: 13–583.

Gottschalk P G and Dunn J R. 2005. The five-parameter logistic: a characterization and comparison with the four-parameter logistic. *Analytical Biochemistry* **343**: 54–65.

Granier C, Aguirrezabal L, Chenu K, Cookson S J, Dauzat M, Hamard P, Thioux J J, Rolland G, Bouchier-Combaud S, Lebaudy A, Muller B, Simonneau T and Tardieu F. 2006. Phenopsis, an automated platform for reproducible phenotyping of plant responses to soil water deficit in *Arabidopsis thaliana* permitted the identification of an accession with low sensitivity to soil water deficit. *New Phytologist* **169**: 623–635.

Gunasena H P M and Harris P M. 1969. The effect of CCC and nitrogen on the growth and yield of the second early potato variety Craig's Royal. *The Journal of Agricultural Science Cambridge* **73**: 245–259.

Gunasena H P M and Harris P M. 1971. The effect of CCC, nitrogen and potassium on the growth and yield of two varieties of potato. *Journal of Agricultural Science Cambridge* **76**: 33.

Heinen M. 1999. Analytical growth equations and their genstat 5 equivalents. *Netherlands Journal of Agricultural Science* **47**: 67–89.

Hughes A P and Freeman P R. 1967. Growth analysis using frequent small harvests. *Journal of Applied Ecology* **4**: 553–560.

Hunt R. 1982. *Plant Growth Curves: The Functional Approach to Plant Growth Analysis*. Edward Arnold, London.

Hunt R. 2003. Growth analysis, individual plants. pp. 588–596. *Encyclopaedia of Applied Sciences* (Eds.) Thomas B D, Murphy J and Murray D. Academic Press, London.

Hunt R, Causton D R, Shipley B and Askew A P. 2002. A modern tool for classical plant growth analysis. *Annals of Botany* **90**: 485–488.

Hunt R and Cornelissen J H. 1997. Components of relative growth rates and their interrelation in 59 temperate plant species. *New Phytologist* **135**: 395–417.

Hunt R and Lloyd P S. 1987. Growth and partitioning. *New Phytologist* **106**: 235–249.

Hunt R and Parsons I T. 1974. A computer program for deriving growth-functions in plant growth-analysis. *Journal of Applied Ecology* **11**: 297–307.

Ifenkwe O P. 1975. *Effects of Row Width and Plant Density on Growth and Development of Two Main Crop Potato Varieties*. PhD Thesis. University College of Wales, Aberystwyth.

Jones O, Maillardet R and Robinson A. 2009. *Introduction to Scientific Programming and Simulation Using R*. Chapman and Hall/CRC, Boca Raton.

Karadavut U, Kayi S A, Palta C and Okur O. 2008. A growth curve application to compare plant heights and dry weights of some wheat varieties. *American Eurasian Journal of Agricultural and Environmental Science* **3**: 888–892.

Leather S R. 2010. Precise knowledge of plant growth stages enhances applied and pure research. *Annals of Applied Biology* **157**: 159–161.

McMahon T A and Bonner J T. 1983. *On Size and Life*. Scientific American Books, NewYork.

Metcalf J C, Rose K E and Rees M. 2003. Evolutionary demography of monocarpic perennials. *Trends in Ecology and Evolution* **18**: 471–480.

Monteith J L. 1977. Climate and the efficiency of crop production in Britain. Phil. *Transaction of Royal Society, London* **281**: 277–294.

Nelissen H, Moloney M and Inzé D. 2014. Translational research: from pot to plot. *Plant Biotechnology Journal* **12**: 277–285.

Nicke A. 1988. Aufbau eines Bestandesentwicklungsmodells. [Construction of a stand development model]. *Beiträge für die Forstwirtschaft* **22**: 71–74.

Paine C E, Marthews T R, Vogt D, Purves D, Rees M, Hector A and Turnbull L A. 2012. How to fir nonlinear plant growth models and calculate growth rates: an update for ecologists. *Methods in Ecology and Evolution* **3:** 245–256.

Pinheiro J and Bates D. 2000. *Mixed-Effect Models in S and S-PLUS*. Springer-Verlag, New York.

Pommerening A and Muszta A. 2015. Methods for modelling plant relative growth rate. *Forest Ecosystem* **2**: 5.

Poorter H. 1989. Plant growth analysis: towards a synthesis of the classical and functional approach. *Phyiologia Plantarum* **75**: 237–244.

Poorter H. 2002. Plant growth and carbon economy. *Encyclopedia of Life Sciences*. Nature Publishing Group, London. Available at htp://www.els.net.

Poorter H and Lewis C. 1986. Testing differences in relative growth rate: a method avoiding curve-fitting and pairing. *Physiologia Plantarum* **67**: 223–226.

Poorter H and Remkes C. 1990. Leaf area ratio and net assimilation rate of 24 wild species differing in relative growth rate. *Oecologia (Berlin)* **83**: 553–559.

Price C and Munns R (Eds.). 2016. Growth analysis: A quantitative approach. *Plants in Action*. 2nd edition. Australian Society of Plant Scientists, New Zealand Society of Plant Biologists, and New Zealand Institute of Agricultural and Horticultural Science, CSIRO, Australia.

Rees M, Osborne C P, Woodward F I and Hulme S P. 2010. Partitioning the components of relative growth rate: how important is plant size variation. *The American Naturalist* **176**: 152–161.

Richards F A. 1959. A flexible growth function for empirical use. *Journal of Experimental Botany* **10**: 290–300.

Rose K E, Atkinson R L, Turnbull L A and Rees M. 2009. The costs and benefits of fast living. *Ecology Letters* **12**: 1379–1384.

Sablowski R and Carnier D M. 2014. Interplay between cell growth and cell cycle in plants. *Journal of Experimental Botany* **65**: 2703–2714.

Sack L, Scoffoni C, McKown A D, Frole K, Rawls M, Havran J C, Tran H and Tran T. 2012. Developmentally based scaling of leaf venation architecture explains global ecological patterns. *Nature Communications* **3**: 837.

Silk W K and Erickson R O. 1979. Kinematics of plant growth. *Journal of Theoretical Biology* **76**: 481–501.

Tessmer O L, Jiao Y, Cruz J A, Kramer D M and Chen J. 2013. Functional approach to high-throughput growth analysis. *BMC Systems Biology* **7**: Available at www.msu.edu/_jinchen/HPGA. Accessed on December 10, 2016.

Turnbull L A, Paul-Victor C, Schmid B and Purves D W. 2008. Growth rates, seed size, and physiology: do small-seeded species really grow faster? *Ecology* **89**: 1352–1363.

Vanclay J K. 1994. *Modelling Forest Growth and Yield*. CAB International, Wallingford.

Vernon A J and Allison J C S. 1963. A method of calculating net assimilation rate. *Nature* **200**: 814–814.

Wang X F, Huang D S, Du J X, Xu H and Heutte L. 2008. Classification of plant leaf images with complicated background. *Applied Mathematics and Computation* **205**: 916–926.

Wenk G. 1994. A yield prediction model for pure and mixed stands. *Forest Ecology and Management* **69**: 259–268.

Wenk G, Antanaitis V and Smelko S. 1990. Waldertragslehre. *Forest Growth and Yield Science*. Deutscher Landwirtschaftsverlag, Berlin.

West C, Briggs G R and Kidd F. 1920. Methods and significant relations in the quantitative analysis of plant growth. *New Phytologist* **19**: 200–207.

Winzeler H, Hunt L A and Mahon J D. 1976. Ontogenetic changes in respiration and photosynthesis in a uniculm barley. *Crop Science* **16**: 786–790.

Wuyts N, Dhondt S and Inze D. 2015. Measurement of growth in view of an integrated analysis of regulatory networks. *Current. Opinion in Plant Biology* **25**: 90–97.

Zeide B. 1993. Analysis of growth equations. *Forest Science* **39**: 594–616.

Zhang X, Hause R J and Borevitz J O. 2012. Natural genetic variation for growth and development revealed by high-throughput phenotyping in *Arabidopsis thaliana. G3: Genes, Genomes and Genetics* **2**: 29–34.

9 Modeling in Horticultural Crops

9.1 INTRODUCTION

Plant growth modeling has become a key modern research tool in almost all areas that are related to plant production, management of natural resources and environmental sciences. The development of plant growth models has progressed enormously during the last two decades mainly because of the advancements made in our understanding of plant processes at molecular and genetic level, availability of modern computational tools, and sharing of information and experiences among biologists, mathematicians and computer scientists. Although interdisciplinary approach has been adopted in plant growth modeling and simulation, still it is one of the most challenging aspects to be encountered so far. Empirical crop and forest growth models, which apply functions fitted to data without considering the physiological processes involved in growth and morphogenesis, continue to be used efficiently for estimating or predicting yield and quality of products. For growth modeling, it is extremely important to understand how plants interact to environmental factors, such as light, temperature, soil characteristics and water availability applied either individually and in combination to influence growth, development and yield in crop plants. It is of paramount importance to take into account various physiological processes involved in growth and development, e.g., water and nutrient uptake, photosynthesis, respiration, photorespiration, hormonal interactions and carbon partitioning (Gifford and Evans, 1981). Recently, the term "process-based model" was introduced to describe those models that considered the interactions between plant functional processes and abiotic factors (Fourcaud et al., 2008).

9.2 PLANT GROWTH MODELING

Plant growth can be regarded as a multi-level process, operating from the cellular to the whole-plant and plant-community level. Cell growth and division constitute the most elementary processes of growth that determines the size and shape of the organ and whole plant. The choice of the model and the level at which growth is measured depend heavily on the purpose for which the growth is being measured. For example, a plant breeder may want to develop a growth model to determine the enhancement of disease resistance, the improvement of yield or product quality. When dealing with very complex traits like crop yield and product quality, an in-depth insight into the physiological and morphological parameters and interaction of these parameters with the growing environment and crop management practices is highly essential to identify a superior genotype. Crop growth models are used to improve the efficiency of a breeding program as model calculations (sensitivity analysis and scenario studies) enable a quantitative analysis of the phenotypic response of modified genetic traits under a range of growing conditions (Boote et al., 2001). Research efforts on the mechanistic understanding of growth, based on measured phenotypic traits, their correlation or causal relation, and their variability in response to the atmospheric and soil environment have delivered enormous amount of data for both modeling purposes and quantitative genetical approaches. Considerable published information is available for a number of crops including cucumber (Marcelis, 1994), tomato (Heuvelink, 1999; Caliskan et al., 2009; Medina-Ruíz et al., 2011), carnation, rose and chrysanthemum (Carvalho et al., 2003).

Crop modeling represents a quantitative way of expressing different phenotypic traits and physiological processes in the plant. The primary aim of mathematical modeling is to study and analyze a

DOI: 10.1201/9781003373889-12

plant system under varying conditions (Mason and Dzierzon, 2006). These models allow for predictions of expression of a phenotypic trait (biomass, height, yield, etc.) under a particular set of growing and environmental conditions. Modeling of phenotypic plant traits and associated physiological processes can help in improving yield and facilitate decision making process to optimize the use of natural resources in crop production (Soltani et al., 2001). The use of mathematical models can save lot of time and money, particularly in the case of perennial crops and if selected trait depends on long-term effects and/or the trait is difficult to monitor (Fraisse et al., 2006). However, in order to create a useful, reliable mathematical model, the modeler must choose suitable mathematical principles and techniques to use and test it against experimental data (Crouch and Haines, 2004). Modern data gathering and processing tools can greatly enhance the ability of biologists to develop useful models to monitor growth, development, yield and product quality in plants. This requires a close interaction between basic crop physiology research and model development tools (Lisson et al., 2005). Plant development involves processes working within efficiencies of scale.

The representation of plant physiological and metabolic processes together with all the related interactions in one model could be a real challenge. Crop growth models do exist for many crops, which are focused on various parameters (functional, quantitative or plant architecture) and processes. They are often distinguished between descriptive and mechanistic models (Pronk et al., 2003; Vazquez-Cruz et al., 2010). Descriptive models do not address underlying mechanisms to the system behavior, whereas mechanistic models incorporate quantitative description of these mechanisms.

Several of the models focus on the description of carbon (C) or nitrogen (N) balance and consider that plant development depends on a change of matter in different plant parts based on uptake (e.g., photosynthesis) and loss (e.g., senescence, abscission, respiration) either within an individual (Heuvelink, 1996, 1999; Marcelis, 1994; Carvalho et al., 2006; Gayler et al., 2008) or in an entire population (Battaglia and Sands, 1998; Gayler et al., 2006; Pretzsch et al., 2008). Since these models do not consider plant morphology (i.e., the geometry and topological connectivity between organs), models linking growth-driven processes with plant morphogenesis (3-D evolution of plant structure over time) were developed over the years (Fourcaud et al., 2008). These models also include interactions of various processes with environmental variations (temperature, soil characteristics and water availability). These models are known as functional-structural (FS) plant models (Godin et al., 2004). Models taking into account the interaction between plant architecture, functions of organs (e.g., root absorption, leaf photosynthesis and sap flow) and growth processes have been developed over the last decade (Perttunen et al., 1998; de Reffye et al., 1999; Godin et al., 2004; Prusinkiewicz, 2004; Vos et al., 2006; Ge´nard et al., 2008). Because of the recent advances in 3-D reconstruction of real plant structures and visual data acquisition, it is now possible to compute plant physiological functioning as an integration of organ-level processes at the whole-plant scale (Godin, 2000; Cheng et al., 2007; Xu et al., 2007; Godin, 2000; Cheng et al., 2007). One of the applications of 3-D virtual plants is in the area of light interception and photosynthesis by the leaves at the plant and community scale (Evers et al., 2007; Rey et al., 2008). However, the success of these models depends on the constant feedback between plant scientists in morphogenesis and physiology (de Reffye et al., 1997; Lambers et al., 2006; Prusinkiewicz and Rolland-Lagan, 2006).

The first ecosystem "models" using the maps were developed in the 16th century, but the one that included stem height and yield of forest stands was developed for the first time in the models in the 19th century (Fourcaud et al., 2008). Lacointe (2000) described the four major carbon-partitioning models currently used in FS models. This use empirically determined allocation coefficients, a description of growth patterns or relationships within the plant, transport-resistance or source–sink relationships in modeling. Le Roux and coworkers reviewed 27 tree growth models that use the main physiological processes involved in C metabolism (Le Roux et al., 2001). Most FS models deal with crop species growing in optimal conditions.

Cell division, elongation and differentiation occur in the meristems during the entire life span of the plant. Meristem cells are the site where several specific cellular activities are performed, which, in turn, encode complex developmental processes in other plant organs (Dupuy et al., 2008). Therefore, the growth can be considered as the consequence of meristematic activity. The location and temporal activity of meristems determine plant morphology and growth rate (Körner, 1991; Sussex and Kerk, 2001; Barthelemy and Caraglio, 2007). Luquet et al. (2006) concluded that plant growth and development depend on both assimilate supply and organogenesis-based demand. Therefore, it is necessary to understand plant development in light of the functioning of meristems in producing new tissues and organs. Dupuy et al. (2008) developed the "cell modeler" software that can simulate cell–cell interactions within meristems and specific tissues, and showed how biophysical and biomechanical processes determine cell production. These models focus on the type and size (strength) of sinks within the plant (stem, root, inflorescence, fruits, etc.), while most traditional crop models consider the environmental effects on source processes (e.g., photosynthesis). The source-driven paradigm lacks effect-based justification (Körner, 1991). Furthermore, since plant's primary production depends on several resources other than C, plant growth rate is not simply determined by the rate of photosynthetic CO_2 uptake per unit leaf area, but also depends on allocation of resources to leaves and the leaf area (the amount of photosynthetically active tissue constructed) and on leaf longevity (how long the photosynthetic tissues are deployed). Moreover, C is used within the plant for both growth and respiration, and the fraction used for respiration varies greatly among plants, increasing with decreasing plant growth rate (Lambers et al., 2008).

Growth is a highly complex phenomenon driven by several processes simultaneously, photosynthesis being only one out of several drivers. Other drivers of growth include the distribution (partitioning) of dry matter to the different plant organs, the feedbacks involved and the plant's capacity for additional resource acquisition, especially under stress conditions, and tissue-specific rates of respiration (Körner, 2006). All these are very important aspects that need to be considered in analyzing and modeling morphological plasticity. Added to this complexity is the location of individual plant within the population or ecosystem and understanding the interactions taking place between and among them. In response to this need, Marks and Lechowicz (2006) developed a plant function model (tree-seedling adaptive design [TAD]) that incorporates 34 key shoot and root traits associated with water, light, C and N, and the multi-task nature of plant organs as well as interactions and trade-offs among traits on whole-plant fitness. According to these authors, the development of such models will therefore allow the investigation of major ecophysiological issues such as how individual plant traits affect plant performance, phenotypic plasticity and competition with neighbors, and determine the potential for alternative designs of similar performance (Marks and Lechowicz, 2006). Körner (1991) developed a list of plant factors, ranging from metabolic factors (i.e., leaf photosynthesis and dark respiration), morphogenetic features (govern plant spatial design and architecture) and environmental factors together determine the C balance during growth. Nutrient factors not only influence respiration and growth but also, directly or indirectly, regulate gene expression and trigger many metabolic and developmental responses that determines growth of both individual organ and plant as a whole (Bijlsma and Lambers, 2000; Yang and Midmore, 2005; Gayler et al., 2008). The effects of nutrients on growth and development are in turn mediated by plant growth regulators (Buck-Sorlin et al., 2008). Therefore, Körner (2003) proposed the hypothesis that growth is not limited only by C supply in many cases (source activity) and provided evidence for the role of sink activity on tree growth, regardless of climatic zones. He also showed that the levels of non-structural carbohydrate (NSC) in tree tissues are a reflection of whether the source activity (photosynthesis) is sufficient to meet the demands of the sink tissues because it has direct bearing on growth. Low NSC pools indicates low source and high sink activity and high NSC concentrations show that photosynthesis meets or exceeds the amount needed for growth (Körner, 2003). Because sink activity is often controlled by finite soil resources and developmental constraints, it is more a stronger driver of growth than the source activity. Molecular and genetic studies, which can help in elucidating the exact relationship between sink and source activity have indicated that sugar signaling of plant assimilate resources to meristems occurs via cell-wall invertases and their coding genes (Dingkuhn et al., 2006). However, whether or not meristem activity is considered as driving force for plant growth processes, the resulting architecture, which is due to primary meristem development, is the key component of many plant growth models.

9.3 STRUCTURE OF THE GROWTH MODEL

Medina-Ruíz et al. (2011) provided a comprehensive review of mathematical modeling in tomato. López et al. (2005) defined the growth and development crop model as follows:

$$\frac{dx}{dt} = f\left(x, u, p\right), \ x\left(t_0\right) = \beta \tag{1}$$

where x is the state vector consisting of n number of crop variables, such as dry weight, number of fruits and foliar area, the input vector (u) deals mainly with the climatic conditions (air temperature, CO_2 concentration, humidity and photosynthetically active radiation [PAR]), the vector (p) represents biological parameters or constants for the model and (β) represents the vector of initial conditions of the state variables. As (f) is a vector of n functions, the model does not have a general analytical solution and must be solved by numerical methods or through computer simulations.

9.4 TYPES OF MODELS

The success of a mathematical model depends on the integration of mathematical principles and techniques used as well as on the validation of the model through testing against actual experimental data (Crouch and Haines, 2004). Models sometimes simplify systems to reduce the datasets required to estimate parameters (Lentz, 1998; Lisson et al., 2005). Depending upon the purpose for which models are designed, different kinds of models are created, some of which are listed below (Murthy, 2003):

- **Statistical models**: These models are used to test the relationship between yield or yield components and weather parameters based on statistical techniques.
- **Mechanistic models**: These models explain both the relationship between weather parameters and yield and mechanisms influencing the dependent variables.
- **Deterministic models**: These models are used to estimate the exact value of the dependent variable using defined coefficients.
- **Stochastic models**: These models define the state of dependent variable at a given rate. For each set of inputs, different outputs are given along with probabilities.
- **Dynamic models**: This model includes time scale as a variable.
- **Static**: Time is not included as a variable. Dependent and independent variables remain constant over a given time scale.
- **Simulation models**: One of the main goals of crop simulation models is to estimate crop production as a function of different input parameters (soil condition, management practices and weather factors).
- **Descriptive model**: A descriptive model defines the behavior of a system in a simple manner. The model does not reflect the mechanisms underlining the phenomena but consists of one or more mathematical equations.
- **Explanatory model**: This includes quantitative description of the mechanisms and processes that cause the behavior of the system. A system is analyzed, and its processes and mechanisms are quantified separately to build the model.

Out of the above three model types, descriptive, mechanistic and explanatory models have been described below.

9.4.1 Descriptive Models

In the past, tedious statistical analyses were used to analyze and interpret the data. This approach, apart from time-consuming and vulnerable to human errors, is expensive (Van Ittersum et al., 2003a). Initial crop models focused on functions designed to describe and analyze plant growth without analyzing the underlying physiological processes (Renton et al., 2005). Descriptive models involve relatively short computing time and use a limited number of variables (Mirschel et al., 2004). The main drawback of descriptive model is related to its inability to simulate plant adaptability and response to different conditions. Also, an entirely new model will have to be built to accommodate extended data set or additional parameter (Renton et al., 2005). Generally, descriptive models describe only descriptive relationships between response and predictor variables. Essentially, descriptive models are most often associated with curve fitting and regression and are not governed by biological processes (Domijan et al., 2006).

9.4.2 Mechanistic Models

These models are used for testing the hypotheses and synthesizing knowledge on complex systems based on physiological processes, such as photosynthesis, assimilation and respiration (Brainard and Bellinder, 2004; Thornley and Cannell, 2000). These models are widely used to investigate the impacts of weather factors on crop growth, productivity and product quality (Tao et al., 2009; Rico-García, 2011). Since photosynthetically active radiation (PAR) has profound impact evapotranspiration and photosynthesis, they are used as one of the variables in photosynthesis-based models (Dutilleul et al., 2007; Sentelhas and Gillespie, 2008). Carbon-rich compounds are distributed according to organ demands through assimilation and mobilization of carbon. Partitioning of carbon into various compounds (i.e., sugars, other fruit compounds and respired CO_2) can also be taken into account. These processes are nearly always described in terms of a set of differential equations. Determination of leaf area index will be required to estimate interception of solar radiation and biomass production (Soltani et al., 2006). Although respiration is one of the main energy sources in growing plants, it has not been sufficiently considered in several crop models (Kuretz et al., 2003). It is possible to predict product composition as a function of climatic variables using the mechanistic models (Heredia and Andres, 2008).

9.4.3 Explanatory Models

Explanatory crop growth models provide a tool to represent and combine knowledge in a generic way (Challa et al., 1994). These models were first developed nearly 40 years ago by Prof. C. T. de Wit in Wageningen, The Netherlands (Bouman et al., 1996; Van Ittersum et al., 2003b). Yin and Van Laar (2005) provided a new approach in the GECROS (Genotype × Environment interaction on CROP growth Simulator) model for examining plant responses (biomass and protein production) to environmental and genotypic characteristics. Several research groups have shown that explanatory crop models can be helpful in quantitative-trait loci (QTL) analyses of complex traits, thereby improving breeding efficiency and enhancing breeding method by design (Yin et al., 2000; Reymond et al., 2003; Messina et al., 2006).

9.5 CROP GROWTH MODELS – CASE STUDIES

For illustration purpose, mathematical models developed for tomato (widely grown vegetable) and chrysanthemum (ornamental crop) are describe in the following.

9.5.1 A Case Study of Tomato

Jones and coworkers developed the dynamic tomato growth and yield model (TOMGRO), a mechanistic model consisting of 69 state variables used to compute development and production in tomato plants (Jones et al., 1991). This model includes input vector of climatic variables (air temperature, solar radiation [PAR] and CO_2 concentration) and the output vector state variable (number of leaves, number of new branches, number of fruits, dry weight of leaves, dry weight of new branches, dry weight of fruits and foliar area) and uses a source-demand function to identify carbohydrates for the growth of different parts of the plant. The latest version of this model consists of 574 state variables for simulating the development of fruit, with each fruit having specific positions on the clusters. A separate model has been incorporated for analyzing fruit development separately from the growth in biomass (Kenig and Jones, 1997). Later on, these authors proposed a relatively simple model with the advantage of having only five state variables (number of nodes on main stem, index leaf area, total plant weight, fruit weight and weight of ripe fruit) (Jones et al., 1999; Ramírez-Arias, 2005). This simplified version was validated with data from several experiments.

De Koning (1994) developed a model of 300 state variables to predict the distribution of dry matter with tomatoes grown in greenhouses. He estimated the number of fruits by predicting the initiation, abortion and harvesting of individual fruit and potential growth rate. The dry-matter distribution in the model found to be proportional to potential growth rate was used to predict demand for assimilates. The prediction of the number of fruits per cluster, however, was not predicted properly acceptably accurate (Tap, 2000).

TOMSIM is another tomato growth model for predicting growth and development of tomato. The production of dry matter is predicted by a sub-module based on the estimation of photosynthesis (Ramírez-Arias, 2005).

9.5.2 CHRYSANTHEMUM

The model described in this chapter, the CHRYSIM1.0, consists of modules for greenhouse radiation transmission, radiation interception by the crop, leaf and canopy photosynthesis, dry-matter production and dry-matter partitioning among plant organs (roots, stem, leaves and flowers). The model calculations are as follows:

- Interception of radiation and canopy gross photosynthesis are calculated for a multilayered uniform canopy.
- Leaf gross photosynthesis is calculated with the biochemical model.
- Net assimilates production results from the difference between canopy gross photosynthesis and maintenance respiration.
- Maintenance respiration is calculated as a function of dry weights of the different plant organs, temperature and crop relative growth rate according to Heuvelink (1999).
- Assimilate partitioning is determined by fixed ratios as determined by crop developmental stage.
- Computation of leaf area increase.
- Daily global radiation was model input.
- A greenhouse transmittance for diffuse radiation of 70%, 71% or 75.6%, respectively, was assumed in the three simulation studies.
- Hourly values of greenhouse temperature and CO_2 concentration were also used as model input.

REFERENCES

Barthelemy D and Caraglio Y. 2007. Plant architecture: a dynamic, multilevel and comprehensive approach to plant form, structure and ontogeny. *Annals of Botany* **99**: 375–407.

Battaglia M and Sands P J. 1998. Process-based-forest productivity models and their application in forest management. *Forest Ecology and Management* **102**: 13–32.

Bijlsma R J and Lambers H. 2000. A dynamic whole-plant model of integrated metabolism of nitrogen and carbon. 2. Balanced growth driven by C fluxes and regulated by signals from C and N substrate. *Plant and Soil* **220**: 71–87.

Boote K J, Kropff M J and Bindraban P S. 2001. Physiology and modelling of traits in crop plants: implications for genetic improvement. *Agricultural System* **70**: 395–420.

Bouman B A M, Van Keulen H, Van Laar H H, et al. 1996. The 'School of de Wit' crop growth simulation models: a pedigree and historical overview. *Agricultural Systems* **52**: 171–198.

Brainard D C and Bellinder R R. 2004. Assessing variability in fecundity of *Amaranthus powellii* using a simulation model. *Weed Research* **44**: 203–217.

Buck-Sorlin G, Hemmerling R, Kniemeyer O, Burema B and Kurth W. 2008. A rule-based model of barley morphogenesis, with special respect to shading and gibberellic acid signal transduction. *Annals of Botany* **101**: 1109–1123.

Caliskan O, Odabas M S and Cirak C. 2009. The modeling of the relation among the temperature and light intensity of growth in *Ocimum basilicum* L. *Journal of Medicinal Plant Research* **3**: 965–977.

Carvalho S M P, Dhonukshe V B and Heuvelink E. 2003. Interactive effects of duration of long-day period and plant density on external quality of cut chrysanthemum. *Acta Horticulturae* **624**: 335–340.

Carvalho S M P, Heuvelink E, Harbinson J and van Kooten O. 2006. Role of sink–source relationships in chrysanthemum flower size and total biomass production. *Physiologia Plantarum* **128**: 263–273.

Challa H, Heuvelink E and Leutscher K J. 1994. Modelling in the search for balance between inputs and outputs in greenhouse cultivation. pp. 187–195. *Plant Production on the Threshold of a New Century: Proceedings of the International Conference at the Occasion of the 75th Anniversary of the Wageningen Agricultural University, Wageningen, Netherlands* (Eds.) Struik P C, Vredenberg W J, Renkema J A, et al. Kluwer Academic Publishers, Dordrecht, The Netherlands.

Cheng Z L, Zhang X P and Chen B Q. 2007. Simple reconstruction of tree branches from a single range image. *Journal of Computer Science and Technology* **22**: 846–858.

Crouch R and Haines C. 2004. Mathematical modeling: transitions between real world and mathematical model. *International Journal of Mathematics, Education, Science and Technology* **35**: 197–206.

De Koning A N M. 1994. *Development and Dry Matter Distribution in Glasshouse Tomato: A Quantitative Approach.* PhD Thesis. Wageningen Agricultural University, Wageningen. The Netherlands.

de Reffye P, Fourcaud T, Blaise F, Barthélémy D and Houllier F. 1997. A functional model of tree growth and tree architecture. *Silva Fennica* **31**: 297–311.

de Reffye P, Houllier F and Blaise F. 1999. Modelling plant growth and architecture. pp. 70–72. *Advances in Environment and Ecological Modelling* (Eds.) Houllier F and Blaise F. Elsevier, Paris and The Netherlands.

Dingkuhn M, Luquet D, Clément-Vidal A, Tambour L, Kim H K and SongY H. 2006. Is plant growth driven by sink regulation? Implications for crop models, phenotyping approaches and ideotypes. *Proceedings of the Frontis Workshop on Functional – Structural Plant Modelling in Crop Production* (Eds.) Spiertz J H J, Struick P C and Laar H H. Wageningen University and Research Centre, Wageningen.

Domijan K, Jorgensen M and Reid J. 2006. Semi-mechanistic modeling in nonlinear regression: a case of study. *Australian New Zealand Journal of Statistics* **48**: 373–392.

Dupuy L, Mackenzie J, Rudge T and Haseloff J. 2008. A system for modelling cell–cell interactions during plant morphogenesis. *Annals of Botany* **101**: 1255–1265.

Dutilleul P, Han L and Smith D L. 2007. Plant light interception can be explained via computed tomography scanning: demonstration with pyramidal cedar (*Thuja occidentalis*, Fastigiata). *Annals of Botany* **101**: 19–23.

Evers J B, Vos J, Chelle M, Andrieu B, Fournier C and Struik P C. 2007. Simulating the effects of localized red: far-red ratio on tillering in spring wheat (*Triticum aestivum*) using a three-dimensional virtual plant model. *New Phytologist* **176**: 325–336.

Fourcaud T, Zhang X, Stokes A, Lambers H and Körner C. 2008. Plant growth modelling and applications: the increasing importance of plant architecture in growth models. *Annals of Botany* **101**: 1053–1063.

Fraisse C W, Breuer N E, Zierden D, Bellow J G, Paz J, Cabrera V E, García y García A, Ingram T, Hatch U, Hoogenboom G, Jones J W and O'Brien J J. 2006. AgClimate: a climate forecast information system for agricultural system for agricultural risk management in the southeastern USA. *Computers and Electronics in Agriculture* **53**: 13–27.

Gayler S, Grams T E E, Heller W, Treutter D and Priesack E. 2008. A dynamical model of environmental effects on allocation to carbon-based secondary compounds in juvenile trees. *Annals of Botany* **101**: 1089–1098.

Gayler S, Grams T E E, Kozovits A R, Winkler J B, Luedemann G and Priesack E. 2006. Analysis of competition effects in mono- and mixed cultures of juvenile beech and spruce by means of the plant growth simulation model PLATHO. *Plant Biology* **8**: 503–514.

Génard M, Dauzat J, Franck N, Lescourret F, Moitrier N, Vaast P and Vercambre G. 2008. Carbon allocation in fruit trees: from theory to modelling. *Trees-Structure and Function*. https://doi.org/10.1007/s00468-007-0176-5.

Gifford R M and Evans L T. 1981. Photosynthesis, carbon partitioning, and yield. *Annual Review of Plant Physiology* **32**: 485–509.

Godin C. 2000. Representing and encoding plant architecture: a review. *Annals of Forest Science* **57**: 413–438.

Godin C, Hanan J, Kurth W, Lacointe A, Takenaka A, Prusinkiewicz P, et al. (Eds.). 2004. *Proceedings of the 4th International Workshop on Functional – Structural Plant Models*. UMR AMAP, Montpellier, France.

Heredia A and Andres A. 2008. Mathematical equations to predict mass fluxes and compositional changes during osmotic dehydration of cherry tomato halves. *Dryland Technology* **26**: 873–883.

Heuvelink E. 1996. Dry matter partitioning in tomato: validation of a dynamic simulation model. *Annals of Botany* **77**: 71–80.

Heuvelink E. 1999. Evaluation of a dynamic simulation model for tomato crop growth and development. *Annals of Botany* **83**: 413–422.

Jones J W, Dayan E, Allen L H, van Keulen H and Challa H. 1991. A dynamic tomato growth and yield model (TOMGRO). *Transactions of American Society for Agricultural and Biological Engineers* **34**: 663–672.

Jones J W, Kenig A and Vallejos C E. 1999. Reduced state-variable tomato growth model. *Transactions of American Society for Agricultural and Biological Engineers* **42**: 255–265.

Kenig A and Jones J W. 1997. TOMGRO v3.0 A dynamic model of tomato growth and yield. Ch. II-5. *Optimal Environmental Control for Indeterminate Greenhouse Crops* (Eds.) Seginer I, Jones J W, Gutman P and Vallejos C E. BARD Research Report No. IS-1995–91RC. Agricultural Engineering Dept., Technion, Haifa, Israel.

Körner C. 1991. Some often-overlooked plant characteristics as determinants of plant growth – a reconsideration. *Functional Ecology* **5**: 162–173.

Körner C. 2003. *Alpine Plant Life*. 2nd edition. Springer, Heidelberg, Germany.

Körner C. 2006. Plant CO_2 responses: an issue of definition, time and resource supply. *New Phytologist* **172**: 393–411.

Kuretz V K, Drozdov S N, Popov E G, Dembo E D, Khilkov N I and Trofimova S A. 2003. Effect and aftereffect of respiration of intact plants. *Russian Journal of Plant Physiology* **50**: 308–312.

Lacointe A. 2000. Carbon allocation among tree organs: a review of basic processes and representation in functional – structural tree models. *Annals of Forest Science* **57**: 521–533.

Lambers H, Chapin F S I and Pons T L. 2008. *Plant Physiological Ecology*. 2nd edition. Springer, New York.

Lambers H, Shane M W, Cramer M D, Pearse S J and Veneklaas E J. 2006. Root structure and functioning for efficient acquisition of phosphorus: matching morphological and physiological traits. *Annals of Botany* **98**: 693–713.

Le Roux X, Lacointe A, Escobar-Gutiérrez A and Le Dizès S. 2001. Carbon-based models of individual tree growth: a critical appraisal. *Annals of Forest Science* **58**: 469–506.

Lentz W. 1998. Model applications in horticulture: a review. *Scientia Horticulturae* **74**: 151–174.

Lisson S N, Inman N G, Robertson M J and Keating B A. 2005. The historical and future contribution of crop physiology and modeling research to sugarcane production systems. *Field Crop Research* **92**: 321–335.

López I L, Rojano A and Ramírez A. 2005. Greenhouse horticultural mathematical models: transcending the dynamic behavior of crops. *Revista Chapingo Series Horticulture* **11**: 257–267.

Luquet D, Dingkuhn M, Kim H, Tambour L and Clement-Vidal A. 2006. EcoMeristem, a model of morphogenesis and competition among sinks in rice. 1. Concept, validation and sensitivity analysis. *Functional Plant Biology* **33**: 309–323.

Marcelis L F M. 1994. A simulation model for dry matter partitioning in cucumber. *Annals of Botany* **74**: 43–52.

Marks C O and Lechowicz M J. 2006. A holistic tree seedling model for the investigation of functional trait diversity. *Ecological Modelling* **193**: 141–181.

Mason E G and Dzierzon H. 2006. Application of modeling to vegetation management. *Canadian Journal of Forestry Research* **36**: 2505–2514.

Medina-Ruíz C A, Mercado-Luna I A, Soto-Zarazúa G M, Torres-Pacheco I and Rico-Garcia E. 2011. Mathematical modelling of tomato plant: a review. *African Journal of Agricultural Research* **6**: 6745–6749.

Messina C D, Jones J W, Boote K J, et al. 2006. A gene-based model to simulate soybean development and yield responses to environment. *Crop Science* **46**: 456–466.

Mirschel W, Schultz A, Wenkel K O, Wieland R and Poluektov R A. 2004. Crop growth modeling on different spatial scales: a wide spectrum of approaches. *Archives of Agronomy and Soil Science* **50**: 329–343.

Murthy Radha Krishna V. 2003. Crop growth modelling and its application in agricultural meteorology. *Satellite Remote Sensing and GIS Applications in Agricultural Meteorology*: 235–261.

Perttunen J, Sievanen R and Nikinmaa E. 1998. LIGNUM: a model combining the structure and the functioning of trees. *Ecological Modelling* **108**: 189–198.

Pretzsch H, Grote R, Reineking B, Rötzer T and Seifert S. 2008. Models for forest ecosystem management: a European perspective. *Annals of Botany* **101**: 1065–1087.

Pronk A A, Heuvelink H and Challa H. 2003. Dry mass production and leaf area development of field-grown ornamental conifers: measurements and simulation. *Agricultural Systems* **78**: 337–353.

Prusinkiewicz P. 2004. Modeling plant growth development. *Current Opinion in Plant Biology* **7**: 79–83.

Prusinkiewicz P and Rolland-Lagan A G. 2006. Modeling plant morphogenesis. *Current Opinion in Plant Biology* **9**: 83–88.

Ramírez-Arias J A. 2005. *Hierarchical Multiobjective Control for Greenhouse Crops*. PhD Thesis. Universidad de Almería, España.

Renton M, Hanan J and Burrage K. 2005. Using the canonical modelling approach to simplify the simulation of function in functional – structural plant models. *New Phytologist* **166**: 845–857.

Rey H, Dauzat J, Chenu K, Barczi J F, Dosio G A and Lecoeur J. 2008. Using a 3-D virtual sunflower to simulate light capture at organ, plant and plot levels: contribution of organ interception, impact of heliotropism and analysis of genotypic differences. *Annals of Botany* **101**: 1139–1151.

Reymond M, Muller B, Leonardi A, et al. 2003. Combining quantitative trait loci analysis and an ecophysiological model to analyze the genetic variability of the responses of maize leaf growth to temperature and water deficit. *Plant Physiology* **131**: 664–675.

Rico-García E. 2011. Mathematical modelling on tomato plants: a review. *African Journal of Agricultural Research* **6**: 6745–6749.

Sentelhas P C and Gillespie T J. 2008. Estimating hourly net radiation for leaf wetness duration using the Penman-Monteith equation. *Theoretical and Applied Climatology* **91**: 205–215.

Soltani A, Khooie F R, Ghassemi-Golezani K and Moghaddam M. 2001. A simulation study of chickpea crop response to limited irrigation in a semiarid environment. *Agricultural Water Management* **49**: 225–237.

Soltani A, Robertson M J, Mohammad-Nejad Y and Rahemi-Karizaki A. 2006. Modeling chickpea growth and development: leaf production and senescence. *Field Crop Research* **99**: 14–23.

Sussex I M and Kerk N M. 2001. The evolution of plant architecture. *Current Opinion in Plant Biology* **4**: 33–37.

Tao F, Zhang Z, Liu J and Yokozawa M. 2009. Modelling the impacts of weather and climate variability on crop productivity over a large area: a new super-ensemble-based probabilistic projection. *Agriculture and Forestry Meteorology* **149**: 1266–1278.

Tap F. 2000. *Economics-Based Optimal Control of Greenhouse Tomato Crop Production*. PhD Thesis. Wageningen Agricultural University, Wageningen, The Netherlands.

Thornley J H M and Cannell M G R. 2000. Modeling the components of plant respiration: representation and realism. *Annals of Botany* **85**: 55–67.

van Ittersum M K, Leffelaar P A, Van Keulen H, et al. 2003b. On approaches and applications of the Wageningen crop models. *European Journal of Agronomy* **18**: 201–234.

van Ittersum M K, Leffelar P A, Van Keulen H, Kropff M J, Bastiaans L and Goudriaan J. 2003a. Developments in modeling crop growth, cropping systems in the Wageningen School. *NJAS-Wagen. Journal of Life Sciences* **50**: 239–247.

Vazquez-Cruz M A, Torres-Pacheco I, Miranda-ópez R, Cornejo-Pérez O, Osornio-Rios A, Romero-Troncoso R and Guevara-González R. 2010. Potential of mathematical modeling in fruit quality. *African Journal of Biotechnology* **9**: 260–267.

Vos J, Marcelis L F M, de Visser P H B, Struick P C and Evers J B. (Eds.). 2006. *Proceedings of the Frontis Workshop on Functional – Structural Plant Modelling in Crop Production*. University and Research Centre, Wageningen, The Netherlands.

Xu H, Gossett N and Chen B. 2007. Knowledge and heuristic-based modeling of laser-scanned trees. *ACM Transactions on Graphics* **26**: 19.

Yang Z J and Midmore D J. 2005. Modelling plant resource allocation and growth partitioning in response to environmental heterogeneity. *Ecological Modelling* **181**: 59–77.

Yin X and Van Laar H H. 2005. *Crop Systems Dynamics*. Wageningen Academic Publishers, The Netherlands.

Yin X Y, Kropff M J, Goudriaan J, et al. 2000. A model analysis of yield differences among recombinant inbred lines in barley. *Agronomy Journal* **92**: 114–120.

Part IV

Response of Plants to Environmental Stimuli

Beginning from Darwin's "The power of movements in plants" theory, using the model plant systems, scientists have made tremendous efforts to shed light on the identity of key signaling events associated with phototropism. These efforts have greatly expanded our understanding of this phenomenon including photoreceptor's identity and their possible role in perceiving and transporting signals, role of auxin as the major growth regulator in eliciting phototropic responses, possible involvement of other phytohormones, and identification of signaling components and mechanisms. Genetic analyses have indicated the involvement of multiple auxin transporter proteins in addition to a number of important regulatory components and factors. However, there is lot more to be done to achieve the goal of fully elucidating all of the molecular components, from the reception of light to the expression of phototropic response. There have also been significant efforts made by several research groups to gain a better understanding of the molecular mechanisms involved in photoperiodic changes occurring within the leaf and transfer of the signal that induces changes in the shoot apex. The current knowledge suggests that plants monitor photoperiods in the leaf, and transmit that information coded within an elusive signal dubbed florigen to the apex to effect changes the developmental program. Recent advances indicate that the core of the day-length measurement mechanism lies in the circadian regulation of CONSTANS (CO) expression and the subsequent photoperiodic induction of the expression of FLOWERING LOCUS T (FT) gene, which might encode a major component of florigen.

Plants have a highly complex mechanism by which they distinguish a few days or weeks of cold treatment and flower after several weeks of that cold exposure even under nonconductive conditions. Research has shown that the cold measurement is related to the VIN3 in Arabidopsis, which is only expressed after several weeks of cold exposure. Furthermore, VIN3 is only expressed in the cold and becomes undetectable after plants are returned to warm conditions. It is believed that during prolonged cold exposure, the levels of certain factor(s) slowly decline or increase until a threshold is reached. This allows the plants to acquire competence to flower. Currently, it is not known how plants measure the duration of cold and what is the link between increase or decrease of certain factors and acquisition of competence. Therefore, the challenge is to understand the nature of the measurement of cold duration at a molecular level.

DOI: 10.1201/9781003373889-13

Woody plants are pruned to maintain a desired size and shape and to promote a certain type of growth. Ornamental plants are pruned to improve the aesthetic quality of the plant, but fruit trees are pruned to improve fruit quality by encouraging proper balance between vegetative (wood) and reproductive (fruiting) growth. Annual pruning of fruit trees always reduces yield but enhances fruit quality. Pruning improves light penetration into the canopy, and light is required for flower-bud development, fruit set and growth, and red color development. Pruning also makes the canopy more open and improves pest control by allowing better spray penetration into the tree; air movement throughout the canopy is increased, which improves drying conditions and reduces severity of many diseases. It is important to understand the role of source–sink relationships in promoting growth and crop productivity. For this, it is important to consider the nature and extent of control exerted by sources and sinks on various components of plant growth and also the exact mechanisms involved in these relationships. An integrated view of the interactions between sinks and sources is needed at the whole-plant scale to identify factors and/or physiological processes that limit plant growth and development.

10 Photoperception and Signaling of Tropic and Nontropic Responses Environment Signals

10.1 HISTORICAL PERSPECTIVE

Plants are sensitive to different environmental cues. One of the most remarkable among them is plant's response to the direction of the incoming light, which is now known as phototropism. The earliest proof for the existence of phototropism was provided by Giambattista della Porta who described plant phototropism as a "rejoicing" response of cucumber plants to the sun (della Porta, 1569). This was followed by the observation of Bacon et al. (1627), who considered phototropism as simple mechanical consequences of wilting in plants. On the other hand, Thomas Browne suggested that plants move to get away from bad effects of neighboring plants (Browne, 1658). Robert Sharrock proved that plants' movement was not a simple mechanical consequence of wilting but rather is triggered by growth (Sharrock, 1672). Later on, Ray et al. (1686) considered phototropism to be a consequence of temperature gradient with the side closer to the window being cooler than the side that is away from the window (Ray et al., 1686). However, Henri Louis DuHamel suggested that it was the light and not the temperature that was causing the bending response (mentioned in Whippo and Hangarter, 2006). Frank was the first to recognize that phototropism and gravitropism are driven by a common process (Whippo and Hangarter, 2006). Darwin (1880) conducted a series of experiments to show how plants respond to external stimuli and to gain understanding of some general principles governing their growth and life. The theory of circumnutation (circular or elliptical movements of stem and tips of plant) proposed by him and his son, Frank, to explain such movements in plants provided the basis for the study on the tropic responses of plants. In his book, *The Power of Movements in Plants*, Darwin hypothesized that plants could grow differentially and directionally in response to external stimuli, such as light and gravity. He also demonstrated that the part of the plant that perceives the stimuli is different from the part that responds to that stimulus. He further reported that the "influence" moves from the site of the stimulus perception to the site of response, where bending actually occurs (Holland et al., 2009). Darwin reported that the site of photoreception (shoot tip) is different from the location of curvature and the substance produced in the shoot tip is responsible for inducing curvature in the lower region. Further proof for the existence of Darwin's "influence" came from the experiments conducted by Boysen-Jensen (1911), who suggested that the "influence" moves from the tip of the plant toward the base in the unlit side of the plant, and that the directional and differential movement of the "influence" is critical for the bending response (Holland et al., 2009). Paal (1918) also provided proof for the involvement of the substance produced in the tip of plant. These findings led Cholodny (1927) and Went (1926, 1928) to independently suggest that light-mediated lateral distribution of a plant growth hormone in the shaded side of the seedling is responsible for differential growth in the phototropic curvature. According to the Choldony–Went theory, an asymmetric accumulation of auxin caused by the tropic stimulus is responsible for the differential growth response leading to curvature of shoot tips. Kogl and Haagen-Smit (1931) named this substance as auxin (derived from the Greek word "auxein" meaning to grow). However, according to Shen-Miller, Gordon and Shen-Miller, the phototropism

DOI: 10.1201/9781003373889-14

is caused by light-mediated inhibition of polar transport of auxin on the illuminated side of the seedlings (Shen-Miller and Gordon, 1966; Shen-Miller et al., 1969). Naqvi (1972) proposed that the production of abscisic acid in the illuminated portion of seedlings induced unequal transport of auxin and caused the curvature on the side away from light. Additional evidences for differential auxin transport are emerging from the studies on model system using *Arabidopsis thaliana*. Although these studies conclusively proven the involvement of auxin in phototropism, the exact mechanism of transport and signaling of auxin is still not clear (Whippo and Hangarter, 2006).

Once it was clear that phototropism was caused by light, the attention was focused on another important area related to the perception of a unilateral light source by the illuminated seedlings. Von Wiesner (1878) was the first scientist to confirm that blue/violet light was most effective in inducing phototropism. This was later corroborated by Blaauw (1909) and Johnston (1934), who demonstrated that wavelength around 480 nm and 440 nm is involved in phototropism. Because the phototropism action spectra coincided with the absorption spectra of carotenoids, and carotenoid's concentration was the highest in the phototropism sensitive area of the seedling, carotenoids were originally thought to be the phototropism photoreceptors. A series of studies conducted by Briggs and coworkers eventually identified a membrane-bound protein (120-kD), whose phosphorylation and activation is influenced by blue light in a fashion that resembled that of the phototropic response (Gallanger et al., 1988; Short and Briggs, 1990; Reymond et al., 1992a, 1992b; Short et al., 1992). They further demonstrated that the microsomes in phototropism mutant JK224 contained low levels of this protein. Subsequently, these researchers identified two domains of this 120-kD protein, LOV 1 (light, oxygen and voltage) and LOV 2, which bind to flavin chromophores with spectral properties resembling that of phototropism (Huala et al., 1997). This protein was later renamed as phototropin (Christie et al., 1999). The studies then focused on understanding the mechanism of light reception.

The current understanding of phototropism suggests that the incoming blue light is perceived by membrane-associated photoreceptors (phototropins) and triggers the formation of a lateral auxin gradient. Once the auxin gradient is established, signaling networks is triggered to control asymmetrical cell elongation and cell growth leading to bending of the hypocotyl toward a light source. Now several excellent, in-depth reviews covering various aspects of phototropism research are available in the literature (Briggs, 2006, 2014; Whippo and Hangarter, 2006; Holland et al., 2009; Sakai and Haga, 2012).

10.2 LIGHT AS A STIMULUS

10.2.1 Photoreception and Signaling

Plants cope with the changes in their environment by integrating the signals of light quality and quantity, modifying overall growth characteristics from seed germination to reproduction (Kami et al., 2010; Pedmale and Liscum, 2007; Pedmale et al., 2010; Chen and Chory, 2011). Although many photoreceptors, signaling pathways and processes they regulate have been identified, the components associated with the induction, modulation and establishment of phototropic responses in higher plants have not been fully understood (Christie and Murphy, 2013). Plant's ability to reorient organs toward (positive phototropism) or away from (negative phototropism) a unidirectional light source has attracted the attention of several researchers for nearly two centuries (Holland et al., 2009). According to Cholodny and Went (Went and Thimann, 1937), tropisms result from the lateral redistribution of auxin in response to stimuli. In the case of stem phototropism, auxins are redistributed from side closest to the incident light to the shaded side leading to differential rate of cell elongation in the shaded and lit portion of the organ and, thus, curve toward the light source (Went and Thimann, 1937; Esmon et al., 2006; Holland et al., 2009). Despite over 100 years of concerted efforts by several research groups, photoreceptors responsible for sensing directional light remained elusive until when Briggs identified the phototropins (Short and Briggs, 1990; Reymond et al., 1992a; Liscum and Briggs, 1995; Huala et al., 1997; Christie et al., 1998). Like the work of Cholodny and Went, the findings of Briggs and his colleagues represented classical research on phototropism.

Arabidopsis thaliana is used as a model system for understanding light perception and the signaling components involved in phototropism (Liscum et al., 2014; Sakai and Haga, 2012). Both root and hypocotyl phototropisms in *Arabidopsis* are regulated by phototropin receptors (Kutschera and Briggs, 2012). Since their discovery, considerable progress has been made in understanding how these photoreceptors are activated by blue light. According to Christie and coworkers, all flowering plants contain two phototropins (phot1 and phot2) (Christie et al., 2014). Phot1 functions primarily as the photoreceptor for root phototropism (Kutschera and Briggs, 2012) and hypocotyl phototropism over a broad range of blue light intensities (Christie et al., 2014). Owing largely to light-mediated increases in protein abundance, the action of phot2 in hypocotyl phototropism is limited to high light intensities (Christie et al., 2014; Sakai et al., 2001). Arabidopsis phototropins also elicit hypocotyl curvature to UV-B wavelengths (Vandenbussche and Van Der Straeten, 2014). Intriguingly, phototropin-deficient seedlings still exhibit phototropism toward UV-B (Vandenbussche and Van Der Straeten, 2014). Vandenbussche and Van Der Straeten (2014) reported that the phototropin-independent response is mediated by the newly identified UV-B photoreceptor, UV RESISTANCE LOCUS 8 (UVR8). The growth mechanism underlying UVR8-dependent hypocotyl phototropism may be distinct from phototropin-mediated events triggering hypocotyl curvature toward blue light (Vandenbussche and Straeten, 2014). Phototropic responses in the gametophytes of the Japanese maidenhair fern, *Adiantium capillis-veneris*, are regulated by both red and blue light (Kawai et al., 2003). A dual red-/blue-light-absorbing phytochrome–phototropin receptor chimera designated "neochrome" is responsible for this response in *Adiantum* fern (and other *polypodiaceous* ferns) and serves to provide enhanced light sensitivity in shaded habitats (Kawai et al., 2003). Recently, Li and coworkers reported that ferns have acquired neochrome from nonvascular hornworts by horizontal gene transfer (Li et al., 2014). While mosses lack neochrome, phototropic response to red and blue light in *Physcomitrella patens* is regulated by direct interactions between phytochrome and phototropin photoreceptors (Jaedicke et al., 2012). Hypocotyl phototropism in angiosperms is not induced by red light; instead, sensing of red light by the phytochromes is known to enhance this response (Goyal et al., 2013). Parks and coworkers found that the curvature enhancement is best when red light is given 1–2 h prior to the directional blue light treatment (Janoudi et al., 1992) and is dependent on phytochrome A (phyA) (Parks et al., 1996). Mutants lacking phyB, phyC, phyD and phyE appear to exhibit normal phototropic response, suggesting that phyA is the primary phytochrome modulating phototropism in Arabidopsis (Strasser et al., 2010). This response has been attributed to the attenuation of phot1 internalization from the plasma membrane. Yet this enhancement response is still apparent in seedlings, where phot1 is constitutively anchored to the plasma membrane (Preuten et al., 2015). Alternatively, phytochrome-regulated expression of key signaling components associated with auxin transport is likely to be involved (Goyal et al., 2013; Haga et al., 2014). Blue light sensing by the cryptochromes is also reported to impact phototropism in Arabidopsis. Hypocotyl curvature is reduced in the cry1cry2 double mutant, implying that cryptochromes either indirectly control the process by modulating blue light-regulated growth or through more direct ways, which is still remain to be fully understood (Goyal et al., 2013; Whippo and Hangarter, 2003).

Phototropins are plasma membrane-bound proteins comprising an amino-terminal light-sensing domain and a carboxy-terminal serine/threonine kinase domain (Liscum et al., 2014). The amino-terminal region contains two specialized LOV1 and LOV2, which function as binding sites for the blue-light-absorbing chromophore flavin mononucleotide (FMN). In the absence of light, FMN is bound non-covalently within the LOV domain, absorbing strongly in the blue region (l max – 450 nm). Irradiation of the LOV domain results in the formation of a covalent adduct between the C (4a) carbon of the FMN chromophore and the sulfur atom of a nearby, conserved cysteine residue (Christie et al., 2014). Adduct formation occurs rapidly (within microseconds), absorbing maximally at 390 nm. This photochemical reaction is fully reversible in darkness and can be completely abolished by mutation of cysteine (Christie et al., 2014). Although both LOV domains can function as light sensors, phototropin kinase activity is regulated primarily by LOV2 (Christie et al., 2014; Han et al., 2013). In darkness, LOV2, together with adjacent peptide sequences, act as repressor of phototropin kinase activity. Photoactivation of LOV2 leads to protein conformational changes that relieve this repression activity, resulting in receptor autophosphorylation and an initiation of phototropic signaling (Christie et al., 2014). A detailed description of the photochemical and biochemical properties of phototropin blue light receptors can be found in the review article by Christie et al. (2014).

In Arabidopsis, several researchers have shown that phototropin-mediated curvature results from the increased elongation at the shaded side combined with a decrease in elongation at the irradiated side (Orbovic and Poff, 1992; Preuten et al., 2013; Yamamoto et al., 2014; Christie et al., 2011). The importance of auxin in perception, transport and signaling has been studied extensively by Sakai and Haga (2012). Inoue and coworkers have demonstrated the importance of protein kinase activity and phosphorylation within the activation loop of the phototropin kinase domain (Inoue et al., 2008, 2011). According to these authors, phosphorylation of signaling components leads to the formation of an auxin gradient. This regulated auxin transport is a likely mechanism controlling phototropism (Rakusova et al., 2014). Young leaves (and cotyledons) are considered a major source of auxin, which is then transported down toward the lower parts of the plant, such as hypocotyls and roots (Tao et al., 2008; Spalding, 2013). Sakai and Haga (2012) and Christie and coworkers (Christie et al., 2011) have shown the involvement of two main families of transporters in auxin transport: long-PIN-FORMED efflux carriers (PIN1–4, PIN7 in *Arabidopsis*) and several multi-drug resistance transporters from the ATP-binding cassette B (ABCB) class, most notably ABCB19. Studies by Ding and coworkers have shown that PIN3 is laterally redistributed in the endodermal cells of etiolated seedlings following phototropic stimulation (Ding et al., 2011). Furthermore, it was shown that PIN activity is regulated by phosphorylation by two protein kinase families from the AGCVIII class: PINOID (PID) and D6 PROTEIN KINASES (D6PKs) (Zourelidou et al., 2014; Barbosa and Schwechheimer, 2014). Although phototropins also belong to the same AGCVIII protein kinase family (Barbosa and Schwechheimer, 2014), there is currently no evidence to show the direct phosphorylation of PINs by the phototropins (Ding et al., 2011). Willige et al. (2013) also reported light-regulated PIN phosphorylation to represent an important regulatory mechanism. However, this idea remains conjecture at this time and how light-activated phototropin kinase activity would lead to such an event remains unknown. Furthermore, the importance of PID and members of this group of kinases in the regulation of phototropism appears to be restricted to rather specific conditions. A more direct regulation of auxin efflux by phot1 has been proposed to inhibit polar auxin transport and indirectly promote a lateral flux of auxin, favoring phototropism in the upper region of the hypocotyl (Christie et al., 2011). Phot1 also directly phosphorylates PHYTOCHROME KINASE SUBSTRATE 4 (PKS4), a post-translational modification that was suggested to inhibit the activity of this positive regulator of phototropism (Demarsy et al., 2012). All these studies suggest that phototropin activation leads to auxin gradient formation and microtubule reorganization both of which are required for regulated anisotropic growth (Christie et al., 2011). However, this hypothesis has been questioned in the past and requires further evaluation (Gao et al., 2015; Shibaoka, 1994).

10.3 PHOTOTROPINS

In higher plants, there are two phototropins (phot1 and phot2) (Huala et al., 1997; Jarillo et al., 2001; Suetsugu and Wada, 2013). Under low-intensity blue light (BL), phot1 has been shown to be the primary receptor controlling phototropism, whereas phot1 and phot2 act redundantly under moderate to high BL conditions (Liscum and Briggs, 1995; Sakai et al., 2000, 2001). In addition to their roles in phototropism, the phots mediate and/or regulate a number of other BL responses, such as stomatal opening and photosynthetic gas exchange (Kinoshita et al., 2001), cotyledon and leaf blade expansion, flattening and positioning in response to BL (Sakamoto and Briggs, 2002; Ohgishi et al., 2004; Takemiya et al., 2005; Inoue et al., 2008; Han et al., 2013), and stimulation of chloroplast positioning under low light conditions to optimize absorption (Sakai et al., 2001). Under high-intensity light, phot2 activation causes chloroplasts to move away from the irradiated edge of the cell to avoid photodamage (Jarillo et al., 2001; Kagawa et al., 2001; Sakai et al., 2001).

10.4 PHOTOTROPIC CONTROL OF DEVELOPMENTAL PROCESSES

10.4.1 Seed Germination

Seeds are able to remain dormant in a dry state for extended periods, germinating only when favorable conditions occur (Bentsink and Koornneef, 2002). Borthwick and coworkers were the first to show that red light stimulates germination in lettuce seeds, and that this induction can be inhibited by subsequent exposure to far-red light. They also showed that lettuce seeds can be sequentially exposed to red and far-red light with the germination response being determined by the final light treatment. These observations led to the purification of the photoreceptor responsible, later termed phytochrome. In Arabidopsis, germination of dormant seeds is controlled by light, temperature and time of storage in the dried state (Koornneef and Karssen, 1994). Light-dependent germination of *Arabidopsis* seed is mediated entirely by phytochrome (Casal and Sánchez, 1998). Analysis of mutants that lack phyA (*phyA*) or phyB (*phyB*) has shown the importance of these phytochromes in the regulation of germination. In *Arabidopsis*, red/far-red reversible (LFR) germination is largely mediated through phyB. Germination is also induced by low quantities of red or far-red light (VLFR) or continuous far-red light (far-red HIR) all mediated through phyA (Casal and Sánchez, 1998). In recent experiments using *phyC*, *phyD* and *phyE* mutants, Hennig and coworkers have demonstrated the role of *phyE* in Arabidopsis seed germination (Hennig et al., 2002).

10.4.2 Seedling Establishment

Following seed germination, light signals act to direct and inhibit hypocotyl extension, while promoting the opening and expansion of cotyledons. The concomitant synthesis of chlorophyll, chloroplast development and opening of stomata enable plants to initiate photosynthetic activity and become photoautotrophic. Transfer of dark-grown seedlings to white light results in an inhibition of hypocotyl growth and the opening of cotyledons, a process termed "de-etiolation." The analysis of mutants deficient in multiple photoreceptors has enabled the assessment of response to different wavelengths of light.

10.4.3 Leaf Development

While under low-light conditions, chloroplasts accumulate on the upper surface of the palisade mesophyll cells to maximize photosynthesis, under strong-light conditions, chloroplasts line up perpendicular to the direction of the light source to minimize photo damage caused by excess light (Briggs and Christie, 2002; Kasahara et al., 2002). Kagawa and Wada (2002) showed that treatment with blue light is sufficient to bring about chloroplast relocation. It was initially thought that Phot2 was the major photoreceptor for chloroplast relocation (Briggs and Christie, 2002). However, subsequent analysis showed that *photl/phot2* double mutants show a much more severe phenotype than single mutants alone (Sakai et al., 2001; Kasahara et al., 2002). Like chloroplast relocation, *photl/phot2* regulates blue-light-induced opening of stomata. Both phototropins show redundancy in regulating blue-light-mediated chloroplast relocation and stomatal opening, with *Photl* and *Phot2* showing some differences in their sensitivity to light intensity.

10.4.4 R:FR Ratio and Shade Avoidance

When grown in close proximity to one another, constraints on photosynthetic productivity can lead to increased competition between individuals for this key resource. Plants have therefore evolved two principle adaptive strategies (shade tolerance and shade avoidance) to enhance their survival in such conditions. Successful shade avoidance requires plants to detect the impending threat of shade and initiate avoidance responses before actual shading occurs (Ballaré, 1999). Plants perceive the presence of neighboring vegetation as a reduction in the ratio of red to far-red wavelengths in the light reflected from, or transmitted through green tissues. Because photosynthetic pigments, chlorophylls and carotenoids absorb light in the visible spectrum radiation reflected from, or transmitted through plant canopy will have relatively higher proportion of far-red wavelengths (Smith, 1982). Changes in R:FR ratio are detected by plants as a change in the relative proportions of Pr and Pfr. The reduction in R:FR leads to a marked acceleration of stem and petiole elongation, often at the expense of leaf and storage organ development. Other effects of relatively higher FR include reduction in leaf area, thinner leaves in dicot plants, reduced angle from the stem and increased apical dominance (McLaren and Smith, 1978: Smith and Whitelam, 1997). Additionally, reduced branching in dicots and reduced tillering in grasses was observed at higher FR levels (Casal et al., 1986).

10.4.5 Photoperiodic Regulation of Flowering

In plants, the day length is an important signal for initiation of flowering. It is thought that flowering is induced in long-day species or repressed in short-day species, when a light signal coincides with a sensitive phase of the circadian clock (Roden et al., 2002). *Arabidopsis*, a facultative long-day plant, flowers earlier under long days but will eventually flower even under short days. In fact, under laboratory conditions, *Arabidopsis* will flower after just a single long-day exposure (Mouradov et al., 2002). *phyA* and *cry2* having only small effects on circadian clock entrainment, mutations in both these genes have marked effects on the photoperiodic control of flowering (Johnson et al., 1994; Guo et al., 1998). Therefore, phyA and cry2 do not affect flowering-time responses by influencing clock entrainment but by modulating other aspects of the flowering response (Samach and Coupland, 2000).

10.5 ROLE OF LIGHT AS AN EXTERNAL FACTOR

10.5.1 Photoperiodic Control of Growth and Development

During their life cycle, the flowering plants pass through distinct developmental phases starting with germination and ending with seed formation. The transition between these phases is dependent on genetic programs that are triggered and often regulated by both environmental and endogenous stimuli (He et al., 2004). Between the stage of germination and the time when they become competent to flower and reproduce, the shoots pass through a phase of vegetative growth, which is characterized by rapid increase in their photosynthetic capacity, size and mass. This stage can be further divided into a juvenile and an adult vegetative phase. The plants in the juvenile phase are insensitive to environmental inductive signals. This is followed by an adult vegetative phase where plants are receptive to external inductive cues for flowering. This shift from adult vegetative stage to reproductive stage is marked by a transition to flowering (Bäurle and Dean, 2006; McDonald and Kwong, 2005). Flowering commences from "floral induction signal" which induces "floral evocation" followed by "floral initiation," "flower development" and eventually "anthesis" stages (Hopkins and Huner, 2004; McDonald and Kwong, 2005). Flowering, which represents an important adaptation in the life cycle of the plant to seasonal changes in the environment, determines reproductive success (Bäurle and Dean, 2006; Kim et al., 2009).

Based on the inductive stimuli, flowering follows one of the four pathways, namely, photoperiod, autonomous, vernalization and gibberellin-induced pathways (Bäurle and Dean, 2006; Lee and Lee,

2010; Sung and Amasino, 2004). A common set of genes have been reported to define the signaling pathways in flowering (Putterill et al., 2004). Photoperiod and vernalization and/or their interactions are reported as main external factors influencing flowering responses and behaviors (Jung and Müller, 2009; Kim et al., 2009; Mouradov et al., 2002; Searle and Coupland, 2004). Manipulation of growing environments along with the use of new technologies (e.g., genetic modifications, hybridization, grafting techniques and plant growth regulators) have been successfully used to develop effective production guidelines and novel varieties for commercial production in many horticultural crops (Wilkie et al., 2008).

The mechanism by which plants monitor environmental cues and trigger the initiation of flowering at the right time has been a subject of tremendous interest for several decades. The ability to sense and respond to changes in day length is known as photoperiodism and is a widespread phenomenon in both plants and animals. Numerous plant species flower during the time of the year when the days are longer than nights and for several years scientists believed that the accumulation of photosynthetic products synthesized during long days was responsible for flowering. The importance of day length in controlling seasonal responses was originally proposed by Tournois and Klebs in the early 1900s. According to these researchers, the duration rather than the quantity of light is a major factor controlling the plant development (Tournois, 1912, 1914; Klebs, 1913). Later in the 1920s, Garner and Allard (1920) found that a mutant variety of tobacco (Maryland Mammoth) continued to grow vegetatively to about 5 m in height without flowering under the prevailing conditions of summer but readily flowered in the greenhouse under natural light conditions during the winter. Based on these observations, Garner and Allard by covering plants grown during the long days found that the artificial short days forced the plants to produce flower. Therefore, Garner and Allard concluded that the length of the day was the determining factor in flowering and subsequently, confirmed this hypothesis in several other species and conditions. Thus, their studies laid the foundations for the subsequent research on photoperiodic responses in several plants.

10.6 PLANT'S RESPONSE TO THE LENGTH OF THE DAY

Based on photoperiodic response, plants are classified into three groups: (i) long-day plants (LDP or qualitative LDP) in which flowering occurs under long days or is promoted by daily light periods longer than a critical day length in a 24-h cycle; (ii) short-day plants (SDP or qualitative SDP), in which flowering is produced under short days or it is accelerated by day lengths shorter than the critical threshold; and (iii) day-neutral plant (DNP), which are not influenced by day lengths. Flowering in DNPs is typically under autonomous regulation, which means that the internal developmental factors control the flower production.

Several models have been put forward to explain the basis of the photoperiodic responses in plants. One of the simple models proposed that the gradual accumulation of a substance is required to trigger a transition to reproductive phase. Increasing the length of darkness would either promote (in SD plants) or inhibit (in LD plants) the accumulation of such a product required for flower induction. According to this model (called as hourglass model), once the threshold day length is attained, further increments would not have further influence on flowering. However, there is mounting evidence for the involvement of some sort of circadian system (circadian clock) in plants that measures the changes in day length. The floral response fluctuates rhythmically under these conditions with maximum response achieved every time the total length is 24 h or multiples of it, whereas the response will be much lower at intermediate cycle lengths (lengths less than 24 h) (Thomas and Vince-Prue, 1997). The presence or absence of light at specific times of the day rather than the duration of the periods of light or darkness has greater influence of flower induction in plants. The effect of night breaks on flowering explains this phenomenon. These observations strongly support that the circadian clock generates a rhythm with a period close to 24 h and is responsive to light only at a particular phase of the cycle.

The LDPs can effectively measure the increase in the day length in spring or early summer and delay flowering until the critical day length (CDL) is reached. Many temperate crops behave this way. In contrast, SDPs flower in fall or winter when the days shorten below the critical day length (e.g., Chrysanthemum). However, plants exhibit several adaptations for avoiding the ambiguity of day-length signal. One is the coupling of a temperature requirement to a photoperiodic response. Certain plant species do not respond to photoperiod until after a cold period (vernalization or overwintering) has occurred. Other plants avoid seasonal ambiguity by distinguishing between shortening and lengthening days. For example, long-short-day plants (LSDPs) flower only after a sequence of long days followed by short days (e.g., *Bryophyllum, Kalanchoe* and *Cestrum nocturnum*). They flower either in the late summer or fall, when the days are shortening. Similarly, short-long-day plants (SLDPs) flower only after a sequence of short days followed by long days (e.g., *Campanula medium* and *Echeveria harmsii*). These plants flower naturally in early spring in response to lengthening days.

A common misconception is that SDPs only flower in short photoperiods and LDPs only in long photoperiods; in fact, some SDPs (*Xanthium strumarium*) have long critical day length (CDL) as much as 15.5 h in a 24 h cycle and will essentially be able to flower under LDs of 15 h. Similarly, there are some LDPs (*Lolium perenne* and *Lolium temulentum*) with low CDLs and are able to flower in short days (SDs) of 9 h (Thomas and Vince-Prue, 1997). Plants in which flowering can only occur in the inducing photoperiod have an obligate response, whereas plants in which flowering is promoted by LDs or SDs, but which can still flower in the other photoperiod, have been known to have a facultative response. As in some tropical species, the difference between inductive and noninductive photoperiods could be as short as 30 min, making it difficult to measure time very accurately (Borchert et al., 2005). This is important as small errors in measurement of the photoperiod can result in premature or delayed flowering. Plants are able to measure time by means of an endogenous time-keeping mechanism called the circadian clock. The CDL varies with age and the prevailing environmental conditions. For example, the CDL in *Hyoscyamus niger* becomes shorter with lower night temperatures. Similarly, CDL of seedlings of *Pharbitis nil* is shorter than that of adult plants (Thomas and Vince-Prue, 1997). Sometimes, the response to photoperiod is used in conjunction with responses to other environmental cues to create a particular developmental life cycle. For example, biennial plants will not flower in the first year of growth despite encountering inductive photoperiods because they require an exposure to cold temperature for a certain period of time (vernalization) to flower and set seed in the following spring or early summer. There are also several known cases where other environmental factors such as vernalization, high temperature, high irradiance or low nitrogen can substitute for photoperiodic induction of flowering, or where the response to photoperiod may be modulated or even repressed by other environmental factors (Bernier and Perilleux, 2005).

The molecular basis of the interactions between the photoperiodic pathway and other pathways that affect flowering is still not fully clear. It is possible that the genes controlling in one pathway can sometimes be regulated or modified by other pathways. Thus, one should always consider the photoperiodic pathway as a component of an interacting network of pathways regulating flowering rather than in isolation. While flowering is only one of many responses that plants have to photoperiod, other responses such as tuberization, bud break and the onset of cold hardiness or dormancy also rely on the basic photoperiodic detection mechanism and may have common features with the control of flowering.

The ability of plants to coordinate certain developmental processes to particular times of the year when environmental conditions are more favorable allows distinct advantages. Timing reproduction to spring allows sufficient time for vulnerable young offspring to fully develop to tolerate harsh conditions of winter. The photoperiod is the duration of light and darkness in a daily cycle of 24 h. It is a constant 12 h light and 12 h dark in the equator, but as you move away from the equator toward either of the earth's poles, because of the tilt of the earth's axis toward the sun, the length of the light and dark periods change to become unequal divisions of the 24 h cycle. As we get to the earth's poles, the photoperiods become either 24 h of light or

24 h darkness. The annual rotation of the earth around the sun also leads to changes throughout the year (except at the equator), with day lengths becoming longer in summer and shorter in winter. The length of the day reaches annual maximum for particular latitude on summer solstice (in the Northern Hemisphere it is around June 20/21), whereas it is the shortest on winter solstice (around December 21/22). Since the annual cycle of variation in photoperiod is consistent from year to year, it is considered as a more reliable indicator of the time of year than temperature fluctuations. Coordinated responses to changing photoperiodic observed in the plants allows synchronous flowering of plants of the same species to maximize cross pollination. It also enables an organism to anticipate variations in environmental conditions predicted to occur around the same time each year and prepare themselves in advance. For example, shortening day length in autumn is used as a cue by many perennial plants to induce cold hardiness and bud dormancy in anticipation of the freezing winter temperatures later on in the season. The ability of desert plants to respond to long photoperiods enables them to enter into a dormant state during the arid summer months and protect themselves from harsh weather conditions. Similarly, plants can use short-day signals to induce flowering in early spring and enable them to complete seed production (Thomas and Vince-Prue, 1997).

10.7 MEASUREMENT OF PHOTOPERIOD

Even though day length and night length are perfectly correlated in the 24 h cycle, plants have a mechanism to measure the length of either the night or the day. Hamner and Bonner (1938) showed that long nights induced flowering in SDP even if these were coupled with a long day, but short days did not induce flowering if the night was also short. Additionally, flowering can be prevented if the long nights were interrupted with a light pulse (a night break), supporting the importance of the dark period. On the other hand, LDPs are generally, less responsive to night break treatments (Thomas and Vince-Prue, 1997). LDPs require longer night interruptions. Far-red (FR) light at the end of the light period promote flowering in LDPs, but flowering is greatly reduced or prevented if plants are exposed to red light after FR (Thomas, 1998; Thomas and Vince-Prue, 1997). In contrast, the light quality given at different parts of the light period has little effect on flowering in SDPs, although some SDPs show a weak response to light quality during the day. For this reason, plants can be classified as either predominantly dark dominant (SDPs) or light dominant (LDPs).

Three basic models (external coincidence, internal coincidence and extended model) have been proposed to measure the length of the day or night. The external coincidence model is based on an internal rhythm that coincides with an external signal (light) under certain conditions and then induces flowering. This is one of the popular models used for measuring the length of the day or night (Bünning, 1936; Pittendrigh, 1966). The internal coincidence model is based on the coincidence of two internal rhythms (Pittendrigh, 1972). Both these models rely on a circadian clock in controlling the internal rhythms. In the external coincidence model, the phase of the rhythm that is sensitive to light (called the photoperiodic response rhythm) is set by light (Lumsden, 1998). This rhythm is constant from the end of the light period in most SDPs. Therefore, Lumsden (1998) postulated that rhythm is suspended in light if the light period is long enough, and then released at the onset of darkness. Based on the evidence on the persistence of rhythmicity during light in SDPs, Lumsden (1996) suggested an extended model in which the circadian clock is not completely arrested in the light but continues to oscillate around a light limit cycle. In this model, the cycle still has a period of 24 h, but will only occupy a limited part of possible phase states. The rhythm will be released at slightly different phases depending on the phase in the light limit cycle at dusk. However, LDPs are generally not responsive to short night breaks, but show a rhythm in the sensitivity to far-red light during the day. Based on this response of LDPs it is suggested that the photoperiodic response rhythm is reset at dawn but the rhythm continues to run in light (Thomas and Vince-Prue, 1997; Lumsden, 1998). This rhythm establishes a phase of sensitivity to far-red light.

10.8 PLANT'S COMPETENCE TO RESPOND TO PHOTOPERIOD

In plants, timing of reproduction, including flowering, tuberization and bulbing, is often controlled by photoperiod (Garner and Allard, 1920; Thomas and Vince-Prue, 1997). Additionally, the annual growth cycle in perennials, especially in temperate areas, is mainly controlled by photoperiod (Wareing, 1956; Dormling, 1973; Vince-Prue, 1975; Eriksson et al., 1978). Several decades of physiological studies on these photoperiodic responses have revealed striking similarities in these responses (Thomas and Vince-Prue, 1997). Although there are some common components in these mechanisms, significant differences have been reported in the literature. These differences are in the way the light signals are detected by the leaves, the entrainment of circadian rhythms, and the production and transmittal of a mobile signal throughout the plant. During the last 10–15 years, a significant progress has been made to understand the molecular mechanisms controlling photoperiodic induction of flowering, tuberization and growth cessation using model species, such as *Arabidopsis thaliana*, potato and aspen (Searle and Coupland, 2004; Imaizumi and Kay, 2006; Turck et al., 2008). Jackson (2009) and Lagercrantz (2009) provided an excellent review and described the present state of the knowledge on various mechanisms involved in plant's response to photoperiod.

The competence to respond to inductive photoperiods changes during plant development. It has been shown that floral induction does not occur until a certain developmental stage has been reached, thus ensuring that the plant has sufficient resources to sustain flowering and subsequent fruit production. Plants in the juvenile phase are unable to respond to an inductive stimulus that would be sufficient to induce flowering in an adult plant. The juvenile phase can be as short as a few days in herbaceous species or extend to several years as in woody tree species (Hackett, 1985). The length of the juvenile phase and thus, plant's ability to respond to inductive photoperiod is influenced by light integral, temperature, photoperiod and gibberellic acid (Hackett, 1985; Chien and Sussex, 1996; Telfer et al., 1997; Adams et al., 1999, 2001). In many plants, transition to adult phase is associated not only with the acquisition of the competence to flower but also with external changes such as alterations in leaf shape as in ivy (*Hedera helix*) and maize (*Zea mays*), and the development of abaxial trichomes as in adult *Arabidopsis* plants (Poethig, 1990; Bongard-Pierce et al., 1996; Telfer et al., 1997). Numerous mutants with altered juvenile phase lengths have been identified through the use of these phenotypical markers, such as *teopod* (*tp*) and *early phase change* (*epc*) mutants of maize, which have extended and shortened juvenile phases, respectively (Poethig, 1988; Dudley and Poethig, 1993; Vega et al., 2002). Interestingly, studies on the *teopod* mutants, where sectors of wild-type tissue were created in *tp1* and *tp2* mutants, indicate that the *TP1* and *TP2* genes affect juvenility non-cell-autonomously (Dudley and Poethig, 1993). While most mutants have been found to have an altered length of the juvenile phase, the rice *moril* mutant is unable to undergo the juvenile to adult phase transition at all and, as a result, did not flower even if grown in inducing SD photoperiods (Asai et al., 2002).

Several researchers have identified genes including *HASTY* (*HST*), *ZIPPY* (*ZIP*), *SERRATE* (*SE*) and *SQUINT* (*SQN*) that are responsible for determining the length of the juvenile phase in *Arabidopsis* (Clarke et al., 1999; Berardini et al., 2001; Bollman et al., 2003; Hunter et al., 2003). Mutations in all of these genes result in a shortened juvenile phase, indicating that the function of these genes is to maintain the length of the juvenile phase. Grigg et al. (2005) established the involvement of microRNAs (miRNAs) and trans-acting small interfering RNAs (ta-siRNAs) in controlling the length of the juvenile phase. They concluded that the *SE* is known to act in miRNA gene silencing pathway. Fahlgren et al. (2006) and Hunter et al. (2006) have shown that *ZIP* encodes an ARGONAUTE protein, which is required for the production and/or stability of ta-siRNAs. Similarly, Park et al. (2005) concluded that *HST* is involved in the synthesis or stability of some miRNAs. Furthermore, Peragine et al. (2004) and Xie et al. (2005) have demonstrated the role of other genes such as *SUPPRESSOR OF GENE SILENCING 3* (*SGS3*), *RNADEPENDENT POLYMERASE 6* (*RDR6*) and *DICER-LIKE 4* (*DCL4*) that are involved in gene silencing. The mutants for these genes were also found to have a shortened juvenile phase. Like *ZIP*, these three genes were found to be involved in the biosynthesis of ta-siRNAs and, as with *ZIP*, their mutant phenotypes are mainly related to the juvenile – adult phase change, rather than the highly pleiotropic phenotypes of miRNA biosynthesis mutants such as *hst* (Bollman et al., 2003). Based on these observations, Willmann and Poethig (2005) concluded that ta-siRNAs are likely to have a more restricted role in plant development than miRNAs. Since the ta-siRNAs target a gene that promotes the adult state and represses the juvenile state, any disruption in the biosynthesis of ta-siRNAs as observed in *zip*, *sgs3*, *rdr6* and *dcl4* mutants will prevent the degradation of transcripts of the target gene. This will lead to shortening of the juvenile phase and a more rapid transition to the adult state (Bäurle and Dean, 2006). The identities of the miRNAs involved in the production of the ta-siRNAs, and the identities of the ta-siRNAs and their target genes involved in the control of juvenility are currently being thoroughly investigated. MiR390 has been shown to be involved in the production of the TAS3 family of ta-siRNAs which target the mRNAs of several *AUXIN RESPONSE FACTOR* (*ARF*) genes, including *ARF3* (Fahlgren et al., 2006). It has been shown by regulating the transcript levels of the *ARF3* gene. TAS3 ta-siRNAs affect the length of the juvenile phase. Thus, *ARF3* is one of the target genes involved in the control of juvenility. Overexpression of miR156 has been shown to extend the juvenile phase, in the most part, through its downregulation of the SBP-box gene *SQUAMOSA PROMOTER BINDING PROTEIN-LIKE 3* (*SPL3*) (Wu and Poethig, 2006). It has recently been shown that *pny pnf* double mutants prevent the activation of *LFY* by FLOWERING LOCUS T (FT) (Kanrar et al., 2008). In plants that require a vernalization treatment, the competence of the shoot apex to respond to inducing signals (inductive photoperiods) is affected by vernalization through the regulation of a repressor of flowering (e.g., FLOWERING LOCUS C [FLC] in *Arabidopsis*). High levels of FLC repress *FT* expression and also prevent induction at the apex (Searle et al., 2006). *FLC* expression is regulated by the autonomous and vernalization flowering pathways and *FRIGIDA* (*FRI*) gene. The vernalization and autonomous pathways act to reduce the levels of *FLC* expression, thereby relieving the repression at the apex and increasing its competence to be induced by other pathways such as the photoperiodic pathway (Mouradov et al., 2002; Henderson et al., 2003; Bäurle and Dean, 2006).

10.9 PHOTOPERIODIC RESPONSE PATHWAY

The ability to respond to photoperiod requires a mechanism that will detect the changes in the day length. In facultative LDP (e.g., *Arabidopsis*), the flowering occurs earlier under LDs than in SDs. Studies conducted by a number of research groups have shown that the mechanism involves the interaction of light signals perceived by photoreceptors (e.g., phytochromes, cryptochromes and the blue light receptor F-box proteins ZEITLUPE [ZTL] and FLAVIN-BINDING KELCH REPEAT F-BOX 1 [FKF1]) with the components of the circadian clock, and CONSTANS (CO) gene and protein whose rhythmic expression is driven by the circadian clock (Samach et al., 2000; Wigge et al., 2005; Yamaguchi et al., 2005). The CONSTANS (CO) protein is a major regulator of photoperiodic flowering, and it directly induces the expression of the floral integrator gene *FT* and the closely related to TWIN SISTER OF FT (TSF). The CO protein is expressed at very low levels, and its level is the limiting factor in the induction of flowering by photoperiod. This was evident from the study conducted by Robson and coworkers wherein reducing CO levels by half in heterozygous plants delays flowering (Robson et al., 2001). The photoperiodic pathway regulates the levels of the CO protein during the course of the day. According to Valverde et al. (2004), the levels of CO protein show increasing trend between about 10 h and about 16 h after dawn (i.e., toward the end of an LD). It was evident from the fact that flowering could be induced by altering the light/dark regime or *CO* expression, such that high levels of *CO* expression occurred in the light period in SDs (Roden et al., 2002; Yanovsky and Kay, 2002).

10.9.1 CIRCADIAN CLOCK

The ability of plants to detect light and measure the day length is based on the existence of a circadian clock. The plant circadian system consists of three components, namely, input pathways that provide temporal information to the clock, the central oscillator mechanism responsible for driving rhythms with a period close to 24 h, and a number of output pathways that regulate metabolic and developmental processes using the temporal information provided by the clock (Jarillo et al., 2006). The light perceived by photoreceptors represents the main input pathway to the clock; the pace of the clock is reset by light every day allowing the progressive adjustment of the clock to the time of dawn, so that the mechanism of the oscillator remains synchronized with external cycles of light and dark (Franklin et al., 2005; Jiao et al., 2007). Plants have evolved an array of photoreceptors to detect light over a large range of fluence rates and wavelengths, including the PHYTOCHROMES (PHY), which absorb in the red and far-red regions of the spectrum, and the CRYPTOCHROMES (CRY), PHOTOTROPINS (PHOT), and the ZEITLUPE (ZTL)/LOV KELCH PROTEIN 2 (LKP2)/ FLAVIN-BINDING KELCH REPEAT F-BOX 1 (FKF1) family, all of which absorb blue and UV-A light (Yanovsky and Kay, 2003; Jiao et al., 2007). The participation of the circadian clock in the control of biological activities allows plants to anticipate and adapt to periodic environmental changes, maximizing their opportunities to survive successfully (Mas, 2005; McClung, 2006; Hotta et al., 2007). The control of flowering by day length determines seasonal patterns of flowering and is regulated by one or more of these output branches of the clock. The core oscillator determines the daily rhythms in output genes, which set the light sensitive phase for triggering the floral transition when plants are exposed to appropriate photoperiodic conditions.

The circadian clock involves several interconnected negative feedback loops (Gardner et al., 2006; McClung, 2006; Hotta et al., 2007). The clock controls several responses that need to be coordinated during the daily cycle. The basic negative feedback loop consists of the TIMING OF CAB EXPRESSION 1 (TOC1) gene whose product positively regulates two partially redundant Myb transcription factors, LATE ELONGATED HYPOCOTYL (LHY) and CIRCADIAN CLOCK ASSOCIATED 1 (CCA1). LHY and CCA1 proteins then send feedback to negatively regulate the expression of TOC1 through binding to an evening element in its promoter (Alabadi et al., 2001). The stability of the TOC1 protein is also regulated, in this case by ZTL, which targets TOC1 for degradation by the 26S proteasome (Más et al., 2003). Other feedback loops involve the PSEUDO RESPONSE REGULATOR (PRR), GIGANTEA (GI) and LUX ARRHYTHMO (LUX) genes (Gardner et al., 2006; McClung, 2006).

10.10 PHOTOPERIODIC CONTROL OF PLANT DEVELOPMENTAL PROCESSES

10.10.1 Induction of Flowering by Light

The main functions of light signal in the photoperiodic control of flowering are as follows:

(i) Since the internal clock is programmed to a 22- to 29-h cycle, it could get out of phase with the normal day/night cycles within a few days without resting. Correct entrainment of the clock is important as it sets the phase of expression of clock-regulated genes such as CO, which are outputs from the clock, in relation to the daily light/dark cycle. The red-light acts through the phytochromes – phyA, phyB, phyD and phyE – and the blue light acts through ZTL and the cryptochromes cry1 and cry2 to entrain the clock (Somers et al., 1998; Devlin and Kay, 2000; Kim et al., 2007b). This is achieved by inducing the expression of genes (LHY, CCA1 and PRR9) that are key components of the clock (Wang and Tobin, 1998; Kim et al., 2003; Farré et al., 2005) and enhancing the stability of ZTL by promoting its interaction with another clock component (GI).

(ii) Promotion of the blue-light-dependent interaction between FKF1 and GI which is necessary for the degradation of a transcriptional repressor of CO called CYCLING DOF FACTOR 1 (CDF1) (Sawa et al., 2007), thereby promoting *CO* expression. It is proposed that CDF1 is bound to the *CO* promoter and inhibits *CO* transcription in the first part of the day. GI and FKF1, which are produced later in the day, form a complex in a blue-light-dependent manner, which binds to CDF1, enabling FKF1 to target CDF1 for degradation by the 26S proteasome, thus relieving the repression of *CO* and allowing its expression toward the end of an LD (Sawa et al., 2007).

(iii) Studies by Valverde et al. (2004) have shown that while red light acting through phyB promotes the degradation of *CO* by the proteasome, the far-red and blue light acting through phyA and the cryptochromes, respectively, increase the stability of CO. The phyB-mediated degradation predominates during the morning, whereas this is antagonized toward the end of the day by the action of phyA and the cryptochromes, resulting in the stabilization of CO at the end of an LD. In the dark, the CO protein is degraded by the proteasome by the CONSTITUTIVE PHOTOMORPHOGENIC 1 (COP1) protein, a RING finger ubiquitin ligase that regulates the stability of transcription factors involved in the plant's response to light (Osterlund et al., 2000; Holm et al., 2001; Seo et al., 2003).

Several research groups (Onouchi et al., 2000; Samach et al., 2000; Yamaguchi et al., 2005; Suàrez-Lόpez et al., 2001; Yanovsky and Kay, 2002) have attempted to elucidate the mechanism for light-dependent regulation of *CO* at both the transcriptional and posttranscriptional levels. According to these authors, CO protein accumulates at high levels during the light period, causing a strong induction of the *FT* gene, which ultimately results in flowering. In SDs, the *CO* expression does not rise to high enough levels during the daytime. There is no GI/FKF1-induced daytime peak of *CO* expression because the proteins are not present in sufficient amounts to relieve the CDF1-mediated repression of *CO*. Therefore, *FT* expression is not induced and subsequent flowering does not occur. In LDPs, CO protein levels will rise to high enough levels above CDL to induce *FT* gene and flowering. For detailed discussion of how *CO* promotes flowering can be found in publications by Onouchi et al. (2000), Samach et al. (2000), Yamaguchi et al. (2005), Takada and Goto (2003) and An et al. (2004).

Jung et al. (2007) established that the overexpression miR172 causes extremely early flowering even in a *co* mutant background. It induces flowering through the alleviation of the repression of *FT* by TOE1, which is one of the target genes of miR172. The expression of miR172 increases with plant age until flowering, and the transcript levels of its target genes *TOE1*, *TOE2*, *SMZ* and *SNZ* decrease with age in a complementary fashion (Jung et al., 2007). These authors have further established dual role for *GI* in photoperiodic control of flowering, regulating both the *CO*- and miR172-mediated induction of *FT* expression. He suggested that both pathways are required for the promotion of flowering in LDs and disruption of either can result in late flowering in LDs.

10.10.2 Movement of Photoperiodic Signals in Plants

The graft-transmissible flower-inducing signal moves from induced leaves through the phloem to the apex (Thomas and Vince-Prue, 1997). It has been shown that there is a nonselective loading of proteins of up to 67 kDa from companion cells into the phloem sieve elements (Stadler et al., 2005), and hence, the small 20-kDa FT protein can easily enter into the phloem. Several researchers have also shown that the FT protein can also move intracellularly from the end of the vasculature into the SAM (Corbesier et al., 2007; Jaeger and Wigge, 2007; Mathieu et al., 2007; Tamaki et al., 2007), and even across graft unions (Corbesier et al., 2007; Lin et al., 2007). Several authors have also established that mRNAs can move through the phloem throughout the plant (Kim et al., 2001; Haywood et al., 2005; Banerjee et al., 2006). When the FT protein arrives in the apex, it interacts with the b-ZIP transcription factor FLOWERING LOCUS D (FD) to form a transcriptional complex that activates the meristem identity gene *AP1* (Abe et al., 2005; Wigge et al., 2005). FT is also known to upregulate *SOC1* expression in the SAM (Yoo et al., 2005). SOC1 forms a complex with another MADS-box protein, AGAMOUS-LIKE 24 (AGL24), which translocates it to the nucleus where it binds the *LFY* promoter to induce *LFY* expression (Lee et al., 2008). LFY induces the expression of *AP1* and vice versa.

Other compounds such as gibberellins, cytokinins, sucrose, nitrate and glutamine are also transported from leaves to the apex. These compounds alter the rate of cell division in the SAM (Bernier and Perilleux, 2005). Photoperiodic induction in *Sinapis* and *Xanthium* leads to increased export of sucrose and cytokinin from the leaf. Increasing cell division in the SAM can cause early flowering (Cockcroft et al., 2000). However, whether these compounds are part of the inducing signal *per se*, or whether their transport to the apex is an early event following induction, still requires to be explained.

The influence of photoperiod on flowering is moderated by other factors such as temperature and the developmental age of the plant. In Arabidopsis, Searle et al. (2006) have shown that FLC is a general repressor of flowering. It acts in both the phloem to inhibit *FT* and *SOC1* expression, and in SAM to inhibit *FD* and *SOC1* expression, thereby affecting the competence of the SAM to respond to the FT protein. The levels of *FLC* expression are controlled by vernalization and autonomous pathways, act to reduce *FLC* expression, thereby relieving the repression of flowering and allowing flowering to be induced by other pathways such as the photoperiodic pathway (Mouradov et al., 2002; Henderson et al., 2003; Bäurle and Dean, 2006).

10.10.3 Hormonal Control of Flowering

The hormone GA induces flowering in SDs under noninductive photoperiod. According to Blázquez and Weigel, GA acts to directly induce *LFY* expression via a domain in the *LFY* promoter that is different to the one required for photoperiodic induction of *LFY* (Blázquez and Weigel, 2000). This domain binds the GAMYB transcription factor which in turn is regulated by an miRNA, miR159 (Achard et al., 2004). In some species, GA is inhibitory for flowering (Roberts et al., 1999), and abscisic acid (ABA) delays flowering. Stress conditions also affect flowering time (Martínez et al., 2004), whereas nitric oxide represses *CO* and *GI* and elevates *FLC* expression to inhibit flowering (He et al., 2004). Thus, the induction of flowering by photoperiod is influenced by several factors.

10.10.4 Photoperiodic Control of Flowering LDPs

The mechanisms outlined above primarily describe the photoperiodic flowering response in the LDP. However, different photoperiodic responses have been reported among members of the same family (Hayama et al., 2007).

10.10.5 Tuberization

Tuberzation is a highly regulated process in wild relatives of potatoes and in some modern-day cultivars. It is promoted by SD and low night temperatures. In fact, tuberization is the plant's way of surviving from season to season. In most commercial varieties, the requirement for SDs for tuberization has largely been selected out. Most research on tuberization is done on such photoperiod-sensitive potatoes. It is now clear that tuberization is not under the control of a single signal. Jan van den Berg and Elmer Ewing by analyzing large progenies raised from crosses between potatoes with diverse genetic makeup, and degree or time of tuberization, concluded that tuberization is under the control of several genes and is regulated by a balance of several factors.

Tuberization in potato is promoted by SDs and inhibited by LDs. Tuberization can also be prevented by a night break, and at the same time, the inhibitory effect of night break with red light can be reversed by a far-red light treatment (Batutis and Ewing, 1982). Like flowering, tuberization is a major developmental switch from vegetative growth to reproductive growth involving several physiological changes. Stems become shorter and thicker at the stolon ends, and leaves become larger and broader with increased tendency for flower-bud abortion and accelerated senescence (Steward et al., 1981). There is also evidence that some of the molecular components involved in the photoperiodic control of flowering (e.g., phyB) may also play a role in the photoperiodic control of tuberization (Jackson et al., 1996). It has also been shown that phyB regulates the production of a graft-transmissible signal (Jackson et al., 1998). Several studies have concluded that both phyA and cryptochromes also play a role in day-length perception in potato (Heyer et al., 1995) and are involved in entraining the circadian clock in potato (Yanovsky et al., 2000). Martinez-Garcia et al. (2002) suggested a possible role for *CO* in the tuberization response. In some lines of *S. tuberosum* ssp. *andigena*, while the tuberization is under photoperiodic control, the flowering is not pointing to the involvement of different mobile signals in these processes (Jackson and Prat, 1996; Jackson et al., 2000).

10.10.6 Bud Growth

The onset of bud set and growth cessation which precedes dormancy in trees has been shown to be under photoperiodic control. It is promoted by falling day length in autumn. According to Böhlenius et al. (2006), *PtFT1* inhibits this process and overexpression of *PtFT1* prevents growth cessation and bud set even after an extended period in SDs, whereas RNAi lines with decreased levels of *PtFT1* show an enhanced response. It is important to consider both the photoperiod-insensitive phase and the photoperiod-sensitive phase together rather than as separate developmental processes. In some cases, the control mechanism involved in the regulation of growth cessation and bud set has been shown to be very similar to that controlling flowering, with PtFT1 acting as a repressor rather than an inducer, while in others different mechanisms are involved. Lot more research needs to be done to elucidate the photoperiodic control mechanisms involved in bud set and cessation of growth.

REFERENCES

Abe M, Kobayashi Y, Yamamoto S, Daimon Y, Yamaguchi A, Ikeda Y, Ichinoki H, Notaguchi M, Goto K and Araki T. 2005. FD, a bZIP protein mediating signals from the floral pathway integrator FT at the shoot apex. *Science* **309**: 1052–1056.

Achard P, Herr A, Baulcombe D C and Harberd N P. 2004. Modulation of floral development by a gibberellin-regulated microRNA. *Development* **131**: 3357–3365.

Adams S R, Pearson S and Hadley P. 2001. Improving quantitative flowering models through a better understanding of the phases of photoperiod sensitivity. *Journal of Experimental Botany* **52**: 655–662.

Adams S R, Pearson S, Hadley P and Patefield W M. 1999. The effects of temperature and light integral on the phases of photoperiod sensitivity in *Petunia hybrida*. *Annals of Botany* **83**: 263–269.

Alabadi D, Oyama T, Yanovsky M J, Harmon F G, Más P and Kay S A. 2001. Reciprocal regulation between TOC1 and LHY/CCA1 within the Arabidopsis circadian clock. *Science* **293**: 880–883.

An H, Roussot C, Suárez-López P, Corbesier L, Vincent C, Piñeiro M, Hepworth S, Mouradov A, Justin S, Turnbull C, et al. 2004. CONSTANS acts in the phloem to regulate a systemic signal that induces photoperiodic flowering of *Arabidopsis*. *Development* **131**: 3615–3626.

Asai K, Satoh N, Sasaki H, Satoh H and Nagato Y. 2002. A rice heterochronic mutant, mori1, is defective in the juvenile-adult phase change. *Development* **129**: 265–273.

Bacon F, Rawley W and Cecil T. 1627. *Sylva Sylvarum: Or A Natural History in Ten Centuries*. John Haviland and Augustine Mathewes, London.

Ballaré C L. 1999. Keeping up with the neighbors: phytochrome sensing and other signaling mechanisms. *Trends in Plant Science* **4**: 97–102.

Banerjee A K, Chatterjee M, Yu Y, Suh S G, Miller W A and Hannapel D J. 2006. Dynamics of a mobile RNA of potato involved in a long-distance signaling pathway. *Plant Cell* **18**: 3443–3457.

Barbosa I C and Schwechheimer C. 2014. Dynamic control of auxin transport-dependent growth by AGCVIII protein kinases. *Current Opinion in Plant Biology* **22**: 108–115.

Batutis E J and Ewing E E. 1982. Far-red reversal of red-light effect during long night induction of potato (*Solanum tuberosum* L.) tuberization. *Plant Physiology* **69**: 672–674.

Bäurle I and Dean C. 2006. The timing of developmental transitions in plants. *Cell* **125**: 655–664.

Bentsink L and Koornneef M. 2002. Seed dormancy and germination. The Arabidopsis Book. *American Society of Plant Biologists*. Available at www.aspb.org/publications/arabidopsis/.

Berardini T Z, Bollman K, Sun H and Poethig R S. 2001. Regulation of vegetative phase change in *Arabidopsis thaliana* by cyclophilin 40. *Science* **291**: 2405–2407.

Bernier G and Perilleux C. 2005. A physiological overview of the genetics of flowering time control. *Plant Biotechnology Journal* **3**: 3–16.

Blaauw A H. 1909. Die perzeption des lichtes. *Recueil des Travaux Botaniques Neerlandais* **5**: 209–372.

Blázquez M A and Weigel D. 2000. Integration of floral inductive signals in Arabidopsis. *Nature* **404**: 889–892.

Böhlenius H, Huang T, Charbonnel-Campaa L, Brunner A M, Jansson S, Strauss S H and Nilsson O. 2006. CO/FT regulatory module controls timing of flowering and seasonal growth cessation in trees. *Science* **312**: 1040–1043.

Bollman K M, Aukerman M J, Park M Y, Hunter C, Berardini T Z and Poethig R S. 2003. HASTY, the Arabidopsis ortholog of exportin 5/MSN5, regulates phase change and morphogenesis. *Development* **130**: 1493–1504.

Bongard-Pierce D K, Evans M M S and Poethig R S. 1996. Heteroblastic features of leaf anatomy in maize and their genetic regulation. *International Journal of Plant Sciences* **157**: 331–340.

Borchert R, Renner S S, Calle Z, Navarrete D, Tye A, Gautier L, Spichiger R and von Hildebrand P. 2005. Photoperiodic induction of synchronous flowering near the Equator. *Nature* **433**: 627–629.

Boysen-Jensen P. 1911. La transmission de l'irritation phototropique dans l'avena. *Bulletin Academie des Sciences et Lettres de Montpellier* **3**: 1–24.

Briggs W R. 2006. Blue-UV-A receptors: historical overview. pp 171–197. *Photomorphogenesis in Plants and Bacteria.* 2nd edition (Eds.) Schäfer E and Nagy F. Springer, Dordrecht, The Netherlands.

Briggs W R. 2014. Phototropism: some history, some puzzles, and a look ahead. *Plant Physiology* **164**: 13–23.

Briggs W R and Christie J M. 2002. Phototropins 1 and 2: versatile plant blue-light receptors. *Trends in Plant Science* **7**: 204–210.

Browne T. 1658. *The Garden of Cyrus.* Golden Cockerel Press, London.

Bünning E. 1936. Die endogene tagesrhythmik als grundlage der photoperiodischen reaktion. *Berichte der Deutschen Botanischen Gesellschaft* **54**: 590–607.

Casal J J and Sánchez R A. 1998. Phytochromes and seed germination. *Seed Science Research* **8**: 317–329.

Casal J J, Sanchez R A and Deregibus V A. 1986. The effect of plant density on tillering: the involvement of R:FR ratio and the proportion of radiation intercepted per plant. *Environmental and Experimental Botany* **26**: 365–371.

Chen M and Chory J. 2011. Phytochrome signaling mechanisms and the control of plant development. *Trends in Cell Biology* **21**: 664–671.

Chien J C and Sussex I M. 1996. Differential regulation of trichome formation on the adaxial and abaxial leaf surfaces by gibberellins and photoperiod in *Arabidopsis thaliana* (L.) Heynh. *Plant Physiology* **111**: 1321–1328.

Cholodny N. 1927. Wuchshormone und tropismem bei den planzen. *Biologisches Zentralblatt* **47**: 604–626.

Christie J M, Blackwood L, Petersen J and Sullivan S. 2014. Plant flavoprotein photoreceptors. *Plant Cell Physiology* **56**: 401–413.

Christie J M and Murphy A S. 2013. Shoot phototropism in higher plants: new light through old concepts. *American Journal of Botany* **100**(1): 35–46.

Christie J M, Reymond P, Powell G K, Bernasconi P, Raibekas A A, Liscum E and Briggs W R. 1998. *Arabidopsis* NPH1: a flavoprotein with the properties of a photoreceptor for phototropism. *Science* **282**: 1698–1701.

Christie J M, Salomon M, Nozue K, Wada M and Briggs W R. 1999. LOV (light, oxygen, or voltage) domains of the blue-light photoreceptor phototropin (nph1): binding sites for the chromophore flavin mononucleotide. *Proceedings of the National Academy of Sciences* **96**: 8779–8783.

Christie J M, Yang H, Richter G L, Sullivan S, Thomson C E, Lin J, Titapiwatanakun B, Ennis M, Kaiserli E and Lee O R. 2011. phot1 inhibition of ABCB19 primes lateral auxin fluxes in the shoot apex required for phototropism. *PLoS Biology* **9**: e1001076.

Clarke J H, Tack D, Findlay K, Van Montagu M and Van Lijsebettens M. 1999. The SERRATE locus controls the formation of the early juvenile leaves and phase length in Arabidopsis. *Plant Journal* **20**: 493–501.

Cockcroft C E, den Boer B G, Healy J M and Murray J A. 2000. Cyclin D control of growth rate in plants. *Nature* **405**: 575–579.

Corbesier L, Vincent C, Jang S, Fornara F, Fan Q, Searle I, Giakountis A, Farrona S, Gissot L, Turnbull C., et al. 2007. FT protein movement contributes to long-distance signaling in floral induction of Arabidopsis. *Science* **316**: 1030–1033.

Darwin C. 1880. *The Power of Movement in Plants.* John Murray Publishers, London.

della Porta G. 1569. *Magiæ Naturalis*. Rouille, London.

Demarsy E, Schepens I, Okajima K, Hersch M, Bergmann S, Christie J, Shimazaki K, Tokutomi S and Fankhauser C. 2012. Phytochrome Kinase Substrate 4 is phosphorylated by the phototropin 1 photoreceptor. *EMBO Journal* **31**: 3457–3467.

Devlin P F and Kay S A. 2000. Cryptochromes are required for phytochrome signaling to the circadian clock but not for rhythmicity. *Plant Cell* **12**: 2499–2509.

Ding Z, Galvan-Ampudia C S, Demarsy E, Langowski L, Kleine-Vehn J, Fan Y, Morita M T, Tasaka M, Fankhauser C, Offringa R and Friml J. 2011. Light-mediated polarization of the PIN3 auxin transporter for the phototropic response in Arabidopsis. *Nature Cell Biology* **13**: 447–452.

Dormling I. 1973. Photoperiodic control of growth and growth cessation in Norway spruce seedlings. *IUFRO Division 2 Symposium on Dormancy in Trees*. Kornik.

Dudley M and Poethig R S. 1993. The heterochronic teopod1 and teopod2 mutations of maize are expressed non-cell-autonomously. *Genetics* **133**: 389–399.

Eriksson G, Ekberg I, Dormling I, Matérn B and von Wettstein D. 1978. Inheritance of bud-set and bud-flushing in Picea abies (L.) Karst. *Theoretical and Applied Genetics* **52**: 2–19.

Esmon C A, Tinsley A G, Ljung K, Sandberg G, Hearne L B and Liscum E. 2006. A gradient of auxin and auxin-dependent transcription precedes tropic growth responses. *Proceedings of National Academy of Science, USA* **103**: 236–241.

Fahlgren N, Montgomery T A, Howell M D, Allen E, Dvorak S K, Alexander A L and Carrington J C. 2006. Regulation of AUXIN RESPONSE FACTOR3 by TAS3 ta-siRNA affects developmental timing and patterning in Arabidopsis. *Current Biology* **16**: 939–944.

Farré E M, Harmer S L, Harmon F G, Yanovsky M J and Kay S A. 2005. Overlapping and distinct roles of PRR7 and PRR9 in the Arabidopsis circadian clock. *Current Biology* **15**: 47–54.

Franklin K A, Larner V S and Whitelam G C. 2005. The signal transducing photoreceptors of plants. *International Journal of Developmental Biology* **149:** 653–664.

Gallanger S, Short T S, Ray P M, Pratt L H and Briggs W R. 1988. Light-mediated changes in two proteins found associated with plasma membrane fraction from pea stem sections. *Proceedings of National Academy of Sciences, United States of America* **85**: 8003–8007.

Gao Y, Zhang Y, Zhang D, Dai X, Estelle M and Zhao Y. 2015. Auxin binding protein 1 (ABP1) is not required for either auxin signaling or *Arabidopsis* development. *Proceedings of National Academy of Science, USA* **112**: 2275–2280.

Gardner M J, Hubbard K E, Hotta C T, Dodd A N and Webb A A R. 2006. How plants tell the time. *Biochemical Journal* **397**: 15–24.

Garner W W and Allard H A. 1920. Effect of the relative length of day and night and other factors of the environment of growth and reproduction in plants. *Journal of Agricultural Research* **18**: 553–603.

Goyal A, Szarzynska B and Fankhauser C. 2013. Phototropism: at the crossroads of light-signaling pathways. *Trends in Plant Science* **18**: 393–401.

Grigg S P, Canales C, Hay A and Tsiantis M. 2005. SERRATE coordinates shoot meristem function and leaf axial patterning in Arabidopsis. *Nature* **437**: 1022–1026.

Guo H, Yang H, Mockler T C and Lin C. 1998. Regulation of flowering time by Arabidopsis photoreceptors. *Science* **279**: 1360–1363.

Hackett W P. 1985. Juvenility, maturation and rejuvenation in woody plants. *Horticultural Reviews* **7**: 109–155.

Haga K, Hayashi K and Sakai T. 2014. PINOID AGC kinases are necessary for phytochrome-mediated enhancement of hypocotyl phototropism in *Arabidopsis*. *Plant Physiology* **166**: 1535–1545.

Hamner K C and Bonner J. 1938. Photoperiodism in relation to hormones as factors in floral initiation and development. *Botanical Gazette* **100**: 388–431.

Han I S, Cho H Y, Moni A, Lee A Y and Briggs W R. 2013. Investigations on the photoregulation of chloroplast movement and leaf positioning in *Arabidopsis*. *Plant and Cell Physiology* **54**: 48–56.

Hayama R, Agashe B, Luley E, King R and Coupland G. 2007. A circadian rhythm set by dusk determines the expression of FT homologs and the short-day photoperiodic flowering response in Pharbitis. *Plant Cell* **19**: 2988–3000.

Haywood V, Yu T S, Huang N C and Lucas W J. 2005. Phloem long-distance trafficking of GIBBERELLIC ACID-INSENSITIVE RNA regulates leaf development. *Plant Journal* **42**: 49–68.

He Y, Doyle M R and Amasino R M. 2004. PAF1-complex-mediated histone methylation of FLOWERING LOCUS C chromatin is required for the vernalization-responsive, winter-annual habit in Arabidopsis. *Genes & Development* **18**: 2774–2784.

Henderson I R, Shindo C and Dean C. 2003. The need for winter in the switch to flowering. *Annual Review of Genetics* **37**: 371–392.

Hennig L, Stoddart W M, Dieterle M, Whitelam G C and Schäfer E. 2002. Phytochrome E controls light-induced germination of *Arabidopsis. Plant Physiology* **128:** 194–200.

Heyer A G, Mozley D, Landschutze V, Thomas B and Gatz C. 1995. Function of phytochrome a in potato plants as revealed through the study of transgenic plants. *Plant Physiology* **109**: 53–61.

Holland J J, Robers D and Liscum E. 2009. Understanding phototropism: from Darwin to today – Darwin review. *Journal of Experimental Botany* **60**: 1969–1978.

Holm M, Hardtke C S, Gaudet R and Deng X W. 2001. Identification of a structural motif that confers specific interaction with the WD40 repeat domain of Arabidopsis COP1. *EMBO Journal* **20**: 118–127.

Hopkins W G and Huner N P A (Eds.). 2004. *Introduction to Plant Physiology*. John Wiley and Sons, Inc., London.

Hotta C T, Gardner M J, Hubbard K E, Baek S J, Dalchau N, Suhita D, Dodd A N and Webb A A R. 2007. Modulation of environmental responses of plants by circadian clocks. *Plant, Cell and Environment* **30**: 333–349.

Huala E, Oeller P W, Liscum E, Han I S, Larsen E and Briggs W R. 1997. Arabidopsis NPH1: a protein kinase with a putative redox-sensing domain. *Science* **278**: 2120–2123.

Hunter C, Sun H and Poethig R S. 2003. The Arabidopsis heterochronic gene ZIPPY is an ARGONAUTE family member. *Current Biology* **13**: 1734–1739.

Hunter C, Willmann M R, Wu G, Yoshikawa M, de la Luz Gutierrez-Nava M and Poethig S R. 2006. Trans-acting siRNA-mediated repression of ETTIN and ARF4 regulates heteroblasty in Arabidopsis. *Development* **133**: 2973–2981.

Imaizumi T and Kay S A. 2006. Photoperiodic control of flowering: not only by coincidence. *Trends in Plant Science* **11**: 550–558.

Inoue S, Kinoshita T, Matsumoto M, Nakayama K I, Doi M and Shimazaki K. 2008. Blue light-induced autophosphorylation of phototropin is a primary step for signaling. *Proceedings of National Academy of Science, USA* **105**: 5626–5631.

Inoue S, Matsushita T, Tomokiyo Y, Matsumoto M, Nakayama K I, Kinoshita T and Shimazaki K. 2011. Functional analyses of the activation loop of phototropin2 in Arabidopsis. *Plant Physiology* **156**: 117–128.

Jackson S D. 2009. Plant responses to photoperiod. *New Phytologist* **181**: 517–531.

Jackson S D, Heyer A, Dietze J and Prat S. 1996. Phytochrome B mediates the photoperiodic control of tuber formation in potato. *Plant Journal* **9**: 159–166.

Jackson S D, James P E, Carrera E, Prat S and Thomas B. 2000. Regulation of transcript levels of a potato gibberellin 20-oxidase gene by light and phytochrome B. *Plant Physiology* **124**: 423–430.

Jackson S D, James P E, Prat S and Thomas B. 1998. Phytochrome B affects the levels of a graft-transmissible signal involved in tuberisation. *Plant Physiology* **124**: 423–430.

Jackson S D and Prat S. 1996. Control of tuberisation in potato by gibberellins and phytochrome B. *Physiologia Plantarum* **98**: 29–32.

Jaedicke K, Lichtenthäler A L, Meyberg R, Zeidler M and Hughes J. 2012. A phytochrome – phototropin light signaling complex at the plasma membrane. *Proceedings of the National Academy of Sciences* **109**: 12231–12236.

Jaeger K E and Wigge P A. 2007. FT protein acts as a long-range signal in Arabidopsis. *Current Biology* **17**: 1050–1054.

Janoudi A K, Konjevic R, Apel P and Poff K L. 1992. Time threshold for second positive phototropism is decreased by a preirradiation with red light. *Plant Physiology* **99**: 1422–1425.

Jarillo J A, Gabrys H, Capel J Alonso J M, Ecker J R and Cashmore A R. 2001. Phototropin-related NPL1 controls chloroplast relocation induced by blue light. *Nature* **410**: 952–954.

Jarillo J A and Pineiro M. 2006. The molecular basis of photoperiodism. *Biological Rhythms Research* **37**: 353–380.

Jiao Y, Lau O and Deng X W. 2007. Light-regulated transcriptional networks in higher plants. *National Review of Genetics* **8**: 217–230.

Johnson E, Bradley M, Herberd N P and Whitelam G C. 1994. Photoresponses of light-grown *phyA* mutants of *Arabidopsis. Plant Physiology* **105**: 141–149.

Johnston E S. 1934. Phototropic sensitivity in relation to wavelength. *Smithsonian Miscellaneous Collections* **92**: 1–17.

Jung C and Müller A E. 2009. Flowering time control and applications in plant breeding. *Trends in Plant Science* **14**: 563–573.

Jung J H, Seo Y H, Seo P J, Reyes J L, Yun J, Chua N H and Park C M. 2007. The GIGANTEA-regulated microRNA172 mediates photoperiodic flowering independent of CONSTANS in Arabidopsis. *Plant Cell* **19**: 2736–2748.

Kagawa T, Sakai T, Suetsugu N, Oikawa K, Ishiguro S, Kato T, Tabata S, Okada K and Wada M. 2001. Arabidopsis NPL1: a phototropin homolog controlling the chloroplast high-light avoidance response. *Science* **291**: 2138–2141.

Kagawa T and Wada M. 2002. Blue-light induced chloroplast relocation. *Plant Cell Physiology* **43**: 367–371.

Kami C, Lorrain S, Hornitschek P and Fankhauser C. 2010. Light-regulated plant growth and development. *Current Topics in Developmental Biology* **91**: 29–66.

Kanrar S, Bhattacharya M, Arthur B, Courtier J and Smith H M. 2008. Regulatory networks that function to specify flower meristems require the function of homeobox genes PENNYWISE and POUND-FOOLISH in Arabidopsis. *Plant Journal* **54**: 924–937.

Kasahara M, Kagawa T, Oikawa K, Suetsugu N, Miyao M and Wada M. 2002. Chloroplast avoidance movement reduces photodamage in plants. *Nature* **420**: 829–832.

Kawai H, Kanegae T, Christensen S, Kiyosue T, Sato Y, Imaizumi T, Kadota A and Wada M. 2003. Responses of ferns to red light are mediated by an unconventional photoreceptor. *Nature* **421**: 287–290.

Kim D-H, Doyle M R, Sung S and Amasino R M. 2009. Vernalization: winter and the timing of flowering in plants. *Annual Review of Cell and Developmental Biology* **25**: 277–299.

Kim J Y, Song H R, Taylor B L and Carre I A. 2003. Light-regulated translation mediates gated induction of the Arabidopsis clock protein LHY. *EMBO Journal* **22**: 935–944.

Kim M, Canio W, Kessler S and Sinha N. 2001. Developmental changes due to long-distance movement of a homeobox fusion transcript in tomato. *Science* **293**: 287–289.

Kim W Y, Fujiwara S, Suh S S, Kim J, Kim Y, Han L, David K, Putterill J, Nam H G and Somers D E. 2007b. ZEITLUPE is a circadian photoreceptor stabilized by GIGANTEA in blue light. *Nature* **449**: 356–360.

Kinoshita T, Doi M, Suetsugu N, Kagawa T, Wada M and Shimazaki K. 2001. Photl and phot2 mediate blue light regulation of stomatal opening. *Nature* **414**: 656–660.

Klebs G. 1913. Über das Verhältnis der aussenwelt zur entwicklung der pflanze. *Heidelberger Academie der Wissenschaften* **5**: 1–47.

Kogl F and Haagen-Smits A. 1931. I. Mitteilung uber pflanzliche wachstumsstoffe. Uber die vhemie des euchsstoffs. *Proceedings Koninklijke Nederlandse Akademie van Wetenschappen* **34**: 1411–1416.

Koornneef M and Karssen C M. 1994. Seed dormancy and germination. pp. 313–334. *Arabidopsis*. (Eds.) Meyerowitz E M and Somerville C R. Harbor Laboratory Press, Cold Spring Harbor, NY.

Kutschera U and Briggs W R. 2012. Root phototropism: from dogma to the mechanism of blue light perception. *Planta* **235**: 443–452.

Lagercrantz U. 2009. At the end of the day: a common molecular mechanism for photoperiod response in plants. *Journal of Experimental Botany* **60**: 2501–2515.

Lee J and Lee I. 2010. Regulation and function of SOC1, a flowering pathway integrator. *Journal of Experimental Botany* **61**: 2247–2254.

Lee J, Oh M, Park H and Lee I. 2008. SOC1 translocated to the nucleus by interaction with AGL24 directly regulates leafy. *Plant Journal* **55**: 832–843.

Li F W, Villarreal J C, Kelly S, Rothfels C J, Melkonian M, Frangedakis E, Ruhsam M, Sigel E M, Der J P and Pittermann J. 2014. Horizontal transfer of an adaptive chimeric photoreceptor from bryophytes to ferns. *Proceedings of National Academy of Science, USA* **111**: 6672–6677.

Lin M, Belanger H, Lee Y, VarkonyiGasic E, Taoka K-C, Miura E, XoconostleCazares B, Gendler K, Jorgensen R and Phinney B. 2007. FLOWERING LOCUS T protein may act as the long-distance florigenic signal in the Cucurbits. *Plant Cell* **19**: 1488–1506.

Liscum E, Askinosie S K, Leuchtman D L, Morrow J, Willenburg K T and Coats D R. 2014. Phototropism: growing towards an understanding of plant movement. *Plant Cell* **26**: 38–55.

Liscum E and Briggs W R. 1995. Mutations in the NPH1 locus of Arabidopsis disrupt the perception of phototropic stimuli. *Plant Cell* **7**: 473–485.

Lumsden P J. 1996. A limit cycle model for the circadian clock in short-day plants. *Flowering Newsletter* **21**: 42–47.

Lumsden P J. 1998. Photoperiodic induction in short-day plants. pp 151–165. *Biological rhythms and photoperiodism in plants* (Eds.) Lumsden P J and Millar A J. BIOS Scientific Press, Oxford.

Martínez C, Pons E, Prats G and León J. 2004. Salicylic acid regulates flowering time and links defense responses and reproductive development. *Plant Journal* **37**: 209–217.

Martinez-Garcia J F, Virgos-Soler A and Prat S. 2002. Control of photoperiod regulated tuberization in potato by the Arabidopsis flowering-time gene CONSTANS. *Proceedings of the National Academy of Sciences, USA* **99**: 15211–15216.

Mas P. 2005. Circadian clock signaling in Arabidopsis thaliana: from gene expression to physiology and development. *The International Journal of Developmental Biology* **49**: 491–500.

Más P, Kim W Y, Somers D E and Kay S A. 2003. Targeted degradation of TOC1 by ZTL modulates circadian function in *Arabidopsis thaliana. Nature* **426**: 567–570.

Mathieu J, Warthmann N, Kuttner F and Schmid M. 2007. Export of FT protein from phloem companion cells is sufficient for floral induction in Arabidopsis. *Current Biology* **17**: 1055–1060.

McClung C R. 2006. Plant circadian rhythms. *The Plant Cell* **18**: 792–803.

McDonald M B and Kwong F Y. (Eds). 2005. *Flower Seeds Biology and Technology*. CABI Publishing, Oxfordshire.

McLaren J S and Smith H. 1978. The function of phytochrome in the natural environment. VI. The growth and development of *Rumex obtensifolius* under simulated canopy light environments. *Plant, Cell and Environment* **1**: 61–67.

Mouradov A, Cremer F and Coupland G. 2002. Control of flowering time: interacting pathways as a basis for diversity. *Plant Cell* **14** (Suppl): S111–S130.

Naqvi S M. 1972. Possible role of abscisic acid in phototropism. *Zeitschrift fur Pflanzenphysiologie* **67**: 454–456.

Ohgishi M, Saji K, Okada K and Sakai T. 2004. Functional analysis of each blue light receptor, cry1, cry2, phot1, and phot2, by using combinatorial multiple mutants in Arabidopsis. *Proceedings of National Academy of Sciences, United States of America* **101**: 2223–2228.

Onouchi H, Igeño M I, Périlleux C, Graves K and Coupland G. 2000. Mutagenesis of plants overexpressing CONSTANS demonstrates novel interactions among Arabidopsis flowering-time genes. *Plant Cell* **12**: 885–900.

Orbovic V and Poff K L. 1992. Growth distribution during phototropism of *Arabidopsis thaliana* seedlings. *Plant Physiology* **103**: 157–163.

Osterlund M T, Hardtke C S, Wei N and Deng X W. 2000. Targeted destabilization of HY5 during light-regulated development of Arabidopsis. *Nature* **405**: 462–466.

Paal A. 1918. Uber phototropische reizleitung. *Jahr. Wiss. Botany* **58**: 406–458.

Park M Y, Wu G, Gonzalez-Sulser A, Vaucheret H and Poethig R S. 2005. Nuclear processing and export of microRNAs in Arabidopsis. *Proceedings of the National Academy of Sciences, USA* **102**: 3691–3696.

Parks B M, Quail P H and Hangarter R P. 1996. Phytochrome a regulates red-light induction of phototropic enhancement in Arabidopsis. *Plant Physiology* **110**: 155–162.

Pedmale U V, Celaya R B and Liscum E. 2010. Phototropism: mechanisms and outcomes. *The Arabidopsis Book* **8**: e0125.

Pedmale U V and Liscum E. 2007. Regulation of phototropic signaling in *Arabidopsis* via phosphorylation state changes in the phototropin 1-interacting protein NPH3. *Journal of Biological Chemistry* **282**: 19992–20001.

Peragine A, Yoshikawa M, Wu G, Albrecht H L and Poethig R S. 2004. SGS3 and SGS2/SDE1/RDR6 are required for juvenile development and the production of trans-acting siRNAs in Arabidopsis. *Genes and Development* **18**: 2368–2379.

Pittendrigh C S. 1966. The circadian oscillation in Drosophila pseudoobscura pupae: a model for the photoperiodic clock. *Zeitschrift fu¨r Pflanzenphysiologie* **54**: 275–307.

Pittendrigh C S. 1972. Circadian surfaces and the diversity of possible roles of circadian organization in photoperiodic induction. *Proceedings of the National Academy of Sciences, United States of America* **69**: 2734–2737.

Poethig R S. 1988. Heterochronic mutations affecting shoot development in maize. *Genetics* **119**: 959–973.

Poethig R S. 1990. Phase change and the regulation of shoot morphogenesis in plants. *Science* **250**: 923–930.

Preuten T, Blackwood L, Christie J and Fankhauser C. 2015. Lipid anchoring of *Arabidopsis* phototropin 1 to assess the functional significance of receptor internalisation: should I stay or should I go? *New Phytologist* **206**: 1038–1050.

Preuten T, Hohm T, Bergmann S and Fankhauser C. 2013. Defining the site of light perception and initiation of phototropism in *Arabidopsis*. *Current Biology* **23**: 1934–1938.

Putterill J, Laurie R and Macknight R. 2004. It's time to flower: the genetic control of flowering time. *Bioessays* **26**: 363–373.

Rakusova H, Fendrych M and Friml J. 2014. Intracellular trafficking and PIN-mediated cell polarity during tropic responses in plants. *Current Opinion in Plant Biology* **23C**: 116–123.

Ray J, Camel G J and Tournefort J P. 1686. *Historia Plantarum*. Henricum Faithorne, London.

Reymond P, Short T W, Briggs W R and Poff K L. 1992a. Blue light activates a specific protein kinase in higher plants. *Plant Physiology* **100**: 655–661.

Reymond P, Short T W, Briggs W R and Poff K L. 1992b. Light induced phosphorylation of a membrane protein plays an early role in signal transduction for phototropism in *Arabidopsis thaliana*. *Proceedings of National Academy of Sciences, United States of America* **89**: 4718–4721.

Roberts A, Blake P, Lewis R, Taylor J and Dunstan D. 1999. The effect of gibberellins on flowering in roses. *Journal of Plant Growth Regulation* **18**: 113–119.

Robson F, Costa M M R, Hepworth S R, Vizir I, Piñeiro M, Reeves P H, Putterill J and Coupland G. 2001. Functional importance of conserved domains in the flowering-time gene CONSTANS demonstrated by analysis of mutant alleles and transgenic plants. *Plant Journal* **28**: 619–631.

Roden L C, Song H R, Jackson S, Morris K and Carre I A. 2002. Floral responses to photoperiod are correlated with the timing of rhythmic expression relative to dawn and dusk in Arabidopsis. *Proceedings of the National Academy of Sciences, USA* **99**: 13313–13318.

Sakai T and Haga K. 2012. Molecular genetic analysis of phototropism in *Arabidopsis*. *Plant and Cell Physiology* **53**: 1517–1534.

Sakai T, Kagawa T, Kasahara M, Swartz T E, Christie J M, Briggs W R, Wada M and Okada K. 2001. Arabidopsis nph1 and npl1: blue light receptors that mediate both phototropism and chloroplast relocation. *Proceedings of National Academy of Sciences, United States of America* **98**: 6969–6974.

Sakai T, Wada T, Ishiguro S and Okada K. 2000. RPT2. A signal transducer of the phototropic response in Arabidopsis. *Plant Cell* **12**: 225–236.

Sakamoto K and Briggs W R. 2002. Cellular and subcellular localization of phototropin 1. *Plant Cell* **14**: 1723–1735.

Samach A and Coupland G. 2000. Time measurement and the control of flowering in plants. *Bioessays* **22**: 38–47.

Samach A, Onouchi H, Gold S E, Ditta G S, Schwarz-Sommer Z, Yanofsky M F and Coupland G. 2000. Distinct roles of CONSTANS target genes in reproductive development of Arabidopsis. *Science* **288**: 1613–1616.

Sawa M, Nusinow D A, Kay S A and Imaizumi T. 2007. FKF1 and GIGANTEA complex formation is required for day-length measurement in Arabidopsis. *Science* **318**: 261–265.

Searle I and Coupland G. 2004. Induction of flowering by seasonal changes in photoperiod. *The EMBO Journal* **23**: 1217–1222.

Searle I, He Y H, Turck F, Vincent C, Fornara F, Kröber S, Amasino R A and Coupland G. 2006. The transcription factor FLC confers a flowering response to vernalization by repressing meristem competence and systemic signaling in Arabidopsis. *Genes and Development* **20**: 898–912.

Seo H S, Yang J Y, Ishikawa M, Bolle C, Ballesteros M L and Chua N H. 2003. LAF1 ubiquitination by COP1 controls photomorphogenesis and is stimulated by SPA1. *Nature* **423**: 995–999.

Sharrock R. 1672. *The History of the Propagation and Improvement of Vegetables*. W. Hall, Oxford.

Shen-Miller J, Cooper P and Gordon S A. 1969. Phototropism and photoinhibition of basipolar transport of auxin in oat coleoptiles. *Plant Physiology* **44**: 491–496.

Shen-Miller J and Gordon S A. 1966. Hormonal relations in the phototropic response. III. The movement of C14-labeled and endogenous indole acetic acid in phototropically stimulated Zea coleoptiles. *Plant Physiology* **41**: 59–65.

Shibaoka H. 1994. Plant hormone-induced changes in the orientation of cortical microtubules – alterations in the cross-linking between microtubules and the plasma-membrane. *Annual Review of Plant Physiology* **45**: 527–544.

Short T W and Briggs W R. 1990. Characterization of a rapid, blue light-mediated change in detectable phosphorylation of a plasma membrane protein from etiolated pea (*Pisum sativum* L.) seedlings. *Plant Physiology* **92**: 179–185.

Short T W, Porst M and Briggs W R. 1992. A photoreceptor system regulating in vivo and in vitro phosphorylation of a pea plasma-membrane protein. *Photochemistry and Photobiology* **55**: 773–781.32.

Smith H. 1982. Light quality, photoperception, and plant strategy. *Annual Review of Plant Physiology* **33**: 481–518.

Smith H and Whitelam G C. 1997. The shade avoidance syndrome: multiple responses mediated by multiple phytochromes. *Plant, Cell and Environment* **20**: 840–844.

Somers D E, Devlin P F and Kay S A. 1998. Phytochromes and cryptochromes in the entrainment of the Arabidopsis circadian clock. *Science* **282**: 1488–1490.

Spalding E P. 2013. Diverting the downhill flow of auxin to steer growth during tropisms. *American Journal of Botany* **100**: 203–214.

Stadler R, Wright K M, Lauterbach C, Amon G, Gahrtz M, Feuerstein A, Oparka K J and Sauer N. 2005. Expression of GFP-fusions in Arabidopsis companion cells reveals nonspecific protein trafficking into sieve elements and identifies a novel post-phloem domain in roots. *Plant Journal* **41**: 319–331.

Steward F, Moreno V and Roca W. 1981. Growth form and composition of potato plants as affected by environment. *Annals of Botany* **48**(Suppl. 2): 45.

Strasser B, Sánchez-Lamas M, Yanovsky M J, Casal J J and Cerdán P D. 2010. *Arabidopsis taliana* life without phytochromes. *Proceedings of National Academy of Sciences, United States of America* **107**: 4776–4781.

Suàrez-López P, Wheatley K, Robson F, Onouchi H, Valverde F and Coupland G. 2001. CONSTANS mediates between the circadian clock and the control of flowering in Arabidopsis. *Nature* **410**: 1116–1120.

Suetsugu N and Wada M. 2013. Evolution of three LOV blue light receptor families in green plants and photosynthetic stramenopiles: phototropin, ZTL/FKF1/LKP2 and aureochrome. *Plant Cell Physiology* **54**: 8–23.

Sung S B and Amasino R M. 2004. Vernalization in *Arabidopsis thaliana* is mediated by the PHD finger protein VIN3. *Nature* **427**: 159–164.

Takada S and Goto K. 2003. TERMINAL FLOWER 2, an Arabidopsis homolog of HETEROCHROMATIN PROTEIN 1, counteracts the activation of FLOWERING LOCUS T by CONSTANS in the vascular tissues of leaves to regulate flowering time. *Plant Cell* **15**: 2856–2865.

Takemiya A, Inoue S, Doi M, Kinoshita T and Shimazaki K. 2005. Phototropins promote plant growth in response to blue light in low light environments. *Plant Cell* **17**: 1120–1127.

Tamaki S, Matsuo S, Wong H L, Yokoi S and Shimamoto K. 2007. Hd3a protein is a mobile flowering signal in rice. *Science* **316**: 1033–1036.

Tao Y, Ferrer J L, Ljung K, Pojer F, Hong F, Long J A, Li L, Moreno J E, Bowman M E and Ivans L J. 2008. Rapid synthesis of auxin via a new tryptophan-dependent pathway is required for shade avoidance in plants. *Cell* **133**: 164–176.

Telfer A, Bollman K M and Poethig R S. 1997. Phase change and the regulation of trichome distribution in *Arabidopsis thaliana. Development* **124**: 645–654.

Thomas B. 1998. Photoperiodism: an overview. pp 151–165. *Biological Rhythms and Photoperiodism in Plants* (Eds.) Lumsden P J and Millar A J. BIOS Scientific Publishers, Oxford.

Thomas B and Vince-Prue D. 1997. *Photoperiodism in Plants*. Academic Press, London.

Tournois J. 1912. Influence de la lumiére sur la florasion du houblon japonais et du chanvre déterminées par des semis haitifs. *Comptes Rendus de l'Academie desSciences, Paris* **155**: 297–300.

Tournois J. 1914. *Etudes sur la sexualite du houblon*. Doctoral Dissertation. Verlag nicht ermittelbar.

Turck F, Fornara F and Coupland G. 2008. Regulation and identity of florigen: FLOWERING LOCUS T moves center stage. *Annual Review of Plant Biology* **59**: 573–594.

Valverde F, Mouradov A, Soppe W, Ravenscroft D, Samach A and Coupland G. 2004. Photoreceptor regulation of CONSTANS protein in photoperiodic flowering. *Science* **303**: 1003–1006.

Vandenbussche F and Van Der Straeten D. 2014. Differential accumulation of ELONGATED HYPOCOTYL5 correlates with hypocotyl bending to ultraviolet-B light. *Plant Physiology* **166**: 40–43.

Vega S H, Sauer M, Orkwiszewski J A J and Poethig R S. 2002. The early phase change gene in maize. *Plant Cell* **14**: 133–147.

Vince-Prue D. 1975. *Photoperiodism in Plants*. McGraw-Hill, London.

von Wiesner J. 1878. *Die Heliotropischen Erscheinungen im Pflanzenreiche: Eine Physiologische Monographie*. Kaiserlich-königlichen Hof- und Staatsdruckerei, Wien, Germany.

Wang Z Y and Tobin E M. 1998. Constitutive expression of the CIRCADIAN CLOCK ASSOCIATED 1 (CCA1) gene disrupts circadian rhythms and suppresses its own expression. *Cell* **93**: 1207–1217.

Wareing P F. 1956. Photoperiodism in woody plants. *Annual Review of Plant Physiology* **7**: 191–214.

Went F W. 1926. On growth accelerating substances in the coleoptile of *Avena sativa. Proceedings Koninklijke Nederlandse Akademie van Wetenschappen* **30**: 10–19.

Went W F. 1928. Wuchsstoff und Wachstum. *Recl. Trav. Bot. Neerland.* **25**: 1–116.

Went F W and Thimann K V. 1937. *Phytohormones.* Macmillan Company, New York.

Whippo C W and Hangarter R P. 2003. Second positive phototropism results from coordinated co-action of the phototropins and cryptochromes. *Plant Physiology* **132**: 1499–1507.

Whippo C W and Hangarter R P. 2006. Phototropism: bending towards enlightenment – Historical prospective. *The Plant Cell* **18**: 1110–1119.

Wigge P A, Kim M C, Jaeger K E, Busch W, Schmid M, Lohmann J U and Weigel D. 2005. Integration of spatial and temporal information during floral induction in Arabidopsis. *Science* **309**: 1056–1059.

Wilkie J D, Sedgley M and Olesen T. 2008. Regulation of floral initiation in horticultural trees. *Journal of Experimental Botany* **59**: 3215–3228.

Willige B C, Ahlers S, Zourelidou M, Barbosa I C, Demarsy E, Trevisan M, Davis P A, Roelfsema M R, Hangarter R and Fankhauser C. 2013. D6PK AGCVIII kinases are required for auxin transport and phototropic hypocotyl bending in Arabidopsis. *Plant Cell* **25**: 1674–1688.

Willmann M R and Poethig R S. 2005. Time to grow up: the temporal role of small RNAs in plants. *Current Opinion in Plant Biology* **8**: 548–552.

Wu G and Poethig R S. 2006. Temporal regulation of shoot development in *Arabidopsis thaliana* by miR156 and its target SPL3. *Development* **133**: 3539–3547.

Xie Z X, Allen E, Wilken A and Carrington J C. 2005. DICER-LIKE 4 functions in trans-acting small interfering RNA biogenesis and vegetative phase change in *Arabidopsis thaliana. Proceedings of the National Academy of Sciences, USA* **102**: 12984–12989.

Yamaguchi A, Kobayashi Y, Goto K, Abe M and Araki T. 2005. TWIN SISTER of FT (TSF) acts as a floral pathway integrator redundantly with FT. *Plant and Cell Physiology* **46**: 1175–1189.

Yamamoto K, Suzuki T, Aihara Y, Haga K, Sakai T and Nagatani A. 2014. The phototropic response is locally regulated within the topmost light-responsive region of the Arabidopsis thaliana seedling. *Plant and Cell Physiology* **55**: 497–506.

Yanovsky M J and Kay S A. 2002. Molecular basis of seasonal time measurement in Arabidopsis. *Nature* **419**: 308–312.

Yanovsky M J, Izaguirre M, Wagmaister J A, Gatz C, Jackson S D, Thomas B and Casal J J. 2000. Phytochrome A resets the circadian clock and delays tuber formation under long days in potato. *Plant Journal* **23**: 223–232.

Yanovsky M J and Kay S A. 2003. Living by the calendar: how plants know when to flower. *Nature Reviews Molecular Cell Biology* **4**: 265–276.

Yoo S K, Chung K S, Kim J K, Lee J H, Hong S M, Yoo S J, Yoo S Y, Lee J S and Ahn J H. 2005. Constans activates suppressor of overexpression of constans 1 through flowering locus T to promote flowering in Arabidopsis. *Plant Physiology* **139**: 770–778.

Zourelidou M, Absmanner B, Weller B, Barbosa I C, Willige B C, Fastner A, Streit V, Port S A, Colcombet J, de la Fuente and van Bentem S. 2014. Auxin efflux by PIN-FORMED proteins is activated by two different protein kinases, D6 PROTEIN KINASE and PINOID. *eLife*. http://doi.org/10.7554/eLife.02860.

11 Vernalization of Horticultural Plants

11.1 INTRODUCTION

Plants continuously perceive the changes in the environmental conditions and align their developmental transitions to seasonal changes in environmental conditions, such as light, temperature and water. Plants used these environmental cues to regulate the transition to flowering. Vernalization is the process wherein plants are induced to flower by extended exposure to cold temperature (Amasino, 2004). Several studies have been conducted to understand how this transition from vegetative to reproductive phase is brought about in plants (Bernier et al., 1981a, 1981b; Halevy, 1985–1989; Bernier, 1988; Kinet, 1993; Finnegan et al., 2001; Amasino, 2004; Song et al., 2012). Plants perceive photoperiod and irradiance through mature intact leaves. In contrast, temperature signals are perceived by all plant parts, although vernalizing temperature is mainly perceived by the shoot apex (Lang, 1986). Like light, temperature also plays a vital role in a number of physiological processes including flowering. The perception of temperature cues by plants is believed to involve a range of sensors, including chromatin-based mechanisms (Kumar and Wigge, 2010). Vernalization occurs during the cold treatment; however, flowering takes place only after several weeks or even months later when other conditions (photoperiods and ambient temperatures) are favorable for flowering (Henderson and Dean, 2004; Kim et al., 2009). While petkus rye (short-day plant) was the first to be used for experimentation, the other plants used for experimentation include *Hyoscyamus niger* (long-day plant), *Triticum aestivum* (CV Winter wheat), *Lunaria biennsis*, *Lolium perennae*, *Beta vulgaris*, *Daucus carota* and *Brassica oleracea*. Most of our current knowledge about the molecular mechanisms in vernalization response was obtained from studies with *Arabidopsis thaliana*. Except for few plants, the optimum vernalizing temperature ranges from 5°C to 8°C (Engelen-Eigles and Erwin, 1997; Wiebe, 1990). The vernalization is completed in 12 (*Allium cepa* L. and *Daucus carota* L.) to 20 weeks (*Brassica oleracea capitata*) (Wiebe, 1990). The vernalization requirement in plants may be either obligatory (plants stay vegetative without cold exposure) or facultative (vernalization has only an additive effect on flower induction) (Wiebe, 1990; Michaels and Amasino, 2000; Engelen-Eigles and Erwin, 1997).

In Arabidopsis, FLOWERING LOCUS C (FLC) gene, which encodes an MADS domain protein, has been shown to be the key to the vernalization response (Michaels and Amasino, 1999, 2001; Sheldon et al., 1999), although other genes may also be involved in this response. Vernalization is very effective at the seed stage or seedling stage. In some plants, even the embryos can be successfully verbalized. While dry seeds cannot be vernalized, the imbibed seeds are responsive to vernalizing temperatures. Additionally, certain plants cannot be vernalized without the presence of minimum number of leaves suggesting that a certain degree of photosynthetic activity is needed to respond to cold treatment. This condition is referred to as "Ripeness to Flowering" (Song et al., 2012). Embryos separated from endosperm fail to be vernalized, but can be effectively vernalized if carbohydrates like sucrose solution or endosperm tissue is supplied to embryo indicating that an energy source is a requirement for vernalization to happen. The meristematic tissues in the shoot apex and leaf are the most sensitive site for vernalization.

11.2 VERNALIZATION PATHWAYS

Using the mutants of *Arabidopsis* with impaired vernalization response together with subsequent analysis of the corresponding genes and changes in chromatin biochemistry, scientists have

DOI: 10.1201/9781003373889-15

successfully characterized the process of vernalization into three phases. These are setting the FLC expression level before exposure to cold, cold-induced FLC silencing and epigenetic silencing of FLC after re-exposure to warm temperatures.

The seedlings from previously vernalized parents require vernalization to flower (Sheldon et al., 2008; Choi et al., 2011). This prevents premature flowering in the autumn. Several regulators have been shown to establish the initial level of FLC expression in *Arabidopsis* (Crevillén and Dean, 2011). This is achieved through two pathways: the first, upregulation of transcription through the conserved RNA polymerase-associated factor 1 complex (Paf1C) and FRIGIDA (FRI) as well as downregulation by the autonomous pathway (Koornneef et al., 1991; Oh et al., 2004; Johanson et al., 2000; Kim et al., 2009).

Swiezewski and coworkers proposed that multiple cold-induced regulatory steps each with different time scales are involved in vernalization (Swiezewski et al., 2009). The transcription of FLC decreases relatively rapidly with the reduction of nascent FLC transcripts within the first two to three weeks of cold exposure. Within a similar timeframe, a cold-dependent upregulation of noncoding antisense transcripts, referred to as COOLAIR, is initiated downstream of the poly-A site of the sense transcript. The addition of COOLAIR promoter sequences to a reporter gene is sufficient to confer silencing of the reporter by cold exposure (Swiezewski et al., 2009). COOLAIR enhances the cold-induced downregulation of FLC expression which has the FLC sense terminator and COOLAIR promoter replaced by an alternative terminator and does not produce antisense transcripts. However, disruption of COOLAIR expression by insertional mutagenesis does not disrupt vernalization in plants that have been exposed to four weeks of cold temperatures (Helliwell et al., 2011).

Significant changes occur at the FLC locus when plants are exposed to warm conditions after prolonged exposure to cold temperatures. Within just a few days after the transfer to warm temperatures, the PHD – PRC2 complex is detected across the whole FLC locus (De Lucia et al., 2008; Shcherban and Salina, 2013). The associated histone modification, H3K27me3, also increases substantially across the whole gene (Angel et al., 2011; De Lucia et al., 2008; Finnegan and Dennis, 2007).

Moisture and oxygen are required for effective vernalization. The seeds or embryos should possess at least 40–50% water in their cells for vernalization to be effective. Thr nrrf for oxygen is substantiated by the fact that carbohydrates are required for effective vernalization.

Meristematic cells in stem apex and leaves have been shown to produce a substance called "vernalin" following the exposure to vernalizing temperature (Kim et al., 2009). This substance is graft transmissible and capable of diffusion and induce flowering. Attempts to isolate and identify the components of vernalin have so far failed. Whether the vernalin is the same as florigin or a precursor of florigin is also not yet clearly established.

11.2.1 Devernalization

After sufficient vernalization, if the seeds or seedlings are subjected to warm temperature (35–40°C), the plants that develop from such treatment fail to flower. The negating effect of higher temperatures is called "devernalization." Nevertheless, if the vernalized plants are maintained at sufficiently low temperatures for a long period of time, devernalization does not happen. This may be due to the putative vernalin might have already moved and acted upon the genetic material and committed it for flower formation. However, devernalized plants can be revernalized by subjecting the same seedling or seed again for another period of cold treatment. Again, if vernalization is prolonged too long, the seedlings lose the effect of vernalization and lose their potentiality to flower.

11.3 HORMONAL CONTROL

Many plants which require cold treatment also require proper photoperiodic treatment for the induction of flowers. In such plants, gibberellins not only substitute vernalization and photoperiodic treatment. But same gibberellins have no effect in some long-day and cold-requiring plant species, such as petkus rye. In order for plants to respond to GA treatment, they should possess a cluster of leaves (rosette). The effect of GA on plants like Henbane has been attributed to its effect on the elongation of internodes at which time GA also promotes and probably elaborates the factors required for the induction flowers. Like cold treatment, GA also induces VER203 expression. Hence, GA can partially substitute cold treatment to initiate flowering in plants that require cold treatment.

11.4 MATHEMATICAL MODELING OF THE VERNALIZATION

Angel et al. (2011) attempted to elucidate the quantitative aspects of vernalization using mathematic modeling. This model assumes that the histone modifications are a key functional element of epigenetic memory and the overall histone modification status characterizes the transcriptional state of a locus, where H3K27me3 correlates with repression. The histone modifications of a certain type tend to recruit protein complexes that lead to the addition of further modifications in a "reading-and-writing" mechanism. Any histone modifications that are lost in the biological processes or DNA replication are replaced by long-range positive feedback of antagonistic histone modifications. A number of models have been developed for field crops (Clarkson and Russell, 1979; Ellis et al., 1989; Habekotté, 1997; Liu, 2007; McMaster et al., 2008) and for *Arabidopsis* (Chew et al., 2012). These models have used linear (Ellis et al., 1989), exponential (Clarkson and Russell, 1979; Larsen, 1988), dent-like (Habekotté, 1997; McMaster et al., 2008), sigmoidal (Liu, 2007) or a combination of beta and broken linear functions (Chew et al., 2012). The use of nonlinear functions typically allows elucidation of the relationships among temperature and the level of vernalization saturation and the duration required for complete vernalization (Liu, 2007; McMaster et al., 2008).

11.5 MOLECULAR BASIS OF VERNALIZATION

Several authors associated epigenetic changes in gene expression such as imprinting with changes in DNA methylation (Tilghman, 1999; Grossniklaus et al., 1998; Russinova and de Vries, 2000). They suggested that vernalization may involve changes in the pattern of DNA methylation, and plants with reduced methylation levels show a number of pleiotropic effects including ectopic expression of floral organ-identity genes such as *AGAMOUS* which are known to induce flowering when ectopically expressed (Dennis et al., 1997; Finnegan et al., 1996, 1998; Mizukami and Ma, 1997). As noted by Finnegan et al. (1998), the promotive effects of vernalization and demethylation on flowering are additive acting on separate flowering pathways. Thus, changing patterns of DNA methylation may contribute to the epigenetic changes in gene expression that are likely to be responsible for the vernalized state, but more research is required to establish the molecular basis of this phenomenon. In many species, the plant hormone gibberellin (GA) plays a role in the regulation of flowering, although evidences suggest that GA may not be directly involved in the vernalization pathway. Using the grafting studies, Lang (1986) demonstrated the possible existence of a transmissible vernalization hormone (vernalin).

REFERENCES

Amasino R. 2004. Vernalization, competence and epigenetic memory of winter. *The Plant Cell* **16**: 2553–2559.

Angel A, Song J, Dean C and Howard M. 2011. A Polycomb-based switch underlying quantitative epigenetic memory. *Nature* **476**: 105–108.

Bernier G. 1988. The control of floral evocation and morphogenesis. *Annual Review of Plant Physiology and Plant Molecular Biology* **39**: 175–219.

Bernier G, Kinet J-M and Sachs R M. 1981a. *The Physiology of Flowering*. Volume I. CRC Press, Boca Raton.

Bernier G, Kinet J-M and Sachs R M. 1981b. *The Physiology of Flowering*. Volume II. CRC Press, Boca Raton.

Chew Y H, Wilczek A M, Williams M, Welch S M, Schmitt J and Halliday K J. 2012. An augmented Arabidopsis phenology model reveals seasonal temperature control of flowering time. *New Phytologist* **194**: 654–665.

Choi K, Kim J, Hwang H J, Kim S, Park C, Kim S Y and Lee I. 2011. The FRIGIDA complex activates transcription of FLC, a strong flowering repressor in *Arabidopsis* by recruiting chromatin modification factors. *Plant Cell* **23**: 289–303.

Clarkson N M and Russell J S. 1979. Effect of temperature on the development of two annual medics. *Australian Journal of Agricultural Research* **30**: 909–916.

Crevillén P and Dean C. 2011. Regulation of the floral repressor gene FLC: the complexity of transcription in a chromatin context. *Current Opinion in Plant Biology* **14**: 38–44.

De Lucia F, Crevillen P, Jones A M, Greb T and Dean C. 2008. A PHD polycomb repressive complex 2 triggers the epigenetic silencing of FLC during vernalization. *Proceedings of National Academy Science, USA* **105**: 16831–16836.

Dennis E S, Bilodeau P, Burn J, Finnegan E J, Genger R, Helliwell C, Kang B J, Sheldon C C and Peacock W J. 1997. Methylation controls the low temperature induction of flowering in Arabidopsis. pp. 97–103. *Control of Plant Development, Genes and Signals* (Eds.) Greenland A J, Meyerowitz E M and Steer M. Portland Press Ltd, London.

Ellis R H, Summerfield R J, Roberts E H and Cooper J P. 1989. Environment control of flowering in barley (*Hordeum vulgare*). III. Analysis of potential vernalization responses, and methods of screening germplasm for sensitivity to photoperiod and temperature. *Annals of Botany* **63**: 687–704.

Engelen-Eigles G and Erwin J E. 1997. A model plant for vernalization. *Scientia Horticulturae* **70**: 197–202.

Finnegan E J, Bagnall D, Helliwell C, Rouse D, Sheldon C, Tagede M, Peacock W J and Denis E. 2001. Vernalization. pp. 1–5. *Encyclopedia of Life Science*. Nature Publishing Group, Macmillan Publishers Ltd., London.

Finnegan E J and Dennis E S. 2007. Vernalization-induced trimethylation of histone H3 lysine 27 at FLC is not maintained in mitotically quiescent cells. *Current Biology* **17**: 1978–1983.

Finnegan E J, Genger R K, Kovac K, Peacock W J and Dennis E S. 1998. DNA methylation and the promotion of flowering by vernalization. *Proceedings of the National Academy of Science United States of America* **95**: 5824–5829.

Finnegan E J, Peacock W J and Dennis E S. 1996. Reduced DNA methylation in Arabidopsis thaliana results in abnormal plant development. *Proceedings of the National Academy of Science United States of America* **93**: 8449–8454.

Grossniklaus U, Vielle-Calzada J P, Hoeppner M A and Gagliano W B. 1998. Maternal control of embryogenesis by *MEDEA*, a polycomb group gene in Arabidopsis. *Science* **280**: 446–450.

Habekotté B. 1997. A model of the phenological development of winter oilseed rape (Brassica napus L.). *Field Crops Research* **54**: 127–136.

Halevy A H. 1985–1989. *Handbook of Flowering*. CRC Press, Boca Raton, LA.

Helliwell C A, Robertson M, Finnegan E J, Buzas D M and Dennis E S. 2011. Vernalization-repression of *Arabidopsis* FLC requires promoter sequences but not antisense transcripts. *PLOS ONE* **6**: e21513.

Henderson I R and Dean C. 2004. Control of *Arabidopsis* flowering: the chill before the bloom. *Development* **131**: 3829–3838.

Johanson U, West J, Lister C, Michaels S, Amasino R M and Dean C. 2000. Molecular analysis of *FRIGIDA*, a major determinant of natural variation in *Arabidopsis* flowering time. *Science* **290**: 344–347.

Kim D H, Doyle D M, Sung S and Amasino R M. 2009. Vernalization: winter and the timing of flowering in plants. *Annual Review of Cell and Developmental Biology* **25**: 277–299.

Kinet J. 1993. Environmental, chemical, and genetic control of flowering. *Horticultural Reviews* **15**: 279–333.

Koornneef M, Hanhart C J and Veen J H. 1991. A genetic and physiological analysis of late flowering mutants in *Arabidopsis thaliana. Molecular and General Genetics* **229**: 57–66.

Kumar S V and Wigge P A. 2010. H2A.Z-containing nucleosomes mediate the thermosensory response in *Arabidopsis*. *Cell* **140**: 136–147.

Lang A. 1986. Hyoscyamus niger. pp. 144–186. *Handbook of Flowering* (Ed.) Halevy A H. CRC Press, Boca Raton.

Larsen R U. 1988. A prediction model of the development rate of the pot plant Senecio × hybridus Hyl: a case study. *Acta Horticulturae* **230**: 381–388.

Liu D L. 2007. Incorporating vernalization response functions into an additive phenological model for reanalysis of the flowering of annual pasture legumes. *Field Crops Research* **101**: 331–342.

McMaster G S, White J W, Hunt L A, Jamieson P D, Dhillon S S and Ortiz-Monasterio J I. 2008. Simulating the influence of vernalization, photoperiod and optimum temperature on wheat development rates. *Annals of Botany* **102**: 561–569.

Michaels S D and Amasino R M. 1999. FLOWERING LOCUS C encodes a novel MADS domain protein that acts as a repressor of flowering. *Plant Cell* **11**: 949–950.

Michaels S D and Amasino R M. 2000. Memories of winter: vernalization and the competence to flower. *Plant, Cell and Environment* **32**: 1145–1153.

Michaels S D and Amasino R M. 2001. Loss of FLOWERING LOCUS C activity eliminates the late-flowering phenotype of FRIGIDA and autonomous pathway mutations but not responsiveness to vernalization. *Plant Cell* **13**: 935–941.

Mizukami Y and Ma H. 1997. Determination of Arabidopsis floral meristem identity by *AGAMOUS*. *The Plant Cell* **9:** 393–408.

Oh S, Zhang H, Ludwig P and van Nocker S. 2004. A mechanism related to the yeast transcriptional regulator Paf1c is required for expression of the Arabidopsis FLC/MAF MADS box gene family. *Plant Cell* **16**: 2940–2953.

Russinova E and de Vries S. 2000. Parental contribution to plant embryos. *Plant Cell* **12**: 461–464.

Shcherban' A B and Salina E A. 20**13**. Epigenetic regulation of expression of vernalization genes**.** *Tsitolgiia* **55**: 234–237.

Sheldon C C, Burn J E, Perez P P, Metzger J and Edwards J A. 1999. The FLF MADS box gene: repressor of flowering in *Arabidopsis* regulated by vernalization and methylation. *Plant Cell* **11**: 445–458.

Sheldon C C, Hills M J, Lister C, Dean C, Dennis E S and Peacock W J. 2008. Resetting of FLOWERING LOCUS C expression after epigenetic repression by vernalization. *Proceedings of National Academy of Science, USA* **105**: 2214–2219.

Song J, Angel A, Howard M and Dean C. 2012. Vernalization – cold-induced epigenetic switch. *Journal of Cell Science* **125**: 3723–3731.

Swiezewski S, Liu F, Magusin A and Dean C. 2009. Cold-induced silencing by long antisense of an Arabidopsis polycomb target. *Nature* **462**: 799–802.

Tilghman S M. 1999. The sins of the fathers and mothers: genomic imprinting in mammalian development. *Cell* **96**: 185–193.

Wiebe H J. 1990. Estimation of raising temperature at time of bolting of Chinese Cabbage. *Acta Horticulturae* **267**: 297–304.

12 Pruning and Training

12.1 INTRODUCTION

Pruning is the removal of unwanted growths plant parts to induce desired new growths. Fruit trees are pruned for several reasons:

- Pruning establishes the basic tree structure and limits the tree size.
- Pruning opens spaces in the tree canopy for sunlight to penetrate and facilitate photosynthesis and the production of carbohydrates and energy that drive the growth of roots, shoots and fruits.
- Pruning reduces branch crowding, reduces the amount of foliage inside the canopy and enables better air circulation that reduces humidity and allows efficient canopy management strategy (CMS).
- Branch angles between a vertical trunk and lateral shoots are stronger as they approach a 90° degree angle. Narrow branch angles are typically weak, and they may break under the weight of developing fruit or branch.
- Dormant pruning is an invigorating process that stimulates vegetative growth while summer pruning is a devigorating process that reduces growth because it involves removing part of the functional leaf area of the tree.
- Pruning is done to maintain the right balance between vegetative growth, flower production and fruiting. Dormant pruning is a reliable crop load management tool. It can be used to control alternate bearing in some fruit crops.
- By removing the dead woods and infected parts, pruning prevents the spread of disease or insect damage.

Training is practiced in horticultural crops to develop desired tree framework or structure with or without pruning of certain plant parts of the plant. It is undertaken to admit more sunlight and air to the center of the plant and expose maximum leaf area for light interception. It is also done to limit the growth and spread of the plant to facilitate trouble-free cultural practices, such as spraying harvesting and pollination.

The main functions of training include

- admitting sunlight and air to the center of the plant and at the same time, expose as many leaves as possible to sun;
- directing growth of the plants in such way that cultural operations can be performed both efficiently and at lower cost;
- protecting the tree from sunburn and wind damage;
- balanced distribution of fruit bearing branches on the main limb or stem.

The major differences between pruning and training are shown in Table 12.1.

DOI: 10.1201/9781003373889-16

TABLE 12.1
Differences between Training and Pruning

Trait	Training	Pruning
Function	Provides main structure or shape to the plant. Determines the general character and even details of the plant's outline and of its branching and framework.	Corrects or maintains tree structure and induces flowering and fruiting in some plants.
Impact	Does not significantly affect physiological processes in plants	Reduces carbohydrate levels and hormonal balance in shoots as well as alters physiological processes.
Types	Includes summer training and summer pruning as well as dormant pruning.	Includes summer pruning, dormant pruning and root pruning.
Parts affected	Training will reduce cutting by directing the growth and involves both pruning and other means (bending)	Involves removal of shoots/roots

12.2 PRUNING TYPES

12.2.1 Summer Pruning

Sometimes, light pruning is done in the summer to allow the light into the tree canopy or to remove excess canopy. Timing and amount of shoot removal are important, since it can impact carbohydrate supply to developing fruit and also direct sun exposure of fruit that could lead to sunburn. Several researchers reported summer pruning to reduce vegetative growth, improve light penetration, alter levels of endogenous growth hormones and shoot carbohydrates, change plant physiological processes, increase the number of flower buds, alter fruit maturation and enhance fruit quality (Hyden and Emerson, 1976; Ritter, 1983; Ferree, 1988; Utermark, 1977; Marini, 1985; Sharif Hossain et al., 2006). These authors have suggested that the summer pruning is less invigorating than dormant pruning. Recently, summer pruning has attracted much attention as a potential method of controlling tree size and reducing pruning costs in orchard operations (Taylor and Ferree, 1984). Sharif Hossain et al. (2006) demonstrated that summer pruning in mid-July after fruit harvest was a useful practice for maintaining slender spindle bush type of peach trees, especially when grafted on vigorous rootstocks.

12.2.2 Mechanical Pruning

Recently, more sophisticated pruning machines have been developed for commercial vineyards. Their guiding systems permit removal of greater amounts of the pruning, although follow-up by hand pruners is always required. Switching to mechanical pruning has significantly reduced costs and improved quality.

12.2.3 Root Pruning

The main purpose of root pruning is to remove part of the root system to reduce the absorption of water and nutrients, as well as alter the hormone balance within the plant body. This results in an overall reduction in shoot growth and altered physiological processes. Root pruning practiced four to six weeks before flowering was found to be most effective (Saure, 2007). Root pruning is effective when topsoil is shallow. Distance from the trunk and pruning depth is influenced by the level of tree vigor and root distribution. Pruning can be carried out on either one or both sides of a tree – but one-sided pruning (alternating each year) is considered least risky. Water stress induced by root pruning is temporary and trees recover shortly, but water stress can significantly slow down shoot growth. Water stress can be overcome by irrigating the trees immediately after root pruning although excess irrigation can lead to increased vegetative growth and reduced flowering.

Responsiveness of trees to root pruning changes with the age of the tree and crop load. Shoot growth is affected more by root pruning in older apple trees with high crop load than in young trees with no or low crop load (Schupp and Ferree, 1987; Schupp et al., 1992; Ferree and Knee, 1997). Hoad and Abbott (1986) attributed this to slow recovery of the root system to high crop load. Furthermore, favorable soil moisture conditions and higher soil fertility may negate the growth inhibiting effects of root pruning, while low soil moisture levels and poor soils increase the negative impact of root pruning (Ferree, 1992; Miller, 1995).

12.3 TYPES OF PRUNING CUTS

12.3.1 Thinning Cut

Also known as selective cutting or drop-crotching, thinning cut involves the complete removal of a branch back to the point of its origin on the main stem or to another lateral branch to stimulate new growth in many growing points. Although the overall general shape of the tree is maintained, the thinning cuts reduce the number of branches to allow air and sunlight into the interior of plant.

12.3.2 Heading Cut

It involves removal of only the tips of the branches to promote the growth of lower buds on the stem. It improves the size, shape and compactness of the plant by refocusing on growth and encouraging the sprouting of lateral buds on the branch. Heading cuts influence flowering and fruiting with light heading cuts usually producing morea greater number of stems and dense foliage. Although the number of flowers and fruits produced are more, the fruits may be smaller in size and of lower quality. Severe heading cuts, on the other hand, promote fewer numbers of flowers and fruits, but they will be of better quality.

12.3.3 Bench Cut

It is a special type of heading cut where terminal section of a branch is removed just above a side branch. Bench cuts are sometimes necessary to redirect upright-growing branches to the outside but should be avoided if possible because waterspouts will develop on the flat branch section below the bench cut. In contrast, removal of slow growing, weak, unproductive and underhanging branches or shorts is referred to as "thin wood pruning."

12.4 TRAINING SYSTEMS

In horticultural plants, training starts from the time the seedlings are planted in the field. It's more efficient to train plants in the first year itself by placing limbs in a position, rather than pruning them later. Growth in fruit trees is regulated by plant hormones, such as cytokinins produced in the roots that transported to the shoot tips to break dormancy. When the apical buds begin to grow, they produce auxin which moves down through gravity and inhibits the growth of lateral buds. This phenomenon is known as "apical dominance." The best way to overcome apical dominance is to train branches early to encourage them to grow more horizontal and remove unwanted upright shoots.

12.4.1 Training Systems

Basic training systems are explained in the following.

12.4.1.1 Central Leader Training

After choosing a vigorous central shoot at the time of planting, all competing upright shoots around it are removed. Limbs are spread so as not to compete with the central leader. It is now an outdated system.

12.4.1.2 Open-Center Training

Three or four shoots are chosen in the first year to form the main architecture, and competing limbs are removed. Other systems have developed from this basic tree architectural system.

12.4.1.3 Espalier Training

The tree structure is created in two dimensions by using trellis systems. The main trunk is tied to the wires. Buds are selected to form branches, and shoots are fastened to training wires. Trellised systems are desirable because they would produce the best quality fruits. Application of growth retardants or dwarfing rootstocks allows these training systems more feasible.

12.4.1.4 Palmette (Trellised or Untrellised)

This system is suitable for wide spaced tree crops. A trellised palmette would be advisable with wires at 50, 100, 150 and 200 cm high.

12.4.1.5 Free-Standing Central Leader

This is an inexpensive and simple system to develop. Disadvantages are that trees can become fairly tall and, because it is untrellised, limb breakage and fruit rubbing to each other can be a problem. Spacing will vary with cultivar and soil depth.

12.5 TREE CANOPY MANAGEMENT

Fruit crops flower and produce fruits naturally without human intervention. In an unmanaged tree canopy, almost two-thirds of the total leaf area occurs in the uppermost 30% of the total tree canopy is exposed to 60–100% of total sunlight, whereas nearly 70% of the leaf area in the tree remains overcrowded and shaded all the time. Consequently, the majority of the flowering and fruiting occurs in the uppermost 30% of the tree canopy, while the growth, flowering, fruiting and fruit quality in the lower portions of the tree are severely restricted by poor-light conditions. Apart from causing a significant reduction in crop productivity, fruits at the top are difficult to reach to carry out cultural operations like spraying and harvesting. Therefore, it is important to manage the tree canopy to strike a proper balance between vegetative growth and fruiting and promote uniform distribution of vegetative and fruiting branches. These problems can be achieved by following an appropriate and effective canopy management strategy (Krajewski and Krajewski, 2011; Krajewski and Pittaway, 2000). Tree canopy management will also help in developing strong tree architecture that will support heavy crop loads while improving productivity and fruit quality in the long term. There have been several excellent reviews of physiological basis of canopy management strategies (CMSs) by Krajewski (1996), Krajewski and Pittaway (2000), Bevington et al. (2002) and Krajewski and Krajewski (2011). CMS is applied through pruning, training and following appropriate horticultural practices (precision irrigation practices and balanced nutrition to balance vegetative and reproductive growths). Pruning alters the apical dominance of shoot tips, reduces the number of budbuds on the tree and removes dead, infected, out-of-place and week branches. It also improves the distribution of light within the canopy lighting, leading to better distribution of sun light throughout the canopy and reduce shading of leaves to promote bud growth, flowering, fruiting, fruit size, fruit pigmentation and tree growth (Krajewski, 1996; Krajewski and Pittaway, 2000). Pruning also improves the flow of sap, which contains water, nutrients, photosynthetic assimilate and growth hormones from different parts of the plant body to the developing fruits. The developing fruits in positions within the canopy where light conditions are optimal and sap flow is strong will also be of high quality (Krajewski and Pittaway, 2000, 2011; Bevington et al., 2002).

The aim of any CMS is to strike a balance between vegetative and fruiting wood, with uncrowded bearing wood situated conveniently within the relative shelter of the well-lit, adequately ventilated inner canopy and low enough to harvest from the ground. To do that, it is important to apply CMS to trees of all ages and appreciate the relation between how young trees are trained and how they are pruned when mature. The benefits of pruning are most often negated by over-irrigation and over-fertilization with especially nitrogen and potassium.

12.6 SOURCE–SINK RELATIONSHIP

A plant consists of both autotrophic organs that produce sugars, amino acids and other growth promoting substances and heterotrophic organs that utilize these substances. The efficiency of these organisms depends on their ability to capture and convert light energy into chemical energy (photosynthesis) and later translocate, store and utilize that chemical energy (De Jong and Ryugo, 1998). The autotrophic organs in the plant (typically mature leaves) are known as sources, while non-photosynthetic heterotrophic organs (fruits, roots and tubers) and immature leaves are known as sinks (Taiz and Zeiger, 2006). The partitioning and allocation of photosynthetic products is closely connected to plant growth since the export of carbohydrate from photosynthesizing leaves provides serves as energy source for the substrate for the growth and maintenance of non-photosynthetic tissues (Osorio et al., 2014). Marschner (2012) pointed out that the genotype and environment interaction strongly influences this source–sink relationships. Modern fruit growers try to manipulate the source–sink relationship to ensure adequate fruit production, uniformity in fruit production (regular bearing) and quality (Gil, 2006; Nerson, 2009). Biomass or crop yield, to a large extent, depends on the whole-plant photosynthesis over the entire growing season, and hence, greenhouse growers enrich the greenhouse environment with supplemental carbon dioxide to enhance photosynthetic rates. However, measurement of photosynthetic rates once or twice during the growing season will not provide sufficient information on biomass production or crop yield because yield is determined by both the availability of photosynthetic assimilate and the way in which the photosynthetic assimilates are partitioned throughout the plant. Although light and CO_2 are considered as key limiting factors determining the assimilate production in source tissues (photosynthetic rate) and, ultimately, the plant growth, the ability of the sink tissues to utilize photosynthetic assimilates is also crucial for plant growth (Foyer and Paul, 2001). Therefore, to fully comprehend the relationship between photoassimilate partitioning and plant growth, it is necessary to consider three important factors: (i) production of photoassimilates (source capacity), (ii) transport of photoassimilates and (iii) utilization of photoassimilates by sink organs (sink strength).

12.6.1 Source

The photosynthesis, a process by which the sun energy is captured by green tissues (chlorophyll in leaves) and used for the synthesis of carbohydrates and other compounds, occurs mainly in leaves (Kozlowski and Pallardy, 1997). Dejong and Ryugo (1998) and Friedrich and Fischer (2000) showed the influence of various factors on photosynthesis in fruit species. Photosynthesis rates in some of the fruit crops are shown in Table 12.2. Plant growth and crop yield depend largely on the ready steady supply of carbohydrates and nutrients (Oliveira and Priestley, 1988).

A part of the photosynthetic assimilates produced in the leaves are utilized by young developing leaves. The remaining quantities are transported to other organs which depend on them (heterotrophic). The leaves that are in close proximity to developing fruits exhibit increased photosynthetic rates as compared to the remaining leaves of the tree (Urban et al., 2003). In other species, the calyx with its photosynthetic capacity plays an important role in the production and translocation of carbohydrates during early fruit development (Fischer and Lüdders, 1997). Leaf drop, whether natural or caused by environmental factors such as temperature, storms, hail, harmful chemicals, drought, salinity and oxygen deficiency, has a profound negative effect on source–sink relationships (Fischer, 2011). According to Fischer (2011), optimal leaf area for favorable growth and quality in several fruit species is 200 cm^2 per 100 g of fresh fruit mass, although grapes require twice as much this value (Table 12.3). Similarly, the number of leaves per fruit also influences carbohydrate

TABLE 12.2
Maximum Photosynthetic Rates in Selected Fruit Crops

Crop	Photosynthetic rate (μmol CO_2 fixed m^{-2} s^{-1})	Reference
Orange	9.9 ± 9.9 ± 1.6	Flore and Lakso (1989)
Peach	13.3 ± 13.3 ± 3.8	Flore and Lakso (1989)
Avocado	4.8 ± 2.4	Flore and Lakso (1989)
Pear	20.2	Flore and Lakso (1989)
Grape	12.4 ± 1.4	Flore and Lakso (1989)
Blue berry	12.7 ± 7.4	
Apple	5.7 ± 5.6	
Strawberry	13.9 ± 2.9	
Gooseberry	8.3	Melgarejo and Fischer (2013)
Hazel nut	12.2	Hampson et al. (1996)

TABLE 12.3
Leaf to Fruit Ratio for Optimal Fruit Production

Crops	Number leaves/fruit	Reference
Mango	30–60	Fischer (2005)
Pineapple	6–8	Malézieux et al. (2003)
Plum	15–25	Fischer (2005)
Peach	15–30	Fischer (2005)
Apple	15–35	Fischer (2005)
Pear	25–40	Fischer (2005)
Grape	7–8	Sauer and Baumann (2007)
Satsuma Mandarin	25	Kubo et al. (2001)

Source: Adapted from Léchaudel et al. (2004); *Acta Horticulturae* 645(645): 429–433.

accumulation in fruits. For instance, in mango, 100 leaves per fruit resulted in maximum dry weight accumulation (Léchaudel et al., 2004).

Fruits accumulate large quantities of starch, sucrose or hexose sugars during their development (Kozlowski and Pallardy, 1997). Accumulation of carbohydrate in fruits depends on the cultivar, leaf:fruit ratio, and growing conditions (Friedrich and Fischer, 2000). While a high fruit load decreases the distribution of assimilates to the roots and other plant organs, the lack of assimilates have negative effects on fruit production in the following years (Lenz, 2009). Poor accumulation of reserves inhibits flower induction, leading to alternate bearing in persimmon, citrus, mango and apple (Ojima et al., 1985; Goldschmidt and Golomb, 1982; Lenz, 2009). Expanded leaves near the fruit have been shown to exhibit increased photosynthetic rates (Urban et al., 2003).

12.6.2 Translocation of Carbohydrates

The distribution (partitioning) of carbohydrates (CH) determines the amounts and patterns of plant growth and yield (Lakso and Flore, 2003). The transport direction and volume depend on sink position and their relative strength (Friedrich and Fischer, 2000). Carbohydrates and other organic substances are translocated by an osmotically generated pressure gradient between source and sink organs (Taiz and Zeiger, 2006). Carbohydrates in the form of sucrose supplied to the fruit are generally from leaf photosynthesis and plant carbohydrate reserves (Friedrich and Fischer, 2000). The upper recently matured fully expanded leaves export CH to young leaves and meristem, whereas the basal leaves send photosynthates predominantly to the roots (Taiz and Zeiger, 2006). According to Kozlowski and Pallardy (1997), the rate of translocation of photosynthates from the sources (mainly leaves) to the sinks (mainly fruits) influences the rate of photosynthesis.

12.6.3 Sink Strength

Carbohydrate partitioning within a tree is not a genetically programmed process but is a result of the unique combination of competing organs and their relative abilities to compete for limited carbohydrates (Lakso and Flore, 2003). The degree of competition among various sinks depends on the organ's metabolic status and the distance from the CH source. Leaves closest to the fruit have a dominating photosynthetic activity (Herold, 1980). Shoots and roots of young trees receive considerable amounts of CH but tends to decline with tree age and with heavy fruit loads (Lakso and Flore, 2003; Lenz, 2009). The percentage of total DM partitioned to the harvested portion (e.g., the fruit) is known as harvest index (Lakso and Flore, 2003).

REFERENCES

Bevington K B, Falivene S, Moulds G and Krajewski A J. 2002. Pruning citrus for export Fruit size. *Citrus Market Development Group (CMDG) Final Report Project 50*. New South Wales Department of Agriculture, Australia.

DeJong T M and Ryugo K. 1998. Carbohydrate assimilation, translocation, and utilization. pp. 109–114. *Walnut Production Manual* (Ed.) Ramos D E. University of California ANR Publication No. 3373. Available at http://anrcatalog.ucdavis.edu/Nuts/3373.aspx.

Ferree D C. 1988. Role of rootstocks and spur-type scions for controlling vegetative growth of apple and peach trees. *HortScience* **23**: 464–467.

Ferree D C. 1992. Time of root pruning influences vegetative growth, fruit size, biennial bearing, and yield of 'Jonathan' apple. *Journal of American Society for Horticultural Science* **117**: 198–202.

Ferree D C and Knee M. 1997. Influence of root pruning an rootstock on growth and performance of 'Golden Delicious' apple. *HortScience* **32**: 645–648.

Fischer G. 2005. Aspectos de la fisiología aplicada de los frutales promisorios en cultivo y poscosecha. *Review of Comalfi* **32**: 22–34.

Fischer G. 2011. La relación hoja/fruto en especies frutícolas. pp. 40–53. *Proceedings of 4th Colombian Congress of Horticulture*. Mosquera, Palmira, Colombia.

Fischer G and Lüdders P. 1997. Developmental changes of carbohydrates in cape gooseberry (*Physalis peruviana* L.) fruits in relation to the calyx and the leaves. *Agronomy Colombia* **14**: 95–107.

Flore J A and Lakso A N. 1989. Environmental and physiological regulation of photosynthesis in fruit crops. *Horticultural Reviews* **36**: 111–157.

Foyer C H and Paul M J. 2001. Source-sink relationships. *Encyclopedia of Life Sciences*. Nature Publishing Group. Available at www.els.net.

Friedrich G and Fischer M. 2000. *Physiologische Grundlagen des Obstbaues*. Ulmer Verlag, Stuttgart, Alemania.

Gil G F. 2006. *Fruticultura: La producción de fruta de climas templados y subtropical y uva de vino*. Ediciones Universidad Católica de Chile, Santiago.

Goldschmidt E E and Golomb A. 1982. The carbohydrate balance of alternate bearing in citrus trees and the significance of reserves for flowering and fruiting. *Journal of American Society for Horticultural Science* **107**: 206–208.

Hampson C R, Azarenko A N and Potter J R. 1996. Photosynthetic rate, flowering, and yield component alteration in hazelnut in response to different Light Environments. *Journal of American Society for Horticultural Science* **121**: 1103–1111.

Herold A. 1980. Regulation of photosynthesis by sink activity – the missing link. *New Phytologist* **86**: 131–144.

Hoad G V and Abbott D L. 1986. Hormonal control of growth and reproductive development in apple. pp. 87–92. *Regulation of Photosynthesis in Fruit Trees* (Eds.) Lakso A N and Lenz F. Special Report NewYork Agriculture. Experimental Station. Cornell University, Ithaca, NY.

Hyden R A and Emerson F H. 1976. The case of summer pruning. *American Fruit Grower* **95**: 27–28.

Kozlowski T T and Pallardy S G. 1997. *Physiology of Woody Trees*. 2nd edition. Academic Press, San Diego, CA.

Krajewski A J. 1996. Pruning of citrus in Southern Africa: a Hacker's guide. *Citrus Journal* **6**: 19–23.

Krajewski A J and Krajewski S A. 2011. Canopy management of sweet orange, grapefruit, lemon, lime and mandarin trees in the tropics: principles, practices and commercial experiences. *Proceedings of the First International Symposium on Tropical Horticulture* **894**: 65–76.

Krajewski A J and Pittaway T. 2000. Manipulation of Citrus flowering and fruiting by pruning. pp. 357–360. *Proceedings of 9th International Citrus Congress*. International Organization of Citrus Virologists, Orlando, FL.

Kubo T, Hohjo I and Hirat S. 2001. Sucrose accumulation and its related enzyme activities in the juice sacs of Satsuma mandarin fruit from trees with different crop loads. *Scientia Hortiulturae* **91**: 215–225.

Lakso A N and Flore J A. 2003. Carbohydrate partitioning and plant growth. pp. 21–30. *Concise Encyclopedia of Temperate Tree Fruit* (Eds.) Baugher T A and Singh S. Food Products Press, New York.

Léchaudel M, Jannoyer M and Génard M. 2004. Effects of the leaf:fruit ratio on growth and partitioning of water and dry matter in mango fruit. *Acta Horticulturae* **645**: 429–433.

Lenz F. 2009. Fruit effects on the dry matter and carbohydrate distribution in apple tree. *Acta Horticulturae* **835**: 21–38.

Malézieux E, Cote F and Bartholomew D P. 2003. Crop environment, plant growth and physiology. pp. 69–107. *The Pineapple: Botany, Production and Uses* (Eds.) Bartholomew S P, Paull R E and Rohrbach K G. CABI Publishing, Wallingford.

Marini R P. 1985. Vegetative growth, yield and fruit quality of peach as influenced by dormant pruning, summer pruning and summer topping. *Journal of American Society of Horticultural Science* **110**: 133–139.

Marschner, P. (Ed.). 2012. *Marschner's Mineral Nutrition of Higher Plants*. 3rd edition. Elsevier, Oxford.

Melgarejo L M and Fischer G. 2013. Ecophysiological response of cape gooseberry (*Physalis peruviana* L.) to the environmental conditions of Bogotá Plateau. https://doi.org/10.17584/rcch.2012v6i2.1980.

Miller S S. 1995. Root pruning and trunk scoring have limited effect on young bearing apple trees. *HortScience* **30**: 981–984.

Nerson H. 2009. Source-sink relationships and their effects on fruit growth and quality of casaba melon (Cucumis melo L. var Inodorous). *Fruit, Vegetable and Cereal Science and Biotechnology* **3**: 44–47.

Ojima M, Dall'Orta F A C, Barbosa W, Tombolata A F C and Rigitano O. 1985. Alternate bearing in the persimmon cultivar Pomelo (IAC 6–22). *Bragantia* **44**: 481–486.

Oliveira C M and Priestley C A. 1988. Carbohydrate reserves in deciduous fruit trees. *Horticultural Reviews* **10**: 403–430.

Osorio S, Ruan Y-L and Fernie A R. 2014. An update on source-to-sink carbon partitioning in tomato. *Frontiers in Plant Science* **5**: 516/1–516/11.

Ritter C M. 1983. Summer pruning of peach trees. *The Mountainer Grower* **448**: 16–19.

Sauer E and Baumann A. 2007. *Qualitätsoptimierung durch Entblätterung*. Available at www.lwg.bayern.de/weinbau/berufsausbildung_qualifizierung/18385/.

Saure M C. 2007. Root pruning – a poorly understood management practice in fruit trees. *International Journal of Fruit Science* **7**: 43–56.

Schupp J R and Ferree D C. 1987. Effect of root pruning at different growth stages on growth and fruiting of apple trees. *HortScience* **22**: 387–390.

Schupp J R, Ferree D C and Warrington I J. 1992. Interactions of root pruning and deblossoming on growth, development and yield of 'Golden Delicious' apple. *Journal of Horticultural Science* **67**: 465–480.

Sharif Hossain A B M, Mizutani F, Onguso J M, El-Shereif A R and Rutto K L. 2006. Effect of summer pruning on shoot growth and fruit quality in peach trees trained as slender spindle bush type. *Memoirs of the Faculty of Agriculture. Ehime University* **51**: 9–13.

Taiz L and Zeiger E. 2006. *Plant Physiology*. 4rd edition. Sinauer Associates, Sunderland, MA.

Taylor B H and Ferree D C. 1984. The influence of summer pruning and cropping on growth and fruiting of apple. *Journal American Society of Horticultural Science* **109**: 17–24.

Urban L, Le Roux X, Sinoquet H, Jaffuel S and Jannoyer M. 2003. A biochemical model of photosynthesis for mango leaves: evidence for the effect of fruit on photosynthesis capacity of nearby leaves. *Tree Physiology* **23**: 289–300.

Utermark H. 1977. Summer pruning to control growth and maintain fruiting in mature apple trees. *Compact Fruit Trees* **10**: 86–90.

Part V

Hormonal Control of Plant Growth and Development

Plant hormones are naturally occurring substances within the plant tissues, which are active at very low concentrations, and have a key role in controlling plant growth and development. They govern cell division, elongation and differentiation and play an important role in regulating all facets of plant life, such as seed growth, flowering, sex of flowers, fruit setting, growth and ripening, and senescence. Every living cell in a plant has potential to produce plant hormones which move freely through the plant tissues and bind to a specific receptor protein. Generally, plant hormones are interchange within the plant by using different types of movements such as (i) cytoplasmic streaming within cells and slow diffusion of ions and molecules between cells and (ii) through vascular tissues, such as phloem or sieve tubes transporting hormones from one part of the plant to another.

Usually, plant cells respond to hormones at specific stages during their life cycle. They require hormones at specific times and sites during the growth. Therefore, hormone production occurs usually at sites of active growth within the meristems. In order to regulate the internal hormonal levels and moderate their effects, plants use different pathways to regulate the amount of chemicals used in the biosynthesis of hormones. Subsequently, they are moved to other parts of the plant, stored in cells to be released later or cannibalize already-formed hormones by conjugating them with carbohydrates, amino acids or peptides. Furthermore, plant hormones have the ability to interact either synergistically or antagonistically in order to induce a particular response. Based on the chemical structure, phytohormones have been categorized as auxins, gibberellins, cytokinins, abscisic acid, ethylene, brassinosteroids and jasmonates. This part describes the structure, biosynthetic and metabolic pathways and physiological roles of various hormones at different stages of plant growth and development.

DOI: 10.1201/9781003373889-17

13 Auxin

13.1 INTRODUCTION

The word auxin is derived from the Greek word "auxein" meaning to grow. At the cellular level, auxins are essential for cell growth, affecting both cell division and cellular expansion. Auxins have the ability to promote axial elongation (shoots), lateral expansion (root) or isodiametric expansion (fruit growth) depending on the specific types of tissue (Woodward and Bartel, 2005; Teale et al., 2006). Auxins are essential for the establishment and maintenance of polarity in organized tissues, maintenance of apical dominance and mediation of tropisms in whole plants (Friml et al., 2003). However, in some cases, auxins promote cellular expansion without cell division (coleoptile growth).

13.2 AUXINS BIOSYNTHESIS

Five naturally occurring (endogenous) auxins, indole-3-acetic acid (IAA), *4-chloroindole-3-acetic acid* (4-ClIAA), *phenylacetic acid* (PAA), *indole-3-butyric acid* (IBA) and *indole-3-propionic acid* (IPA) have been identified in plants (Simon and Petrášek, 2011; Ludwig-Müller, 2011). Major synthetic auxins include 1-naphthaleneacetic acid (NAA), 2,4-dichlorophenoxyacetic acid (2,4-D), 2, 4, 5-Trichlorophenoxyacetic acid (2,4,5-T), 4-Amino-3,5,6-trichloropicolinic acid (picloram) and 2-Methoxy-3,6-dichlorobenzoic acid (dicamba) (Simon and Petrášek, 2011). IAA, the most abundant naturally occurring auxin in plants, exists in both free and conjugated forms (Zhao et al., 2013). Although plants share evolutionary conserved core mechanisms for auxin biosynthesis, different plant species may have distinct strategies and modifications to optimize IAA biosynthesis (Zhao et al., 2010). Generally, auxin biosynthesis can be divided into two general categories: (i) *de novo* auxin biosynthesis (Cheng et al., 2013; Xu and Huang, 2014) and (ii) release from auxin conjugates (Mano and Nemoto, 2012; Brumos et al., 2014; Zhao et al., 2013; Tivendale et al., 2014). *De novo* auxin biosynthesis has been further divided into two groups: (i) tryptophan (Trp)-dependent and (ii) Trp-independent biosynthesis. Trp-dependent auxin biosynthesis is essential for embryogenesis, seedling growth, flower development, vascular pattern formation as well as for other developmental processes (Stepanova et al., 2008; Tao et al., 2008), and it is biosynthesized from Trp through four suggested routes, such as indole-3-acetaldoxime (IAOx), indole-3-pyruvic acid (IPyA), indole-3-acetamide (IAM) and tryptamine (TAM) according to their key intermediates (Woodward and Bartel, 2005).

13.2.1 TAA/YUC Pathway (Tryptophan Aminotransferase)

The YUCCA (YUC) flavin-containing monooxygenase (FMO) has been identified as a key auxin biosynthesis gene from an activation-tagging screen for long hypocotyl mutants in Arabidopsis (Zhao et al., 2001; Kasahara, 2016; Cheng et al., 2006). It has been demonstrated through various physiological and molecular studies that yuc1D is an auxin overproduction mutant (Zhao et al., 2001). Later, YUC has been reported be a member of a gene family with 11 genes in the Arabidopsis genome and all of YUC genes contribute in auxin biosynthesis (Cheng et al., 2006, 2007). Further detailed investigation using various yuc mutant combinations have confirmed that YUC genes are essential for almost all of the major developmental processes including embryogenesis, seedling growth, vascular initiation and patterning, flower development and plant architecture (Cheng et al., 2006, 2007). The biochemical mechanisms of YUC-mediated auxin biosynthesis showed that YUC enzymes use NADPH and molecular oxygen to catalyze the oxidative decarboxylation of

DOI: 10.1201/9781003373889-18

IPA to generate IAA (Mashiguchi et al., 2011; Dai et al., 2013). Three sequential chemical steps are involved in YUC6-catalyzed conversion of IPA into IAA such as (i) reduction of FAD to FADH2 using electrons from NADPH, (ii) binding of molecular oxygen to FADH2 to form the C4a-(hydro) peroxyl flavin intermediate and (iii) the reaction of the C4a intermediate with IPA to produce IAA from decarboxylation of Indole 3-pyruvic acid or IPA (Dai et al., 2013). Moreover, IPA as a substrate, YUC6, can also catalyze the decarboxylation of phenyl-pyruvate (PPA) to produce phenylacetic acid (PAA) (Dai et al., 2013), proposing that YUC enzymes do not have strict substrate specificities. Physiological, biochemical and genetic studies clearly demonstrated that the YUC family of flavin-containing monooxygenases plays a key role in auxin biosynthesis (Gao and Zhao, 2014). Tryptophan Aminotransferase of Arabidopsis (TAA) family of aminotransferases has been reported to have a role in auxin biosynthesis. However, TAA1 is one of the main auxin biosynthesis enzymes (Stepanova et al., 2008; Tao et al., 2008; Yamada et al., 2009). For example, YUC genes from rice (Yamamoto et al., 2007; Fujino et al., 2008; Abu-Zaitoon et al., 2012), maize (Gallavotti et al., 2008), tomato (Expósito Rodríguez et al., 2011), petunia (Tobeña-Santamaria et al., 2002), strawberry (Liu and Lin, 2012) and other species (Kim et al., 2012; Cheol Park et al., 2013) have been functionally characterized, and they all participate in auxin biosynthesis.

13.2.2 Other Trp-Dependent Auxin Biosynthesis Pathways

Trp is metabolized into several other indolic compounds, which show their role in auxin biosynthesis. However, the physiological roles of indolic compounds other than IPA in auxin biosynthesis are still unclear.

13.2.2.1 Indole-3-Acetamide (IAM) Pathway

Detectable amount of IAM has been reported from Arabidopsis, maize, rice and tobacco. This is the key intermediate in the bacterial auxin biosynthesis pathway characterized in *Agrobacterium* and *Pseudomonas* (Yamada et al., 1985; Romano et al., 1995; Sugawara et al., 2009; Novák et al., 2012). Generally, in plant pathogenic bacteria, Trp is oxidized by the iaaM Trp-2-monooxygenase to IAM that is subsequently hydrolyzed by iaaH to produce IAA. It has been suggested that IAM may be synthesized from IAA as a way to control free IAA levels and conversion of IAA into IAM may be accomplished using mechanisms similar to glutamine biosynthesis (Gao and Zhao, 2014). Group of hydrolases, which are homologous to the bacterial hydrolase iaaH, plays a role in converting IAM into IAA (Pollmann et al., 2006; Hoffmann et al., 2010). However, it is still not clear whether IAM is an important auxin biosynthesis intermediate in plants because IAM-deficient mutants have not been identified.

13.2.2.2 Tryptamine (TAM) Pathway

TAM was a proposed substrate for the YUC flavin monooxygenases, which have now been shown to catalyze the conversion of IPA into IAA in vitro and in vivo (Zhao et al., 2001; Kim et al., 2012). However, all of the flavin-containing monooxygenases form the C4a-(hydro) peroxyl flavin intermediates and able to ensure both nucleophilic and electrophilic reactions, depending on the reaction conditions. However, until now, there has not been sufficient evidence to show whether TAM is an important intermediate in auxin biosynthesis. Biosynthesis and metabolism of TAM is also not well understood until now.

13.2.2.3 Indole-3-Acetonitrile (IAN) Pathway

IAN is known to encourage adventitious root development and inhibit primary root elongation. The conversion of IAN into IAA is catalyzed by nitrilases. And inactivation of nitrilase genes leads to resistance to exogenous IAN, but the nitrilase mutants do not display distinct developmental defects observed in known auxin signaling and transport mutants (Bartel and Fink, 1994; Normanly et al.,

1997). Arabidopsis genome contains four copies of the nitrilase gene. However, developmental consequences of disrupting all four nitrilase genes have not yet been investigated. Therefore, it is still debatable whether IAN plays a significant role in auxin biosynthesis (Sugawara et al., 2009).

13.2.2.4 Indole-3-Acetadehyde (IAAld) Pathway

IAAld is produced from IPA by IPA decarboxylases in some IAA-producing bacteria (Carreno-Lopez et al., 2000), and IAAld can be further oxidized into IAA by aldehyde oxidases. However, inactivation of Arabidopsis aldehyde oxidases does not disturb auxin homeostasis, indicating that IAAld does not contribute significantly to *de novo* auxin biosynthesis (Mashiguchi et al., 2011). Nevertheless, IAAld can also be oxidized by aldehyde dehydrogenases, which have not been characterized in Arabidopsis.

13.2.2.5 Indole-3-Acetaldoxime (IAO_X) Pathway

IAOx has been identified in Arabidopsis and related species (Mashiguchi et al., 2011). In Arabidopsis, CYP79B2 and CYP79B3 convert Trp directly into IAOx (Hull et al., 2000; Zhao et al., 2002). IAOx is also a precursor for indolic glucosinolate biosynthesis, and when the genes encoding glucosinolate biosynthesis enzymes are mutated, more IAOx is fluxed into IAA biosynthesis, causing overproduction of auxin phenotypes (Boerjan et al., 1995; Bak and Feyereisen, 2001). Therefore, it seems plausible that CYP79B2 and CYP79B3 are the only genes responsible for producing IAOx in Arabidopsis.

13.2.3 Trp-Independent IAA Biosynthetic Pathway

The existence of a Trp-independent IAA biosynthetic pathway has been suggested based on the analysis of Trp auxotrophic mutants (Wright and Cohen, 1991; Normanly et al., 1993; Tivendale et al., 2014). A number of studies strongly suggested the existence of the Trp-independent IAA biosynthetic pathway that might branch from IGP and/or indole. Wang et al. (2015) reported that Arabidopsis indole synthase mutant synthesized the auxin through spatially and temporally regulated pathway that contributes significantly to the establishment of the apical–basal axis, which profoundly affects the early embryogenesis in Arabidopsis, and these results explained the Trp-independent auxin biosynthetic pathway and its functions in regulating plant growth and development. However, the molecular basis, biological functions and spatiotemporal regulation of the Trp-independent IAA biosynthetic pathway have remained unclear.

13.3 AUXIN CONJUGATION AND DEGRADATION PATHWAY

Although, auxin level regulation in plant cell is operated mainly by homeostatic mechanism during *de novo* synthesis, the levels of IAA can be weakened by conjugation, mainly to amino acids and sugars, and also by degradation (Normanly, 2010; Rosquete et al., 2012). Various genes that are involved in IAA conjugation and IAA conjugate hydrolysis have been identified (Woodward and Bartel, 2005; Ludwig-Müller, 2011). Several research groups have reported that IAA can be conjugated and stored in seeds and could be used for early seedling growth. 2-oxoindole-3-acetic acid (oxIAA) and oxIAA glucose (oxIAA-Glc) have been reported to be major degradation products of IAA (Östin et al., 1998; Kai et al., 2007; Novák et al., 2012).

A number of research groups have studied the localization of protein and the presence of specific peptides in the enzymes involved in shikimate and L-Trp biosynthesis pathways and concluded that auxin biosynthesis is localized in plastids (Woodward and Bartel, 2005; Tzin and Galili, 2010; Mano and Nemoto, 2012). The pathways downstream of L-Trp are believed to be localized to the cytosol (Woodward and Bartel, 2005; Mano and Nemoto, 2012). Kriechbaumer et al. (2012) reported that YUC4 enzyme is localized both in the cytosol and in cytosolic side of the membrane.

However, the available information on the localization/concentration of IAA and its metabolites as well as the enzymes involved in IAA metabolism in other subcellular compartments like Vacuoles and apoplast is not sufficient to draw definite conclusions.

13.4 AUXIN METABOLISM DURING PLANT GROWTH AND DEVELOPMENT

It has now been well established that both root and shoot tissues are capable of synthesizing and metabolizing IAA, although at different rates. Therefore, these tissues can function as both source and sink for IAA (Ljung et al., 2001). In order to regulate intracellular concentrations of IAA, the degradation and transport mechanisms need to be kept under strict control. The current understanding of major pathways of IAA metabolism and transport from source to sink tissues is described in the following.

13.4.1 Auxin Metabolism in Root

Auxin homeostasis has profound influence on root architecture (Jones and Ljung, 2012). Ljung et al. (2005) reported that several genes involved in auxin biosynthesis in the root apical meristem, where the rate of IAA biosynthesis is high. Local auxin biosynthesis combined with polar auxin transport is involved in forming the auxin gradient within the root apex and plays an important role in polarity planning and hair positioning (Ljung et al., 2005; Petersson et al., 2009; Ikeda and Ohme-Takagi, 2009). Besides this, local IAA biosynthesis also affects a number of processes including cell division, differentiation and elongation, lateral root initiation and elongation within the root (Overvoorde et al., 2010; Jones and Ljung, 2012). Initiation, development and outgrowth of lateral root primordia (LRP) have been reported to be strongly influenced by local auxin biosynthesis in the root apex (Bhalerao et al., 2002). Exogenous application of IAA has been found effective in root, inducing in stem cuttings of *Malus hupehensis* (Zhang et al., 2017).

13.4.2 Auxin Metabolism in Other Organs

In Arabidopsis, YUCCA genes have been shown to play critical roles during embryogenesis, leaf, flower, vascular and fruit development (Cheng et al., 2006, 2007; Eklund et al., 2010; Wang et al., 2011). However, in rice, maize and potato, YUCCA genes are important for normal plant development and resistance to different stress conditions (Yamamoto et al., 2007; Gallavotti et al., 2008; Bernardi et al., 2012; Kim et al., 2012). Furthermore, IAA biosynthesis is strongly upregulated by soluble sugars, which is mediated via the phytochrome interacting factor (PIF) transcriptional regulators and has an impact on hypocotyl elongation during early seedling growth and development (Sairanen et al., 2012; Lilley et al., 2012).

13.5 AUXIN TRANSPORT IN PLANTS

Auxin (IAA) is mainly transported from the source tissues such as young leaves and flowers by an unregulated bulk flow in the mature phloem especially over long distance. It can also be directly transported from cell to cell by forming local maxima and creating a gradient, which is known as polar auxin transport (PAT) (Revalska et al., 2015). PAT is maintained by the coordinated action of efflux (PIN) and auxin influx (AUX/LAX) carrier proteins.

13.5.1 Influx Carriers

Four auxin influx carriers have been identified in Arabidopsis, and each one of them has its own function. Additionally, some homologs have also been identified in other plants. AUX1 and LAX3 have been reported to mediate IAA uptake (Yang et al., 2006; Swarup et al., 2008).

13.5.2 Efflux Carriers

Studies on several allelic root mutants (*agravitropic 1* [*agr1*], *wavy roots 6* [*wav6*], *ethylene-insensitive root 1* [*eir1*] and the floral mutant *pin-formed1* [*pin1*]) of Arabidopsis led to the identification of auxin efflux carrier candidates (Bell and Maher, 1990; Okada and Shimura, 1990; Romano et al., 1995). These mutants show decreased PAT in shoots and roots. The corresponding *PIN1* gene encodes a plant-specific protein with two trans membrane regions separated by a hydrophilic loop (Gälweiler et al., 1998). The PIN1, PIN2, PIN3, PIN4 and PIN7 proteins are localized at the plasma membrane, where they act as auxin efflux carriers (Petrášek et al., 2006; Mravec et al., 2008), and some of them have been functionally characterized.

13.6 OTHER ENDOGENOUS AUXINS

4-Chloroindole-3-acetic acid (4-Cl-IAA), 4-chloroindole-3-acetic acid is found in certain higher plants and is more active in some bioassays than the IAA (Reinecke et al., 1999). It has been reported that some species of legumes such as *Lens culinaris*, *Lathyrus latifolius*, *Vicia faba* and *Pisum sativum* seeds are the rich source of 4-Cl-IAA. Ozga et al. (2009) reported 4-Cl-IAA to move from young seeds into the pod and help in pod elongation. A recent study showed that 4-Cl-IAA is produced from 4-chlorotryptophan by two aminotransferases, TRYPTOPHAN AMINOTRANSFERASE RELATED1 and TRYPTOPHAN AMINOTRANSFERASE RELATED2 and synthesized via the novel intermediate 4-chloroindole-3-pyruvic acid in *Pisum sativum* seeds (Tivendale et al., 2014)

PHENYLACETIC ACID (PAA)

Phenylacetic acid is a non-indolic active auxin, which is present at physiologically relevant levels in several higher plant species (Wightman and Rauthan, 1974; Wightman and Lighty, 1982; Ludwig-Müller and Cohen, 2002). It contains a phenyl functional group and a carboxylic acid functional group with honey-like dour. However, the biological significance of PAA is not fully understood.

INDOLE-3-BUTYRIC ACID (IBA)

IBA has been identified in a number of plant species. IBA is found in conjugated forms, although it is present significantly at lower levels than IAA (Ludwig-Müller et al., 1993; Ludwig-Müller, 2000). IBA and IAA can be interconverted, which has led to the suggestion that IBA may act as a precursor to IAA (Bartel et al., 2001). IBA was found to be the most efficient PGR for root induction in various species such as *Ficus hawaii* and Syzygium *javanica* (Siddiqui and Hussain, 2007; Paul and Aditi, 2009).

INDOLE-3-PROPIONIC ACID (IPA)

IPA is a plant hormone with a number of functions including cell division, elongation, autonomal loss of leaves and the formation of buds, roots, flowers and fruit. It is sometimes used in tissue culture.

13.7 PHYSIOLOGICAL ROLE OF AUXINS

Auxins play an important role in determining several physiological activities in plants including cell elongation, apical dominance, root initiation, prevention of abscission, parthenocarpy, respiration, callus formation, vascular differentiation, weed control, enzymatic activity, flower initiation and increased metabolic activity. These functions have been briefly explained in this section.

13.7.1 Cell Elongation

Auxin stimulates the cell elongation in shoot. Various theories have been proposed to explain its role in cell elongation: (i) increasing the osmotic solutes in the cells, (ii) reducing the wall pressure, (iii) increasing the permeability of cells to water, (iv) enhancing cell-wall synthesis and (v) inducing the synthesis of specific DNA-dependent mRNA and specific enzymatic proteins (Majda and Roberts, 2018; Ma and Li, 2019).

13.7.2 Apical Dominance

The apical dominance was first discovered by Skoog and Thimann (1934), in which the apical bud suppresses the outgrowth of lateral buds, or lateral buds start developing only when the apical bud is removed. Removal of the apical bud (or decapitation) from the shoot did not decrease the auxin concentration in lateral buds, but there was, in fact, a manifold increase in auxin concentration even after a few hours of decapitation (Gocal et al., 1991). Three theories of apical dominance, namely, the direct, diversion and indirect were proposed in the 1930s to explain the role of auxin synthesized in the shoot apex in inhibiting axillary bud outgrowth, and thus, the shoot branching. After studying the direct and diversion theories of apical dominance in detail, they are replaced with the current auxin transport canalization and second messenger theories, respectively, although these theories still do not fully explain the phenomenon of apical dominance. The indirect theory explains the apical dominance as auxin-induced stem growth indirectly inhibiting bud outgrowth. However, how a growing stem might inhibit bud outgrowth is not fully understood. Recent findings indicate that the bud dormancy (quiescence or paradormancy) in response to intrinsic and environmental factors in diverse species is linked to enhanced growth of the main shoot and reduced sugar level in the buds. Since a growing stem is a strong sink for sugars, and sugar is indispensable for shoot branching, the indirect theory of apical dominance might be related to auxin-induced stem growth that inhibits bud outgrowth by diverting sugars away from buds. Advances in the effect of source–sink status on dormancy and outgrowth of axillary buds can eventually improve our understanding of apical dominance and shoot branching in plants (Kebrom, 2017).

The ratio of auxin to cytokinins has been shown to be one of the key factors involved in apical dominance. When the shoot is intact, the auxin in the shoot apex helps to maintain high concentration of ABA in lateral buds so that their growth is suppressed. By removing the shoot apex, the concentration of ABA in lateral buds decreases so that lateral buds grow rapidly.

13.7.3 Root Initiation and Development

Reports suggest that many genes involved in the synthesis of auxin are expressed in roots and that root-generated auxin contributes to the maintenance of the gradients and maxima required for normal root development (Ljung et al., 2005; Ikeda and Ohme-Takagi, 2009; Petersson et al., 2009). Therefore, the control of the auxin concentration in each cell is essential for normal root growth and development. Auxins not only are involved in the process of root hair initiation but also play an important role in the elongation of root hairs (Knox et al., 2003; Fischer et al., 2006; Rahman et al., 2002; Lee and Cho, 2006; Cho et al., 2007). Exogenous application of auxin to *A. thaliana* increased the length of root hairs (Pitts et al., 1998; Rahman et al., 2002). Furthermore, when the intracellular auxin concentration of developing hair cells was artificially changed, the length of the resulting root hairs showed a positive correlation with the auxin concentration (Lee and Cho, 2006; Cho et al., 2007). However, higher concentration of auxin inhibits the elongation of root but increases the number of lateral roots. Application of auxin especially IBA to the base of the stem cutting promotes early and extensive rooting. This fact has been widely utilized in asexual propagation of economically useful plants propagated by cuttings.

13.7.4 Prevention of Abscission

Abscission is a process in which vegetative and reproductive organs are detached from the mother plant at a specific location (abscission zone) in response to developmental, environmental, hormonal and molecular cues (Estornell et al., 2013). Abscission has significant role in plant development. For example, shedding the leaves in autumn minimizes water loss in the winter; separating seeds from fruits permits propagation, and rapidly detaching damaged or infected organs is a defense strategy against pest or disease attack. Endogenous auxins play an important role in controlling premature abscission of leaves and fruits. It operates by suppressing enzymatic (cellulase and polygalacturonase) degradation of cell-wall components (pectins, celluloses and hemicelluloses). Thus, leaves and fruits must produce auxin continuously to prevent the formation of the abscission zone which cuts off their nutrient and water supply. However, auxins have often been reported either to delay or to induce fruit abscission depending upon the time of application and, hence, may operate as growth hormones or as abscising agents.

13.7.5 Parthenocarpy

Auxin and gibberellin are the most common phytohormones used to trigger artificial parthenocarpic fruit or seed production. Auxin can encourage the formation of parthenocarpic fruits (i.e., the formation of seedless fruits without fertilization). This phenomenon is common under natural conditions and, in such cases, the concentration of auxins in the ovaries has been found to be higher than in the ovaries of plants which produce fruits only after fertilization. In the latter case, the concentration of the auxin in ovaries increases after pollination and fertilization.

Auxin synthesis within unpollinated ovary has been achieved by expressing the *iaaM* gene under control of the *DefH9* (Deficiens Homolog 9) (Rotino et al., 1997). Indole-acetamide is converted by either spontaneous or enzymatic hydrolysis to indole-3-acetic acid (IAA) or ammonia within plant cells. The *DefH9-iaaM* construct has been used to induce parthenocarpy in several other crop plants (Rotino et al., 1997; Ficcadenti et al., 1999; Mezzetti et al., 2004; Yin et al., 2006). Furthermore, parthenocarpy can also be achieved by modifying auxin signaling. Auxin-dependent transcriptional regulation is controlled by the *Aux/IAA* and *AUXIN RESPONSE FACTOR* (*ARF*) gene families (Leyser, 2006; Guilfoyle and Hagen, 2007). Manipulation of the activity of specific Aux/IAA and ARF genes has now been reported to affect fruit development. For example, genetic alterations of ARF8 function induced parthenocarpy in *Arabidopsis* (Vivian-Smith et al., 2001; Goetz et al., 2006; Goetz et al., 2007).

13.7.6 Respiration

Auxin stimulates respiration, and there is a strong correlation between auxin-induced growth and an increased respiration rate. It has been reported that auxin increases the rate of respiration indirectly through increased supply of ADP (Adenosine diphosphate) or by rapidly utilizing the ATP in the expanding cells (French and Beevers, 1953).

13.7.7 Callus Formation

In tissue culture, IAA promotes cell division and elongation leading to the development of callus. Cell division and growth of most types of plant cells cultured *in vitro* require an external source of auxin (Petrášek et al., 2002). 2,4-dichlorophenoxyacetic acid (2,4-D) is the most widely used auxin for inducing somatic embryogenesis in tissue culture (Pellegrineschi et al., 2004; Vikrant and Rashid, 2003).

13.7.8 Vascular Differentiation

Auxin has been correlated with vascular differentiation in plants. Auxin transport is associated with the vascular system, and therefore, auxin induces vascular differentiation (Sachs, 1991).

13.7.9 Eradication of Weeds

Some of the synthetic auxins, especially 2,4-D (2,4-dichlorophenoxyacetic acid), act as herbicides and are used for the eradication of weeds. Spraying of this auxin at higher concentration induces the root cells to divide very rapidly. As a result, the roots become distorted and sieve tubes get closed and, consequently, the plant dies. Broad-leaved dicots are more sensitive to 2,4-D than monocots, thereby selectively killing dicot weeds without affecting the monocots.

13.7.10 Flower Initiation

Relation between auxin and flower development was first established when the auxin transport mutant *pin1* was isolated and characterized (Okada et al., 1991; Gälweiler et al., 1998). Auxin accumulation in groups of cells in the periphery of the shoot apical meristem specifies the site of leaf or floral primordium initiation (Reinhardt et al., 2003). Besides this, auxin also regulates other aspects of flower development such as floral organ initiation, growth and patterning, and later events that ensure reproductive success of the mature flower (Nemhauser et al., 2004; Cheng and Zhao, 2007; Sundberg and Østergaar, 2009). The *pin1* mutant showed pleiotropic defects through all stages of development, but the most striking phenotype was that the inflorescence of *pin1* often does not have any flowers indicating that auxin plays an essential role in initiating flower primordia. PIN1 encodes a transmembrane protein that has recently been shown to localize with polarity in cells and to participate in auxin efflux (Gälweiler et al., 1998; Petrášek et al., 2006; Wiśniewska et al., 2006). In some plants such as *Ananas sativus*, flowering continues throughout the year. In such cases, spraying of dilute solution of the synthetic auxins (like 2, 4-D and NAA) initiates simultaneous flowering in the whole crop.

REFERENCES

Abu-Zaitoon Y M, Bennett K, Normanly J and Nonhebel H M. 2012. A large increase in IAA during development of rice grains correlates with the expression of tryptophan aminotransferase OsTAR1 and a grain-specific YUCCA. *Physiologia Plantarum* **146**: 487–499.

Bak S and Feyereisen R. 2001. The involvement of two P450 enzymes, CYP83B1 and CYP83A1, in auxin homeostasis and glucosinolate biosynthesis. *Plant Physiology* **127**: 108–118.

Bartel B and Fink G R. 1994. Differential regulation of an auxin-producing nitrilase gene family in Arabidopsis thaliana. *Proceedings of the National Academy of Sciences United States of America* **91**: 6649–6653.

Bartel B, LeClere S, Magidin M and Zolman B. 2001. Inputs to the active indole-3-acetic acid pool: de novo synthesis, conjugate hydrolysis, and indole-3-butyric acid b-oxidation. *Journal of Plant Growth Regulation* **20**: 198–216.

Bell C J and Maher E P. 1990. Mutants of Arabidopsis thaliana with abnormal gravitropic responses. *Molecular and General Genetics MGG* **220**: 289–293.

Bernardi J, Lanubile A, Li Q B, Kumar D, Kladnik A, Cook S D and Chourey P S. 2012. Impaired auxin biosynthesis in the defective endosperm18 mutant is due to mutational loss of expression in the ZmYuc1 gene encoding endosperm-specific YUCCA1 protein in maize. *Plant Physiology* **160**: 1318–1328.

Bhalerao R P, Eklöf J, Ljung K, Marchant A, Bennett M and Sandberg G. 2002. Shoot-derived auxin is essential for early lateral root emergence in Arabidopsis seedlings. *The Plant Journal* **29**: 325–332.

Boerjan W, Cervera M T, Delarue M, Beeckman T, Dewitte W, Bellini C and Inzé D. 1995. Superroot, a recessive mutation in Arabidopsis, confers auxin overproduction. *The Plant Cell* **7**: 1405–1419.

Brumos J, Alonso J M and Stepanova A N. 2014. Genetic aspects of auxin biosynthesis and its regulation. *Physiologia Plantarum* **151**: 3–12.

Carreno-Lopez R, Campos-Reales N, Elmerich C and Baca B E. 2000. Physiological evidence for differently regulated tryptophan-dependent pathways for indole-3-acetic acid synthesis in *Azospirillum brasilense*. *Molecular and General Genetics* **264**: 521–530.

Cheng Y, Dai X and Zhao Y. 2006. Auxin biosynthesis by the YUCCA flavin monooxygenases controls the formation of floral organs and vascular tissues in Arabidopsis. *Genes & Development* **20**: 1790–1799.

Cheng Y, Dai X and Zhao Y. 2007. Auxin synthesized by the YUCCA flavin monooxygenases is essential for embryogenesis and leaf formation in Arabidopsis. *The Plant Cell* **19**: 2430–2439.

Cheng Y and Zhao Y. 2007. A role for auxin in flower development. *Journal of Integrative Plant Biology* **49**: 99–104.

Cheng Z J, Wang L, Sun W, Zhang Y, Zhou C, Su Y H and Cheng Y. 2013. Pattern of auxin and cytokinin responses for shoot meristem induction results from the regulation of cytokinin biosynthesis by AUXIN RESPONSE FACTOR3. *Plant Physiology* **161**: 240–251.

Cheol Park H, Cha J Y and Yun D J. 2013. Roles of YUCCAs in auxin biosynthesis and drought stress responses in plants. *Plant Signaling & Behavior* **8**(6): 337–349.

Cho M, Lee S H and Cho H T. 2007. P-Glycoprotein4 displays auxin efflux transporter – like action in Arabidopsis root hair cells and tobacco cells. *The Plant Cell* **9**: 3930–3943.

Dai X, Mashiguchi K, Chen Q, Kasahara H, Kamiya Y, Ojha S, DuBois J, Ballou D and Zhao Y. 2013. The biochemical mechanism of auxin biosynthesis by an arabidopsis YUCCA flavin-containing monooxygenase. *Journal of Biological Chemistry* **288**: 1448–1457.

Eklund D M, Ståldal V, Valsecchi I, Cierlik I, Eriksson C, Hiratsu K and Sundberg E. 2010. The Arabidopsis thaliana STYLISH1 protein acts as a transcriptional activator regulating auxin biosynthesis. *The Plant Cell* **22**: 349–363.

Estornell L H, Agustí J, Merelo P, Talón M and Tadeo F R. 2013. Elucidating mechanisms underlying organ abscission. *Plant Science* **199**: 48–60.

Expósito-Rodríguez M, Borges A A, Borges-Pérez A and Pérez J A. 2011. Gene structure and spatiotemporal expression profile of tomato genes encoding YUCCA-like flavin monooxygenases: the ToFZY gene family. *Plant Physiology and Biochemistry* **49**: 782–791.

Ficcadenti N, Sestili S, Pandolfini T, Cirillo C, Rotino G L and Spena A. 1999. Genetic engineering of parthenocarpic fruit development in tomato. *Molecular Breeding* **5**: 463–470.

Fischer D G, Hart S C, Rehill B J, Lindroth R L, Keim P and Whitham T G. 2006. Do high-tannin leaves require more roots? *Oecologia* **149**: 668–675.

French R C and Beevers H. 1953. Respiratory and growth responses induced by growth regulators and allied compounds. *American Journal of Botany* **12**: 660–666.

Friml J, Vieten A, Sauer M, Weijers D, Schwarz H, Hamann T and Jürgens G. 2003. Efflux-dependent auxin gradients establish the apical–basal axis of Arabidopsis. *Nature* **426**: 147–153.

Fujino K, Matsuda Y, Ozawa K, Nishimura T, Koshiba T, Fraaije M W and Sekiguchi H. 2008. NARROW LEAF 7 controls leaf shape mediated by auxin in rice. *Molecular Genetics and Genomics* **279**: 499–507.

Gallavotti A, Yang Y, Schmidt R J and Jackson D. 2008. The relationship between auxin transport and maize branching. *Plant Physiology* **147**: 1913–1923.

Gälweiler L, Guan C, Müller A, Wisman E, Mendgen K, Yephremov A and Palme K. 1998. Regulation of polar auxin transport by AtPIN1 in Arabidopsis vascular tissue. *Science* **282**: 2226–2230.

Gao Y and Zhao Y. 2014. Auxin biosynthesis and catabolism. pp. 21–38. *Auxin and Its Role in Plant Development* (Eds.) Zazímalová E, Petrášeky J and Benková E. Springer, Vienna.

Gocal G F, Pharis R P, Yeung E C and Pearce D. 1991. Changes after decapitation in concentrations of indole-3-acetic acid and abscisic acid in the larger axillary bud of Phaseolus vulgaris L. cv Tender Green. *Plant Physiology* **95**: 344–350.

Goetz M, Hooper L C, Johnson S D, Rodrigues J C M, Vivian-Smith A and Koltunow A M. 2007. Expression of aberrant forms of AUXIN RESPONSE FACTOR8 stimulates parthenocarpy in Arabidopsis and tomato. *Plant Physiology* **145**: 351–366.

Goetz M, Vivian-Smith A, Johnson S D and Koltunow A M. 2006. AUXIN RESPONSE FACTOR8 is a negative regulator of fruit initiation in Arabidopsis. *The Plant Cell* **18**: 1873–1886.

Guilfoyle T J and Hagen G. 2007. Auxin response factors. *Current Opinion in Plant Biology* **10**: 453–460.

Hoffmann M, Hentrich M and Pollmann S. 2010. Auxin-Oxylipin Crosstalk: relationship of AntagonistsF. *Journal of Integrative Plant Biology* **53**: 429–445.

Hull A K, Vij R and Celenza J L. 2000. Arabidopsis cytochrome P450s that catalyze the first step of tryptophan-dependent indole-3-acetic acid biosynthesis. *Proceedings of the National Academy of Sciences* **97:** 2379–2384.

Ikeda M and Ohme-Takagi M. 2009. A novel group of transcriptional repressors in Arabidopsis. *Plant and Cell Physiology* **50**: 970–975.

Jones B and Ljung K. 2012. Subterranean space exploration: the development of root system architecture. *Current Opinion in Plant Biology* **15**: 97–102.

Kai K, Horita J, Wakasa K and Miyagawa H. 2007. Three oxidative metabolites of indole-3-acetic acid from Arabidopsis thaliana. *Phytochemistry* **68**: 1651–1663.

Kasahara H. 2016. Current aspects of auxin biosynthesis in plants. *Bioscience Biotechnology and Biochemistry* **80**: 34–42.

Kebrom T H. 2017. A growing stem inhibits bud outgrowth – the overlooked theory of apical dominance. *Frontiers in Plant Science* **8**: 1874 https://doi.org/10.3389/fpls.2017.01874.

Kim S Y, Lee J, Eshed-Williams L, Zilberman D and Sung Z R. 2012. EMF1 and PRC2 cooperate to repress key regulators of Arabidopsis development. *PLoS Genetics* **8**: e1002512.

Knox K, Grierson C S and Leyser O. 2003. AXR3 and SHY2 interact to regulate root hair development. *Development* **130**: 5769–5777.

Kriechbaumer V, Wang P, Hawes C and Abell B M. 2012. Alternative splicing of the auxin biosynthesis gene YUCCA4 determines its subcellular compartmentation. *The Plant Journal* **70**: 292–302.

Lee S H and Cho H T. 2006. PINOID positively regulates auxin efflux in Arabidopsis root hair cells and tobacco cells. *The Plant Cell* **18**: 1604–1616.

Leyser O. 2006. Dynamic integration of auxin transport and signalling. *Current Biology* **16**: 24–433.

Lilley J L S, Gee C W, Sairanen I, Ljung K and Nemhauser J L. 2012. An endogenous carbon-sensing pathway triggers increased auxin flux and hypocotyl elongation. *Plant Physiology* **160**: 2261–2270.

Liu C J and Lin J Y. 2012. Anti-inflammatory and anti-apoptotic effects of strawberry and mulberry fruit polysaccharides on lipopolysaccharide-stimulated macrophages through modulating pro-/anti-inflammatory cytokines secretion and Bcl-2/Bak protein ratio. *Food and Chemical Toxicology* **50**: 3032–3039.

Ljung K, Bhalerao R P and Sandberg G. 2001. Sites and homeostatic control of auxin biosynthesis in Arabidopsis during vegetative growth. *The Plant Journal* **28**: 465–474.

Ljung K, Hull A K, Celenza J, Yamada M, Estelle M, Normanly J and Sandberg G. 2005. Sites and regulation of auxin biosynthesis in Arabidopsis roots. *The Plant Cell* **17**: 1090–1104.

Ludwig-Müller J. 2000. Indole-3-butyric acid in plant growth and development. *Plant Growth Regulation* **32**: 219–230.

Ludwig-Müller J. 2011. Auxin conjugates: their role for plant development and in the evolution of land plants. *Journal of Experimental Botany* **62**: 1757–1773.

Ludwig-Müller J and Cohen J D. 2002. Identification and quantification of three active auxins in different tissues of *Tropaeolum majus*. *Physiologia Plantarum* **115**: 320–329.

Ludwig-Müller J, Sass S, Sutter E G, Wodner M and Epstein E. 1993. Indole-3-butyric acid in Arabidopsis thaliana. *Plant Growth Regulation* **13**: 179–187.

Ma L. and Li G. 2019. Auxin-dependent cell elongation during the shade avoidance response. *Frontiers in Plant Science* **10**: 914. https://doi.org/10.3389/fpls.2019.00914.

Majda M. and Roberts S. 2018. The role of auxin in cell wall expansion. *International Journal of Molecular Sciences* **19**: 951–970.

Mano Y and Nemoto K. 2012. The pathway of auxin biosynthesis in plants. *Journal of Experimental Botany* **63**: 2853–2872.

Mashiguchi K, Tanaka K, Sakai T, Sugawara S, Kawaide H, Natsume M and McSteen P. 2011. The main auxin biosynthesis pathway in Arabidopsis. *Proceedings of the National Academy of Sciences* **108**: 18512–18517.

Mezzetti B, Landi L, Pandolfini T and Spena A. 2004. The defH9-iaaM auxin-synthesizing gene increases plant fecundity and fruit production in strawberry and raspberry. *BMC Biotechnology* **4**: 4.

Mravec J, Kubeš M, Bielach A, Gaykova V, Petrášek J, Skůpa P and Friml J. 2008. Interaction of PIN and PGP transport mechanisms in auxin distribution-dependent development. *Development* **135**: 3345–3354.

Nemhauser J L, Mockler T C and Chory J. 2004. Interdependency of brassinosteroid and auxin signaling in Arabidopsis. *PLoS Bio*logy **2**: e258.

Normanly J. 2010. Approaching cellular and molecular resolution of auxin biosynthesis and metabolism. *Cold Spring Harbor Perspectives in Biology* **2**: a001594.

Normanly J, Cohen J D and Fink G R. 1993. Arabidopsis thaliana auxotrophs reveal a tryptophan-independent biosynthetic pathway for indole-3-acetic acid. *Proceedings of the National Academy of Sciences* **90**: 10355–10359.

Normanly J, Grisafi P, Fink G R and Bartel B. 1997. Arabidopsis mutants resistant to the auxin effects of indole-3-acetonitrile are defective in the nitrilase encoded by the NIT1 gene. *The Plant Cell* **9**: 1781–1790.

Novák O, Hényková E, Sairanen I, Kowalczyk M, Pospíšil T and Ljung K. 2012. Tissue-specific profiling of the Arabidopsis thaliana auxin metabolome. *The Plant Journal* **72**: 523–536.

Okada K and Shimura Y. 1990. Reversible root tip rotation in Arabidopsis seedlings induced by obstacle-touching stimulus. *Science* **250:** 274.

Okada N, Sasakawa C, Tobe T, Talukder K A, Komatsu K and Yoshikawa M. 1991. Construction of a physical map of the chromosome of Shigella flexneri 2a and the direct assignment of nine virulence-associated loci identified by Tn5 insertions. *Molecular Microbiology* **5**: 2171–2180.

Östin A, Kowalyczk M, Bhalerao R P and Sandberg G. 1998. Metabolism of indole-3-acetic acid in Arabidopsis. *Plant Physiology* **118**: 285–296.

Overvoorde P, Fukaki H and Beeckman T. 2010. Auxin control of root development. *Cold Spring Harbor Perspectives in Biology* **2**: a001537.

Ozga J A, Reinecke D M, Ayele B T, Ngo P, Nadeau C and Wickramarathna A D. 2009. Developmental and hormonal regulation of gibberellin biosynthesis and catabolism in pea fruit. *Plant Physiology* **150:** 448–462.

Paul R and Aditi C. 2009. IBA and NAA of 1000 ppm induce more improved rooting characters in air-layers of waterapple (*Syzygium javanica* L.). *Bulgarian Journal of Agricultural Science* **15:** 123–128.

Pellegrineschi A, Brito R M, McLean S and Hoisington D. 2004. Effect of 2, 4-dichlorophenoxyacetic acid and NaCl on the establishment of callus and plant regeneration in durum and bread wheat. *Plant Cell, Tissue and Organ Culture* **77**: 245–250.

Petersson S V, Johansson A I, Kowalczyk M, Makoveychuk A, Wang J Y, Moritz T and Ljung K. 2009. An auxin gradient and maximum in the Arabidopsis root apex shown by high-resolution cell-specific analysis of IAA distribution and synthesis. *The Plant Cell* **21**: 1659–1668.

Petrášek J, Březinová A, Holík J and Zažimalová E. 2002. Excretion of cytokinins into the cultivation medium by suspension-cultured tobacco cells. *Plant Cell Reports* **21**: 97–104.

Petrášek J, Mravec J, Bouchard R, Blakeslee J J, Abas M, Seifertová D and Dhonukshe P. 2006. PIN proteins perform a rate-limiting function in cellular auxin efflux. *Science* **312**: 914–918.

Pitts R J, Cernac A and Estelle M. 1998. Auxin and ethylene promote root hair elongation in Arabidopsis. *The Plant Journal* **16**: 553–560.

Pollmann S, Mülle A and Weiler E W. 2006. Many roads lead to "auxin": of nitrilases, synthases, and amidases. *Plant Biology* **8**: 326–333.

Rahman A, Hosokawa S, Oono Y, Amakawa T, Goto N and Tsurumi S. 2002. Auxin and ethylene response interactions during Arabidopsis root hair development dissected by auxin influx modulators. *Plant Physiology* **130**: 1908–1917.

Reinecke H, Zhang M, Bartosek T and Murry C E. 1999. Survival, integration, and differentiation of cardio-myocyte grafts. *Circulation* **100**: 193–202.

Reinhardt D, Pesce E R, Stieger P, Mandel T, Baltensperger K, Bennett M and Kuhlemeier C. 2003. Regulation of phyllotaxis by polar auxin transport. *Nature* **426**: 255–260.

Revalska M, Vassileva V, Zechirov G and Iantcheva A. 2015. Is the auxin influx carrier LAX3 essential for plant growth and development in the model plants *Medicago truncatula, Lotus japonicus* and *Arabidopsis thaliana*? *Biotechnology & Biotechnological Equipment* **29**: 786–797.

Romano C P, Robson P R H, Smith H, Estelle M and Klee H. 1995. Transgene-mediated auxin overproduction in Arabidopsis: hypocotyl elongation phenotype and interactions with the hy6–1 hypocotyl elongation and axrl auxin-resistant mutants. *Plant Molecular Biology* **27**: 1071–1083.

Rosquete M R, Barbez E and Kleine-Vehn J. 2012. Cellular auxin homeostasis: gatekeeping is housekeeping. *Molecular Plant* **5**: 772–786.

Rotino G L, Perri E, Zottini M, Sommer H and Spena A. 1997. Genetic engineering of parthenocarpic plants. *Nature Biotechnology* **15**: 1398–1401.

Sachs T. 1991. Cell polarity and tissue patterning in plants. *Development* **113**: 83–93.

Sairanen I, Novák O, Pěnčík A, Ikeda Y, Jones B, Sandberg G and Ljung K. 2012. Soluble carbohydrates regulate auxin biosynthesis via PIF proteins in Arabidopsis. *The Plant Cell* **24**: 4907–4916.

Siddiqui M I and Hussain S A. 2007. Effect of indole butyric acid and types of cuttings on root initiation of Ficus hawaii. *Sarhad Journal of Agriculture* **23**: 919.

Simon S and Petrášek J. 2011. Why plants need more than one type of auxin. *Plant Science* **180**: 454–460.

Skoog F and Thimann K V. 1934. Further experiments on the inhibition of the development of lateral buds by growth hormone. *Proceedings of the National Academy of Sciences* **20**: 480–485.

Stepanova A N, Robertson-Hoyt J, Yun J, Benavente L M, Xie D Y, Doležal K and Alonso J M. 2008. TAA1-mediated auxin biosynthesis is essential for hormone crosstalk and plant development. *Cell* **133**: 177–191.

Sugawara S, Hishiyama S, Jikumaru Y, Hanada A, Nishimura T, Koshiba T and Kasahara H. 2009. Biochemical analyses of indole-3-acetaldoxime-dependent auxin biosynthesis in Arabidopsis. *Proceedings of the National Academy of Sciences* **106**: 5430–5435.

Sundberg E and Østergaard L. 2009. Distinct and dynamic auxin activities during reproductive development. *Cold Spring Harbor Perspectives in Biology* **1**: a001628.

Swarup K, Benková E, Swarup R, Casimiro I, Péret B, Yang Y and Levesque M P. 2008. The auxin influx carrier LAX3 promotes lateral root emergence. *Nature Cell Biology* **10**: 946–954.

Tao Y, Ferrer J L, Ljung K, Pojer F, Hong F, Long J A and Cheng Y. 2008. Rapid synthesis of auxin via a new tryptophan-dependent pathway is required for shade avoidance in plants. *Cell* **133**: 164–176.

Teale W D, Paponov I A and Palme K. 2006. Auxin in action: signalling, transport and the control of plant growth and development. *Nature Reviews Molecular Cell Biology* **7**: 847–859.

Tivendale N D, Ross J J and Cohen J D. 2014. The shifting paradigms of auxin biosynthesis. *Trends in Plant Science* **19**: 44–51.

Tobeña-Santamaria R, Bliek M, Ljung K, Sandberg G, Mol J N, Souer E and Koes R. 2002. FLOOZY of petunia is a flavin mono-oxygenase-like protein required for the specification of leaf and flower architecture. *Genes & Development* **16**: 753–763.

Tzin V and Galili G. 2010. New insights into the shikimate and aromatic amino acids biosynthesis pathways in plants. *Molecular Plant* **3**: 956–972.

Vikrant T and Rashid A. 2003. Somatic embryogenesis or shoot formation following high 2,4-D pulse – treatment of mature embryos of Paspalum scrobiculatum. *Biologia Plantarum* **46**: 297–300.

Vivian-Smith A, Luo M, Chaudhury A and Koltunow A. 2001. Fruit development is actively restricted in the absence of fertilization in Arabidopsis. *Development* **128**: 2321–2331.

Wang C, Liu C, Roqueiro D, Grimm D, Schwab R, Becker C and Weigel D. 2015. Genome-wide analysis of local chromatin packing in Arabidopsis thaliana. *Genome Research* **25**: 246–256.

Wang X, Wang H, Wang J, Sun R, Wu J, Liu S and Huang S. 2011. The genome of the mesopolyploid crop species *Brassica rapa*. *Nature Genetics* **43**: 1035–1039.

Wightman F and Lighty D L. 1982. Identification of phenylacetic acid as a natural auxin in the shoots of higher plants. *Physiologia Plantarum* **55**: 17–24.

Wightman F and Rauthan B S. 1974. Evidence for the biosynthesis and natural occurrence of the auxin, phenylacetic acid, in shoots of higher plants. pp. 15–27. In *Plant Growth Substances; Proceedings of the International Conference*. Hirokawa Publishing Co., Tokyo.

Wiśniewska J, Xu J, Seifertová D, Brewer P B, Růžička K, Blilou I and Friml J. 2006. Polar PIN localization directs auxin flow in plants. *Science* **312**: 883–883.

Woodward A W and Bartel B. 2005. Auxin: regulation, action, and interaction. *Annals of Botany* **95**: 707–735.

Wright A D and Cohen J D. 1991. Indole-3-acetic acid biosynthesis in the mutant maize orange pericarp, a tryptophan auxotroph. *Science* **254**: 998.

Xu L and Huang H. 2014. Genetic and epigenetic controls of plant regeneration. *Current Topics in Developmental Biology* **108**: 1–33.

Yamada M, Greenham K, Prigge M J, Jensen P J and Estelle M. 2009. The TRANSPORT INHIBITOR RESPONSE2 gene is required for auxin synthesis and diverse aspects of plant development. *Plant Physiology* **151**: 168–179.

Yamada T, Palm C J, Brooks B and Kosuge T. 1985. Nucleotide sequences of the Pseudomonas savastanoi indoleacetic acid genes show homology with Agrobacterium tumefaciens T-DNA. *Proceedings of the National Academy of Sciences* **82**: 6522–6526.

Yamamoto Y, Kamiya N, Morinaka Y, Matsuoka M and Sazuka T. 2007. Auxin biosynthesis by the YUCCA genes in rice. *Plant Physiology* **143**: 1362–1371.

Yang Y, Hammes U Z, Taylor C G, Schachtman D P and Nielsen E. 2006. High-affinity auxin transport by the AUX1 influx carrier protein. *Current Biology* **16**: 1123–1127.

Yin Z, Malinowski R, Ziółkowska A, Sommer H, Plcader W and Malepszy S. 2006. The DefH9-iaaM-containing construct efficiently induces parthenocarpy in cucumber. *Cellular and Molecular Biology Letters* **11**: 279–290.

Zhang W, Fan J, Tan Q, Zhao M and Cao F. 2017. Mechanisms underlying the regulation of root formation in Malus hupehensis stem cuttings by using exogenous hormones. *Journal of Plant Growth Regulation* **36**: 174–185.

Zhao S Q, Hu J, Guo L B, Qian Q and Xue H W. 2010. Rice leaf inclination2, a VIN3-like protein, regulates leaf angle through modulating cell division of the collar. *Cell Research* **20**: 935–947.

Zhao S Q, Xiang J J and Xue H W. 2013. Studies on the rice leaf INCLINATION1 (LC1), an IAA – amido synthetase, reveal the effects of auxin in leaf inclination control. *Molecular Plant* **6**: 174–187.

Zhao Y, Christensen S K, Fankhauser C, Cashman J R, Cohen J D, Weigel D and Chory J. 2001. A role for flavin monooxygenase-like enzymes in auxin biosynthesis. *Science* **291**: 306–309.

Zhao Y, Hull A K, Gupta N R, Goss K A, Alonso J, Ecker J R, Normanly J, Chory J and Celenza J L. 2002. Trp-dependent auxin biosynthesis in Arabidopsis: involvement of cytochrome P450s CYP79B2 and CYP79B3. *Genes and Development***16**: 3100–3112.

14 Gibberellins

14.1 INTRODUCTION

Gibberellins (GAs) comprise a large group of diterpenoid hormones that are essential for many developmental processes in plants, including seed germination, stem elongation, root elongation, leaf expansion, trichome development, pollen maturation and induction of flowering (Achard and Genschik, 2009). They encourage organ growth through enhancement of cell elongation and, in some cases, cell division. GA was first identified in the pathogenic fungus *Gibberella fujikuroi*, which causes the foolish-seedling disease of rice (Yabuta and Sumiki, 1938). More than 130 gibberellins have been identified so far in plants, fungi and bacteria. However, only a few GAs such as GA_1, GA_3, GA_4 and GA_7 are derived from a basic diterpenoid carboxylic acid skeleton and commonly have a C_3-hydroxyl group. It has been reported that developing seeds contain a very high concentrations of GA, as well as a broad range of different structural forms. However, vegetative and early reproductive tissues also contain a limited amount of GAs. In contrast, the endosperm, cotyledons and testa in the advanced stage of seed development contain large amount of active and inactive GAs, but their role is unclear. Generally, those seeds also contain high levels of GA-inactivating activity to prevent premature germination and abnormal seedling growth (Ross et al., 1995). GAs promote certain developmental switches between seed dormancy and germination, juvenile and adult growth phases, and vegetative and reproductive development (Hedden and Thomas, 2012). They normally act in response to certain developmental and environmental cues that may regulate GA biosynthesis, turnover (deactivation), perception and signal transduction.

14.2 GIBBERELLIN BIOSYNTHESIS

GAs are produced from geranyl geranyl diphosphate (GGDP) by a complex pathway involving three cell compartments and three classes of enzymes (Hedden and Phillips, 2000). They are biosynthesized in apical root and shoot tissues, developing seeds and fruits and young leaves of developing apical buds. Quantitative analysis of dwarf plant bioassay revealed the presence of GA in active growing tissues such as shoot apices, young leaves and flowers (Jones and Phillips, 1966; Kobayashi et al., 1988). Presence of GAs in xylem and phloem exudates indicates a long-distance transport of GAs (Hoad, 1995). It has been reported that active transport of GAs and their intermediates was supported by grafting experiments (Proebsting et al., 1992). Usually, GAs are synthesized via the terpenoid pathway requiring three enzymes such as terpene synthase (TPSs), cytochrome P450 monooxygenase (P450s) and 2-oxoglutarate dependent dehydrogenase (2 ODDs) from GGDP in plants (Gupta and Chakrabarty, 2013). *ent*-copalyl diphosphate synthase (CPS) and *ent*-kaurene synthase (KS) are the terpene synthase mostly located in plastids and involved in the conversion of GGDP to tetracyclic hydrocarbon intermediate *ent*-kaurene (Aach et al., 1997; Helliwell et al., 2001). First, *ent*-kaurene oxidase (KO) present in the outer membrane of plastid catalyzes the sequential oxidation on C-19 to produce *ent*-kaurenoic acid, and then *ent-kaurenoic* acid oxidase (KAO) that is present in endoplasmic reticulum converts *ent*-kaurenoic acid into GA (Helliwell et al., 2001; Nelson et al., 2009). Bioactive GA_4 is converted from GA_{12} through oxidations of C-20 and GA_3 by GA 20-oxidase (GA20ox) and GA 3-oxidase (GA3ox), respectively. GAs are mostly inactivated by conjugation, epoxidation, methylation and 2-oxidation (Schneider and Schliemann, 1994; Zhu et al., 2006; Varbanova et al., 2007). C19-GA_2-oxidation is considered to be the primary inactivation pathway (Rieu et al., 2008). Previous study reported that GA biosynthesis is regulated

DOI: 10.1201/9781003373889-19

by phytochrome in germinating *L. sativa* and *A. thaliana* seeds, in which genes encodes for GA_3-oxidases (Yamauchi et al., 2004). However, regulation of bioactive GA levels also involves feedback and feed-forward regulation of the expression of 2ODDs (Yamaguchi, 2008).

14.2.1 GA-Biosynthetic Pathway

In higher plants, GAs are primarily produced from the methylerythritol phosphate pathway. The hydrocarbon intermediate ent-kaurene is produced from GGPP (trans-geranylgeranyl diphosphate) in proplastids (Aach et al., 1997; Kasahara et al., 2002). It has been postulated that ent-kaurene synthesis competes with that of other GGPP-derived metabolites for a common pool of GGPP (Fray et al., 1996). Ent-kaurene synthesis from GGPP is synthesized by the two-step via ent-copalyl diphosphate (i) the reactions catalyzed by separate enzymes, ent-copalyl diphosphate synthase (CPS), a class II (proton-initiated) cyclase and ent-kaurenesynthase (KS), a class I (initiated by phosphate ionization) cyclase, respectively in higher plants and (ii) the conversion of ent-copalyl diphosphate to ent-kaurene (MacMillan and Beale, 1999). Previous studies reported that CPS and KS are encoded by a single gene in Arabidopsis but in others two or more CPS- and KS-like genes are expressed (Sakamoto et al., 2004; Toyomasu et al., 2009; Keeling, 2010). ent-Kaurene is relatively volatile and has been found to interchange with the external environment and therefore, it may function as a mediator of plant-to-plant communication (Otsuka et al., 2004). Meanwhile, regulation of GA biosynthesis occurs mainly at later stages of the pathway therefore, exogenous ent-kaurene is impossible to have a major influence on GA content and the resulting growth (Fleet et al., 2003). Conversion of ent-kaurenein to bioactive forms involves the action of membrane-associated P450s (cytochrome P450 monooxygenases) and soluble ODDs (2-oxoglutarate-dependent dioxygenases) (Hedden and Thomas, 2012).

The formation of GA_{12}, which is the common precursor for all GAs in plants, involves six oxidative steps, catalyzed by two monooxygenases, ent-kaurene oxidase (KO) belonging to the CYP701A P450 clade and ent-kaurenoic acid oxidase (KAO) belonging to CYP88A (Hedden and Phillips, 2000; Mizutani and Ohta, 2010). These enzymes are localized in both endoplasmic reticulum and the plastid (Helliwell et al., 2001). The three-step KO-catalyzed oxidation of ent-kaurene to ent-kaurenoic acid via ent-kaurenol and ent-kaurenal has been shown to involve repeated hydroxylations on C-19, with the initial step rate limiting and the intermediates retained at the enzyme's active site (Morrone et al., 2010). However, the oxidation of ent-kaurenoic acid to GA_{12} by KAO occurs in three steps via the intermediates ent-7α-hydroxykaurenoic acid and GA_{12}-aldehyde, requiring successive oxidations at C7β, C-6 β and C-7 (Hedden, 1997).

The nature of the enzymatic activity responsible for 13-hydroxylation, through which GA_{12} is converted into GA_{53}, is unclear because both P450 and ODD 13-hydroxylase activities have been detected in higher plants (Sponsel and Hedden, 2004). GA_{12} and GA_{53} are oxidized on C-20 by GA_{20}ox (GA $_{20}$-oxidase), an ODD, converting these substrates in parallel pathways into GA_9 and GA_{20}, respectively, and this whole process involves three to four steps depending on the mechanism for the loss of C-20 (Hedden, 1997). However, mechanism by which C-20 is lost is still not clear, although it seems that the formation of a free radical on C-10 that responds with the C-19 carboxylic acid forms the lactone (Ward et al., 1997). Ward and coworkers (Ward et al., 2002) reported that C-20 is lost directly as CO_2 that requires the existence of an intermediate between the C20 aldehyde and the C19 γ-lactone product. However, no candidate intermediate has been identified so far suggesting that it remains bound at the enzyme active site (Hedden, 1997).

The formation of biologically active hormones involves the 3β-hydroxylation of GA9 and GA20 to GA4 and GA1, respectively. This final step is catalyzed by the enzyme ODD GA3ox (GA 3-oxidase). This reaction tends to be highly region specific, with a single product being produced in shoots of dicotyledonous species, whereas in addition to GA_1, GA_3 is formed from GA_{20} via the intermediate GA_5 by the action of a single GA_3ox in monocotyledons (Martin et al., 1997; Williams et al., 1998; Itoh et al., 2001; Appleford et al., 2006). In some dicotyledonous species, developing seeds exhibit relatively low levels of GA3oxs, which are region-specific. These enzymes possess not only 3β-hydroxylase activity but also additional functionalities such as 2β-hydroxylase and 2,3-desaturase activities (Smith et al., 1990; Lange et al., 1997; Itoh et al., 2001; Ward et al., 2010). Unsaturated GAs (C-2,3-) may be further oxidized by GA3oxs to the 2,3-epoxide or to the 3β-hydroxyl,2-unsaturated products. Consequently, GA_5 may be converted into GA6 or GA_3, and this latter conversion involves abstraction of the 1β-H, rearrangement of the 2,3-double bond to the 1,2-position and then addition of OH at C-3β (Albone et al., 1990). Generally, the pathway of GA biosynthesis could be divided into three stages, and each of them is accomplished in a different cellular compartment.

14.2.1.1 Stage I. Formation of Terpenoid Precursors and Ent-Kaurene in Plastids

GAs are biosynthesized from a 5-C precursor isopentenyl diphosphate (IPP), and the IPP may be synthesized in either plastids or cytosol. From IPP, 10-C (GPP), 15-C (FPP) and 20-C (GGPP) precursors of terpenoids are formed by condensation of 5-C units (IPP). After the formation of GGPP, the pathway becomes specific for GAs. GGPP is converted by two cyclization reactions through copalyl pyrophosphate into ent-kaurene. These reactions are catalyzed by the enzymes cyclases that are located in proplastids and not in mature chloroplasts, and in-fact, it constitutes the first step that is specific for GAs. This step of GA biosynthesis is inhibited by compounds such as Amo-1618, Phosphon D and CCC.

14.2.1.2 Stage II. Oxidations to Form GA_{12} and GA_{53} on ER Through GA_{12} Aldehyde

The ent-kaurene is transported from plastids to endoplasmic reticulum (ER). Now, a methyl group on ent-kaurene at 19th-carbon position is oxidized to carboxylic group which is followed by contraction of ring B from 6-C to 5-C ring structure to form GA_{12}-aldehyde. GA_{12}-aldehyde is subsequently oxidized to give GA_{12}, which is precursor to all other GAs in plants. Hydroxylation of GA_{12} at C-13 results in the formation of GA_{53}. The enzymes catalyzing the above oxidation reactions are monooxygenases, which are located on ER and utilize cytochrome P450 in these reactions. Activity of these enzymes is inhibited by paclobutrazol and other inhibitors before GA_{12}-aldehyde.

14.2.1.3 Stage III. Formation of All Other GAs From GA_{12} or GA_{53} in Cytosol

All other steps in the biosynthesis of GAs from GA_{12} or GA_{53} are carried out in cytosol by soluble enzymes called dioxygenases. These enzymes require molecular O_2 and 2-oxoglutarate as cosubstrates and use ferrous iron (Fe^{+2}) and ascorbic acid as cofactors. Activity of these enzymes is inhibited by cyclohexanetriones. Environmental factors such as temperature and photoperiod are known to affect biosynthesis of gibberellins.

14.3 EVOLUTION OF GIBBERELLIN BIOSYNTHESIS AND SIGNAL TRANSDUCTION

The availability of genome sequences has driven interest in the evolution of GA production and signaling (Hedden and Sponsel, 2015). The presence of functional GA biosynthesis and signaling pathways in lycophyte (*Selaginella moellendorffii*) indicates that they are evolved in vascular plants (Hirano et al., 2007; Vandenbussche et al., 2007; Yasumura et al., 2007). GA signaling regulates sporulation in *Selaginella moellendorffii*, which indicates that the pathway evolved to regulate GAMYB, which is involved in reproductive development (Aya et al., 2011). The role of GA in growth responses of higher plants may have occurred through modifications to DELLA that extended the range of transcription factors with which it can interact (Hedden and Sponsel, 2015). Gibberellin production has also been evolved in some fungal and bacterial species, but it seems to have occurred independently of that in plants (Hedden et al., 2002).

14.4 GENES ENCODING ENZYMES FOR GA BIOSYNTHESIS AND CATABOLISM

A number of GA-deficient dwarf mutants such as *ga1*, *ga2*, *ga3*, *ga4* and *ga5* have been isolated from Arabidopsis (Koornneef and van der Veen, 1980). By cloning corresponding genes, it was confirmed that GA_1 encodes CPS, *GA2* encodes KS, GA_3 encodes KO, GA_4 encodes GA_3ox1 and GA_5 encodes $GA_{20}ox1$ (Sun and Kamiya, 1994; Yamaguchi et al., 1998a; Helliwell et al., 1999; Chiang et al., 1995; Phillips et al., 1995). Furthermore, Hedden and Phillips (2000) and Yamaguchi (2008) have successfully isolated gene-encoding enzymes involved in most steps of GA metabolic pathways (except for 13-hydroxylation) (Hedden and Phillips, 2000; Yamaguchi, 2008). It has been reported that CPS, KS and KO are each encoded by a single gene in most of the plant species. Synthesis of *ent*-kaurene which is the first committed intermediate in the GA pathway has been shown to be catalyzed by CPS and KS (Silverstone et al., 1997).

Actively growing tissues such as shoot apices, root tips, developing anthers and seeds contain higher amounts of *CPS* mRNA. The expression pattern of *KS* is similar to that of *CPS*, but the overall amount of *KS* mRNA is much higher than *CPS*, which suggests that synthesis of *ent*-kaurene is primarily controlled by the expression and location of CPS (Yamaguchi et al., 1998a). Overexpression of CPS causes a dramatic increase in *ent*-kaurene accumulation confirming that CPS play an important role in controlling the location and the activity of the early stages of GA biosynthesis (Fleet et al., 2003). However, $GA_{20}ox$ and GA_3ox are encoded by a small gene family with each gene showing a tissue-specific expression pattern (Thomas et al., 1999; Mitchum et al., 2006; Rieu et al., 2008). Some of these genes are expressed in all stages of plant development, whereas some have specific expression at a particular phase of plant development (Rieu et al., 2008).

14.4.1 Gibberellin Signal Transduction

Regulatory elements of the GA signal transduction pathway have been identified by using (i) screening for mutants with altered GA sensitivity, (ii) identification of transcriptional regulators of the GA responsive genes and (iii) identifying differentially expressed genes. Such approaches have revealed both positive and negative regulators of GA transduction (Jacobsen et al., 1995; Olszewski et al., 2002; Sun and Gubler, 2004). Thomas and coworkers have described the response of various GA-response genes in different plants (Thomas et al., 2005).

14.4.2 DELLA Proteins in Gibberellin Signaling

Hormone-mediated control of plant growth and development involves both synthesis and response of GAs (Nakajima et al., 2006). A bioactive GA binds to a GA receptor such as Gibberellin insensitive dwarf 1 (GID1) and, subsequently, forms a GA-GID 1 complex, which triggers the degradation of DELLAs. This mechanism is also called de-repression of GA (Gallego-Giraldo et al., 2014; Fukazawa et al., 2014). This process requires a specific ubiquitin E3 ligase complex ($SCF^{SLY1/GID2}$) in Arabidopsis and rice (Sun, 2010). In recent years, positive (SLEEPY1 [SLY1], GAMYB, and PICKLE) and negative (short internode [SHI], SPINDLY [SPY]) factors have been reported to regulate GA signaling in higher plants (Sun and Gubler, 2004; Ogas et al., 1997; Swain et al., 2002; McGinnis et al., 2003; Wang et al., 2015).

DELLA proteins are required for GA regulation, whereas a GRAS domain is required for its function. The DELLA genes are members of the GRAS (GAI-RGA and Scarecrow) family (Pysh et al., 1999). The C-terminal GRAS domain contains sequences similar to those found in metazoan signal transducers and activators of transcription (STAT) factors including two leucine heptad repeats (LHR) and SH2-like domain (Peng et al., 1999). GAI-RGA and Scarecrow proteins contain a variable N-terminal domain. The N-terminal domain of the DELLA subfamily is defined by the consensus DELLA and VHYNP amino acid sequences (Thomas et al., 2005). Deletions,

N-terminal truncations and amino acid substitutions within the DELLA domain have been shown to result in a semi-dominant GA-insensitive dwarf phenotype (Boss and Thomas, 2002; Dill et al., 2001; Peng et al., 1999).

14.4.3 Regulation of GA Metabolism in Light

Light quantity, quality and photoperiod are the most important factors that regulate all aspects of plant growth and development. Light exerts profound influence by causing changes in concentration and/or sensitivity to GAs (Kamiya and García-Martínez, 1999; Olszewski et al., 2002; Yamaguchi and Kamiya, 2000). It has been reported that the response of seeds to light is mediated by an increase in GA biosynthesis (Yamaguchi and Kamiya, 2000). Arabidopsis seeds have an absolute requirement for both GAs and red light. In addition, red light enhances the sensitivity of the seed to GA and promotes germination (Hilhorst and Karssen, 1988). Germination studies in Arabidopsis and lettuce have provided definite evidence that phytochrome upregulates GA biosynthesis by promoting GA_3-oxidation (Yamaguchi et al., 1998b; Toyomasu et al., 1993, 1998).

14.4.4 Gibberellins Transport in Plants

The GA transport was investigated by Kato (1958) and confirmed that equal movement of GA in the acropetal and basipetal directions. On the basis of the colocation of genes encoding GA-biosynthetic enzymes and signaling components, it has been concluded that GAs are synthesized at their site of action in shoot apices and stamens of rice (Kaneko et al., 2003). Long-distance transport of GAs from leaves to shoot apex has been implicated in floral initiation in the shoot apex, promotion of elongation and increase in secondary growth of the stem (King, 2012; Dayan et al., 2012). Recently, grafting experiments with Arabidopsis mutants provided clear evidence that GA_{12} is the main mobile form in both the xylem and phloem (Regnault et al., 2015). Gibberellins were present in both phloem and xylem exudates (Hoad and Bowen, 1968). Transport of gibberellins in plants through phloem is similar to those of carbohydrates and other organic solutes. The GA transport is non-polar and may also occur in xylem due to lateral movement between the two vascular tissues, i.e., xylem and phloem. The gibberellins are not translocated in plant as free molecules but probably in their bound form as gibberellin-glycosides (Chiba et al., 2015).

14.5 GIBBERELLIN CONJUGATES

Ester and ether GA conjugates have been identified in a number of plant species using glucose as the predominant conjugate partner (Schneider and Schliemann, 1994). Some GA conjugates were shown to have biological activity in bioassays, but this was believed to be due to hydrolysis to the aglycone in the assay system (Sembdner et al., 1976). Evidence for the reversibility of GA conjugate formation enhances the possibility of their having a storage role, providing a mechanism for rapid release of bioactive GA (Schneider et al., 1992). Recently, ectopic expression of a β-glucosidase in transplastomic tobacco resulted in increased levels of several hormones including GA_1 and GA_4 suggesting that plastids may act as a store for hormone conjugates (Jin et al., 2012).

14.6 DEACTIVATION OF GIBBERELLINS

GAs are inactivated to regulate homeostasis and allow a rapid reduction in GA concentration when required. Several mechanisms are involved in deactivation of gibberellin, such as introduction of a 2-β-hydroxyl group into a GA markedly reduces its biological activity. Conversion of active GA_1 into inactive GA_8 is one such example. Another example is conversion of GA_{20} to GA_{29}. Conversion of free GAs into their bound forms, such as gibberellin-glycosides or gibberellin-glycosyl ethers or esters also inactivates GAs, especially in mature seeds. Formation of Gibberethione from GA_3 by the oxidation of its 3-P-OH group followed by addition of mercaptopyruvic acid or L-cysteine to the resulting a P-unsaturated ketone leads to inactivation.

14.6.1 Seed Reserve Mobilization by Aleurone Cells

Aleurone cells are a highly specialized tissue that differentiates from peripheral endosperm cells during seed development (Hooley, 1994). These cells are commonly used for studying the cellular and molecular biology of GA-action. Aleurone cells synthesize and secrete a variety of hydrolytic enzymes that participate in the systematic breakdown of starch and protein reserves in the endosperm to provide nutrients for the growing embryonic seedling immediately after seed germination. Aleurone cells possess the majority of the seed's stored reserves of myo-inositol, phosphorous and mineral (K^+ and Mg^{2+}) that are released into the endosperm following germination and provide the carbohydrate for cell-wall synthesis and phosphorous and essential cations for emerging seedlings (Hooley, 1994). Reserve mobilization by aleurone cells appears to be coordinated to a large extent by GA produced by the embryo (Fincher, 1989; Jones and Jacobsen, 1991).

14.7 PHYSIOLOGICAL ROLE OF GIBBERELLINS IN PLANTS

14.7.1 Seed Dormancy and Germination

Breaking of seed dormancy is controlled by several environmental factors (light, temperature and moisture) as well as by the endogenous growth regulating hormones (GA and ABA). Normally, GA increases the growth potential of embryo and also induces hydrolytic enzymes. Therefore, it stimulates the seed germination (Kucera et al., 2005; Finch-Savage and Leubner-Metzger, 2006). Embryonic GA released during seed germination triggers weakening of seed cover by stimulating gene expression that is involved in cell expansion and modification (Finkelstein et al., 2008). GAs stimulate the production of hydrolytic enzyme (ά-amylase) in the aleurone layer of germinating cereal grains (Seo et al., 2009). In cereal grains, the stored food reserves of the endosperms are broken down into soluble sugars, amino acids and other products that are transported to the growing embryo. After physiological and biochemical characterization, it was confirmed that the bioactive GAs are produced in the embryo and transported to aleurone layer where it triggers the expression of ά-amylase (Gubler et al., 1995). These authors concluded that during seed germination, the aleurone layer is unable to synthesize GA but perceive the GA signals.

14.7.2 Stimulation of Growth and Elongation

GA promote cell growth in a variety of tissues, which was confirmed by comparing a GA-deficient mutant phenotype with the wild-type or GA-treated mutant (Hooley, 1994). GA-regulated growth responses include shoot elongation (internodes, hypocotyls, coleoptiles, mesocotyls and epicotyls) and growth of other tissues that occurs during specific growth stages (Hooley, 1994). It is reported that transition from juvenile to adult leaf shape is impacted by GA in a range of plant species (Jones, 1987; Ross et al., 1993). Furthermore, GAs also influence cell growth polarity in root tips and shoot apical meristems and stimulates pollen germination and pollen tube growth following pollination (Nitsch, 1970; Baluška et al., 1993).

GA stimulates stem elongation by promoting cell elongation and cell division (Huttly and Phillips, 1995). Gibberellin treatment causes reorientation of microtubules, which will facilitate axial elongation (Shibaoka, 1994). GA induces enzymes that are responsible for wall loosening and expansion, such as xyloglucan endotransglycosylase/hydrolase (XET or XTH), expansins and pectin methylesterase (PME). Xyloglucan endotransglycosylases split cell-wall xyloglucan polymers endolytically and then rejoin the free ends with another xyloglucan chain (Campbell and Braam, 1999). Expansins disrupt hydrogen bonding in the cell wall and appear to be under GA control in

Arabidopsis and rice (Cosgrove, 2000; Lee et al., 2001; Ogawa et al., 2003). Changes in the expression of a number of genes are now known to accompany GA-induced elongation growth (Shi et al., 1992; Mendu and Silflow, 1993).

GA plays an important role in internode elongation stimulated by cell division and expansion in response to light or dark (photomorphogenesis and skoto-morphogenesis) (Hooley, 1994; Swain and Olszewski, 1996; Ross et al., 1995; Alabadí et al., 2008; Feng et al., 2008; Gallego-Bartolomé et al., 2011). Silverstone et al. (1997) and Li et al. (2013) reported the expression of GA-regulated genes in growing tissues of Arabidopsis and concluded that GA promotes cell elongation through release of DELLA-mediated inhibition of BZR1 transcription factor (Silverstone et al., 1997; Li et al., 2013).

14.7.3 Flowering and Sex Expression

Plant development progresses through distinct phases such as vegetative growth followed by a reproductive phase and eventually seed set and senescence and the changes between these phases are controlled by genetic circuits that integrate endogenous and environmental cues (Huijser and Schmid, 2011). Generally, plants first transit from juvenile vegetative phase to an adult vegetative phase and then transit from a reproductively incompetent to a reproductively competent phase (Poethig, 1990). Gibberellin has been reported to regulate both of these processes (Zimmerman et al., 1985; Reid and Potts, 1986).

The transition from vegetative growth to flowering in long-day or cold-requiring plants and conifers is promoted by the application of GAs (Hooley, 1994). Vernalization and long days promote flowering by stimulating GA biosynthesis (Bernier et al., 1993). Photoperiod also effects floral development in Arabidopsis mutants, where GA can overcome the effect of short days on floral meristem identity and development (Okamuro et al., 1993). GAs may influence the expression of floral homeotic genes such as LFY and AG.

GAs play an important role in controlling male and female fertility (Griffiths et al., 2006). For example, GA-deficient mutants in Arabidopsis and tomato showed abnormal stamen development, while extreme GA deficiency caused female sterility (Goto and Pharis, 1999; Chhun et al., 2007; Hu et al., 2008; Rieu et al., 2008). Because stamens and/or flower receptacles are potential sites for bioactive GA synthesis in Arabidopsis flowers, GAs produced in these organs are transported to promote petal growth (Hu et al., 2008). GA also plays a very important role in pollen germination and pollen tube development, and it has been observed that pollens do not germinate unless rescued by exogenous GA in GA deficit mutants (Rieu et al., 2008). GA in coordination with jasmonic acid relegate the stamen development (filament elongation, anther dehiscence and pollen maturation), whereas the GA alone mediated early anther development (Song et al., 2013).

Gibberellin also modifies sex expression in some plants. GAs tend to suppress the development of staminate flowers while accelerating the development of pistillate flowers in cucumber (Fuchs et al., 1997), Pimenta (Lange et al., 2012) and bitter gourd (Banerjee and Basu, 1992).

14.7.4 Fruit Development

Gibberellins control fruit growth and development in various ways and at different developmental stages (Chudasama and Thake, 2007). Growing fruits are metabolically very active and act as strong sinks for nutrients with hormones possibly modulating the process (Brenner and Cheikh, 1995). Gibberellins are known to influence both cell division and cell enlargement (Adams et al., 1975; Kamijima, 1981). Seeds are rich source of many hormones, which are essential for normal development of fruits (Crane, 1969). Active gibberellins (GA_1 or GA_3) are able to induce fruit set in several horticultural species (Gillaspy et al., 1999; Dorcey et al., 2009). Increased levels of gibberellin in pollinated ovary together with an increased expression of GA-biosynthetic genes enhance fruit set in the parthenocarpic tomato mutants pat, pat2 and pat3/4 (Serrani et al., 2007). However, the application of inhibitors of GA biosynthesis limits fruit growth (Serrani et al., 2008). Auxin and gibberellin act synergistically on fruit growth, suggesting that the two phytohormones interact in regulating the fruit development (Ozga and Reinecke, 2003; Serrani et al., 2008).

14.7.5 Vernalization

Many plants that require cold treatment also require proper photoperiodic treatment for the induction of flowers, without which vernalization does not have any effect. If such plants are treated with gibberellins, they produce flowers without subjecting the plants to cold and photoperiodic treatments (Hazebroek and Metzger, 1990). Therefore, gibberellins substitute not only vernalization but also photoperiodic treatment. It is reported that cold treatment activates VER203 gene expression and GA also induces VER203 expression. Gibberellins are known to overcome both cold treatment and photoperiodic treatment in long-day plants, but it has no effect on short-day plants.

GA-deficient mutants are short and often have retarded flowering without supplemental GA treatment (Rood et al., 1989; Reid, 1990). GA-deficient *Brassica* mutants such as rosette resemble nonvernalized winter annuals suggesting that the failure of non-vernalized plants to bolt and flower could be the result of impaired GA biosynthesis (Rood et al., 1989; Zanewich et al., 1990)

14.7.6 Stress Tolerance

GAs are also actively involved in responding to various stresses such as cold, salt and osmotic stress (DongLei et al., 2013). External application of GA_3 can alleviate adverse effects of high salinity by increasing nutrient uptake (Ashraf, 2002). Furthermore, gibberellic acid increases the germination potential of wheat seeds especially at moderate salinity levels (Radi et al., 2006). Foliar application GA_3 increases salt tolerance by enhancing nutrient uptake in sugarcane (Shomeili et al., 2011). It is reported that application of GA_3 alleviates salt stress by activating enzymes that participate in RNA and protein synthesis (Bejaoui, 1985).

REFERENCES

Aach H, Bode H, Robinson D G and Graebe J E. 1997. ent-Kaurene synthase is located in proplastids of meristematic shoot tissues. *Planta* **202**: 211–219.

Achard P and Genschik P. 2009. Releasing the brakes of plant growth: how GAs shutdown DELLA proteins. *Journal of Experimental Botany* **60**: 1085–1092.

Adams P A, Montague M J, Tepfer M, Rayle D L, Ikuma H and Kaufman P B. 1975. Effect of gibberellic acid on the plasticity and elasticity of Avena stem segments. *Plant Physiology* **56**: 757–760.

Alabadí D, Gallego-Bartolomé J, Orlando L, García-Cárcel L, Rubio V, Martínez C and Blázquez M A. 2008. Gibberellins modulate light signaling pathways to prevent Arabidopsis seedling de-etiolation in darkness. *The Plant Journal* **53**: 324–335.

Albone K S, Gaskin P, MacMillan J, Phinney B O and Willis C L. 1990. Biosynthetic origin of gibberellins A3 and A7 in cell-free preparations from seeds of *Marah macrocarpus* and *Malus domestica*. *Plant Physiology* **94**: 132–142. Appleford N E, Evans D J, Lenton J R, Gaskin P, Croker S J, Devos K M and Hedden P. 2006. Function and transcript analysis of gibberellin-biosynthetic enzymes in wheat. *Planta* **223**: 568–582.

Ashraf M. 2002. Salt tolerance of cotton: some new advances. *Critical Reviews in Plant Sciences* **21**: 1–30.

Aya K, Hiwatashi Y, Kojima M, Sakakibara H, Ueguchi-Tanaka M, Hasebe M and Matsuoka M. 2011. The Gibberellin perception system evolved to regulate a pre-existing GAMYB-mediated system during land plant evolution. *Nature Communications* **2**: 544.

Baluška F, Parker J S and Barlow P W. 1993. A role for gibberellic acid in orienting microtubules and regulating cell growth polarity in the maize root cortex. *Planta* **191**: 149–157.

Banerjee S and Basu P S. 1992. Hormonal regulation of flowering and fruit development: effect of gibberellic acid and ethrel on fruit setting and development of Momordica charantia L. *Biologia Plantarum* **34**: 63.

Bejaoui M. 1985. Interactions between NaCl and some phytohormones on soybean growth. *Journal of Plant Physiology* **120**: 95–110.

Bernier G, Havelange A, Houssa C, Petitjean A and Lejeune P. 1993. Physiological signals that induce flowering. *The Plant Cell* **5**: 1147.

Boss P K and Thomas M R. 2002. Association of dwarfism and floral induction with a grape 'green revolution' mutation. *Nature* **416:** 847–850.

Brenner M L and Cheikh N. 1995. The role of hormones in photosynthate partitioning and seed filling. pp. 649–670. *Plant Hormones: Physiology, Biochemistry, and Molecular Biology*. 2nd edition (Ed.) Davis PJ. Kluwer Academic Publishers, Dordrecht, The Netherlands.

Campbell P and Braam J. 1999. In vitro activities of four xyloglucan endotransglycosylases from Arabidopsis. *The Plant Journal* **18**: 371–382.

Chhun T, Aya K, Asano K, Yamamoto E, Morinaka Y, Watanabe M and Ueguchi-Tanaka M. 2007. Gibberellin regulates pollen viability and pollen tube growth in rice. *The Plant Cell***19**: 3876–3888.

Chiang H H, Hwang I and Goodman H M. 1995. Isolation of the Arabidopsis GA4 locus. *The Plant Cell* **7**: 195–201.

Chiba Y, Shimizu T, Miyakawa S, Kanno Y, Koshiba T, Kamiya Y B and Seo M. 2015. Identification of Arabidopsis thaliana NRT1/PTR FAMILY (NPF) proteins capable of transporting plant hormones. *Journal of Plant Research* **128**: 679–686.

Chudasama R S and Thaker V S. 2007. Relationship between gibberellic acid and growth parameters in developing seed and pod of pigeon pea. *Brazilian Journal of Plant Physiology* **19**: 43–51.

Cosgrove D J. 2000. Loosening of plant cell walls by expansins. *Nature* **407**: 321–326.

Crane J C. 1969. The role of hormones in fruit set and development. *HortScience* **4**: 1969–1970.

Dayan J, Voronin N, Gong F, Sun T P, Hedden P, Fromm H and Aloni R. 2012. Leaf-induced gibberellin signaling is essential for internode elongation, cambial activity, and fiber differentiation in tobacco stems. *The Plant Cell* **24**: 66–79.

Dill A, Jung H S and Sun T P. 2001. The DELLA motif is essential for gibberellin-induced degradation of RGA. *Proceedings of the National Academy of Sciences* **98**: 14162–14167.

DongLei Y, Yao J, Mei C S, Tong X H, Zeng L J, Li Q and Lee C M. 2013. Plant hormone jasmonate prioritizes defense over growth by interfering with gibberellin signaling cascade. *Proceedings of the National Academy of Sciences* **109**: E1192–E1200.

Dorcey E, Urbez C, Blázquez M A, Carbonell J and Perez-Amador M A. 2009. Fertilization-dependent auxin response in ovules triggers fruit development through the modulation of gibberellin metabolism in Arabidopsis. *The Plant Journal* **58**: 318–332.

Feng S, Martinez C, Gusmaroli G, Wang Y, Zhou J, Wang F and Schäfer E. 2008. Coordinated regulation of Arabidopsis thaliana development by light and gibberellins. *Nature* **451**: 475–479.

Fincher G B. 1989. Molecular and cellular biology associated with endosperm mobilization in germinating cereal grains. *Annual Review of Plant Biology* **40**: 305–346.

Finch-Savage W E and Leubner-Metzger G. 2006. Seed dormancy and the control of germination. *New Phytologist* **171**: 501–523.

Finkelstein R, Reeves W, Ariizumi T and Steber C. 2008. Molecular aspects of seed dormancy. *Annual Review of Plant Biology* **59**: 387–415.

Fleet C M, Yamaguchi S, Hanada A, Kawaide H, David C J, Kamiya Y and Sun T P. 2003. Overexpression of AtCPS and AtKS in Arabidopsis confers increased ent-kaurene production but no increase in bioactive gibberellins. *Plant Physiology* **132**: 830–839.

Fray M, Chomet P, Glawischnig E, Stettner C, Grün S, Winklmair A and Simcox K. 1996. Analysis of a chemical plant defense mechanism in grasses. *Science* **277**: 696–699.

Fuchs M, McFerson J R, Tricoli D M, McMaster J R, Deng R Z, Boeshore M L and Gonsalves D. 1997. Cantaloupe line CZW-30 containing coat protein genes of cucumber mosaic virus, zucchini yellow mosaic virus, and watermelon mosaic virus-2 is resistant to these three viruses in the field. *Molecular Breeding* **3**: 279–290.

Fukazawa J, Teramura H, Murakoshi S, Nasuno K, Nishida N, Ito T and Takahashi Y. 2014. DELLAs function as coactivators of GAI-ASSOCIATED FACTOR1 in regulation of gibberellin homeostasis and signaling in Arabidopsis. *The Plant Cell* **26**: 2920–2938.

Gallego-Bartolomé J, Alabadí D and Blázquez M A. 2011. DELLA-induced early transcriptional changes during etiolated development in *Arabidopsis thaliana*. *PLOS ONE* **6**: e23918.

Gallego-Giraldo C, Hu J, Urbez C, Gomez M D, Sun T P and Perez-Amador M A. 2014. Role of the gibberellin receptors GID1 during fruit-set in Arabidopsis. *The Plant Journal* **79**: 1020–1032.

Gillaspy G, Ben-David H and Gruissem W. 1999. Fruits: a developmental perspective. *Plant Cell* **5:** 1439–1451.

Goto N and Pharis R P. 1999. Role of gibberellins in the development of floral organs of the gibberellin-deficient mutant, ga1–1, of Arabidopsis thaliana. *Canadian Journal of Botany* **77**: 944–954.

Griffiths J, Murase K, Rieu I, Zentella R, Zhang Z L, Powers S J and Thomas S G. 2006. Genetic characterization and functional analysis of the GID1 gibberellin receptors in Arabidopsis. *The Plant Cell* **18**: 3399–3414.

Gubler F, Kalla R, Roberts J K and Jacobsen J V. 1995. Gibberellin-regulated expression of a myb gene in barley aleurone cells: evidence for Myb transactivation of a high-pI alpha-amylase gene promoter. *The Plant Cell* **7**: 1879–1891.

Gupta R and Chakrabarty S K. 2013. Gibberellic acid in plant: still a mystery unresolved. *Plant Signaling & Behavior* **8**: e25504.

Hazebroek J P and Metzger J D. 1990. Thermoinductive regulation of gibberellin metabolism in Thlaspi arvense LI Metabolism of [2H]-ent-kaurenoic acid and [14C] gibberellin A12-aldehyde. *Plant Physiology* **94**: 157–165.

Hedden P. 1997. Gibberellin biosynthesis: enzymes, genes and their regulation. *Annual Review of Plant Biology* **48**: 431–460.

Hedden P and Phillips A L. 2000. Gibberellin metabolism: new insights revealed by the genes. *Trends in Plant Science* **5:** 523–530.

Hedden P, Phillips A L, Rojas M C, Carrera E and Tudzynski B. 2002. Gibberellin biosynthesis in plants and fungi: a case of convergent evolution? *Journal Plant Growth Regulation* **20**: 319–331.

Hedden P and Sponsel V. 2015. A century of gibberellin research. *Journal of Plant Growth Regulation* **34**: 740–760.

Hedden P and Thomas S G. 2012. Gibberellin biosynthesis and its regulation. *Biochemical Journal* **444**: 11–25.

Helliwell C A, Poole A, Peacock W J and Dennis E S. 1999. Arabidopsis ent-kaurene oxidase catalyzes three steps of gibberellin biosynthesis. *Plant Physiology* **119**: 507–510.

Helliwell C A, Sullivan J A, Mould R M, Gray J C, Peacock W J and Dennis E S. 2001. A plastid envelope location of Arabidopsis ent-kaurene oxidase links the plastid and endoplasmic reticulum steps of the gibberellin biosynthesis pathway. *The Plant Journal* **28**: 201–208.

Hilhorst H W and Karssen C M. 1988. Dual effect of light on the gibberellin-and nitrate-stimulated seed germination of *Sisymbrium officinale* and *Arabidopsis thaliana. Plant Physiology* **86**: 591–597.

Hirano K, Nakajima M, Asano K, Nishiyama T, Sakakibara H, Kojima M and Banks J A. 2007. The GID1-mediated gibberellin perception mechanism is conserved in the lycophyte Selaginella moellendorffii but not in the bryophyte Physcomitrella patens. *The Plant Cell* **19**: 3058–3079.

Hoad G V. 1995. Transport of hormones in the phloem of higher plants. *Plant Growth Regulation* **16**: 173–182.

Hoad G V and Bowen M R. 1968. Evidence for gibberellin-like substances in phloem exudate of higher plants. *Planta* **82**: 22–32.

Hooley R. 1994. Gibberellins: perception, transduction and responses. *Plant Molecular Biology* **26**: 1529–1555.

Hu J, Mitchum M G, Barnaby N, Ayele B T, Ogawa M, Nam E and Swain S M. 2008. Potential sites of bioactive gibberellin production during reproductive growth in Arabidopsis. *The Plant Cell* **20**: 320–336.

Huijser P and Schmid M. 2011. The control of developmental phase transitions in plants. *Development* **138**: 4117–4129.

Huttly A K and Phillips A L. 1995. Gibberellin-regulated plant genes. *Physiologia Plantarum* **95**: 310–317.

Itoh H, Ueguchi-Tanaka M, Sato Y, Ashikari M and Matsuoka M. 2001. The gibberellin signaling pathway is regulated by the appearance and disappearance of SLENDER RICE1 in nuclei. *The Plant Cell* **14**: 57–70.

Jacobsen S E, Binkowski K A and Olszewski N E. 1995. SPINDLY, a tetratricopeptide repeat protein involved in gibberellin signal transduction in Arabidopsis. *Proceedings of the National Academy of Sciences* **93**: 9292–9296.

Jin S, Zhang X and Daniell H. 2012. Pinellia ternata agglutinin expression in chloroplasts confers broad spectrum resistance against aphid, whitefly, lepidopteran insects, bacterial and viral pathogens. *Plant Biotechnology Journal* **10**: 313–327.

Jones M G. 1987. Gibberellins and the procera mutant of tomato. *Planta* **172**: 280–284.

Jones R L and Jacobsen J V. 1991. Regulation of synthesis and transport of secreted proteins in cereal aleurone. *International Review of Cytology* **126**: 49–88.

Jones R L and Phillips I D J. 1966. Organs of gibberellin synthesis in light-grown sunflower plants. *Plant Physiology* **41**: 1381–1386.

Kamijima O. 1981. Consideration on the mechanism of expression of dwarf genes in rice plants. II. The actions of dwarf genes on cell division and cell elongation in parenchyma of internode. *Japanese Journal of Breeding* **31**: 302–315.

Kamiya Y and García-Martínez J L. 1999. Regulation of gibberellin biosynthesis by light. *Current Opinion in Plant Biology* **2**: 398–403.

Kaneko M, Itoh H, Inukai Y, Sakamoto T, Ueguchi-Tanaka M, Ashikari M and Matsuoka M. 2003. Where do gibberellin biosynthesis and gibberellin signaling occur in rice plants? *The Plant Journal* **35**: 104–115.

Kasahara A, Nukumizu K, Hitoki G, Takata T, Kondo J N, Hara M and Domen K. 2002. Photoreactions on LaTiO2N under visible light irradiation. *The Journal of Physical Chemistry* **106**: 6750–6753.

Kato J. 1958. Studies on the Physiological Effect of Gibberellin II. *Physiologia Plantarum* **11**: 10–15.

Keeling P J. 2010. The endosymbiotic origin, diversification and fate of plastids. *Philosophical Transactions of the Royal Society B: Biological Sciences* **365**: 729–748.

King R W. 2012. Mobile signals in day length-regulated flowering: gibberellins, flowering locus T, and sucrose. *Russian Journal of Plant Physiology* **59**: 479–490.

Kobayashi M, Yamaguchi I, Murofushi N, Ota Y and Takahashi N. 1988. Fluctuation and localization of endogenous gibberellins in rice. *Agricultural and Biological Chemistry* **52**: 1189–1194.

Koornneef M and Van der Veen J H. 1980. Induction and analysis of gibberellin sensitive mutants in *Arabidopsis thaliana* (L.) Heynh. *Theoretical and Applied Genetics* **58**: 257–263.

Kucera B, Cohn M A and Leubner-Metzger G. 2005. Plant hormone interactions during seed dormancy release and germination. *Seed Science Research* **15**: 281–307.

Lange M J P, Knop N and Lange T. 2012. Stamen-derived bioactive gibberellin is essential for male flower development of *Cucurbita maxima* L. *Journal of Experimental Botany*: err448.

Lange T, Robatzek S and Frisse A. 1997. Cloning and expression of a gibberellin 2 beta, 3 beta-hydroxylase cDNA from pumpkin endosperm. *The Plant Cell* **9**: 1459–1467.

Lee Y, Choi D and Kende H. 2001. Expansins: ever-expanding numbers and functions. *Current Opinion in Plant Biology* **4**: 527–532.

Li Z Y, Xu Z S, Chen Y, He G Y, Yang G X, Chen M and Ma Y Z. 2013. A novel role for Arabidopsis CBL1 in affecting plant responses to glucose and gibberellin during germination and seedling development. *PLoS One* **8**(2): e56412.

Macmillan J A K E and Beale M H. 1999. Diterpene biosynthesis. *Comprehensive Natural Products Chemistry* **2**: 217–243.

Martin D N, Proebsting W M and Hedden P. 1997. Mendel's dwarfing gene: cDNAs from the *Le* alleles and function of the expressed proteins. *Proceedings of the National Academy of Sciences* **94**: 8907–8911.

McGinnis K M, Thomas S G, Soule J D, Strader L C, Zale J M, Sun T P and Steber C M. 2003. The Arabidopsis SLEEPY1 gene encodes a putative F-box subunit of an SCF E3 ubiquitin ligase. *The Plant Cell* **15**: 1120–1130.

Mendu N and Silflow C D. 1993. Elevated levels of tubulin transcripts accompany the GA3-induced elongation of oat internode segments. *Plant and Cell Physiology* **34**: 973–983.

Mitchum M G, Yamaguchi S, Hanada A, Kuwahara A, Yoshioka Y, Kato T and Sun T P. 2006. Distinct and overlapping roles of two gibberellin 3-oxidases in Arabidopsis development. *The Plant Journal* **45**: 804–818.

Mizutani M and Ohta D. 2010. Diversification of P450 genes during land plant evolution. *Annual Review of Plant Biology* **61**: 291–315.

Morrone D, Chen X, Coates R M and Peters R J. 2010. Characterization of the kaurene oxidase CYP701A3, a multifunctional cytochrome P450 from gibberellin biosynthesis. *Biochemical Journal* **431**: 337–347.

Nakajima M, Shimada A, Takashi Y, Kim Y C, Park S H, Ueguchi-Tanaka M and Maeda T. 2006. Identification and characterization of Arabidopsis gibberellin receptors. *The Plant Journal* **46**: 880–889.

Nelson D C, Riseborough J A, Flematti G R, Stevens J, Ghisalberti E L, Dixon K W and Smith S M. 2009. Karrikins discovered in smoke trigger Arabidopsis seed germination by a mechanism requiring gibberellic acid synthesis and light. *Plant Physiology* **149**: 863–873.

Nitsch J P. 1970. Hormonal factors in growth and development. *Biochemistry of Fruits and Their Products*. Volume 1. Academic Press, Cambridge, Massachusetts, United States.

Ogas J, Cheng J C, Sung Z R and Somerville C. 1997. Cellular differentiation regulated by gibberellin in the Arabidopsis thaliana pickle mutant. *Science* **277**: 91–94.

Ogawa M, Hanada A, Yamauchi Y, Kuwahara A, Kamiya Y and Yamaguchi S. 2003. Gibberellin biosynthesis and response during Arabidopsis seed germination. *The Plant Cell* **15**: 1591–1604.

Okamuro J K, Den Boer B G and Jofuku K D. 1993. Regulation of Arabidopsis flower development. *The Plant Cell* **5**: 1183.

Olszewski N, Sun T P and Gubler F. 2002. Gibberellin signaling biosynthesis, catabolism, and response pathways. *The Plant Cell* **14**: S61–S80.

Otsuka M, Kenmoku H, Ogawa M, Okada K, Mitsuhashi W, Sassa T and Yamaguchi S. 2004. Emission of ent-kaurene, a diterpenoid hydrocarbon precursor for gibberellins, into the headspace from plants. *Plant and Cell Physiology* **45**: 1129–1138.

Ozga J A and Reinecke D M. 2003. Hormonal interactions in fruit development. *Journal of Plant Growth Regulation* **22**: 73–81.

Peng J, Richards D E, Hartley N M, Murphy G P, Devos K M, Flintham J E and Sudhakar D. 1999. 'Green revolution' genes encode mutant gibberellin response modulators. *Nature* **400**: 256–261.

Phillips A L, Ward D A, Uknes S, Appleford N E, Lange T, Huttly A K and Hedden P. 1995. Isolation and expression of three gibberellin 20-oxidase cDNA clones from Arabidopsis. *Plant Physiology* **108**: 1049–1057.

Poethig R S. 1990. Phase change and the regulation of shoot morphogenesis in plants. *Science* **250**: 923.

Proebsting W M, Hedden P, Lewis M J, Croke, S J and Proebsting L N. 1992. Gibberellin concentration and transport in genetic lines of pea effects of grafting. *Plant Physiology* **100**: 1354–1360.

Pysh L D, Wysocka-Diller J W, Camilleri C, Bouchez D and Benfey P N. 1999. The GRAS gene family in Arabidopsis: sequence characterization and basic expression analysis of the SCARECROW-LIKE genes. *The Plant Journal* **18**: 111–119.

Radi A, Lange T, Niki T, Koshioka M and Lange M J P. 2006. Ectopic expression of pumpkin gibberellin oxidases alters gibberellin biosynthesis and development of transgenic Arabidopsis plants. *Plant Physiology* **140**: 528–536.

Regnault T, Davière J M, Wild M, Sakvarelidze-Achard L, Heintz D, Bergua E C and Achard P. 2015. The gibberellin precursor GA12 acts as a long-distance growth signal in Arabidopsis. *Nature Plants* **1**: 15073.

Reid J B. 1990. Phytohormone mutants in plant research. *Journal of Plant Growth Regulation* **9**: 97–111.

Reid J B and Potts W C. 1986. Internode length in Pisum. Two further mutants, lh and ls, with reduced gibberellin synthesis, and a gibberellin insensitive mutant, lk. *Physiologia Plantarum* **66**: 417–426.

Rieu I, Eriksson S, Powers S J, Gong F, Griffiths J, Woolley L and Phillips A L. 2008. Genetic analysis reveals that C19-GA 2-oxidation is a major gibberellin inactivation pathway in Arabidopsis. *The Plant Cell* **20**: 2420–2436.

Rood S B, Pearce D, Williams P H and Pharis R P. 1989. A gibberellin-deficient Brassica mutant – rosette. *Plant Physiology* **89**: 482–487.

Ross J J, Murfet I C and Reid J B. 1993. Distribution of gibberellins in *Lathyrus odoratus* L. and their role in leaf growth. *Plant Physiology* **102**: 603–608.

Ross J J, Reid J B, Swain S M, Hasan O, Poole A T, Hedden P and Willis C L. 1995. Genetic regulation of gibberellin deactivation in Pisum. *The Plant Journal* **7**: 513–523.

Sakamoto T, Miura K, Itoh H, Tatsumi T, Ueguchi-Tanaka M, Ishiyama K and Miyao A. 2004. An overview of gibberellin metabolism enzyme genes and their related mutants in rice. *Plant Physiology* **134**: 1642–1653.

Schneider G, Jensen E, Spray C R and Phinney B O. 1992. Hydrolysis and reconjugation of gibberellin A20 glucosyl ester by seedlings of Zea mays L. *Proceedings of the National Academy of Sciences* **89**: 8045–8048.

Schneider G and Schliemann W. 1994. Gibberellin conjugates: an overview. *Plant Growth Regulation* **15**: 247–260.

Sembdner G, Borgmann E, Schneider G, Liebisch H W, Miersch O, Adam G and Schrefber K. 1976. Biological activity of some conjugated gibberellins. *Planta* **132**: 249–257.

Seo M, Nambara E, Choi G and Yamaguchi S. 2009. Interaction of light and hormone signals in germinating seeds. *Plant Molecular Biology* **69**(4): 463.

Serrani J C, Ruiz-Rivero O, Fos M and García-Martínez J L. 2008. Auxin-induced fruit-set in tomato is mediated in part by gibberellins. *The Plant Journal* **56**: 922–934.

Serrani J C, Sanjuán R, Ruiz-Rivero O, Fos M and García-Martínez J L. 2007. Gibberellin regulation of fruit set and growth in tomato. *Plant Physiology* **145:** 246–257.

Shi L, Gast R T, Gopalraj M and Olszewski N E. 1992. Characterization of a shoot-specific, GA3-and ABA-Regulated gene from tomato. *The Plant Journal* **2**: 153–159.

Shibaoka H. 1994. Plant hormone-induced changes in the orientation of cortical microtubules: alterations in the cross-linking between microtubules and the plasma membrane. *Annual Review of Plant Biology* **45**: 527–544.

Shomeili M, Nabipour M, Meskarbashee M and Memari H R. 2011. Effects of gibberellic acid on sugarcane plants exposed to salinity under a hydroponic system. *African Journal of Plant Science* **5**: 609–616.

Silverstone A L, Chang C W, Krol E and Sun T P. 1997. Developmental regulation of the gibberellin biosynthetic gene GA1 in *Arabidopsis thaliana. The Plant Journal* **12**: 9–19.

Smith V A, Gaskin P and MacMillan J. 1990. Partial purification and characterization of the gibberellin A20 3β-hydroxylase from seeds of Phaseolus vulgaris. *Plant Physiology* **94**: 1390–1401.

Song Y H, Ito S and Imaizumi T. 2013. Flowering time regulation: photoperiod-and temperature-sensing in leaves. *Trends in Plant Science* **18**: 575–583.

Sponsel V and Hedden V. 2004. Gibberellin biosynthesis and inactivation. pp. 63–94. *Plant Hormones: Biosynthesis, Signal Transduction, Action. Ed 2* (Ed.) Davies P. Kluwer Academic Publishers, Dordrecht, The Netherlands.

Sun T P. 2010. Gibberellin-GID1-DELLA: a pivotal regulatory module for plant growth and development. *Plant Physiology* **154**: 567–570.

Sun T P and Gubler F. 2004. Molecular mechanism of gibberellin signaling in plants. *Annual Review of Plant Biology* **55**: 197–223.

Sun T P and Kamiya Y. 1994. The Arabidopsis GA1 locus encodes the cyclase ent-kaurene synthetase a of gibberellin biosynthesis. *The Plant Cell* **6**: 1509–1518.

Swain S M and Olszewski N E. 1996. Genetic analysis of gibberellin signal transduction. *Plant Physiology* **112**: 11.

Swain S M, Tseng T S, Thornton T M, Gopalraj M and Olszewski N E. 2002. SPINDLY is a nuclear-localized repressor of gibberellin signal transduction expressed throughout the plant. *Plant Physiology* **129**: 605–615.

Thomas D, Guthridge M, Woodcock J and Lopez A. 2005. 14–3–3 protein signaling in development and growth factor responses. *Current Topics in Developmental Biology***67:** 285–303.

Thomas S G, Phillips A L and Hedden P. 1999. Molecular cloning and functional expression of gibberellin 2-oxidases, multifunctional enzymes involved in gibberellin deactivation. *Proceedings of the National Academy of Sciences USA* **96**: 4698–4703.

Toyomasu T, Kaneko A, Tokiwano T, Kanno Y, Kanno Y, Niida R and Dairi T. 2009. Biosynthetic gene-based secondary metabolite screening: a new diterpene, methyl phomopsenonate, from the fungus Phomopsis amygdali. *The Journal of Organic Chemistry* **74**: 1541–1548.

Toyomasu T, Kawaide H, Mitsuhashi W, Inoue Y and Kamiya Y. 1998. Phytochrome regulates gibberellin biosynthesis during germination of photoblastic lettuce seeds. *Plant Physiology* **118**: 1517–1523.

Toyomasu T, Tsuji H, Yamane H, Nakayama M, Yamaguchi I, Murofushi N and Inoue Y. 1993. Light effects on endogenous levels of gibberellins in photoblastic lettuce seeds. *Journal of Plant Growth Regulation* **12**: 85.

Vandenbussche F, Vancompernolle B, Rieu I, Ahmad M, Phillips A, Moritz T and Van Der Straeten D. 2007. Ethylene-induced Arabidopsis hypocotyl elongation is dependent on but not mediated by gibberellins. *Journal of Experimental Botany* **58**(15–16): 4269–4281.

Varbanova M, Yamaguchi S, Yang Y, McKelvey K, Hanada A, Borochov R and Ma C J. 2007. Methylation of gibberellins by Arabidopsis GAMT1 and GAMT2. *The Plant Cell* **19**: 32–45.

Wang C, Liu Y, Li S S and Han G Z. 2015. Insights into the origin and evolution of the plant hormone signaling machinery. *Plant Physiology* **167**: 872–886.

Ward D A, MacMillan J, Gong F, Phillips A L and Hedden P. 2010. Gibberellin 3-oxidases in developing embryos of the southern wild cucumber, Marah macrocarpus. *Phytochemistry* **71**: 17.

Ward J L, Gaskin P, Brown R G S, Jackson G S, Hedden P, Phillips A, Willis C L and Beale M H. 2002. Probing the mechanism of loss of carbon-20 in gibberellin biosynthesis. Synthesis of gibberellin 3a,20-hemiacetal and 19,20-lactol analogues and their metabolism by a recombinant GA 20-oxidase. *Journal of the Chemical Society, Perkin Trans* **1**: 232–241.

Ward J L, Jackson G J, Beale M H, Gaskin P, Hedden P, Mander L N, Phillips A L, Seto H, Talon M and Willis C L. 1997. Stereochemistry of the oxidation of gibberellin 20-alcohols, GA, and GA, to 20-aldehydes by gibberellin 20-oxidases. *Journal of Chemistry SOC Chemistry Communications* **5**: 13–14.

Williams J, Phillips A L, Gaskin P and Hedden P. 1998. Function and substrate specificity of the gibberellin 3β-hydroxylase encoded by the arabidopsis GA4Gene. *Plant Physiology* **117**: 559–563.

Yabuta T and Sumiki Y. 1938. The crystallization of gibberellins A and B. *Journal of the Agricultural Chemical Society of Japan* **14**: 1526.

Yamaguchi S. 2008. Gibberellin metabolism and its regulation. *Annual Review of Plant Biology* **59**: 225–251.

Yamaguchi S and Kamiya Y. 2000. Gibberellin biosynthesis: its regulation by endogenous and environmental signals. *Plant and Cell Physiology* **41**: 251–257.

Yamaguchi S, Smith M W, Brown R G, Kamiya Y and Sun T P. 1998a. Phytochrome regulation and differential expression of gibberellin 3β-hydroxylase genes in germinating Arabidopsis seeds. *The Plant Cell* **10**: 2115–2126.

Yamaguchi S, Sun T P, Kawaide H and Kamiya Y. 1998b. The GA2 locus of Arabidopsis thaliana encodes ent-kaurene synthase of gibberellin biosynthesis. *Plant Physiology* **116**: 1271–1278.

Yamauchi Y, Ogawa M, Kuwahara A, Hanada A, Kamiya Y and Yamaguchi S. 2004. Activation of gibberellin biosynthesis and response pathways by low temperature during imbibition of Arabidopsis thaliana seeds. *The Plant Cell* **16**: 367–378.

Yasumura Y, Crumpton-Taylor M, Fuentes S and Harberd N P. 2007. Step-by-step acquisition of the gibberellin-DELLA growth-regulatory mechanism during land-plant evolution. *Current Biology* **17**: 1225–1230.

Zanewich K P, Rood S B and Williams P H. 1990. Growth and development of Brassica genotypes differing in endogenous gibberellin content. I. Leaf and reproductive development. *Physiologia Plantarum* **79**: 673–678.

Zhu Y, Nomura T, Xu Y, Zhang Y, Peng Y, Mao B and Zhu X. 2006. Elongated uppermost internode encodes a cytochrome P450 monooxygenase that epoxidizes gibberellins in a novel deactivation reaction in rice. *The Plant Cell* **18**: 442–456.

Zimmerman R H, Hackett W P and Pharis R P. 1985. Hormonal aspects of phase change and precocious flowering. pp. 79–115. *Encyclopedia of Plant Physiology* (Eds.) Pharis R P and Reid D M. Springer, New York.

15 Cytokinins

15.1 INTRODUCTION

Cytokinins (N_6-substituted adenine derivatives) refer to the class of plant hormones that promote cell division and are also involved in many other aspects of plant growth and development including shoot initiation and growth, shoot and root branching, leaf expansion, lateral bud growth, floral transition, apical dominance, senescence and photomorphogenesis (Miller et al., 1955; Miller, 1956; Mok, 1994; Mok and Mok, 2001). Nutrient mobilization and biomass distribution have also been reported to be influenced by cytokinins (Takei et al., 2001). Additionally, they are also involved in the maintenance of meristem function and response to environmental stimuli (Werner et al., 2006; Kurakawa et al., 2007; Werner and Schmülling, 2009). The presence of the particular cytokinin can vary greatly among plant species, developmental stages and environmental conditions (Emery et al., 1998; Takei et al., 2001; Quesnelle and Emery, 2007; Hirose et al., 2008; Vyroubalova et al., 2009). The physiological processes are also influenced by various other stimuli, and the outcomes reflect a highly integrated response to these multiple stimuli. Cytokinins are classified depending on the structure as isoprenoid and aromatic cytokinins. Biological activities of both isoprenoid and aromatic cytokinins are qualitatively similar but they may differ quantitatively in different processes (Schmülling, 2004). Isoprenoid cytokinins are dominant and they are either isopentenyl (iP)-type cytokinins (having an isopentenyl N6-side chain) or zeatin-type cytokinins (having a hydroxylated isopentenyl N6-side chain). The side chain of a zeatin-type cytokinin occurs in either cis or trans form depending on which of the two methyl groups is hydroxylated. Usually, cis form is less active as compared to trans form. Recent studies have suggested that cis-zeatin is the dominant cytokinin in various plant species (Emery et al., 2000; Suttle and Banowetz, 2000; Veach et al., 2003; Vyroubalova et al., 2009), although low levels of trans-zeatin can also be found in tRNA (Vreman et al., 1978).

Aromatic cytokinins (e.g., benzyladenine) that have an aromatic benzyl group at N6 occur infrequently, and due to their greater stability, they are often used in tissue culture (Schmülling, 2004). Synthetic cytokinins are the structurally unrelated phenylurea-type cytokinins such as diphenylurea and thidiazuron, and they are highly active but do not occur naturally (Shaw, 1994).

15.2 CYTOKININ BIOSYNTHESIS

Although cytokinins occur mostly as free molecules in plants, they also exist as tRNAs in cytoplasm and chloroplast (George et al., 2008). Roots are by far the main site of cytokinin biosynthesis in plants; however, other actively growing tissues also produce cytokinin (Van Staden and Davey, 1979; Chen et al., 1985). Root tip is considered as the primary site of cytokinin biosynthesis (Ochatt and Power, 1988). The breakdown of tRNA was originally suggested as a possible mechanism for cytokinin synthesis. Two pathways have been suggested for cytokinin biosynthesis: (i) the direct pathway involving the formation of N6-isopentenyladenosine monophosphate (i6AMP) from AMP and dimethylallyl pyrophosphate (DMAPP) followed by hydroxylation of the side chain to form zeatin-type compounds and (ii) the indirect pathway, in which cytokinins are released by turnover of tRNA containing cis-zeatin. Studies with enzyme preparations that convert AMP and DMAPP into i6AMP support the direct biosynthesis. Isopentenyl transferases mediating this conversion have been isolated from the slime mold *Dictyostelium discoideum* and tobacco callus (Chen and Ertl, 1994). The slime mold enzyme has been purified extensively (Ihara et al., 1984). However, the tobacco enzyme, which is highly unstable, easily loses their activity during purification (Chen and Melitz, 1979). Subsequently, studies with Arabidopsis genomic sequence identified a total of nine

 DOI: 10.1201/9781003373889-20

ipt-homologs designated as AtIPT1–AtIPT9 (Takei et al., 2001; Kakimoto, 2001). These authors revealed that AtIPT2 and AtIPT9 encode a putative tRNA-ipt, whereas the other seven AtIPTs formed a distinct clade more closely related to the bacterial ipt/tmr gene. These seven genes promoted the secretion of the cytokinins iP and zeatin, confirming their role in encoding cytokinin biosynthetic genes in *Escherichia coli* (Takei et al., 2001). Isopentenyladenosine-5'-triphosphate (iPTP) and isopentenyladenosine-5′-diphosphate (iPDP) are probably the product of the plant enzyme that can be consequently interconverted to zeatin (Haberer and Kieber, 2002). However, different AtIPT genes show distinct, tissue-specific patterns of expression, which may provide new insights into the sites of cytokinin production (Kakimoto, 2001). Another study in Arabidopsis indicated that an alternative cytokinin biosynthesis pathway may exist in plants (Åstot et al., 2000). These authors suggested the presence of an iPMP-independent pathway, in which ZMP is directly synthesized by IPT from AMP and a yet unidentified side chain precursor. This pathway could be strongly inhibited by mevastatin indicating a terpenoid origin of the hypothetical precursor.

15.2.1 Interconversion

Rapid metabolic interconversion of bases, such as ribosides and ribotides, occurs during cytokinin metabolism. Attachment of ribose or ribose-5′-phosphate at the N9 atom of the adenine ring leads to the formation of ribosides and ribotides, respectively. Interconversion of cytokinins is apparently an important mechanism that regulates the concentration of active compounds, which is catalyzed by the same enzymes that metabolize adenine, adenosine and AMP. The conversion between the cis- and trans-isomers of zeatin is catalyzed by the enzyme cis–trans zeatin isomerase, and zeatin is converted into dihydrozeatin by an NADPH-dependent zeatin reductase (Schmülling, 2004).

15.2.2 Conjugation

Among isoprenoid cytokinins, trans-zeatin and zeatin are most and least active, respectively, in bioassays. In plants, cytokinins exist both as free bases and in the form of nucleosides and nucleotides. Cytokinins can be inactivated by glycosylation of the purine ring or of the side chain (Schmülling, 2004). The cytokinin bases can be further conjugated to glucose at N3-, N7- and N9 of the adenine ring and at the hydroxyl of the side chain. The hydroxyl group of the side chain can also be conjugated to xylose (Turner et al., 1987). The N9 can be modified with L-alanine and C2 with a 2 methylthio group (Entsch et al., 1983; Persson et al., 1994; Pertry et al., 2009). N modifications produces irreversible inactive cytokinins because N-glucosides are resistant to glucosidases (Mok and Mok, 2001; Sakakibara, 2006).

15.3 CYTOKININ SIGNALING AND FUNCTIONS

Cytokinin acts at very low concentrations as both a paracrine local signal in meristematic tissues and as distal signals as in signaling of nutrition availability (Sakakibara, 2006; Galuszka et al., 2008; Werner and Schmülling, 2009). However, the distribution of trans-zeatin-type cytokinins in xylem and isopentenyladenine-type cytokinins in phloem indicates the involvement of selective transport systems (Hirose et al., 2008). Previous studies revealed that some of the purine permeases (PUPs) function as transporters for cytokinin nucleotides (Bürkle et al., 2003; Cedzich et al., 2008). However, according to Frebort and coworkers, PUPs can't transport cytokinins efficiently (Frébort et al., 2011). Hirose et al. (2008) reported the equilibrative nucleoside transporters (ENTs) for cytokinin nucleosides.

The cytokinin signal is carried by a two-component signaling system: (i) histidyl-to-aspartyl system and (ii) His–Asp phosphorelay (Frébort et al., 2011). Membrane-spanning histidine kinase that contains extracellular CHASE (cyclases/histidine kinases-associated sensing extracellular) domain is the first component of the cytokinin signaling cascade, which passes the phosphate to the histidine phosphotransfer proteins that further convey the signal into the nucleus and then to the last component, type-B response regulators. These work as transcription regulators and affect the expression of the cytokinin target genes including type-A response regulators (which inhibit the phosphotransfer from histidine phosphotransfer proteins to the type-B response regulators and thus work as a negative feedback loop) (To et al., 2007).

15.4 CYTOKININ METABOLISM

Enzymes that are involved in the purine metabolism might be responsible for inter conversions among different forms of cytokinins, but their biological significance is still unclear (Mok and Mok, 2001). The adenine ring system can be glucosylated at the N3-, N7- and N9- positions. Glucosyl-conjugates at the N7- and N9- positions and the N7- and N9- modifications will irreversibly inactivate cytokinins, but the precise *in vivo* function of these N-glucosylconjugates is unknown (Mok and Mok, 2001). O-Glucosyl-conjugates of the N6-side chain are a common modification in all plants (Haberer and Kieber, 2002). In *Phaseolus* spp., O-xylosyl-conjugates and the purified corresponding glycosyltransferases show high substrate specificity both to the cytokinins zeatin and dihydrozeatin and to the glycosyl donor (Dixon and Weed, 1989; Mok and Mok, 2001). The two genes zeatin-O-glucosyltransferase (ZOG1) and zeatin-O-xylosyltransferase (ZOX1), which prefer UDP-Glc and UDP-Xyl, have been isolated from *Phaseolus* spp., and the binding site of the UDP-Glc has been mapped by domain swap experiments (Martin et al., 1999a, 1999b; Mok and Mok, 2001).

The cis-zeatin-oglucosyltransferase (cis-ZOG1) has been reported to specifically glucosylates cis-zeatin but not trans-zeatin or dihydrozeatin, suggesting high specificity of cisZOG1 to the cis-isomer. This might specify an important role for cis-zeatin in plants (Martin et al., 2001). O-Glycosylated cytokinins are resistant to the cleavage of the N6-side chain by cytokinin oxidases. Furthermore, these forms can easily be converted into active cytokinins by β-glucosidases (Brzobohaty et al., 1993). Therefore, it is assumed that O-glycosylated cytokinins are inactive, and only stable storage forms play an important role in balancing cytokinin levels.

15.5 CYTOKININ TRANSLOCATION IN PLANTS

Cytokinin moves to target cells by diffusion and/or by a selective transport system (Sakakibara, 2006). Plant cells are capable of absorbing cytokinin nucleobases and nucleosides and tZ-type and iP-type cytokinin from xylem and phloem, respectively, confirming their selective transport (Singh et al., 1998; Lejeune et al., 1994; Takei et al., 2001; Bürkle et al., 2003; Corbesier et al., 2003). The major cytokinin form in the xylem is nucleosides such as tZR (Beveridge et al., 1997; Takei et al., 2001) where the mobility of kinetin riboside exceeds that of BA and kinetin (Hill-Cottingham and Lloyd-Jones, 1968). Therefore, nucleosides are considered the major translocation form of cytokinin. Occurrence of cytokinins in the xylem sap and the identification of the root tip as a major site of cytokinin biosynthesis, it is generally assumed that cytokinins are transported in the xylem to exert their effects on the aerial parts of plants (Haberer and Kieber, 2002). Recently, Kudo et al. (2010) reported that long-distance transport of cytokinins occurs in xylem (mainly as tZ-riboside (tZR) and phloem tissues (mainly iP-type cytokinins). However, short-distance transport of cytokinin across the plasma membrane and into the lumen of the endoplasmic reticulum occurs due to the presence of purine transporters, AtPUP1 (At1g28230) and AtPUP2 (At2g33750) (Bürkle et al., 2003). AtPUP1 isolated in Arabidopsis suggests that AtPUP1 represents a possible cytokinin/purine transporter involved in the translocation of root-derived cytokinins to the aerial plant parts (Gillissen et al., 2000). Plastids are the major subcellular compartments of *de novo* cytokinin biosynthesis (Kasahara et al., 2004; Takei et al., 2001). Therefore, it is assumed that cytokinin is transported from plastids to the cytosol at the intracellular level (Sakakibara, 2006).

15.6 CYTOKININ DEGRADATION

Cytokinin degradation is accomplished through the oxidative cleavage of its N6 side chain that results in the formation of adenine (or its corresponding derivative for N9 substituted cytokinins) and a side chain-derived aldehyde (Brownlee et al., 1975). Cytokinin oxidases, found in several plant species, represent a class of highly diverse proteins (Jones and Schreiber, 1997). The reaction proceeds via cytokinin dehydrogenation to an imine intermediate when an electron acceptor withdraws two electrons from the enzyme's flavin cofactor. A variety of electron acceptors, especially the quinone type, such as 2,3-dimethoxy-5-methyl-1,4-benzoquinone (Q_0), function more efficiently for cytokinin dehydrogenase (CKX) activity (Galuszka et al., 2001). CKX is the only known enzyme to catalyze irreversible inactivation of cytokinins, and its activity has been reported in several plant species (Frébort et al., 2011).

Isolation of pure protein is extremely difficult because of low concentration of CKX in plant tissues. Morris et al. (1999) tried to clone the gene encoding CKX from maize that allowed the identification of CKX genes in other species and their expression in heterologous hosts or overexpression in plants. Presence of a covalently bound FAD molecule has been reported from all known plant CKX enzymes and the cofactor is linked through the 8-methyl group of the isoalloxazine ring to a histidine residue of a well-conserved GHS motif in the enzyme's N-terminal half (Malit et al., 2004). Covalent binding of FAD to protein enables redox catalysis and the oxidative power of flavin (Galuszka et al., 2008). Mature CKX enzymes consist of two domains for FAD and substrate binding (Frébort et al., 2011). Approximately, one-third of all amino acids are highly conserved among different CKX proteins.

15.7 PHYSIOLOGICAL ROLES OF CYTOKININ

Cytokinin controls several physiological activities in plants including cell division, cell enlargement, initiation of interfascicular cambium, morphogenesis, breaking of apical dominance, seed dormancy, delaying of senescence (Richmond-Lang Effect) and promotion of chloroplast development.

15.7.1 Cell Division

Cytokinins regulate cell division activity in embryos and mature plants by altering the size and activity of meristems (Medford et al., 1989; Chaudhury et al., 1993; Werner et al., 2001). However, any defect caused by mutation of the receptor gene *AHK4* (*wol* mutant allele) or knockouts in the *AHK* gene family during cytokinin signal transduction can affect division (Mähönen et al., 2000). An increase in endogenous cytokinin levels during cell division shows overexpression of the KNOTTED homeobox transcription factor homologs (KNAT1 and STM genes) that are important in the regulation of meristem function in transgenic Arabidopsis (Rupp et al., 1999). Similar results were obtained in transgenic tobacco where overexpression of KNAT1 also increase cytokinin level, suggesting a symbiotic relationship between KNAT and the level of cytokinin (Werner et al., 2001). Recent studies have shown that KNOX genes can elevate cytokinin levels and induce the expression of cytokinin biosynthesis genes, AtIPT5, AtIPT7 and ARR5 (cytokinin primary response gene) (Jasinski et al., 2005; Yanai et al., 2005). This indicates that there is a strong relationship between phytohormone cytokinin and KNOX transcription factors in promoting shoot meristem activity and maintenance. Furthermore CRE1 (cytokinin receptor) have been reported to be unable to proliferate callus tissue indicating that cytokinins have a key role in cell division (Inoue et al., 2001). Application of exogenous cytokinin has been shown to induce cell division especially in those tissues where cytokinin is absent. In fact, cytokinin has been implicated with all stages of the cell cycle (Frank and Schmülling, 1999; den Boer and Murray, 2000).

15.7.2 Cell Enlargement

Cell enlargement in cotyledons of dicots, such as mustard, cucumber and sunflower where cotyledons grow as a result of cell expansion during seedling growth, has been shown to be controlled by cytokinins (http://wenku.baidu.com/view/c2d0acc3d5bbfd0a79567313.html). Cytokinin treatment promotes additional cell expansion without increasing their dry weight. Seedlings grown in the light showed greater expansion of cotyledons compared to those grown under dark, although cytokinins was able to promote cotyledon growth in both light- and dark-grown seedlings (Huff and Ross, 1975). It has been shown that cytokinin-stimulated expansion is associated with an increase in the mechanical extensibility of the cell wall in radish cotyledons (Huff and Ross, 1975). Significant cell enlargement was observed in leaf discs cut from etiolated leaves of *Phaseolus vulgaris*, pumpkin cotyledons, tobacco pith cultures, cortical cells of tobacco roots and excised Jerusalem artichoke tissue after kinetin treatment.

15.7.3 Initiation of Interfascicular Cambium

Formation of vascular tissue in plants is a process with broad developmental and physiological consequences (Hejátko et al., 2009; Mähönen et al., 2000; Friml, 2003; Emery et al., 2003; Caño-Delgado et al., 2004; Prigge et al., 2005). Cytokinin is an important regulator for primary vascular growth which promotes the division of procambial cells (Aloni, 1987; Medford et al., 1989; Nieminen et al., 2008). The signal is carried by a two-component signaling system, where a His kinase receptor transfers a phosphate to downstream response regulators, which is a key for cytokinin response (Kim et al., 2006; Hejátko et al., 2009). It has been reported that cytokinin-induced signaling via its receptor Arabidopsis Histidine Kinase4 (AHK4) and the type-B response regulators such as ARR1, ARR10 and ARR12 are essential for formation of procambium in *Arabidopsis thaliana* roots (Scheres et al., 1995; Mähönen et al., 2000; Yokoyama et al., 2007). Hejátko et al. (2009) reported that His kinase cytokinin-independent1 (CKI1) and the cytokinin receptors Arabidopsis Histidine kinase2 (AHK2) and AHK3 are important regulators of vascular tissue development in *Arabidopsis thaliana* shoots. Reducing cytokinin levels in the cambium region has been found to reduce the number of cambium cells in Popular trees by the expression of the Arabidopsis CKX2 gene from a birch CRE1 promoter (Nieminen et al., 2008). However, any alteration in the cytokinin signaling pathway can result in defects in vascular development (Mähönen et al., 2000; Bishopp et al., 2006). Recently, identification of vascular tissue-specific expression of genes that is involved in cytokinin biosynthesis also confirms the role of cytokinin in vascular tissue formation (Zhao et al., 2005).

15.7.4 Morphogenesis

Primary morphogenesis in leaves leads to the initiation of the lamina and formation of different areas of the lamina (e.g., midrib, petiole and leaf base) and marginal structures (e.g., leaflets, lobes and serrations) (Bar and Ori, 2014). However, most of the leaf expansion occurs during secondary morphogenesis when specific tissues have completed their differentiation (Bar and Ori, 2014). Immediately after initiation, leaf primordia undergo growth, morphogenesis and differentiation in a highly flexible process that ultimately gives rise to the final leaf shape. However, this process involves several genes such as YABBY and AIL/PLT linked to the promotion of lamina outgrowth and expansion in *Arabidopsis*, maize and rice (Mizukami and Fischer, 2000; Dai et al., 2007; Sarojam et al., 2010). Balance between morphogenesis and differentiation in leaf development is affected by cytokinin. Increased cytokinin degradation enhances the cell expansion in *Arabidopsis* leaf primordial, which indicate that cytokinin postpones the initiation of cell differentiation (Werner et al., 2001; Holst et al., 2011). Additionally, cytokinin also reported to be involved in the maintenance of extended morphogenetic activity in the tomato leaf margin (Shani et al., 2010). Leaves maintain morphogenetic activity after leaf expansion, which leads to further variability in leaf shape, as seen in the *cla* mutant. It has been reported that GA and cytokinin together control the late morphogenetic activity in tomato (Shani et al., 2010; Yanai et al., 2011). Therefore, the plasticity of leaf shape is attained by tuning the balance between hormones that promote indeterminate state (cytokinin) and hormones that promote differentiation (GA).

15.7.5 Apical Dominance

Apical dominance occurs when the shoot apex inhibits the growth of lateral buds so that the plant can grow vertically. Vertical growth allows them to grow tall and allows to capture more light for photosynthesis. Although apical dominance is mostly controlled by auxin, cytokinins also play a role in initiating the growth of lateral buds. It is reported that cytokinins represent a second messenger for auxin signaling in regulating bud activity because it can be transported acropetally in the xylem sap and enter axillary buds. This will promote the growth of axillary buds (Müller and Leyser, 2011). Application of cytokinins to an axillary bud promotes their growth even in the presence of a growing shoot apex or apically supplied auxin (Wickson and Thimann, 1958; Faiss et al., 1997). Bangerth (1994) reported higher cytokinin levels in xyllen sap of bean and chickpea following decapitation, but this increase could be prevented by application of auxin to the decapitated stump, which indicates that auxin inhibits branching partly by regulating the supply of root-derived cytokinins to the shoot. Similarly, lateral buds of pea plants can be released from apical dominance and even be transformed into dominant shoots by repeatedly treating them with exogenous cytokinins (Li and Bangerth, 2003).

15.7.6 Dormancy and Germination of Seeds

Several research groups have investigated the role of cytokinin in seed germination and dormancy in many plants' species (Prasad et al., 1983; Rao et al., 1976; Kabar, 1998; Chiwocha et al., 2005; Hassani et al., 2009). Cytokinin promotes seed germination by overcoming the inhibitory role imposed by ABA and other inhibitory factors (Mok, 1994). Identification of a gain-of-function mutant gim1 (germination insensitive to ABA mutants) that shows resistance to ABA during seed germination and antagonizing effect of cytokinin on ABA-inhibited seed germination was attributed to the cytokinin-repressed expression of ABI5 (Wang et al., 2011).

15.7.7 Promotion of Chloroplast Development

Besides serving as the main site of photosynthesis, chloroplasts are also involved in synthesis of many primary and secondary metabolites (Lopez-Juez and Pyke, 2005). Chloroplasts develop from proplastids, which are present in the immature cells of plant meristems. The shift from etioplast to chloroplast is part of the de-etiolation process which coincides with the biosynthesis of chlorophyll (von Arnim and Deng, 1996; Tanaka and Tanaka, 2007). Stetler and Laetsch (1965) were the first to recognize the promoting influence of cytokinin in chloroplast development. They found that cytokinin plays an important role in chloroplast differentiation from proplastids. This was confirmed by Parthier (1979), who showed that cytokinins accelerate the differentiation of the prolamellar body into the distinctive lamellar structure during dark-to-light transitions and increased the rate of chlorophyll biosynthesis. Stimulatory effect of cytokinin on chloroplast development was reported to be dependent on an intact prolamellar body and a cytokinin-dependent increased stability of POR enzymes (Kusnetsov et al., 1998). It has been shown that cytokinin encouraged the changes in chloroplast ultrastructure at different time scale after transferring dark-grown Arabidopsis seedlings into the light (Cortleven and Schmülling, 2015). Cytokinins also influence chloroplast number in the presence of light (Boasson and Laetsch, 1969). It regulates plastid division by acting at the level of plastid division 2 (PDV2) by influencing plastid-related nuclear genes (Okazaki et al., 2009). The cytokinin receptor triple mutant and B-type ARR triple mutant (arr 1, 10 and 12) exhibit reduced chlorophyll levels, indicating the positive effect of cytokinin on chloroplast development and maintenance (Riefler et al., 2006; Argyros et al., 2008).

15.7.8 Inhibition of Leaf Senescence

Ageing of leaves is accompanied by the loss of chlorophyll (yellowing) and rapid breakdown of proteins. According to Osborne (1962), higher protein content in kinetin-treated tissue is probably due to the synthesis of greater amounts of proteins than their degradation, which, in turn, may be due to the regulatory action of kinetin on RNA synthesis. Exogenous application of cytokinin retards the chlorophyll loss during leaf senescence (Richmond and Lang, 1957). Increasing cytokinin production delays senescence of leaves, and conversely, reduction in cytokinin level accelerates it (Gan and Amasino, 1995; Ori et al., 1999; Masferrer et al., 2002). Several studies have revealed low levels of cytokinin in leaves to accelerate the onset and progression of senescence (Nooděn and Leopold, 1988; Singh et al., 1992; Nooděn et al., 1990). Similar trends in cytokinin content in xylem sap were shown in sorghum cultivar, where the presence of higher cytokinin levels in xylem sap delayed leaf senescence (Ambler et al., 1992). Transcriptome analyses of Arabidopsis leaf showed a decrease in the expression of cytokinin biosynthesis genes during the senescence (Buchanan-Wollaston et al., 2005; Breeze et al., 2011). Receptor CKI1 (cytokinin independent 1) and the Arabidopsis response regulator (ARR) 2 were found to be the key signals responsible for lowering cytokinin levels before the onset of senescence (Nooděn et al., 1990; Gan and Amasino, 1995; Hwang and Sheen, 2001). Cytokinins are known to mediate the conversion of etioplasts into chloroplasts when etiolated seedlings that were treated with cytokinins are exposed to light as indicated by an increase in the activities of enzymes involved in chloroplast formation (Zubo et al., 2008).

15.7.9 Development and Maturation of Fruits

Hormones play a key role in fruit development (Mariotti et al., 2011). Exogenous application of one hormone can trigger fruit development in some species, whereas a combination of hormones is required to induce fruit development in others, indicating that interaction between hormones is important in fruit set and fruit growth (Crane, 1964; Gillaspy et al., 1993; Vivian-Smith and Koltunow, 1999). Auxin and cytokinin levels have been shown to increase during the seed development until maturity (Yang et al., 2002; Devoghalaere et al., 2012), but removal of the seeds leads to reduced gibberellin biosynthesis in the pericarp (Ozga et al., 1992). Studies by several scientists have shown that cytokinins play a critical role in stimulating cell division during fruit development (Wismer et al., 1995; Srivastava and Handa, 2005; Mariotti et al., 2011). According to Marsch-Martínez et al. (2012), cytokinins play an early proliferation-inducing role in the medial region of the developing gynoecia and a later role during formation of fruit valve margins in Arabidopsis. Werner and coworkers have suggested that cytokinin-deficient fruits of Arabidopsis show non synchronous ripening with less viable seeds compared to normal fruits, suggesting that cytokinin has a role in the regulation of fruit maturation and ripening processes (Werner et al., 2003).

15.7.10 Lateral Root Development

Auxins and cytokinins have antagonistic role in lateral root development. While auxin promotes lateral root development, cytokinin inhibits it (Kieber and Schaller, 2014). Laplaze and coworkers concluded that only xylem-pole pericycle cells were sensitive to inhibition by cytokinin, whereas the young lateral root primordia were insensitive to inhibition by cytokinin (Laplaze et al., 2007). It has been shown that cytokinin reduces the level of PIN1 protein by regulating the endocytic recycling of PIN1 (At1g73590) during lateral root development and transmitting it for lytic degradation in vacuoles (Marhavý et al., 2011). AHP6 pseudo-HPt inhibits cytokinin signaling during early lateral root development (Moreira et al., 2013). AHP6 is essential for the correct orientation of cell divisions at the starting point of lateral root development and is involved in localizing PIN1 (Moreira et al., 2013). Therefore, cytokinins play an important role in regulating auxin flow required for lateral roots development.

15.7.11 Abiotic Stress

Cytokinin concentrations have been shown to decrease in response to prolonged stress (Hubick et al., 1986; Hare et al., 1997; Hansen and Dörffling, 2003; Kudoyarova et al., 2007; Albacete et al., 2008; Ghanem et al., 2008; Ha et al., 2012; O'Brien and Benková, 2013). Conversely, there are also reports that suggest an increase in cytokinin levels in response to short- and prolonged stresses (Hansen and Dörffling, 2003; Pospisilova et al., 2005; Alvarez et al., 2008). Zwack and Rashotte (2015) noticed a transient increase in cytokinin concentrations upon exposure to the stress followed by either complete decrease under constant moderate stress or maintenance of initial increased levels when challenged with more severe conditions. The culture media supplemented with cytokinin improved salt tolerance response in sprouted potato tubers and wheat seedlings (Naqvi et al., 1982; Abdullah and Ahmad, 1990). Similarly, the cytokinin supplemented medium resulted in higher survival of Arabidopsis plants exposed to freezing or dehydrating conditions compared to the medium without cytokinin supplementation (Jeong et al., 2010; Kang et al., 2012). However, spatial, temporal and developmental factors play an important role in determining cytokinin's role in stress signaling.

15.7.12 Plants' Immunity to Pathogens

Cytokinins are known to enhance plant's immunity against diverse pathogens, such as biotrophs, hemi-biotrophs and necrotrophs (Swartzberg et al., 2008; Choi et al., 2010; Argueso et al., 2012; Naseem and Dandekar, 2012). Application of cytokinin reduces the replication of White Clover Mosaic Potex virus and induction of defense gene expression in bean plants (Clarke et al., 1998). Similarly, plants supplemented with cytokinins have been found to upregulate defense-related genes and therefore, promote resistance to the bacterial pathogens, *Pseudomonas syringae* pv. tomato DC3000 (Pst DC3000) (Rashotte et al., 2003). However, there are also reports showing increased growth of pathogens under high cytokinin levels in several plant species (Levin, 1984; Babosha, 2009). Argueso and coworkers have reported the enhancement in plants' defense against pathogens by cytokinins was associated with modulation of expression of defense gene because high concentration of cytokinin resulted in increased SA-mediated defense responses against a virulent oomycete pathogen infection (Argueso et al., 2012).

REFERENCES

Abdullah Z and Ahmad R. 1990. Effect of pre-and post-kinetin treatments on salt tolerance of different potato cultivars growing on saline soils. *Journal of Agronomy and Crop Science* **165**: 94–102.

Albacete A, Ghanem M E, Martínez-Andújar C, Acosta M, Sánchez-Bravo J, Martínez V and Pérez-Alfocea F. 2008. Hormonal changes in relation to biomass partitioning and shoot growth impairment in salinized tomato (*Solanum lycopersicum* L.) plants. *Journal of Experimental Botany* **59**: 4119–4131.

Aloni R. 1987. Differentiation of vascular tissues. *Annual Review of Plant Physiology* **38**: 179–204.

Alvarez S, Marsh E L, Schroeder S G and Schachtman D P. 2008. Metabolomic and proteomic changes in the xylem sap of maize under drought. *Plant, Cell & Environment* **31**: 325–340.

Ambler J R, Morgan P W and Jordan W R. 1992. Amounts of zeatin and zeatin riboside in xylem sap of senescent and nonsenescent sorghum. *Crop Science* **32**: 411–419.

Argueso C T, Ferreira F J, Epple P, To J P, Hutchison C E, Schaller G E and Kieber J J. 2012. Two-component elements mediate interactions between cytokinin and salicylic acid in plant immunity. *PLoS Genet* **8**: e1002448.

Argyros R D, Mathews D E, Chiang Y H, Palmer C M, Thibault D M, Etheridge N and Schaller G E. 2008. Type B response regulators of Arabidopsis play key roles in cytokinin signaling and plant development. *The Plant Cell* **20**: 2102–2116.

Åstot C, Dolezal K, Nordström A, Wang Q, Kunkel T, Moritz T and Sandberg G. 2000. An alternative cytokinin biosynthesis pathway. *Proceedings of the National Academy of Sciences* **97**: 14778–14783.

Babosha A V. 2009. Regulation of resistance and susceptibility in wheat – powdery mildew pathosystem with exogenous cytokinins. *Journal of Plant Physiology* **166**: 1892–1903.

Bangerth F. 1994. Response of cytokinin concentration in the xylem exudate of bean (Phaseolus vulgaris L.) plants to decapitation and auxin treatment, and relationship to apical dominance. *Planta* **194**: 439–442.

Bar M and Ori N. 2014. Leaf development and morphogenesis. *Development* **141**: 4219–4230.

Beveridge C A, Murfet I C, Kerhoas L, Sotta B, Miginiac E and Rameau C. 1997. The shoot controls zeatin riboside export from pea roots. Evidence from the branching mutant rms4. *The Plant Journal* **11**: 339–345.

Bishopp A, Mähönen A P and Helariutta Y. 2006. Signs of change: hormone receptors that regulate plant development. *Development* **133**: 1857–1869.

Boasson R and Laetsch W M. 1969. Chloroplast replication and growth in tobacco. *Science* **166**: 749–751.

Breeze E, Harrison E, McHattie S, Hughes L, Hickman R, Hill C and Zhang C. 2011. High-resolution temporal profiling of transcripts during Arabidopsis leaf senescence reveals a distinct chronology of processes and regulation. *The Plant Cell* **23**: 873–894.

Brownlee B G, Hall R H and Whitty C D. 1975. 3-Methyl-2-butenal: an enzymatic degradation product of the cytokinin, N 6-(Δ2-Isopentenyl) adenine. *Canadian Journal of Biochemistry* **53**: 37–41.

Brzobohaty B, Moore I, Kristoffersen P, Bako L, Campos N, Schell J and Palme K. 1993. Release of active cytokinin by a-glucosidase localized to the maize root meristem. *Science-New York Then Washington* **262**: 1051–1051.

Buchanan-Wollaston V, Page T, Harrison E, Breeze E, Lim P O, Nam H G and Leaver C J. 2005. Comparative transcriptome analysis reveals significant differences in gene expression and signaling pathways between developmental and dark/starvation-induced senescence in Arabidopsis. *The Plant Journal* **42**: 567–585.

Bürkle L, Cedzich A, Döpke C, Stransky H, Okumoto S, Gillissen B and Frommer W B. 2003. Transport of cytokinins mediated by purine transporters of the PUP family expressed in phloem, hydathodes, and pollen of Arabidopsis. *The Plant Journal* **34**: 13–26.

Caño-Delgado A, Yin Y, Yu C, Vafeados D, Mora-García S, Cheng J C and Chory J. 2004. BRL1 and BRL3 are novel brassinosteroid receptors that function in vascular differentiation in Arabidopsis. *Development* **131**: 5341–5351.

Cedzich A, Stransky H, Schulz B and Frommer W B. 2008. Characterization of cytokinin and adenine transport in Arabidopsis cell cultures. *Plant Physiology* **148**: 1857–1867.

Chaudhury A M, Letham S, Craig S and Dennis E S. 1993. ampl-a mutant with high cytokinin levels and altered embryonic pattern, faster vegetative growth, constitutive photomorphogenesis and precocious flowering. *The Plant Journal* **4**: 907–916.

Chen C M and Ertl J R. 1994. Cytokinin biosynthetic enzymes in plants and slime mold. pp. 81–85. *Cytokinins: Chemistry, Activity, and Function*. CRC Press, Boca Raton, FL.

Chen C M, Ertl J R, Leisner S M and Chang C C. 1985. Localization of cytokinin biosynthetic sites in pea plants and carrot roots. *Plant Physiology* **78**(3): 510–513.

Chen C M and Melitz D K. 1979. Cytokinin biosynthesis in a cell-free system from cytokinin-autotrophic tobacco tissue cultures. *FEBS Letters* **107**: 15–20.

Chiwocha S D, Cutler A J, Abrams S R, Ambrose S J, Yang J, Ross A R and Kermode A R. 2005. The etr1-2 mutation in Arabidopsis thaliana affects the abscisic acid, auxin, cytokinin and gibberellin metabolic pathways during maintenance of seed dormancy, moist-chilling and germination. *The Plant Journal* **42**: 35–48.

Choi J, Huh S U, Kojima M, Sakakibara H, Paek K H and Hwang I. 2010. The cytokinin-activated transcription factor ARR2 promotes plant immunity via TGA3/NPR1-dependent salicylic acid signaling in Arabidopsis. *Developmental Cell* **19**: 284–295.

Clarke S F, McKenzie M J, Burritt D J, Guy P L and Jameson P E. 1998. Influence of white clover mosaic potex virus infection on the endogenous cytokinin content of bean. *Plant Physiology* **120**: 547–552.

Corbesier L, Prinsen E, Jacqmard A, Lejeune P, Van Onckelen H, Périlleux C and Bernier G. 2003. Cytokinin levels in leaves, leaf exudate and shoot apical meristem of Arabidopsis thaliana during floral transition. *Journal of Experimental Botany* **54**: 2511–2517.

Cortleven A and Schmülling T. 2015. Regulation of chloroplast development and function by cytokinin. *Journal of Experimental Botany* **66:** 4999–5013.

Crane J C. 1964. Growth substances in fruit setting and development. *Annual Review of Plant Physiology* **15**: 303–326.

Dai M, Hu Y, Zhao Y, Liu H and Zhou D X. 2007. A WUSCHEL-LIKE HOMEOBOX gene represses a YABBY gene expression required for rice leaf development. *Plant Physiology* **144**: 380–390.

den Boer B G and Murray J A. 2000. Triggering the cell cycle in plants. *Trends in Cell Biology* **10**: 245–250.

Devoghalaere F, Doucen T, Guitton B, Keeling J, Payne W, Ling T J and Diak R. 2012. A genomics approach to understanding the role of auxin in apple (Malus × domestica) fruit size control. *BMC Plant Biology* **12**: 7.

Dixon J B and Weed S B. 1989. *Minerals in Soil Environments*. Soil Science Society of America Inc. Soil Science Society of America. doi:10.2136/sssabookser1.2ed.

Emery J F, Floyd S K, Alvarez J, Eshed Y, Hawker N P, Izhaki A, Baum S F and Bowman J L. 2003. Radial patterning of Arabidopsis shoots by class III HD-Zip genes is mediated by miRNA regulation. *Current Biology* **13**: 1768–1774.

Emery R N, Longnecker N E and Atkins C A. 1998. Branch development in *Lupinus angustifolius* L. II. Relationship with endogenous ABA, IAA and cytokinins in axillary and main stem buds. *Journal of Experimental Botany* **49**: 555–562.

Emery R N, Ma Q and Atkins C A. 2000. The forms and sources of cytokinins in developing white lupine seeds and fruits. *Plant Physiology* **123**: 1593–1604.

Entsch B, Parker C W and Letham D S. 1983. An enzyme from lupin seeds forming alanine derivatives of cytokinins. *Phytochemistry* **22**: 375–381.

Faiss M, Zalubìlová J, Strnad M and Schmülling T. 1997. Conditional transgenic expression of the ipt gene indicates a function for cytokinins in paracrine signaling in whole tobacco plants. *The Plant Journal* **12**: 401–415.

Frank M and Schmülling T. 1999. Cytokinin cycles cells. *Trends in Plant Science* **4**: 243–244.

Frébort I, Kowalska M, Hluska T, Frébortová J and Galuszka P. 2011. Evolution of cytokinin biosynthesis and degradation. *Journal of Experimental Botany* **62**: 2431–2452.

Friml J. 2003. Auxin transport – shaping the plant. *Current Opinion in Plant Biology* **6**: 7–12.

Galuszka P, Frébort I, Šebela M, Sauer P, Jacobsen S and Peč P. 2001. Cytokinin oxidase or dehydrogenase? *European Journal of Biochemistry* **268**: 450–461.

Galuszka P, Spíchal L, Kopečný D, Tarkowski P, Frébortová J, Šebela M and Frébort I. 2008. Metabolism of plant hormones cytokinins and their function in signaling, cell differentiation and plant development. *Studies in Natural Products Chemistry* **34**: 203–264.

Gan S and Amasino R M. 1995. Inhibition of leaf senescence by autoregulated production of cytokinin. *Science* **270**: 1986.

George E F, Hall M A and De Klerk G J. 2008. Plant growth regulators II: cytokinins, their analogues and antagonists. pp. 205–226. *Plant Propagation by Tissue Culture*. Springer, Rotterdam, The Netherlands.

Ghanem M E, Albacete A, Martínez-Andújar C, Acosta M, Romero-Aranda R, Dodd I C and Pérez-Alfocea F. 2008. Hormonal changes during salinity-induced leaf senescence in tomato (*Solanum lycopersicum* L.). *Journal of Experimental Botany* **59**: 3039–3050.

Gillaspy G, Ben-David H and Gruissem W. 1993. Fruits: a developmental perspective. *The Plant Cell* **5**: 1439.

Gillissen B, Bürkle L, André B, Kühn C, Rentsch D, Brandl B and Frommer W B. 2000. A new family of high-affinity transporters for adenine, cytosine, and purine derivatives in Arabidopsis. *The Plant Cell* **12**: 291–300.

Ha S, Vankova R, Yamaguchi-Shinozaki K, Shinozaki K and Tran L S P. 2012. Cytokinins: metabolism and function in plant adaptation to environmental stresses. *Trends in Plant Science* **17**: 172–179.

Haberer G and Kieber J J. 2002. Cytokinins. New insights into a classic phytohormone. *Plant Physiology* **128**: 354–362.

Hansen H and Dörffling K. 2003. Root-derived trans-zeatin riboside and abscisic acid in drought-stressed and rewatered sunflower plants: interaction in the control of leaf diffusive resistance? *Functional Plant Biology* **30**: 365–375.

Hare P D, Cress W A and Van Staden J. 1997. The involvement of cytokinins in plant responses to environmental stress. *Plant Growth Regulation* **23**: 79–103.

Hassani S B, Saboora A, Radjabian T and Fallah Husseini H. 2009. Effects of temperature, GA3 and cytokinins on breaking seed dormancy of *Ferula assafoetida* L. *Iranian Journal of Science and Technology* **33**: 75–85.

Hejátko J, Ryu H, Kim G T, Dobešová R, Choi S, Choi S M and Brzobohatý B. 2009. The histidine kinases CYTOKININ-INDEPENDENT1 and ARABIDOPSIS HISTIDINE KINASE2 and 3 regulate vascular tissue development in Arabidopsis shoots. *The Plant Cell* **21**: 2008–2021.

Hill-Cottingham D G and Lloyd-Jones C P. 1968. Relative mobility of some organic nitrogenous compounds in the xylem of apple shoots. *Nature* **220**: 389–390.

Hirose N, Takei K, Kuroha T, Kamada-Nobusada T, Hayashi H and Sakakibara H. 2008. Regulation of cytokinin biosynthesis, compartmentalization and translocation. *Journal of Experimental Botany* **59**: 75–83.

Holst K, Schmülling T and Werner T. 2011. Enhanced cytokinin degradation in leaf primordia of transgenic Arabidopsis plants reduces leaf size and shoot organ primordia formation. *Journal of Plant Physiology* **168**: 1328–1334.

Hubick K T, Taylor J S and Reid D M. 1986. The effect of drought on levels of abscisic acid, cytokinins, gibberellins and ethylene in aeroponically-grown sunflower plants. *Plant Growth Regulation* **4**: 139–151.

Huff A K and Ross C W. 1975. Promotion of radish cotyledon enlargement and reducing sugar content by zeatin and red light. *Plant Physiology* **56**: 429–433.

Hwang I and Sheen J. 2001. Two-component circuitry in Arabidopsis cytokinin signal transduction. *Nature* **413**: 383–389.

Ihara M, Taya Y, Nishimura S and Tanaka Y. 1984. Purification and some properties of D2 – isopentenyl pyrophosphate: 50 AMP D2-isopentenyltransferase from the cellular slime mold Dictyostelium discoideum. *Archive of Biochemistry and Biophysics* **230**: 652–660.

Inoue T, Higuchi M, Hashimoto Y, Seki M, Kobayashi M, Kato T and Kakimoto T. 2001. Identification of CRE1 as a cytokinin receptor from Arabidopsis. *Nature* **409**: 1060–1063.

Jasinski S, Piazza P, Craft J, Hay A, Woolley L, Rieu I and Tsiantis M. 2005. KNOX action in Arabidopsis is mediated by coordinate regulation of cytokinin and gibberellin activities. *Current Biology* **15**: 1560–1565.

Jeong S W, Das P K, Jeoung S C, Song J Y, Lee H K, Kim Y K and Choi G. 2010. Ethylene suppression of sugar-induced anthocyanin pigmentation in Arabidopsis. *Plant Physiology* **154**: 1514–1531.

Jones R J and Schreiber B M. 1997. Role and function of cytokinin oxidase in plants. *Plant Growth Regulation* **23**: 123–134.

Kabar K. 1998. Comparative effects of kinetin, benzyladenine, and gibberellic acid on abscisic acid inhibited seed germination and seedling growth of red pine and arbor vitae. *Turkish Journal of Botany* **22**: 1–6.

Kakimoto, T. 2001. Identification of plant cytokinin biosynthetic enzymes as dimethylallyl diphosphate: ATP/ADP isopentenyltransferases. *Plant and Cell Physiology* **42**(7): 677–685.

Kang N Y, Cho C, Kim N Y and Kim J. 2012. Cytokinin receptor-dependent and receptor-independent pathways in the dehydration response of Arabidopsis thaliana. *Journal of Plant Physiology* **169**: 1382–1391.

Kasahara H, Takei K, Ueda N, Hishiyama S, Yamaya T, Kamiya Y and Sakakibara H. 2004. Distinct isoprenoid origins of cis-and trans-zeatin biosyntheses in Arabidopsis. *Journal of Biological Chemistry* **279**: 14049–14054.

Kieber J J and Schaller G E. 2014. Cytokinins. *The Arabidopsis Book* **12**: e0168.

Kim J Y, Kim S H, Lee H H, Lee K, Ma W, Gong X and Heeger A J. 2006. New Architecture for high-efficiency polymer photovoltaic cells using solution-based titanium oxide as an optical spacer. *Advanced Materials* **18**: 572–576.

Kudo T, Kiba T and Sakakibara H. 2010. Metabolism and long-distance translocation of cytokinins. *Journal of Integrative Plant Biology* **52**: 53–60.

Kudoyarova G R, Vysotskaya L B, Cherkozyanova A and Dodd I C. 2007. Effect of partial root zone drying on the concentration of zeatin-type cytokinins in tomato (*Solanum lycopersicum* L.) xylem sap and leaves. *Journal of Experimental Botany* **58:** 161–168.

Kurakawa T, Ueda N, Maekawa M, Kobayashi K, Kojima M, Nagato Y and Kyozuka J. 2007. Direct control of shoot meristem activity by a cytokinin-activating enzyme. *Nature* **445**: 652–655.

Kusnetsov V, Herrmann R G, Kulaeva O N and Oelmüller R. 1998. Cytokinin stimulates and abscisic acid inhibits greening of etiolated Lupinus luteus cotyledons by affecting the expression of the light-sensitive protochlorophyllide oxidoreductase. *Molecular and General Genetics* **259**: 21–28.

Laplaze L, Benkova E, Casimiro I, Maes L, Vanneste S, Swarup R and Offringa R. 2007. Cytokinins act directly on lateral root founder cells to inhibit root initiation. *The Plant Cell* **19**: 3889–3900.

Lejeune P, Bernier G, Requier M C and Kinet J M. 1994. Cytokinins in phloem and xylem saps of Sinapis alba during floral induction. *Physiologia Plantarum* **90**: 522–528.

Levin S. 1984. Frontal lobe dysfunctions in schizophrenia – II. Impairments of psychological and brain functions. *Journal of Psychiatric Research* **18**: 57–72.

Li C and Bangerth F. 2003. Stimulatory effect of cytokinins and interaction with IAA on the release of lateral buds of pea plants from apical dominance. *Journal of Plant Physiology* **160**: 1059–1063.

Lopez-Juez E and Pyke K A. 2005. Plastids unleashed: their development and their integration in plant development. *International Journal of Developmental Biology* **49**: 557–577.

Mähönen A P, Bonke M, Kauppinen L, Riikonen M, Benfey P N and Helariutta Y. 2000. A novel two-component hybrid molecule regulates vascular morphogenesis of the Arabidopsis root. *Genes & Development* **14**: 2938–2943.

Malit M F, LeDuc D L, Tarun A S, Montes-Bayon M, Meija J and Wu C P. 2004. Overexpression of selenocysteine methyltransferase in *Arabidopsis* and Indian mustard increases selenium tolerance and accumulation. *Plant Physiology* **135**: 377–383.

Marhavý P, Bielach A, Abas L, Abuzeineh A, Duclercq J, Tanaka H and Benková E. 2011. Cytokinin modulates endocytic trafficking of PIN1 auxin efflux carrier to control plant organogenesis. *Developmental Cell* **21**: 796–804.

Mariotti L, Picciarelli P, Lombardi L and Ceccarelli N. 2011. Fruit-set and early fruit growth in tomato is associated with increases in indoleacetic acid, cytokinin, and bioactive gibberellin contents. *Journal of Plant Growth* Regulator **30**: 405–415.

Marsch-Martínez N, Ramos-Cruz D, Irepan Reyes-Olalde J, Lozano-Sotomayor P, Zúñiga-Mayo V M and de Folter S. 2012. The role of cytokinin during Arabidopsis gynoecia and fruit morphogenesis and patterning. *The Plant Journal* **72**: 222–234.

Martin R C, Mok M C, Habben J E and Mok D W. 2001. A maize cytokinin gene encoding an O-glucosyltransferase specific to cis-zeatin. *Proceedings of the National Academy of Sciences* **98**: 5922–5926.

Martin R C, Mok M C and Mok D W. 1999a. A gene encoding the cytokinin enzyme zeatinO-xylosyltransferase of *Phaseolus vulgaris*. *Plant Physiology* **120**: 553–558.

Martín-Cabrejas M A, Jaime L, Karanja C, Downie A J, Parker M L, Lopez-Andreu F J and Waldron K W. 1999b. Modifications to physicochemical and nutritional properties of hard-to-cook beans (*Phaseolus vulgaris* L.) by extrusion cooking. *Journal of Agricultural and Food Chemistry* **47**: 1174–1182.

Masferrer A, Arró M, Manzano D, Schaller H, Fernández-Busquets X, Moncaleán P and Ferrer A. 2002. Overexpression of Arabidopsis thaliana farnesyl diphosphate synthase (FPS1S) in transgenic Arabidopsis induces a cell death/senescence-like response and reduced cytokinin levels. *The Plant Journal* **30**: 123–132.

Medford J I, Horgan R, El-Sawi Z and Klee H J. 1989. Alterations of endogenous cytokinins in transgenic plants using a chimeric isopentenyl transferase gene. *The Plant Cell* **1**: 403–413.

Miller C O. 1956. Similarity of some kinetin and red-light effects. *Plant Physiology* **31**: 318–319.

Miller C O, Skoog F von Saltza M H and Strong M. 1955. Kinetin, a cell division factor from deoxyribonucleic acid. *Journal of American Chemical Society* **77**: 1329–1334.

Mizukami Y and Fischer R L. 2000. Plant organ size control: AINTEGUMENTA regulates growth and cell numbers during organogenesis. *Proceedings of the National Academy of Sciences* **97**: 942–947.

Mok D W and Mok M C. 2001. Cytokinin metabolism and action. *Annual Review of Plant Biology* **52**: 89–118.

Mok M C. 1994. Cytokinins and plant development. *Cytokinins: Chemistry, Activity and Function*: 155–166.

Moreira S, Bishopp A, Carvalho H and Campilho A. 2013. AHP6 inhibits cytokinin signaling to regulate the orientation of pericycle cell division during lateral root initiation. *PLOS ONE* **8**: e56370.

Morris R O, Bilyeu K D, Laskey J G and Cheikh N N. 1999. Isolation of a gene encoding a glycosylated cytokinin oxidase from maize. *Biochemical and Biophysical Research Communications* **255**: 328–333.

Müller D and Leyser O. 2011. Auxin, cytokinin and the control of shoot branching. *Annals of Botany*: mcr069.

Naqvi S S M, Ansari R and Khanzada A N. 1982. Response of salt-stressed wheat seedlings to kinetin. *Plant Science Letters* **26**: 279–283.

Naseem M and Dandekar T. 2012. The role of auxin-cytokinin antagonism in plant-pathogen interactions. *PLoS Pathog* **8**: e1003026.

Nieminen K, Immanen J, Laxell M, Kauppinen L, Tarkowski P, Dolezal K and Bhalerao R. 2008. Cytokinin signaling regulates cambial development in poplar. *Proceedings of the National Academy of Sciences* **105**: 20032–20037.

Noodén L D and Leopold A C (Eds.). 1988. *Senescence and Aging in Plants*. pp. 499–518. Academic Press.

Noodén L D, Singh S and Letham D S. 1990. Correlation of xylem sap cytokinin levels with monocarpic senescence in soybean. *Plant Physiology* **93**: 33–39.

O'Brien J A and Benková E. 2013. Cytokinin cross-talking during biotic and abiotic stress responses. *Frontiers in Plant Science* **4**: 451.

Ochatt S J and Power J B. 1988. An alternative approach to plant regeneration from protoplasts of sour cherry (*Prunus cerasus* L.). *Plant Science* **56**: 75–79.

Okazaki K, Kabeya Y, Suzuki K, Mori T, Ichikawa T, Matsui M and Miyagishima S Y. 2009. The PLASTID DIVISION1 and 2 components of the chloroplast division machinery determine the rate of chloroplast division in land plant cell differentiation. *The Plant Cell* **21**: 1769–1780.

Ori N, Juarez M T, Jackson D, Yamaguchi J, Banowetz G M and Hake S. 1999. Leaf senescence is delayed in tobacco plants expressing the maize homeobox gene knotted1 under the control of a senescence-activated promoter. *The Plant Cell* **11**: 1073–1080.

Osborne D J. 1962. Effect of kinetin on protein & nucleic acid metabolism in Xanthium leaves during senescence. *Plant Physiology* **37**: 595.

Ozga J A, Brenner M L and Reinecke D M. 1992. Seed effects on gibberellin metabolism in pea pericarp. *Plant Physiology* **100**: 88–94.

Parthier B. 1979. Role of phytohormones (cytokinins) in chloroplast development. *Biochemie und Physiologie der Pflanzen* **174**: 173–214.

Persson B C, Esberg B, O'lafsson O and Bjoerk G R. 1994. Synthesis and function of isopentenyl adenosine derivatives in tRNA. *Biochimie* **76**: 1152–1160.

Pertry I, Václavíková K, Depuydt S, Galuszka P, Spíchal L, Temmerman W and Strnad M. 2009. Identification of Rhodococcus fascians cytokinins and their modus operandi to reshape the plant. *Proceedings of the National Academy of Sciences* **106**: 929–934.

Pospisilova J, Vagner M, Malbeck J, Travnickova A and Batkova P. 2005. Interactions between abscisic acid and cytokinins during water stress and subsequent rehydration. *Biologia Plantarum* **49**: 533–540.

Prasad V N, Gupta V N P and Bajracharya D. 1983. Alleviation by gibberellic acid and kinetin of the inhibition of seed germination in Maize (*Zea mays* L.) under submerged conditions. *Annals of Botany* **52**: 649–652.

Prigge M J, Otsuga D, Alonso J M, Ecker J R, Drews G N and Clark S E. 2005. Class III homeodomain-leucine zipper gene family members have overlapping, antagonistic, and distinct roles in Arabidopsis development. *The Plant Cell* **17**: 61–76.

Quesnelle P E and Emery R N. 2007. cis-Cytokinins that predominate in Pisum sativum during early embryogenesis will accelerate embryo growth in vitro. *Botany* **85**: 91–103.

Rao V S, Braun J W and Khan A A. 1976. Promotive effects of organic solvents and kinetin on dark germination of lettuce seeds. *Plant Physiology* **3**: 446–449.

Rashotte A M, Carson S D, To J P and Kieber J J. 2003. Expression profiling of cytokinin action in Arabidopsis. *Plant Physiology* **132**: 1998–2011.

Richmond A E and Lang A. 1957. Effect of kinetin on protein content and survival of detached Xanthium leaves. *Science* **125**: 650–651.

Riefler M, Novak O, Strnad M and Schmülling T. 2006. Arabidopsis cytokinin receptor mutants reveal functions in shoot growth, leaf senescence, seed size, germination, root development, and cytokinin metabolism. *The Plant Cell* **18**: 40–54.

Rupp H M, Frank M, Werner T, Strnad M and Schmülling T. 1999. Increased steady state mRNA levels of the STM and KNAT1 homeobox genes in cytokinin overproducing *Arabidopsis thaliana* indicate a role for cytokinins in the shoot apical meristem. *The Plant Journal* **18**: 557–563.

Sakakibara H. 2006. Cytokinins: activity, biosynthesis, and translocation. *Annual Review of Plant Biology* **57**: 431–449.

Sarojam R, Sappl P G, Goldshmidt A, Efroni I, Floyd S K, Eshed Y and Bowman J L. 2010. Differentiating Arabidopsis shoots from leaves by combined YABBY activities. *The Plant Cell* **22**: 2113–2130.

Scheres B, DiLaurenzio L, Willemsen V, Hauser M T, Janmaat K, Weisbeek P and Benfey P N. 1995. Mutations affecting the radial organization of the *Arabidopsis* root display specific defects throughout the embryonic axis. *Development* **121**: 53–62.

Schmülling T. 2004. Cytokinin. pp. 627–631. *Encyclopedia of Biological Chemistry*. 2nd edition. Reference Module in Biomedical Sciences. Academic Press.

Shani E, Ben-Gera H, Shleizer-Burko S, Burko Y, Weiss D and Ori N. 2010. Cytokinin regulates compound leaf development in tomato. *The Plant Cell* **22**: 3206–3217.

Shaw G. 1994. Chemistry of adenine cytokinins. pp. 15–34. *Cytokinins: Chemistry, Activity and Function* (Eds.) Mok D W S and Mok M C. CRC Press, Boca Raton, FL.

Singh O V, Labana S, Pandey G, Budhiraja R and Jain R K. 1992. Phytoremediation: an overview of metallic ion decontamination from soil. *Applied Microbiology and Biotechnology* **61**: 405–412.

Singh S, Letham D S and Palni L M S. 1998. Cytokinin biochemistry in relation to leaf senescence. VIII. Translocation, metabolism and biosynthesis of cytokinins in relation to sequential leaf senescence of tobacco. *Physiologia Plantarum* **86:** 398–406.

Srivastava A and Handa A K. 2005. Hormonal regulation of tomato fruit development: a molecular perspective. *Journal of Plant Growth Regulation* **24**: 67–82.

Stetler D A and Laetsch W M. 1965. Kinetin-induced chloroplast maturation in cultures of tobacco tissue. *Science* **149**: 1387–1388.

Suttle J C and Banowetz G M. 2000. Changes in cis-zeatin and cis-zeatin riboside levels and biological activity during potato tuber dormancy. *Physiologia Plantarum* **109**: 68–74.

Swartzberg D, Kirshner B, Rav-David D, Elad Y and Granot D. 2008. Botrytis cinerea induces senescence and is inhibited by autoregulated expression of the IPT gene. *European Journal of Plant Pathology* **120**: 289–297.

Takei K, Sakakibara H, Taniguchi M and Sugiyama T. 2001. Nitrogen-dependent accumulation of cytokinins in root and the translocation to leaf: implication of cytokinin species that induces gene expression of maize response regulator. *Plant and Cell Physiology* **42**: 85–93.

Tanaka R and Tanaka A. 2007. Tetrapyrrole biosynthesis in higher plants. *Annual Review of Plant Biology* **58**: 321–346.

To J P, Deruere J, Maxwell B B, Morris V F, Hutchison C E, Ferreira F J, Schaller G E and Kieber J J. 2007. Cytokinin regulates type-A *Arabidopsis* response regulator activity and protein stability via two-component phosphorelay. *The Plant Cell* **19**: 3901–3914.

Turner J E, Mok D W, Mok M C and Shaw G. 1987. Isolation and partial purification of an enzyme catalyzing the formation of O-xylosylzeatin in *Phaseolus vulgaris* embryos. *Proceedings of the National Academy of Sciences* **84**: 3714–3717.

Van Staden J and Davey E J. 1979. The cytokinin content of aseptically cultured pea fruits. *Z Pflanzenphysiol* **87**: 129–135.

Veach Y K, Martin R C, Mok D W, Malbeck J, Vankova R and Mok M C. 2003. O-glucosylation of cis-zeatin in maize. Characterization of genes, enzymes, and endogenous cytokinins. *Plant Physiology* **131**: 1374–1380.

Vivian-Smith A and Koltunow A M. 1999. Genetic analysis of growth-regulator-induced parthenocarpy in Arabidopsis. *Plant Physiology* **121**: 437–452.

Von Arnim A and Deng X W. 1996. Light control of seedling development. *Annual Review of Plant Biology* **47**: 215–243.

Vreman H J, Thomas R, Corse J, Swaminathan S and Murali N. 1978. Cytokinins in tRNA obtained from *Spinacia oleracea* L. leaves and isolated chloroplasts. *Plant Physiology* **61**: 296–306.

Vyroubalová Š, Václavíková K, Turečková V, Novák O, Šmehilová M, Hluska T and Galuszka P. 2009. Characterization of new maize genes putatively involved in cytokinin metabolism and their expression during osmotic stress in relation to cytokinin levels. *Plant Physiology* **151**: 433–447.

Wang Y, Li L, Ye T, Zhao S, Liu Z, Feng Y Q and Wu Y. 2011. Cytokinin antagonizes ABA suppression to seed germination of Arabidopsis by downregulating ABI5 expression. *The Plant Journal* **68**: 249–261.

Werner T, Köllmer I, Bartrina I, Holst K and Schmülling T. 2006. New insights into the biology of cytokinin degradation. *Plant Biology* **8**: 371–381.

Werner T, Motyka V, Laucou V, Smets R, Van Onckelen H and Schmülling T. 2003. Cytokinin-deficient transgenic Arabidopsis plants show multiple developmental alterations indicating opposite functions of cytokinins in the regulation of shoot and root meristem activity. *The Plant Cell* **15**: 2532–2550.

Werner T, Motyka V, Strnad M and Schmülling T. 2001. Regulation of plant growth by cytokinin. *Proceedings of National Academy of Sciences USA* **98**: 10487–10492.

Werner T and Schmülling T. 2009. Cytokinin action in plant development. *Current Opinion in Plant Biology* **12**: 527–538.

Wickson M and Thimann K V. 1958. The antagonism of auxin and kinetin in apical dominance. *Physiologia Plantarum* **11**: 62–74.

Wismer P T, Proctor J T A and Elfving D C. 1995. Benzyladenine affects cell division and cell size during apple fruit thinning. *Journal of the American Society for Horticultural Science* **120**: 802–807.

Yanai O, Shani E, Dolezal K, Tarkowski P, Sablowski R, Sandberg G and Ori N. 2005. Arabidopsis KNOXI proteins activate cytokinin biosynthesis. *Current Biology* **15**: 1566–1571.

Yanai O, Shani E, Russ D and Ori N. 2011. Gibberellin partly mediates LANCEOLATE activity in tomato. *The Plant Journal* **68**: 571–582.

Yang J, Zhang J, Wang Z, Zhu Q and Liu L. 2002. Abscisic acid and cytokinins in the root exudates and leaves and their relationship to senescence and remobilization of carbon reserves in rice subjected to water stress during grain filling. *Planta* **215**: 645–652.

Yokoyama A, Yamashino T, Amano Y I, Tajima Y, Imamura A, Sakakibara H and Mizuno T. 2007. Type-B ARR transcription factors, ARR10 and ARR12, are implicated in cytokinin-mediated regulation of protoxylem differentiation in roots of Arabidopsis thaliana. *Plant and Cell Physiology* **48**: 84–96.

Zhao C, Craig J C, Petzold H E, Dickerman A W and Beers E P. 2005. The xylem and phloem transcriptomes from secondary tissues of the Arabidopsis root-hypocotyl. *Plant Physiology* **138:** 803–818.

Zubo Y O, Yamburenko M V, Selivankina S Y, Shakirova F M, Avalbaev A M, Kudryakova N V and Börner T. 2008. Cytokinin stimulates chloroplast transcription in detached barley leaves. *Plant Physiology* **148:** 1082–1093.

Zwack P J and Rashotte A M. 2015. Interactions between cytokinin signaling and abiotic stress responses. *Journal of Experimental Botany* **66**: 4863–4871.

16 Abscisic Acid

16.1 INTRODUCTION

The environmental factors such as drought, flooding, high and low temperatures, light and salt stress have profound influence on plant growth and development. Any kind of stress imposed by these environmental factors may induce common responses including enhancement in the levels of some plant hormones. Since many kinds of stresses including water, salt, hot and cold temperatures and pests induce ABA synthesis, ABA is recognized as a plant stress hormone. It regulates several important aspects of plant growth and development. Recent studies have demonstrated a pivotal role for ABA in modulation of adaptative responses at the gene level in plants under adverse environmental conditions (Cutler et al., 2010). ABA is also involved in several other physiological processes such as stomatal closure, embryo morphogenesis, development of seeds, and synthesis of storage proteins and lipids, germination, leaf senescence and defense against pathogens (Eckardt, 2002; Finkelstein, 2013). Nevertheless, ABA mediates adaptative plant responses to environmental stresses. Sustainable horticulture under harsh environments requires an understanding of the ways that plants respond to both biotic and abiotic factors.

Abscisic acid (ABA) is a sesquiterpenoid ($C_{15}H_{20}O_4$) with one asymmetric, optically active carbon atom at C-1′. It was discovered independently by several research groups in the 1960s. Originally, it was recognized as a potent growth inhibitor that promotes abscission of fruits and dormancy in woody plants, although its role in these processes is still not clear. It was first identified in abscising cotton fruit (abscisin II) and leaves of dormant sycamore trees (dormin) (Ohkuma et al., 1963; Nambara and Marion-Poll, 2005; Wasilewska et al., 2008; Cutler et al., 2010). ABA indeed act as a growth promoter, especially in ABA-deficient mutants (Sharp et al., 2000; Finkelstein et al., 2002). Besides this, ABA also promotes stomata closure at night when there is no requirement for carbon dioxide or during drought stretches when the plant has to conserve moisture by reducing transpiration (Weyers et al., 1995).

16.2 ABA BIOSYNTHESIS

Abscisic acid (ABA) is a C-15 sesquiterpene originating from isopentenyl pyrophosphate (IPP) (Endo et al., 2014). There are two distinct pathways for ABA biosynthesis, a direct pathway occurring in phytopathogenic fungi and an indirect pathway, which operates in plants (Nambara and Marion-Poll, 2005; Oritani and Kiyota, 2003; Schwartz and Zeevaart, 2010). The biosynthetic origin of isopentenyl diphosphate (IDP) for ABA synthesis in fungi is different from that of plants. The fungi use IDP derived from mevalonate pathway for ABA synthesis, whereas higher plants biosyntesise ABA from IDP derived from non-mevalonate pathway (carotenoid pathway involving cleavage of 9Z-xanthophyll). Contrast to ABA biosynthesis in higher plants, two pathways (direct and carotenoid) have been proposed in the case of fungi.

16.2.1 Direct Pathway

The direct pathway involves the oxidation of ionylideneethanol by oxidase enzyme after cyclization of farnesyl diphosphate (FDP). In this pathway, IPP is synthesized from the mevalonate (MVA) (Newman and Chappell, 1999). Once radiolabeled MVA or farnesyl diphosphate (FDP) was served to ABA-producing fungi, the labels were effectively incorporated into ABA (Neill et al., 1982; Norman et al., 1983). Therefore, it was concluded that fungal ABA is synthesized

DOI: 10.1201/9781003373889-21

from IPP produced from the MVA pathway because all intermediates between FDP and ABA are sesquiterpenes (Oritani and Kiyota, 2003; Nambara and Marion-Poll, 2005). However, the intermediates between FDP and ionylideethanol have not yet been found. The phytopathogenic bacteria, *Botrytis cinerarea* biosynthesises ABA by the direct pathway via ionylideethane. In this pathway, FDP is converted to 2E, 4E, 6E-allofarnesenevia 4,5-didehydrofarnesyl diphosphate or 6E,10E-2,6,10-trimethyl2,6,10-dodecatriene and then to 2Z, 4E, 6E-allofarnesene. Allofarnesene by cyclization and isomerization of gives rise to (R)-2Z,4E-ion, which is combined at C-1, -1, C-1′, and C-4′ during the conversion. This reaction involves P450 inhibitors during oxidation (Norman et al., 1986; Inomata et al., 2004a, 2004b). It is consistent with the finding that ionylideneacetic acid, which have been identified from several fungi, serves as endogenous precursors of ABA in fungi (Oritani and Kiyot, 2003). Inomata et al. (2004a, 2004b) also reported that allofarnesene and ionylideneethane are the endogenous precursors of ABA in *Botrytis cinerea* and *Cercospora cruenta* (Inomata et al., 2004a, 2004b; Endo et al., 2014). Recently, BcCPR1 and BcABA1 genes encoding for P450 oxidoreductase and P450 monooxygenase have been identified in *Botrytis cinerea* (Siewers et al., 2004). The bcaba1 has been identified as ABA biosynthesis gene. The ABA level was reduced in BcCPR1; however, it was completely abolished in the BcABA1 mutant (Siewers et al., 2004). These findings are in consistent with oxidation of ionylideethane by molecular oxygen. This pathway has been found common among ABA-producing fungi (Inomata et al., 2004a).

16.2.2 Indirect Pathway

The indirect pathway uses the methylerythritol phosphate (MEP) as a source of IPP (Endo et al., 2014). In this pathway, FPP continues synthesis to Geranylgeranyl pyrophosphate (GGPP) and C40 carotenoids such as photogene, ζ-carotene, lycopene and β-carotene till Xanthoxin appears (Yang and Feng, 2015). However, sesquiterpenoid ABA is produced from an indirect pathway from the cleavage products of C40 carotenoids in higher plants (Schwartz et al., 2003; Galpaz et al., 2008). It is reported that first MVA polymerizes into C40 carotenoids in the plastid and then C40 carotenoids transformed into Zeaxanthin, which can create some epoxy-carotenoids by cyclization such as 9-cisViolaxanthin or 9-cis-Neoxanthin, and 9-cis-Violaxanthin or 9-cis-Neoxanthin will cleave into Xanthoxin in the plastid (Walton and Li, 1995; Seo and Koshiba, 2002; Cutler et al., 2010; Rodrigo et al., 2013; Ohmiya, 2013). In the final step, Xanthoxin as a C15 framework forms ABA by a series of changes in the cytosol (Seo and Koshiba, 2002; Schwartz et al., 2003; Nambara and Marion-Poll, 2005; Dong et al., 2015). These reactions have been reported to be catalyzed by the transcriptional expression of ABA4 gene in Arabidopsis (North et al., 2007). Xanthoxin is the direct C15 precursor of ABA and it is transported to the cytosol and then, transformed into abscisic aldehyde by the short chain dehydrogenase (SDR), which is encoded by the AtABA2/AtGIN1 genes in Arabidopsis in the biosynthesis pathway (Tan et al., 1997a; Iuchi et al., 2001; Cheng et al., 2002; González-Guzmán et al., 2002). Consequently, abscisic aldehyde is oxidized by aldehyde oxidase (AO) to generate ABA (Seo et al., 2000).

Different enzymes, such as Deoxyoxylulose-5-phosphate synthase (DXS), 3-Hydroxy3-methylglutaryl-coenzyme A reductase (HMGR), Farnesyl diphosphate synthase (FPS), Phytoene Synthase (PSY), Phytoene desaturase (PDS), Zeaxanthin epoxidase (ZEP), Neoxanthin synthase (NXS), Xanthophyll isomerase (XISO) (predicted), 9-cis-epoxycarotenoid dioxygenase (NCED), Short chain dehydrogenase (SDR) and Aldehyde oxidase (AO), are reported to be involved in the synthesis of ABA (Yoshioka et al., 1998; Taylor et al., 2000; Sauter et al., 2001). It has been shown that 9-cis-Violaxanthin or 9-cis-Neoxanthin cleaving into Xanthoxin process is the rate-limiting step in ABA biosynthesis. Therefore, NCED is a key enzyme, which catalyzes this reaction in the ABA biosynthesis pathway in higher plants (Tan et al., 1997a; Iuchi et al., 2001; Messing et al., 2010). Furthermore, nordihydroguaiaretic acid (NDGA), which is an inhibitor of NCED, could effectively inhibit the ABA biosynthesis pathway (Sun et al., 2012).

16.3 ABA CATABOLISM

ABA catabolism occurs mainly by hydroxylation of one of the three methyl groups (C-7′, C-8′ and C-9′) on the ring structure. However, 8′-hydroxylation is believed to serve as the dominant ABA catabolic pathway (Cutler and Krochko, 1999; Okamoto et al., 2009). Consequently, phaseic acid (PA) and dihydrophaseic acid (DPA) are the main ABA catabolitic products of the 8′-hydroxylation pathway (Ng et al., 2016). It is reported that 8′-hydroxy ABA displays considerable biological activity; however, the spontaneous cyclization of 8′-hydroxy ABA into PA causes a significant decline in the activity (Zou et al., 1995). PA is further catabolized into biologically inactive compound, DPA (Ng et al., 2016). Hydroxylation of ABA 8′ is catalyzed by ABA 8′-hydroxylase, a cytochrome P450 encoded by the CYP707A family gene (Kushiro et al., 2004; Saito et al., 2004). Transcript levels of all the family members CYP707A such as *CYP707A1*, *CYP707A2*, *CYP707A3* and *CYP707A4* have been reported to increase during drought stress and are strongly induced upon rehydration with highest induction by *CYP707A3* (Kushiro et al., 2004; Umezawa et al., 2006). It has been shown that CYP707A3 reduces systemic ABA levels in vascular tissues, whereas CYP707A1 catabolizes local ABA pools inside guard cells in response to high humidity (Okamoto et al., 2009). ABA or its metabolites can also be inactivated by conjugation to another molecule, such as ABA glucosyl ester (ABA-GE), which is the main glucose conjugate of ABA. ABA-GE is naturally inactive, and it might function as a storage molecule in the vacuoles to form a releasable pool of ABA (Dietz et al., 2000; Boursiac et al., 2013).

16.4 ABA SIGNALING AND DEGRADATION OF PROTEASOME

Abscisic acid is an important endogenous messenger in plants, and three main components are involved during ABA signaling (i) PYR/PYL/RCAR (an ABA receptor), (ii) type 2C protein phosphatase (PP2C; a negative regulator) and (iii) SNF1-related protein kinase 2 (SnRK2; a positive regulator) (Yoshida et al., 2006; Ma et al., 2009; Park et al., 2009). Ubiquitination and sumoylation play important roles in regulating ABA signaling (Lyzenga et al., 2013). Ubiquitination of the PYR/PYL/RCAR ABA receptors has been reported to cause their degradation in the absence of ABA (Irigoyen et al., 2014). Furthermore, PYL family inhibits the phosphatase activities of ABI1, ABI2 and HAB1 in an ABA-dependent manner (Park et al., 2009; Ma et al., 2009). PYL4 and PYR1 in the proximity of the plasma membrane are regulated by the interaction with a single subunit RING-type E3 ubiquitin ligase (RSL1) (Bueso et al., 2014). ABI3, ABI5, ABFs ABI4 and ATHB6 are the important transcription factors that are involved in proteasome degradation (Finkelstein et al., 2011). ABI3 (abscisic acid-insensitive 3) is a dominant regulator in ABA signaling, which is polyubiquitinated by an ABI3-interacting protein (AIP2) and AIP2 negatively regulate ABA signaling by targeting ABI3 for post-translational destruction (Zhang et al., 2005). It has been shown that ABA induces AIP2 expression during vegetative growth and promotes its accumulation during seed maturation (Cao et al., 2007). However, ABI5 acts in controlling ABA-dependent post-germination growth arrest as well as late seed maturation phase, and ABI5 is strongly controlled by the ubiquitin-*26S* proteasome system (Lopez-Molina et al., 2001). When ABA is absent in cytosol, Keepongoing (KEG) negatively regulates ABA signaling by promoting ABI5 ubiquitination and following degradation by the 26S proteasome (Liu and Stone, 2010; Liu and Stone, 2013). Both ABF1 and ABF3 protein are ubiquitylation substrates of KEG, and their abundance is affected by ABA and the ubiquitin pathway (Chen et al., 2013). It has been reported that ABI4 is essential for proper ABA signaling during seed development and germination because it is subjected to a strict posttranscriptional regulation that involves protein degradation and prevents its accumulation at high levels (Gregorio et al., 2014).

Degradation of proteosomal is regulated by phosphorylation/dephosphorylation mechanisms in response to ABA, which is regulated by SnRK2 kinases belonging to the core ABA signaling complex in which ABA promotes the self-ubiquitination and degradation of KEG after phosphorylation

(Antoni et al., 2011; Vilela et al., 2015). Similarly, CIPK26 interacts with the ABA signaling components (e.g., ABI1, ABI2 and ABI5), and it affects the sensitivity of germinating seeds to the inhibitory effects of ABA and is also targeted by KEG for proteasomal degradation (Lyzenga et al., 2013).

16.5 ABA ACCUMULATION AND TRANSPORT

ABA has been detected in all living tissues from root caps to apical buds including roots, stems, buds, leaves, fruits and seeds as well as in phloem and xylem sap and even in nectar. However, vegetative tissues such as immature leaves and buds contain the highest levels of ABA (Hoad, 1975; Raschke and Zeevaar, 1976; Zeevaart, 1977; Sivakumaran and Hall, 1978). The concentration of ABA in specific plant tissues varies greatly with developmental stages or in response to environmental stress. It is reported that ABA is capable of moving both up and down the stem by vascular tissues (Walton and Li, 1995). Furthermore, ABA translocation among cells, tissues and organs also plays important roles especially during stress conditions (Sah et al., 2016). By addition of a proton to an atom or molecule in ABA, it can diffuse passively across biological membranes (Wilkinson and Davies, 2010; Ng et al., 2016). Across membranes, ABA is transported by ATP-binding cassette (ABC) that contains transporter proteins (Kang et al., 2010; Kuromori et al., 2010). It is reported that *AtABCG25* is responsible for transferring ABA from vascular tissues which are the main site of ABA synthesis in plants (Kuromori et al., 2010). However, AtBCG40 acts as an ABA carrier in plant cells, and AtABCG22, which is essential for stomatal regulation, was proposed as an ABA transporter (Kang et al., 2010; Kuromori et al., 2011).

16.6 PHYSIOLOGICAL ROLE OF ABA

16.6.1 Seed Germination and Dormancy

The transition between seed dormancy and germination signifies a key stage in the plant life cycle (Sah et al., 2016). During early seed development, ABA plays an important role in maintaining embryos in a developmental mode until they are fully formed and have accumulated sufficient reserves, which is essential for germination and seedling establishment (Kermode, 2005). ABA is accumulated in dry seed, and it decreases with seed imbibition, and therefore *de novo* synthesis of ABA during the imbibition of dormant mature seeds contributes to dormancy maintenance (Grappin et al., 2000; Ali-Rachedi et al., 2004; Kermode, 2005; Del Carmen Rodríguez-Gacio et al., 2009). Therefore, amount of seed dormancy is correlated with endogenous ABA levels in imbibed seeds rather than in dry seeds in various species (Grappin et al., 2000; Jacobsen et al., 2002; Ali-Rachedi et al., 2004; Gonai et al., 2004). It is reported that during early seed development, the ABA level is low but increases subsequently around mid-maturation (Kermode, 2005). However, ABA level further declines during late developmental phase (maturation/drying) (Bewley and Black, 1994; Meinke, 1995; Bewley, 1997). These changes in endogenous ABA during different developmental phase have been correlated with the expression pattern of the genes that encode ABA biosynthesis enzymes (Seo and Koshiba, 2002). Besides this, environmental factors such as light, temperature and water availability also play an important role in determining ABA content and sensitivity of the mature seed (Baskin and Baskin, 1998). ABA acts as a positive regulator of physiological dormancy (Finch-Savage and Leubner-Metzger, 2006; Nambara et al., 2010). Previous studies have reported that ABA and GA act antagonistically where ABA inhibits germination by maintaining seed dormancy and GA promotes it (Finch-Savage and Leubner-Metzger, 2006; Kucera et al., 2005).

16.6.2 Root Architecture

ABA alters the root position, angle and length of lateral root, and therefore, it changes the pattern of growth and quiescence in plant roots (Sah et al., 2016). Interaction between the roots and its environment during the plant life plays an important role in determining root organization (Harris, 2015). Levels of ABA generally alter when there is water scarcity or when there is inconsistency in water supply, and therefore, root environment is subsequently affected by ABA-mediated responses (Dodd et al., 2008; Dodd et al., 2010). Furthermore, different stresses (drought, salt, high temperature and osmotic stress) reduce the water potential in the soil. During the stress condition, ABA regulates turgor by decreasing transpiration and increasing water influx into roots (Glinka and Reinhold, 1971). It has been shown that the ratio of root growth to shoot growth is higher in presence of ABA at low water potential as compared to the absence of ABA (Saab et al., 1990).

16.6.3 Fruit Ripening

ABA has been found to be associated with regulation of non-climacteric fruit ripening and stimulation of sugar metabolism and accumulation in fleshy fruits (Davies et al., 1997; Kobashi et al., 1999; Richings et al., 2000; Giovannoni, 2001; Rodrigo et al., 2003; Pan et al., 2005; Zhang et al., 2009a). 9-cis-epoxycarotenoid dioxygenase (NCED), a key enzyme which is involved in ABA biosynthesis, is associated with ripening of several fruits, *Persea americana, Citrus sinensis, Vitis vinifera, Prunus persica and Solanum lycopersicum* (Chernys and Zeevaar, 2007; Rodrigo et al., 2006; Zhang et al., 2009a, 2009b). Besides this, ABA is also found effective in inducing the ripening of non-climacteric fruits of *Fragaria ananassa* (Manning, 1994; Perkins-Veazie, 1995; Jiang and Joyce, 2003). Previous studies reported that both ABA and sucrose control the expression of some genes together (Li et al., 2002; Jossier et al., 2009; Tang et al., 2009). ABA levels in both climacteric and non-climacteric fruits increases sharply at the start of the ripening process (Buesa et al., 1994; Kojima et al., 1995; Kondo and Inoue, 1997). Furthermore, ABA accumulates just before the release of ethylene in climacteric fruits, and application of exogenous ABA increases the production of several metabolites, which are involved in fruit ripening (Çakir et al., 2003; Jeong et al., 2004; Zhang et al., 2009a, 2009b; Giribaldi et al., 2010).

16.6.4 Senescence

Both endogenous and environmental factors play a key role in regulating leaf senescence (van der Graaff et al., 2006; Lim and Nam, 2005). ABA promotes leaf yellowing (break down of chlorophyll), which is an indication of start of leaf senescence process in many species (Ray et al., 1983; He and Jin, 1999; Xue-Xuan et al., 2010; Yang et al., 2003; Fang et al., 2008). Increasing level of endogenous ABA has been correlated with leaf senescence through the expression of several senescence-associated genes (Yang et al., 2002; Xue-Xuan et al., 2010; Finkelstein, 2013). It has been reported that ABA-deficient mutant of *Arabidopsis* displays an altered or delayed senescence (Gan, 2003; Lim and Nam, 2005; Passioura, 2007). Receptor protein kinase 1 (RPK1) is an ABA-inducible membrane-bound receptor kinase which has a regulatory role in ABA-mediated and age-dependent leaf senescence. However, loss-of-function alterations in *RPK1* led to a significant delay in ABA-induced and age-dependent leaf senescence (Lee et al., 2011). Recently, Zhang et al. (2012) reported that senescence-associated gene 113 (*SAG113*) positively regulates ABA-induced leaf senescence by inhibiting stomata closure, which accelerates water loss in senescing leaves.

16.6.5 Stomata Regulation

Stomatal functioning is controlled by guard cell turgor pressure (Lim et al., 2015). Generally, light stimulates stomata opening, whereas ABA and elevated CO_2 levels promote partial or complete closure of stomata (Kim et al., 2010). Under stress (biotic and abiotic), ABA acts as a chemical messenger that induces stomata closure through the activation and inactivation of ion channels by protein kinases and phosphatases (Sokolovski et al., 2005; Garcia-Mata and Lamattina, 2003; Chen et al., 2010). Reduction in gas exchange cause less photosynthate production while decreased transpiration due to stomatal closure reduces water loss from leaves (Suzuki et al., 2013; Mittler and Blumwald, 2015). It has been reported that ABA alteration of guard cell ion transport stimulates stomata closure and reduce water loss during drought conditions (Kim et al., 2010). ABA-activated protein kinase (AAPK) is a positive regulator of ABA-induced stomata closure. It functions by activating plasma membrane anion channels (Li et al., 2000). Anion channels that facilitate anion release from guard cells are triggered by elevated cytosolic Ca^{2+} levels (slow-acting sustained, S-type) and rapid transient (R-type) (Schroeder and Keller, 1992; Schroeder et al., 2001). However, only S-type channel has been found to be accountable for the ABA-mediated stomata closure (Joshi-Saha et al., 2011). Open stomata 1 (OST1) kinase is an AAPK-related protein kinase that facilitates ABA-induced stomata closure in Arabidopsis (Mustilli et al., 2002). It has been reported that activated OST1 kinase binds directly to and phosphorylates the anion channel SLAC1, which interfere with the anion release from the guard cells and promote stomata closure (Geiger et al., 2009, 2010; Lee et al., 2009; Brandt et al., 2012). Continuous efflux of both anions and K^+ from guard cells motivate ions water efflux and contributes toward the loss of guard cell turgor leading to stomata closure (Fan et al., 2004; Munemasa et al., 2015).

ABA also encourages the accumulation of reactive oxygen species (ROS) in guard cells that trigger the closing of stomata (Sah et al., 2016). *AtrbohD*, and *AtrbohF* (NADPH oxidases) have been shown to be responsible for ABA-promoted ROS production in Arabidopsis guard cells (Kwak et al., 2003). ROS production in guard cells is enhanced by ABA treatment, and ROS accumulation is involved in the induction of stomata closure (An et al., 2008; Khokon et al., 2011).

16.6.6 Abiotic and Abiotic Stress

High ABA levels during abiotic stressful conditions are correlated with their adaptation to various stress signals (Swamy and Smith, 1999; Tuteja, 2007). It has been reported that endogenous ABA levels increased under drought and cold stress conditions in many species (Daie and Campbell, 1981; Kannangara and Field, 1983; Henson, 1984; Lalk and Dörffling, 1985; Stewart and Voetberg, 1985; Bensen et al., 1988; Wang et al., 2008; Guóth et al., 2009; Thameur et al., 2011). Increased level of endogenous ABA has also been observed in salt stressed tobacco cells and alfalfa seedlings (Singh et al., 1987). Exogenous application of ABA was found to alleviate salt stress in common bean and potato (Khadri et al., 2006; Etehadnia et al., 2008), enhance drought- and cold-resistance in some species (Flores et al., 1988; Waterland et al., 2010; Du et al., 2013; Yadegari et al., 2014; Wei et al., 2015). ABA induces the expression of several genes that are involved in stress responses and tolerance, such as enzymes for osmoprotectant synthesis (Fujita et al., 2011). Seki et al. (2002) reported that among the ABA-inducible genes, 155 genes (63%) were induced by drought, 133 genes (54%) by high salinity, and 25 genes (10%) by cold treatment in Arabidopsis.

16.6.7 ABA and Defense against Phytopathogens

Besides serving as a path for exchange of gas and water vapor, stomata also forms path for pathogen to enter the plant (Melotto et al., 2006). Since they play a crucial role in the plant immune response, regulation of stomata movement is an early stage of defense mechanism against pathogen infection (Lim et al., 2015)

REFERENCES

Ali-Rachedi S, Bouinot D, Wagner M H, Bonnet M, Sotta B, Grappin P and Jullien M. 2004. Changes in endogenous abscisic acid levels during dormancy release and maintenance of mature seeds: studies with the Cape Verde Islands ecotype, the dormant model of *Arabidopsis thaliana*. *Planta* **219**: 479–488.

An Z, Jing W, Liu Y and Zhang W. 2008. Hydrogen peroxide generated by copper amine oxidase is involved in abscisic acid-induced stomatal closure in *Vicia faba*. *Journal of Experimental Botany* **59**: 815–825.

Antoni R, Rodriguez L, Gonzalez-Guzman M, Pizzio G A and Rodriguez P L. 2011. News on ABA transport, protein degradation, and ABFs/WRKYs in ABA signaling. *Current Opinion in Plant Biology* **14**: 547–553.

Baskin C C and Baskin J M. 1998. *Seeds: Ecology, Biogeography, and, Evolution ofDormancy and Germination*. 2nd edition. Elsevier, Inc., Amsterdam, Netherlands.

Bensen R J, Boyer J S and Mullet J E. 1988. Water deficit-induced changes in abscisic acid, growth, polysomes, and translatable RNA in soybean hypocotyls. *Plant Physiology* **88**: 289–294.

Bewley J D. 1997. Seed germination and dormancy. *The Plant Cell* **9**: 1055.

Bewley J D and Black M. 1994. Seeds. pp. 1–33. *Seeds*. Springer, Berlin, Germany.

Boursiac Y, Léran S, Corratgé-Faillie C, Gojon A, Krouk G and Lacombe B. 2013. ABA transport and transporters. *Trends in Plant Science* **18**(6): 325–333.

Brandt B, Brodsky D E, Xue S, Negi J, Iba K, Kangasjärvi J and Schroeder J I. 2012. Reconstitution of abscisic acid activation of SLAC1 anion channel by CPK6 and OST1 kinases and branched ABI1 PP2C phosphatase action. *Proceedings of the National Academy of Sciences, United States of America* **109**: 10593–10598.

Buesa C, Dominguez M and Vendrell M. 1994. Abscisic acid effects on ethylene production and respiration rate in detached apple fruits at different stages of development. *Revista Espanola de Cienciay Tecnologia de Alimentos* **34**: 495–506.

Bueso E, Rodriguez L, Lorenzo-Orts L, Gonzalez-Guzman M, Sayas E, Muñoz-Bertomeu J and Rodriguez P L. 2014. The single-subunit RING-type E3 ubiquitin ligase RSL1 targets PYL4 and PYR1 ABA receptors in plasma membrane to modulate abscisic acid signaling. *The Plant Journal* **80**: 1057–1071.

Çakir B, Agasse A, Gaillard C, Saumonneau A, Delrot S and Atanassova R. 2003. A grape ASR protein involved in sugar and abscisic acid signaling. *The Plant Cell* **15**: 2165–2180.

Cao Z, Huang B, Wang Q, Xuan W, Ling T, Zhang B and Shen W. 2007. Involvement of carbon monoxide produced by heme oxygenase in ABA-induced stomatal closure in *Vicia faba* and its proposed signal transduction pathway. *Chinese Science Bulletin* **52**: 2365–2373.

Chen L, Zhang L, Li D, Wang F and Yu D. 2013. WRKY8 transcription factor functions in the TMV-cg defense response by mediating both abscisic acid and ethylene signaling in Arabidopsis. *Proceedings of the National Academy of Sciences, United States of America* **110**: E1963–E1971.

Chen Y H, Hu L, Punta M, Bruni R, Hillerich B, Kloss B and Hendrickson W A. 2010. Homologue structure of the SLAC1 anion channel for closing stomata in leaves. *Nature* **467**: 1074–1080.

Cheng S H, Willmann M R, Chen H C and Sheen J. 2002. Calcium signaling through protein kinases. The Arabidopsis calcium-dependent protein kinase gene family. *Plant Physiology* **129**: 469–485.

Chernys J T and Zeevaart J A D. 2007. Characterization of the 9-cis-epoxycarotenoid dioxygenase gene family and the regulation of abscisic acid biosynthesis in avocado. *Plant Physiology* **124**: 343–353.

Cutler A J and Krochko J E. 1999. Formation and breakdown of ABA. *Trends in Plant Science* **4**: 472–478.

Cutler S R, Rodriguez P L, Finkelstein R R and Abrams S R. 2010. Abscisic acid: emergence of a core signaling network. *Annual Review of Plant Biology* **61**: 651–679.

Daie J and Campbell W F. 1981. Response of tomato plants to stressful temperatures increase in abscisic acid concentrations. *Plant Physiology* **67**: 26–29.

Davies C, Boss P K and Robinson S P. 1997. Treatment of grape berries, a nonclimacteric fruit with a synthetic auxin, retards ripening and alters the expression of developmentally regulated genes. *Plant Physiology* **115**: 1155–1161.

Del Carmen Rodríguez-Gacio M, Matilla-Vázquez M A and Matilla A J. 2009. Seed dormancy and ABA signaling: the breakthrough goes on. *Plant Signaling and Behavior* **4**: 1035–1048.

Dietz K J, Sauter A, Wichert K, Messdaghi D and Hartung W. 2000. Extracellular β-glucosidase activity in barley involved in the hydrolysis of ABA glucose conjugate in leaves. *Journal of Experimental Botany* **51**: 937–944.

Dodd I C, Egea G and Davies W J. 2008. Accounting for sap flow from different parts of the root system improves the prediction of xylem ABA concentration in plants grown with heterogeneous soil moisture. *Journal of Experimental Botany* **59**: 4083–4093.

Dodd I C, Egea G, Watts C W and Whalley W R. 2010. Root water potential integrates discrete soil physical properties to influence ABA signaling during partial root zone drying. *Journal of Experimental Botany*: erq195.

Dong Q, Fang Y, Shao Y, Mulligan P, Qiu J, Cao L and Huang J. 2015. Electron-hole diffusion lengths > 175 μm in solution-grown CH3NH3PbI3 single crystals. *Science* **347**: 967–970.

Du H, Liu H and Xiong L. 2013. Endogenous auxin and jasmonic acid levels are differentially modulated by abiotic stresses in rice. *Frontiers in Plant Science* **4**: 397.

Eckardt N A. 2002. Abscisic acid biosynthesis gene underscores the complexity of sugar, stress, and hormone interactions. *The Plant Cell* **14**: 2645–2649.

Endo A, Okamoto M and Koshiba T. 2014. ABA biosynthesis and catabolism pathways. pp. 21–46. *Abscisic Acid: Metabolism Transport and Signaling* (Ed.) Zhang D P. Springer, Berlin, Germany.

Etehadnia M, Waterer D, De Jong H and Tanino K K. 2008. Scion and rootstock effects on ABA-mediated plant growth regulation and salt tolerance of acclimated and unacclimated potato genotypes. *Journal of Plant Growth Regulation* **27**: 125–140.

Fan L M, Zhao Z and Assmann S M. 2004. Guard cells: a dynamic signaling model. *Current Opinion in Plant Biology* **7(5)**: 537–546.

Fang Y, You J, Xie K, Xie W and Xiong L. 2008. Systematic sequence analysis and identification of tissue-specific or stress-responsive genes of NAC transcription factor family in rice. *Molecular Genetics and Genomics* **280**: 547–563.

Finch-Savage W E and Leubner-Metzger G. 2006. Seed dormancy and the control of germination. *New Phytologist* **171**: 501–523.

Finkelstein R R. 2013. Abscisic acid synthesis and response. *The Arabidopsis Book* 11: e0166.

Finkelstein R R, Gampala S S and Rock C D. 2002. Abscisic acid signaling in seeds and seedlings. *The Plant Cell* **14**: S15–S45.

Finkelstein R R, Lynch T J, Reeves W, Petitfils M and Mostachetti M. 2011. Accumulation of the transcription factor ABA-insensitive (ABI) 4 is tightly regulated post-transcriptionally. *Journal of Experimental Botany* **62**: 3971–3979.

Flores A, Grau A, Laurich F and Dörffling K. 1988. Effect of new terpenoid analogues of abscisic acid on chilling and freezing resistance. *Journal of Plant Physiology* **132**: 362–369.

Fujita Y, Fujita M, Shinozaki K and Yamaguchi-Shinozaki K. 2011. ABA-mediated transcriptional regulation in response to osmotic stress in plants. *Journal of Plant Research* **124**: 509–525.

Galpaz N, Wang Q, Menda N, Zamir D and Hirschberg J. 2008. Abscisic acid deficiency in the tomato mutant high-pigment 3 leading to increased plastid number and higher fruit lycopene content. *The Plant Journal* **53**: 717–730.

Gan S. 2003. Mitotic and postmitotic senescence in plants. *Science's SAGE KE* **38**: 7.

García-Mata C and Lamattina L. 2003. Abscisic acid, nitric oxide and stomatal closure – is nitrate reductase one of the missing links? *Trends in Plant Science* **8**: 20–26.

Geiger D, Scherzer S, Mumm P, Marten I A P L, Ache P, Matschi S and Romeis T. 2010. Guard cell anion channel SLAC1 is regulated by CDPK protein kinases with distinct Ca2+ affinities. *Proceedings of the National Academy of Sciences* **107**: 8023–8028.

Geiger D, Scherzer S, Mumm P, Stange A, Marten I, Bauer H and Romeis T. 2009. Activity of guard cell anion channel SLAC1 is controlled by drought-stress signaling kinase-phosphatase pair. *Proceedings of the National Academy of Sciences* **106**: 21425–21430.

Giovannoni J. 2001. Molecular biology of fruit maturation and ripening. *Annual Review of Plant Biology* **52**: 725–749.

Giribaldi M, Gény L, Delrot S and Schubert A. 2010. Proteomic analysis of the effects of ABA treatments on ripening Vitis vinifera berries. *Journal of Experimental Botany* **61**: 2447–2458.

Glinka Z and Reinhold L. 1971. Abscisic acid raises the permeability of plant cells to water. *Plant Physiology* **48**: 103.

Gonai T, Kawahara S, Tougou M, Satoh S, Hashiba T, Hirai N and Yoshioka T. 2004. Abscisic acid in the thermoinhibition of lettuce seed germination and enhancement of its catabolism by gibberellin. *Journal of Experimental Botany* **55**(394): 111–118.

González-Guzmán M, Apostolova N, Belles J M, Barrero J M, Piqueras P and Ponce M R. 2002. The short-chain alcohol dehydrogenase ABA2 catalyzes the conversion of xanthoxin to abscisic aldehyde. *Plant Cell* **14**: 1833–1846.

Grappin P, Bouinot D, Sotta B, Miginiac E and Jullien M. 2000. Control of seed dormancy in *Nicotiana plumbaginifolia*: post-imbibition abscisic acid synthesis imposes dormancy maintenance. *Planta* **210**: 279–285.

Gregorio J, Hernández-Bernal A F, Cordoba E and León P. 2014. Characterization of evolutionarily conserved motifs involved in activity and regulation of the ABA-INSENSITIVE (ABI) 4 transcription factor. *Molecular Plant* **7**: 422–436.

Guóth A, Tari I, Gallé Á, Csiszár J, Pécsváradi A, Cseuz L and Erdei L. 2009. Comparison of the drought stress responses of tolerant and sensitive wheat cultivars during grain filling: changes in flag leaf photosynthetic activity, ABA levels, and grain yield. *Journal of Plant Growth Regulation* **28**: 167–176.

Harris J M. 2015. Abscisic acid: hidden architect of root system structure. *Plants* **4**: 548–572.

He P and Jin J Y. 1999. Relationships among hormone changes, transmembrane Ca2+ flux and lipid peroxidation during leaf senescence in spring maize. *Acta Botanica Sinica* **41**: 1221–1225.

Henson I E. 1984. Inhibition of abscisic acid accumulation in seedling shoots of pearl millet (*Pennisetum americanum* [L.] Leeke) following induction of chlorosis by norflurazon. *Zeitschrift für Pflanzenphysiologie* **114**: 35–43.

Hoad G V. 1975. Effect of osmotic stress on abscisic acid levels in xylem sap of sunflower (*Helianthus annuus* L.). *Planta* **124**: 25–29.

Inomata M, Hirai N, Yoshida R and Ohigashi H. 2004a. The biosynthetic pathway to abscisic acid via ionylideneethane in the fungus Botrytis cinerea. *Phytochemistry* **65**: 2667–2678.

Inomata M, Hirai N, Yoshida R and Ohigashi H. 2004b. Biosynthesis of abscisic acid by the direct pathway via ionylideneethane in a fungus, *Cercospora cruenta. Bioscience, Biotechnology and Biochemistry* **68**: 2571–2580.

Irigoyen M L, Iniesto E, Rodriguez L, Puga M I, Yanagawa Y, Pick E and Rubio V. 2014. Targeted degradation of abscisic acid receptors is mediated by the ubiquitin ligase substrate adaptor DDA1 in Arabidopsis. *The Plant Cell* **26**: 712–728.

Iuchi S, Kobayashi M, Taji T, Naramoto M, Seki M, Kato T and Shinozaki K. 2001. Regulation of drought tolerance by gene manipulation of 9-cis-epoxycarotenoid dioxygenase, a key enzyme in abscisic acid biosynthesis in Arabidopsis. *The Plant Journal* **27**: 325–333.

Jacobsen J V, Pearce D W, Poole A T, Pharis R P and Mander L N. 2002. Abscisic acid, phaseic acid and gibberellin contents associated with dormancy and germination in barley. *Physiologia Plantarum* **115**: 428–441.

Jeong S T, Goto-Yamamoto N, Kobayashi S and Esaka M. 2004. Effects of plant hormones and shading on the accumulation of anthocyanins and the expression of anthocyanin biosynthetic genes in grape berry skins. *Plant Science* **167**: 247–252.

Jiang Y and Joyce D C. 2003. ABA effects on ethylene production, PAL activity, anthocyanin and phenolic contents of strawberry fruit. *Plant Growth Regulation* **39**: 171–174.

Joshi-Saha A, Valon C and Leung J. 2011. Abscisic acid signal off the STARting block. *Molecular Plant* **4**: 562–580.

Jossier M, Bouly J P, Meimoun P, Arjmand A, Lessard P, Hawley S and Thomas M. 2009. SnRK1 (SNF1-related kinase 1) has a central role in sugar and ABA signalling in *Arabidopsis thaliana. The Plant Journal* **59**: 316–328.

Kang J, Hwang J U, Lee M, Kim Y Y, Assmann S M, Martinoia E and Lee Y. 2010. PDR-type ABC transporter mediates cellular uptake of the phytohormone abscisic acid. *Proceedings of the National Academy of Sciences* **107**: 2355–2360.

Kannangara H W and Field R J. 1983. The effect of shade and crop interference on reproduction and growth of seedling' yarrow (*Achillea millefolium* L). *Aspects of Applied Biology* **4**: 147–155.

Kermode A R. 2005. Role of abscisic acid in seed dormancy. *Journal of Plant Growth Regulation* **24**: 319–344.

Khadri M, Tejera N A and Lluch C. 2006. Sodium chloride – ABA interaction in two common bean (*Phaseolus vulgaris*) cultivars differing in salinity tolerance. *Environmental and Experimental Botany* **60**: 211–218.

Khokon M D, Okuma E I J I, Hossain M A, Munemasa S, Uraji M, Nakamura Y and Murata Y. 2011. Involvement of extracellular oxidative burst in salicylic acid-induced stomatal closure in Arabidopsis. *Plant, Cell and Environment* **34**: 434–443.

Kim T H, Böhmer M, Hu H, Nishimura N and Schroeder J I. 2010. Guard cell signal transduction network: advances in understanding abscisic acid, CO_2, and Ca2+ signaling. *Annual Review of Plant Biology* **61**: 561–591.

Kobashi K, Gemma H and Iwahori S. 1999. Sugar accumulation in peach fruit as affected by abscisic acid (ABA) treatment in relation to some sugar metabolizing enzymes. *Journal of the Japanese Society for Horticultural Science* **68**: 465–470.

Kojima K, Yamada Y and Yamamoto M. 1995. Effects of abscisic acid injection on sugar and organic acid contents of citrus fruit. *Journal of the Japanese Society for Horticultural Science* **64**: 17–21.

Kondo S and Inoue K. 1997. Abscisic acid (ABA) and 1-aminocyclopropane-l-carboxylic acid (ACC) content during growth of 'Satohnishiki' cherry fruit, and the effect of ABA and ethephon application on fruit quality. *Journal of Horticultural Science* **72**: 221–227.

Kucera B, Cohn M A and Leubner-Metzger G. 2005. Plant hormone interactions during seed dormancy release and germination. *Seed Science Research* **15**: 281–307.

Kuromori T, Miyaji T, Yabuuchi H, Shimizu H, Sugimoto E, Kamiya A and Shinozaki K. 2010. ABC transporter AtABCG25 is involved in abscisic acid transport and responses. *Proceedings of the National Academy of Sciences* **107**: 2361–2366.

Kuromori T, Sugimoto E and Shinozaki K. 2011. Arabidopsis mutants of AtABCG22, an ABC transporter gene, increase water transpiration and drought susceptibility. *The Plant Journal* **67**(5): 885–894.

Kushiro T, Okamoto M, Nakabayashi K, Yamagishi K, Kitamura S, Asami T and Nambara E. 2004. The Arabidopsis cytochrome P450 CYP707A encodes ABA 8′-hydroxylases: key enzymes in ABA catabolism. *The EMBO Journal* **23**: 1647–1656.

Kwak J M, Mori I C, Pei Z M, Leonhardt N, Torres M A, Dangl J L and Schroeder J I. 2003. NADPH oxidase AtrbohD and AtrbohF genes function in ROS-dependent ABA signaling in Arabidopsis. *The EMBO Journal* **22**: 2623–2633.

Lalk I and Dörffling K. 1985. Hardening, abscisic acid, proline and freezing resistance in two winter wheat varieties. *Physiologia Plantarum* **63**: 287–292.

Lee I C, Hong S W, Whang S S, Lim P O, Nam H G and Koo J C. 2011. Age-dependent action of an ABA-inducible receptor kinase, RPK1, as a positive regulator of senescence in Arabidopsis leaves. *Plant and Cell Physiology* **52**: 651–662.

Lee S C, Lan W, Buchanan B B and Luan S. 2009. A protein kinase-phosphatase pair interacts with an ion channel to regulate ABA signaling in plant guard cells. *Proceedings of the National Academy of Sciences* **106**: 21419–21424.

Li J, Kinoshita T, Pandey S, Ng C K Y, Gygi S P, Shimazaki K-I and Assmann S M. 2002. Modulation of an RNA-binding protein by abscisic-acid-activated protein kinase. *Nature***418:** 793.

Li J, Wang X Q, Watson M B and Assmann S M. 2000. Regulation of abscisic acid-induced stomatal closure and anion channels by guard cell AAPK kinase. *Science* **287**: 300–303.

Lim C W, Baek W, Jung J, Kim J H and Lee S C. 2015. Function of ABA in stomatal defense against biotic and drought stresses. *International Journal of Molecular Sciences* **16**: 15251–15270.

Lim P O and Nam H G. 2005. 2 The molecular and genetic control of leaf senescence and longevity in Arabidopsis. *Current Topics in Developmental Biology* **67**: 50–85.

Liu H and Stone S L. 2010. Abscisic acid increases Arabidopsis ABI5 transcription factor levels by promoting KEG E3 ligase self-ubiquitination and proteasomal degradation. *The Plant Cell* **22**: 2630–2641.

Liu H and Stone S L. 2013. Cytoplasmic degradation of the Arabidopsis transcription factor abscisic acid insensitive 5 is mediated by the RING-type E3 ligase KEEP ON GOING. *Journal of Biological Chemistry* **288**: 20267–20279.

Lopez-Molina L, Mongrand S and Chua N H. 2001. A postgermination developmental arrest checkpoint is mediated by abscisic acid and requires the ABI5 transcription factor in Arabidopsis. *Proceedings of the National Academy of Sciences* **98**: 4782–4787.

Lyzenga W J, Liu H, Schofield A, Muise-Hennessey A and Stone S L. 2013. Arabidopsis CIPK26 interacts with KEG, components of the ABA signalling network and is degraded by the ubiquitin – proteasome system. *Journal of Experimental Botany* **64**: 2779–2791.

Ma Y, Szostkiewicz I, Korte A, Moes D, Yang Y, Christmann A and Grill E. 2009. Regulators of PP2C phosphatase activity function as abscisic acid sensors. *Science* **324**: 1064–1068.

Manning K. 1994. Changes in gene expression during strawberry fruit ripening and their regulation by auxin. *Planta* **194**: 62–68.

Meinke D W. 1995. Molecular genetics of plant embryogenesis. *Annual Review of Plant Biology* **46**: 369–394.

Melotto M, Underwood W, Koczan J, Nomura K and He S Y. 2006. Plant stomata function in innate immunity against bacterial invasion. *Cell* **126**: 969–980.

Messing S A, Gabelli S B, Echeverria I, Vogel J T, Guan J C, Tan B C and Amzel L M. 2010. Structural insights into maize viviparous14, a key enzyme in the biosynthesis of the phytohormone abscisic acid. *The Plant Cell* **22**: 2970–2980.

Mittler R and Blumwald E. 2015. The roles of ROS and ABA in systemic acquired acclimation. *The Plant Cell* **27**: 64–70.

Munemasa S, Hauser F, Park J, Waadt R, Brandt B and Schroeder J I. 2015. Mechanisms of abscisic acid-mediated control of stomatal aperture. *Current Opinion in Plant Biology* **28**: 154–162.

Mustilli A C, Merlot S, Vavasseur A, Fenzi F and Giraudat J. 2002. Arabidopsis OST1 protein kinase mediates the regulation of stomatal aperture by abscisic acid and acts upstream of reactive oxygen species production. *The Plant Cell* **14**: 3089–3099.

Nambara E and Marion-Poll A. 2005. Abscisic acid biosynthesis and catabolism. *Annual Review of Plant Biology* **56**: 165–185.

Nambara E, Okamoto M, Tatematsu K, Yano R, Seo M and Kamiya Y. 2010. Abscisic acid and the control of seed dormancy and germination. *Seed Science Research* **20**: 55–67.

Neill S J, Horgan R, Walton D C and Lee T S. 1982. The biosynthesis of abscisic acid in *Cercospora rosicola*. *Phytochemistry* **21**: 61–65.

Newman J D and Chappell J. 1999. Isoprenoid biosynthesis in plants: carbon partitioning within the cytoplasmic pathway. *Critical Reviews in Biochemistry and Molecular Biology* **34**: 95–106.

Ng L M, Melcher K, Teh B T and Xu H E. 2016. Abscisic acid perception and signaling: structural mechanisms and applications. *Acta Pharmacologica Sinica* **35**: 567–584.

Norman S M, Bennett R D, Poling S M, Maier V P and Nelson M D. 1986. Paclobutrazol inhibits abscisic acid biosynthesis in Cercospora rosicola. *Plant Physiology* **80**: 122–125.

Norman S M, Poling S M, Maier V P and Orme E D. 1983. Inhibition of abscisic acid biosynthesis in Cercospora rosicola by inhibitors of gibberellin biosynthesis and plant growth retardants. *Plant Physiology* **71**: 15–18.

North H M, Almeida A D, Boutin J P, Frey A, To A, Botran L and Marion-Poll A. 2007. The Arabidopsis ABA-deficient mutant aba4 demonstrates that the major route for stress-induced ABA accumulation is via neoxanthin isomers. *The Plant Journal* **50**: 810–824.

Ohkuma K, Lyon J L, Addicott F T and Smith O E. 1963. Abscisin II, an abscission-accelerating substance from young cotton fruit. *Science* **142**: 1592–1593.

Ohmiya A. 2013. Qualitative and quantitative control of carotenoid accumulation in flower petals. *Scientia Horticulturae* **163**: 10–19.

Okamoto M, Tanaka Y, Abrams S R, Kamiya Y, Seki M and Nambara E. 2009. High humidity induces abscisic acid 8′-hydroxylase in stomata and vasculature to regulate local and systemic abscisic acid responses in Arabidopsis. *Plant Physiology* **149**: 825–834.

Oritani T and Kiyota H. 2003. Biosynthesis and metabolism of abscisic acid and related compounds. *Natural Product Reports* **20**: 414–425.

Pan Q-H, Li M-J, Peng C-C, Zhang N, Zou X, Zou K-Q, Wang X-L, Yu X-C, Wang X-F and Zhang D-P. 2005. Abscisic acid activates acid invertases in developing grape berry. *Physiologia Plantarum* **125**: 157–170.

Park S Y, Fung P, Nishimura N, Jensen D R, Fujii H, Zhao Y and Alfred S E. 2009. Abscisic acid inhibits type 2C protein phosphatases via the PYR/PYL family of START proteins. *Science* **324**: 1068–1071.

Passioura J. 2007. The drought environment: physical, biological and agricultural perspectives. *Journal of Experimental Botany* **58**: 113–117.

Perkins-Veazie P M, Huber D J and Brecht J K. 1995. Characterization of ethylene production in developing strawberry fruit. *Plant Growth Regulation* **17**: 33–39.

Raschke K and Zeevaart J A. 1976. Abscisic acid content, transpiration, and stomatal conductance as related to leaf age in plants of *Xanthium strumarium* L. *Plant Physiology* **58**: 169–174.

Ray S, Mondal W A and Choudhuri M A. 1983. Regulation of leaf senescence, grain-filling and yield of rice by kinetin and abscisic acid. *Physiologia Plantarum* **59**: 343–346.

Richings E W, Cripps R F and Cowan A K. 2000. Factors affecting 'Hass' avocado fruit size: carbohydrate, abscisic acid and isoprenoid metabolism in normal and phenotypically small fruit. *Physiologia Plantarum* **109**: 81–89.

Rodrigo M J, Alquézar B, Alós E, Medina V, Carmona L, Bruno M and Zacarías L. 2013. A novel carotenoid cleavage activity involved in the biosynthesis of Citrus fruit-specific apocarotenoid pigments. *Journal of Experimental Botany* **64**: 4461–4471.

Rodrigo M J, Alquezar B and Zacarías L. 2006. Cloning and characterization of two 9-cis-epoxycarotenoid dioxygenase genes, differentially regulated during fruit maturation and under stress conditions, from orange (*Citrus sinensis* L. Osbeck). *Journal of Experimental Botany* **57**: 633–643.

Rodrigo M J, Marcos J F, Alférez F, Mallent M D and Zacarías L. 2003. Characterization of Pinalate, a novel Citrus sinensis mutant with a fruit-specific alteration that results in yellow pigmentation and decreased ABA content. *Journal of Experimental Botany* **54**: 727–738.

Saab I N, Sharp R E, Pritchard J and Voetberg G S. 1990. Increased endogenous abscisic acid maintains primary root growth and inhibits shoot growth of maize seedlings at low water potentials. *Plant Physiology* **93**: 1329–1336.

Sah S K, Reddy K R and Li J. 2016. Abscisic acid and abiotic stress tolerance in crop plants. *Frontiers in Plant Science* **7**.

Saito S, Hirai N, Matsumoto C, Ohigashi H, Ohta D, Sakata K and Mizutani M. 2004. Arabidopsis CYP707As encode (+)-abscisic acid 8′-hydroxylase, a key enzyme in the oxidative catabolism of abscisic acid. *Plant Physiology* **134**: 1439–1449.

Sauter A, Davies W J and Hartung W. 2001. The long-distance abscisic acid signal in the droughted plant: the fate of the hormone on its way from root to shoot. *Journal of Experimental Botany* **52**: 1991–1997.

Schroeder J I and Keller B U. 1992. Two types of anion channel currents in guard cells with distinct voltage regulation. *Proceedings of the National Academy of Sciences* **89**: 5025–5029.

Schroeder J I, Kwak J M and Allen G J. 2001. Guard cell abscisic acid signalling and engineering drought hardiness in plants. *Nature* **410**: 327–330.

Schwartz S H, Qin X and Zeevaart J A. 2003. Elucidation of the indirect pathway of abscisic acid biosynthesis by mutants, genes, and enzymes. *Plant Physiology* **131**: 1591–1601.

Schwartz S H and Zeevaart J A D. 2010. Biosynthesis, signal transduction, action. pp. 137–155. *Plant Hormones: Physiology, Biochemistry and Molecular Biology. Ed 2* (Ed.) Davies P J. Kluwer Academic Publishers, Dordrecht, The Netherlands.

Seki M, Ishida J, Narusaka M, Fujita M, Nanjo T, Umezawa T and Satou M. 2002. Monitoring the expression pattern of around 7,000 Arabidopsis genes under ABA treatments using a full-length cDNA microarray. *Functional and Integrative Genomics* **2**: 282–291.

Seo M and Koshiba T. 2002. Complex regulation of ABA biosynthesis in plants. *Trends in Plant Science* **7**: 41–48.

Seo M, Peeters A J, Koiwai H, Oritani T, Marion-Poll A, Zeevaart J A and Koshiba T. 2000. The Arabidopsis aldehyde oxidase 3 (AAO3) gene product catalyzes the final step in abscisic acid biosynthesis in leaves. *Proceedings of the National Academy of Sciences, United States of America* **97**: 12908–12913.

Sharp R E, LeNoble M E, Else M A, Thorne E T and Gherardi F. 2000. Endogenous ABA maintains shoot growth in tomato independently of effects on plant water balance: evidence for an interaction with ethylene. *Journal of Experimental Botany* **51**: 1575–1584.

Siewers V, Smedsgaard J and Tudzynski P. 2004. The P450 monooxygenase BcABA1 is essential for abscisic acid biosynthesis in Botrytis cinerea. *Applied and Environmental Microbiology* **70**: 3868–3876.

Singh N K, Bracker C A, Hasegawa P M, Handa A K, Buckel S, Hermodson M A and Bressan R A. 1987. Characterization of osmotin A thaumatin-like protein associated with osmotic adaptation in plant cells. *Plant Physiology* **85**: 529–536.

Sivakumaran S and Hall M A. 1978. Effects of age and water stress on endogenous levels of plant growth regulators in *Euphorbia lathyrus* L. *Journal of Experimental Botany* **29**: 195–205.

Sokolovski S, Hills A, Gay R, Garcia-Mata C, Lamattina L and Blatt M R. 2005. Protein phosphorylation is a prerequisite for intracellular Ca2+ release and ion channel control by nitric oxide and abscisic acid in guard cells. *The Plant Journal* **43**: 520–529.

Stewart C R and Voetberg G. 1985. Relationship between stress-induced ABA and proline accumulations and ABA-induced proline accumulation in excised barley leaves. *Plant Physiology* **79**: 24–27.

Sun L, Sun Y, Zhang M, Wang L, Ren J, Cui M and Chen P. 2012. Suppression of 9-cis-epoxycarotenoid dioxygenase, which encodes a key enzyme in abscisic acid biosynthesis, alters fruit texture in transgenic tomato. *Plant Physiology* **158**: 283–298.

Suzuki N, Rivero R M, Shulaev V, Blumwald E and Mittler R. 2013. Abiotic and biotic stress combinations. *New Phytologist* **203**: 32–43.

Swamy P M and Smith B N. 1999. Role of abscisic acid in plant stress tolerance. *Current Science* **76**: 1220–1227.

Tan B C, Schwartz S H, Zeevaart J A and McCarty D R. 1997a. Genetic control of abscisic acid biosynthesis in maize. *Proceedings of the National Academy of Sciences* **94**: 12235–12240.

Tan S, Ducret A, Aebersold R and Gantt E. 1997b. Red algal LHC I genes have similarities with both Chl a/b- and a/c-binding proteins: a 21 kDa polypeptide encoded by LhcaR2 is one of the six LHC I polypeptides. *Photosynthesis Research* **53**: 129–140.

Tang T, Xie H, Wang Y, Lü B and Liang J. 2009. The effect of sucrose and abscisic acid interaction on sucrose synthase and its relationship to grain filling of rice (*Oryza sativa* L.). *Journal of Experimental Botany* **60**: 2641–2652.

Taylor I B, Burbidge A and Thompson A J. 2000. Control of abscisic acid synthesis. *Journal of Experimental Botany* **51**: 1563–1574.

Thameur A, Ferchichi A and López-Carbonell M. 2011. Quantification of free and conjugated abscisic acid in five genotypes of barley (Hordeum vulgare L.) under water stress conditions. *South African Journal of Botany* **77**: 222–228.

Tuteja N. 2007. Abscisic acid and abiotic stress signaling. *Plant Signaling and Behavior* **2**: 135–138.

Umezawa T, Okamoto M, Kushiro T, Nambara E, Oono Y, Seki M and Shinozaki K. 2006. CYP707A3, a major ABA 8′-hydroxylase involved in dehydration and rehydration response in Arabidopsis thaliana. *The Plant Journal* **46**: 171–182.

van der Graaff E, Schwacke R, Schneider A, Desimone M, Flügge U I and Kunze R. 2006. Transcription analysis of Arabidopsis membrane transporters and hormone pathways during developmental and induced leaf senescence. *Plant Physiology* **141**: 776–792.

Vilela B, Nájar E, Lumbreras V, Leung J and Pagès M. 2015. Casein kinase 2 negatively regulates abscisic acid-activated SnRK2s in the core abscisic acid-signaling module. *Molecular Plant* **8**: 709–721.

Walton D C and Li Y. 1995. Abscisic acid biosynthesis and metabolism. pp. 140–157. *Plant Hormones: Physiology, Biochemistry and Molecular Biology. Ed 2* (Ed.) Davies P J. Kluwer Academic Publishers, Dordrecht, The Netherlands.

Wang Q, Guan Y, Wu Y, Chen H, Chen F and Chu C. 2008. Overexpression of a rice OsDREB1F gene increases salt, drought, and low temperature tolerance in both Arabidopsis and rice. *Plant Molecular Biology* **67**: 589–602.

Wasilewska A, Vlad F, Sirichandra C, Redko Y, Jammes F, Valon C and Leung J. 2008. An update on abscisic acid signaling in plants and more. *Molecular Plant* **1**: 198–217.

Waterland N L, Cambell C A, Finer J J and Jones M L. 2010. Abscisic acid application enhances drought stress tolerance in bedding plants. *HortScience* **45**: 409–413.

Wei L, Wang L, Yang Y, Wang P, Guo T and Kang G. 2015. Abscisic acid enhances tolerance of wheat seedlings to drought and regulates transcript levels of genes encoding ascorbate-glutathione biosynthesis. *Frontier in Plant Science* **6**: 458.

Weyers J D, Paterson N W, A'Brook R and Peng Z Y. 1995. Quantitative analysis of the control of physiological phenomena by plant hormones. *Physiologia Plantarum* **95**: 486–494.

Wilkinson S and Davies W J. 2010. Drought, ozone, ABA and ethylene: new insights from cell to plant to community. *Plant, Cell and Environment* **33**: 510–525.

Xue-Xuan X, Hong-Bo S, Yuan-Yuan M, Gang X, Jun-Na S, Dong-Gang G and Cheng-Jiang R. 2010. Biotechnological implications from abscisic acid (ABA) roles in cold stress and leaf senescence as an important signal for improving plant sustainable survival under abiotic-stressed conditions. *Critical Reviews in Biotechnology* **30**: 222–230.

Yadegari L Z, Heidari R, Rahmani F and Khara J. 2014. Drought tolerance induced by foliar application of abscisic acid and sulfonamide compounds in tomato. *Journal of Stress Physiology and Biochemistry* **10**: 327–334.

Yang F W and Feng X Q. 2015. Abscisic acid biosynthesis and catabolism and their regulation roles in fruit ripening. *Phyton, International Journal of Experimental Botany* **84**: 444–453.

Yang J C, Zhang J H, Wang Z Q, Zhu Q S and Liu L J. 2002. Abscisic acid and cytokinins in the root exudates and leaves and their relationship to senescence and remobilization of carbon reserves in rice subjected to water stress during grain filling. *Planta* **215**: 645–652.

Yang J C, Zhang J H, Wang Z Q, Zhu Q S and Liu L J. 2003. Involvement of abscisic acid and cytokinins in the senescence and remobilization of carbon reserves in wheat subjected to water stress during grain filling. *Plant, Cell and Environment* **26**: 1621–1631.

Yoshida T, Nishimura N, Kitahata N, Kuromori T, Ito T, Asami T and Hirayama T. 2006. ABA-hypersensitive germination3 encodes a protein phosphatase 2C (AtPP2CA) that strongly regulates abscisic acid signaling during germination among Arabidopsis protein phosphatase 2Cs. *Plant Physiology* **140**: 115–126.

Yoshioka T, Endo T and Satoh S. 1998. Restoration of seed germination at supraoptimal temperatures by fluridone, an inhibitor of abscisic acid biosynthesis. *Plant and Cell Physiology* **39**: 307–312.

Zeevaart J A. 1977. Sites of abscisic acid synthesis and metabolism in *Ricinus communis* L. *Plant Physiology* **59**: 788–791.

Zhang K, Xia X, Zhang Y and Gan S S. 2012. An ABA-regulated and Golgi-localized protein phosphatase controls water loss during leaf senescence in Arabidopsis. *The Plant Journal* **69**: 667–678.

Zhang M, Leng P, Zhang G and Li X. 2009a. Cloning and functional analysis of 9-cis-epoxycarotenoid dioxygenase (NCED) genes encoding a key enzyme during abscisic acid biosynthesis from peach and grape fruits. *Journal of Plant Physiology* **166**: 1241–1252.

Zhang M, Yuan B and Leng P. 2009b. The role of ABA in triggering ethylene biosynthesis and ripening of tomato fruit. *Journal of Experimental Botany* **60**: 1579–1588.

Zhang X, Garreton V and Chua N H. 2005. The AIP2 E3 ligase acts as a novel negative regulator of ABA signaling by promoting ABI3 degradation. *Genes and Development* **19**: 1532–1543.

Zou J, Abrams G D, Barton D L, Taylor D C, Pomeroy M K and Abrams S R 1995. Induction of lipid and oleosin biosynthesis by (+)-abscisic acid and its metabolites in microspore-derived embryos of Brassica napus L. cv Reston (biological responses in the presence of 8 [prime]-hydroxyabscisic acid). *Plant Physiology* **108**(2): 563–571.

17 Ethylene

17.1 INTRODUCTION

Ethylene (C_2H_4), a gaseous plant hormone, plays an essential role in several aspects of plant growth and development, such as seed germination, root hair development, root nodulation, flower and leaf senescence, leaf abscission, wounding response, fruit ripening, programmed cell death and responsiveness to biotic (pathogen attack) and abiotic (such as hypoxia and chilling) stresses (Johnson and Ecker, 1998; Lin et al., 2009). It is a small hydrocarbon gas naturally occurring in plants and can also be produced by combustion and other processes. Most plant tissues and cell types produce ethylene during their development, but the production rates vary and highest rates being associated with meristematic, stressed or ripening tissues (Abeles et al., 1992; Wang et al., 2002). It was discovered by Gane in 1934.

17.2 HISTORICAL PERSPECTIVE

In 1864, the gas leaked from streetlights showed stunting of growth, twisting of plants and abnormal thickening of stems (known as the triple response) (Salisbury and Ross, 1992). In 1901, a Russian scientist named Dimitry Neljubow reported that the active component in the leaked gas was ethylene (Neljubow, 1901). Later on in 1917, Doubt discovered that ethylene stimulated abscission (Doubt, 1917), and subsequently in 1934, it was shown for the first time that plants synthesize ethylene (Gane, 1934). In 1935, Crocker proposed that ethylene was the plant hormone responsible for fruit ripening as well as inhibition of vegetative tissues (Crocker et al., 1935). Ethylene is now reported to have many other functions as well.

17.3 OCCURRENCE, DISTRIBUTION AND TRANSPORT OF ETHYLENE

All groups of plants such as bacteria, fungi, blue–green algae, bryophytes, pteridophytes, gymnosperms and angiosperms produce ethylene. Various plant organs such as roots, stems, leaves, tubers, bulbs, fruits and seeds are able to synthesize ethylene in higher plants. However, its production varies depending on the type of tissue and stage of development with highest production in ripening fruits. It has been reported that ethylene is biologically active at very low concentration (<1 ppm) and the minimum threshold value for many ethylene-dependent responses in plants is approximately 0.1 ppm. Due to its hydrophobic nature, it can easily pass through plasma membrane into the cell and reach various organs by diffusion through intercellular spaces (Kidd and West, 1945).

17.4 ETHYLENE BIOSYNTHESIS

Ethylene biosynthesis in plants takes place via a methionine-dependent pathway with S-AdoMet as the main precursor. AdoMet, the major methyl donor in plants, is used as a substrate for many biochemical pathways (Adams and Yang, 1979; Yang and Hoffman, 1984; Kende, 1993). Conversion of S-adenosyl methionine (SAM) to 1-aminocyclopropane-1-carboxylic acid (ACC) by ACC synthase (1-aminocyclopropane-1-carboxylic acid synthase, ACS) is the initial step in the ethylene biosynthesis. Ethylene is produced by oxidation of ACC in the presence of ACC oxidase (1-aminocyclopropane-1-carboxylic acid oxidase, ACO). Conversion of S-AdoMet to ACC by ACS is the rate-limiting step in ethylene biosynthesis and this process is encoded by a multigene family, which is regulated by a complex network of developmental and environmental

DOI: 10.1201/9781003373889-22

signals responding to both internal and external stimuli (Johnson and Ecker, 1998). In addition to ACC, ACC synthase (ACS) also produces 5-methylthioadenosine (MTA), which is utilized for the synthesis of new methionine via a modified methionine cycle and preserves the methyl group through every revolution of the cycle utilizing one molecule of ATP for another round of ethylene production (Bleecker and Kende, 2000; Wang et al., 2002). Therefore, ethylene can be synthesized continuously without demanding an increasing pool of methionine. At the same time, the sulfur group of the methionine is also conserved. Finally, ACC is oxidized by ACC oxidase (ACO) to form ethylene, CO_2, and cyanide, which is detoxified to – cyanoalanine by – cyanoalanine synthase (-CAS, EC 4.4.1.9) to prevent toxicity of accumulated cyanide during high rates of ethylene synthesis. Phosphorylation of ACS2 and ACS6 by MPK6 (Mitogen-activated protein kinases 6) leads to the accumulation of ACS protein and therefore, it elevates the levels of cellular ACS activity and ethylene production (Liu and Zhang, 2004). Expression of the ACS genes is regulated by a variety of signals that actives ACC synthase and maintains it at low levels, which confirm that ethylene biosynthesis is tightly controlled. E4, a putative methionine sulphoxide reductase protein and ER69 a putative cobalamine-independent methionine synthase gene has been reported to play possible role in the methionine cycle (Zegzouti et al., 1999). However, both positive and negative feedback regulation of ethylene biosynthesis have been reported in different plant species (Kende, 1993; Nakatsuka et al., 1998; Barry et al., 2000). Positive feedback regulation of ethylene biosynthesis is a characteristic feature of ripening fruits and senescing flowers in which exposure to exogenous ethylene or propylene results in a large increase in ethylene production due to the induction of ACS and ACO enzymes, which are encoded by small multigene families and their expression is differentially regulated by various developmental, environmental and hormonal signals (Kende, 1993; Zarembinski and Theologis, 1994; Barry et al., 2000; Llop-Tous et al., 2000).

17.5 ETHYLENE SIGNAL PERCEPTION AND TRANSDUCTION

After its synthesis, ethylene is perceived and transduced through transduction machinery to trigger specific biological responses in plants. Ethylene signal transduction pathway was identified from a simple genetic screen using ethylene effect on dark-grown seedlings, which is known as the triple response. Triple response is characterized by inhibition of hypocotyl and root elongation, a thickened hypocotyl and an exaggerated apical hook in *Arabidopsis thaliana*. Several ethylene-insensitive mutants have been identified such as etrl and etr2 (ethylene response), ein2, ein3, ein4, ein5 and ein6 (ethylene insensitive), hlsl (hookless) and eirl (ethylene-insensitive root) (Bleecker et al., 1988; Guzman and Ecker, 1990; Chang et al., 1993; Roman et al., 1995; Chao et al., 1997; Alonso et al., 1999). Furthermore, mutants that exhibited a triple response in the absence of ethylene such as ctrl (constitutive triple response) and ranl (responsive to antagonist) have also been identified (Kieber et al., 1993; Hirayama et al., 1999; Woeste and Kieber, 2000). Using the genetic and molecular analyses of these mutants, Chen and coworkers attempted to define a pathway for ethylene signal transduction after the initial hormone perception to transcriptional regulation (Chen et al., 2005). They reported that ethylene is perceived by a family of five membrane-bound receptors (ETRl, ETR2, ERSl, ERS2 and EIN4) that have similarity to two-component regulators from bacteria (Bleecker, 1999; Schaller and Kieber, 2002). However, loss-of function (LOF) mutations in any single ethylene receptor have little or no effect upon seedling growth, consistent with functional overlap within the receptor family (Hua and Meyerowitz, 1998).

Plants with multiple LOF mutations in the receptors show a constitutive ethylene response suggesting that these receptors are negative regulators of ethylene signaling (Hua and Meyerowitz, 1998). Most receptors are localized to the plasma membrane, analysis of the ethylene receptor ETRl supported its localization to the endoplasmic reticulum (ER), and such location is compatible with the ready diffusion of ethylene in both aqueous and lipid environments (Abeles et al., 1992; Chen et al., 2002). CTR1, a Raf-like ser/thr kinase, which is similar to a mitogen-activated protein kinase kinase (MAPKKK), has been identified as a downstream component in the signaling pathway, suggesting the involvement of an MAP-kinase like signaling cascade in the regulation of ethylene signaling (Kieber et al., 1993; Huang et al., 2003; Chen et al., 2005). CTR1 is a negative regulator of ethylene signaling, whereas EIN2 positively regulates the ethylene signaling pathway by altering ion concentrations (Alonso et al., 1999). However, there is no evidence so far that confirms ion transport activity of EIN2, which is a small family of transcription factors that includes EIN3 and various EIN3 like (EIL) proteins (Roman et al., 1995; Chao et al., 1997). LOF mutations in EIN3 cause partial ethylene insensitivity, which can be released by expression of EIN3, EIL1 or EIL2, indicating the involvement of at least these two EILs together with EIN3 in the mediation of ethylene response because EIN3/EIL family is involved in a regulatory cascade and stimulation of the transcription of other transcription factors such as ERF1 (ethylene-response factor) (Alonso et al., 2003; Chao et al., 1997). Normally, ethylene receptors activate the kinase activity of CTR1 in the absence of ethylene. Therefore, CTR1 actively suppresses the downstream responses of EIN2 and the EIN3/EIL transcription factors that remain inactive. However, after ethylene binding, the receptors no longer activate CTR1, which allows for activation of EIN2 and the establishment of ethylene responses (Chen et al., 2005). Mechanism of the ethylene signal transduction pathway is well explored in both dicots and monocots, although some differences are observed (Adams-Phillips et al., 2004; Klee, 2004).

17.6 PHYSIOLOGICAL ROLES OF ETHYLENE

17.6.1 Seed Germination

Generally, ethylene production in seeds starts immediately after the onset of imbibition, and it increases with the progress of germination. However, the highest production has been shown to be associated with the radicle protrusion through the seed coat and its production varies with species (KeÇpczyński and KeÇpczyńska, 1997; Matilla, 2000; Siriwitayawan et al., 2003; El-Maarouf-Bouteau et al., 2014). Several researchers have reported that exogenous application of ethylene or ethephon (ethylene-releasing compound) improves germination by breaking primary-, secondary- and thermo-dormancy (KeÇpczyński and KeÇpczyńska, 1997; Matilla, 2000; Matilla and Matilla-Vázquez, 2008; Arc et al., 2013). Corineau and coworkers have listed the plants in which primary-, secondary- and thermo-dormancy is broken by ethylene and ethylene-releasing compounds (ethephon, ACC) (Corbineau et al., 2014). It promotes germination of non-dormant seeds when they were incubated at high temperature, osmotic stress, hypoxia and salinity (Abeles, 1986; Gallardo et al., 1991; Khan et al., 2009; Silva et al., 2014).

Ethylene is also found to be effective in breaking embryo dormancy in apple, beechnut and sunflower (Sińska and Lewandowska, 1991; Corbineau et al., 1990; Calvo et al., 2004).

17.6.2 Root Hair Formation and Elongation

Ethylene stimulates adventitious root formation in many plant species (Zimmerman and Hitchcock, 1933). Recent studies have shown that ethylene promotes, inhibits or has no effects on rooting (Coleman et al., 1980; Mudge and Swanson, 1978). Ethylene precursor, ACC1, has been found to promote root hair initiation (Masucci and Schiefelbein, 1996). However, the role of ethylene in root hair formation is still controversial because the ethylene-insensitive mutants (*etr1* and *ein2*) maintained normal root hair numbers (Masucci and Schiefelbein, 1996). Similarly, exogenous application of ethylene, cytokinin and auxin together with environmental factors can influence the root hair elongation (Pitts et al., 1998; Schiefelbein, 2000).

17.6.3 Nodule Formation

Several studies demonstrated inhibitory effects of ethylene on nodule formation in legumes (Grobbelaar et al., 1971; Goodlass and Smith, 1979; Lee and LaRue, 1992). However, Tamimie and Timko (2003) reported increased nodulation when plants were sprayed with ethylene biosynthesis inhibitors such as AVG and AOA (aminooxyacetic acid). Ethylene-insensitive mutants of *Medicago truncatula* (sickle mutants) produced higher number of nodules compared to its wild-type relative (Penmetsa and Cook, 1997).

17.6.4 Flower and Leaf Senescence

Plant senescence is a developmentally regulated and genetically programmed process that ultimately leads to the death of a particular organ or whole plant (Tripathi and Tuteja, 2007). Ethylene biosynthesis occurs in senescing leaves and in ripening fruits, and hence, it acts as a strong promoter of senescence in many plants (Jing et al., 2005; Lim et al., 2005). Pollination is known to trigger a large number of changes in flowers such as petal and anther senescence and ovary maturation to contain developing seeds (O'Neill, 1997). Flower senescence is mediated by pollination induced ethylene generation in various plant species and once ethylene is initiated, it becomes autocatalytic and accelerates the senescence process (Woodson and Lawton, 1988). Various ethylene receptors and downstream genes also play an active role in the progression of senescence (Müller et al., 2000; Shibuya et al., 2002). A model for regulatory pathways for flower senescence was proposed by Tripathi and Tuteja (2007). A leaf and flower senescence is described in detail in Chapter 22.

17.6.5 Fruit Ripening

Fruits are classified into non-climacteric (whose maturation does not depend on ethylene such as citrus, grapes, cherries, raspberries, strawberries and cashew) and climacteric fruits (such as apples, apricots, avocados, bananas, blackberries, kiwi, plums, peaches, pears and tomatoes where maturation and ripening depend on ethylene production) (Abeles et al., 1992). The role of various hormones including ethylene in fruit development and ripening is explained in detail in Chapter 21.

17.6.6 Stress Tolerance

Besides regulating the plant growth and development, ethylene has a vital role in biotic and abiotic stress responses in plants (Abeles et al., 1992). Generally, ethylene accumulation increases in plants that are subjected to environmental stresses (Gamalero and Glick, 2012). Ethylene level and its direct precursor ACC could be obviously induced by salinity and other abiotic stresses in many plant species (Morgan and Drew, 1997). It has been shown that halophytes such as *Cakile maritima* and *Thellungiella salsuginea* accumulated more ACC in both leaves and roots under high salinity (Ellouzi et al., 2014). Application of ethylene or ACC could improve plant tolerance to high salinity by enhancing the expression of reactive oxygen species (ROS) (Cao et al., 2007). It has been reported that ethylene production is quickly stimulated under elevated salinity and other stress conditions (Morgan and Drew, 1997). Recent studies have shown that ACS and S-AdoMet (ethylene precursor) play a major role in plant's response to biotic and abiotic stresses (Bleecker and Kende, 2000; Tao et al., 2015).

Low temperature is a primary environmental factor that limits plant growth and development; therefore, plant evolves various physiological and biochemical processes in response to cold stress (Thomashow, 1999; Shi et al., 2012). Although role of ethylene in cold stress tolerance varies with species, endogenous ethylene levels were shown to change under cold stress (Sun et al., 2016). Ethyle has been reported to enhance cold tolerance in *Lycopersicon esculentum* but to decrease it in tobacco *Nicotiana tabacum* (Zhang and Huang, 2010; Ciardi et al., 1997). Therefore, the role of ethylene on cold tolerance is species dependent (Kazan, 2015).

REFERENCES

Abeles F B. 1986. Role of ethylene in *Lactuca sativa* cv Grand Rapids' seed germination. *Plant Physiology* **81**: 780–787.

Abeles F B, Morgan P and Saltveit Jr M. 1992. Regulation of ethylene production by internal, environmental and stress factors Ethyl. *Plant Biology* **2**: 56–119.

Adams D O and Yang S F. 1979. Ethylene biosynthesis: identification of 1-aminocyclopropane-1-carboxylic acid as an intermediate in the conversion of methionine to ethylene. *Proceedings of the National Academy of Sciences, United States of America* **76**: 170–174.

Adams-Phillips L, Barry C and Giovannoni J. 2004. Signal transduction systems regulating fruit ripening. *Trends in Pant Science* **9**: 331–338.

Alonso J M, Hirayama T, Roman G, Nourizadeh S and Ecker J R. 1999. EIN2, a bifunctional transducer of ethylene and stress responses in *Arabidopsis. Science* **284**: 2148–2152.

Alonso J M, Stepanova A N, Leisse T J, Kim C J, Chen H, Shinn P and Gadrinab, C. 2003. Genome-wide insertional mutagenesis of *Arabidopsis thaliana. Science* **301**: 653–657.

Arc E, Sechet J, Corbineau F, Rajjou L and Marion-Poll A. 2013. ABA crosstalk with ethylene and nitric oxide in seed dormancy and germination. *Frontiers in Plant Science* **4**: 63.

Barry C S, Llop-Tous M I and Grierson D. 2000. The regulation of 1-aminocyclopropane-1-carboxylic acid synthase gene expression during the transition from system-1 to system-2 ethylene synthesis in tomato. *Plant Physiology* **123**: 979–986.

Bleecker A B. 1999. Ethylene perception and signaling: an evolutionary perspective. *Trends in Plant Science* **4**: 269–274.

Bleecker A B, Estelle M A, Somerville C and Kende H. 1988. Insensitivity to ethylene conferred by a dominant mutation in Arabidopsis thaliana. *Science* **241**: 1086–1090.

Bleecker A B and Kende H. 2000. Ethylene: a gaseous signal molecule in plants. *Annual Review of Cell and Developmental Biology* **16**: 1–18.

Calvo A P, Nicolas C, Nicolas G and Rodríguez D. 2004. Evidence of a cross-talk regulation of a GA 20-oxidase (FsGA20ox1) by gibberellins and ethylene during the breaking of dormancy in *Fagus sylvatica* seeds. *Physiologia Plantarum* **120**: 623–630.

Cao W H, Liu J, He X J, Mu R L, Zhou H L, Chen S Y and Zhang J S. 2007. Modulation of ethylene responses affects plant salt-stress responses. *Plant Physiology* **143**: 707–719.

Chang C, Kwok S F, Bleecker A B and Meyerowitz E M. 1993. Arabidopsis ethylene-response gene ETR1: similarity of product to two-component regulators. *Science-New York Then Washington* **262**: 539–539.

Chao Q, Rothenberg M, Solano R, Roman G, Terzaghi W and Ecker J R. 1997. Activation of the ethylene gas response pathway in Arabidopsis by the nuclear protein ETHYLENE-INSENSITIVE3 and related proteins. *Cell* **89**: 1133–1144.

Chen H, Qualls R G and Miller G C. 2002. Adaptive responses of *Lepidium latifolium* to soil flooding: biomass allocation, adventitious rooting, aerenchyma formation and ethylene production. *Environmental and Experimental Botany* **48**: 119–128.

Chen Y F, Etheridge N and Schaller G E. 2005. Ethylene signal transduction. *Annals of Botany* **95**: 901–915.

Ciardi J A, Tieman D M, Jones J B and Klee H J. 1997. Reduced expression of the tomato ethylene receptor gene LeETR4 enhances the hypersensitive response to *Xanthomonas campestris* pv. *vesicatoria. Molecular Plant-Microbe Interactions* **14**: 487–495.

Coleman W K, Huxter T J, Reid D M and Thorpe T A. 1980. Ethylene as an endogenous inhibitor of root regeneration in tomato leaf discs cultured in vitro. *Physiologia Plantarum* **48**: 519–525.

Corbineau F, Bagniol S and Côme D. 1990. Sunflower (*Helianthus annuus* L.) seed dormancy and its regulation by ethylene. *Israel Journal of Botany* **39**: 313–325.

Corbineau F, Xia Q, Bailly C and El-Maarouf-Bouteau H. 2014. Ethylene, a key factor in the regulation of seed dormancy. *Frontiers in plant Science* **5**: 113486.

Crocker W, Hitchcock A E and Zimmerman P W. 1935. Similarities in the effects of ethylene and the plant auxins. *Contributions of Boyce Thompson Institute* **7**: 231–248.

Doubt S L. 1917. The response of plants to illuminating gas. *Botanical Gazette* **63**: 209–224.

Ellouzi H, Hamed K B, Hernández I, Cela J, Müller M, Magné C and Munné-Bosch S. 2014. A comparative study of the early osmotic, ionic, redox and hormonal signaling response in leaves and roots of two halophytes and a glycophyte to salinity. *Planta* **240**: 1299–1317.

El-Maarouf-Bouteau H A Y A T, Sajjad Y, Bazin J, Langlade N, Cristescu S M, Balzergue S and Bailly C. 2014. Reactive oxygen species, abscisic acid and ethylene interact to regulate sunflower seed germination. *Plant, Cell & Environment* **38**: 364–374.

Gallardo M, del Mar Delgado M, Sánchez-Calle I M and Matilla A J. 1991. Ethylene production and 1-aminocyclopropane-1-carboxylic acid conjugation in thermoinhibited *Cicer arietinum* L. seeds. *Plant Physiology* **97**: 122–127.

Gamalero E and Glick B R. 2012. Ethylene and abiotic stress tolerance in plants. pp. 395–412. *Environmental Adaptations and Stress Tolerance of Plants in the Era of Climate Change*. Springer, New York.

Gane R. 1934. Production of ethylene by some ripening fruits. *Nature* **134**: 1008.

Goodlass G and Smith K A. 1979. Effects of ethylene on root extension and nodulation of pea (*Pisum sativum* L.) and white clover (*Trifolium repens* L.). *Plant and Soil* **51:** 387–395.

Grobbelaar N, Clarke B and Hough M C. 1971. The nodulation and nitrogen fixation of isolated roots of *Phaseolus vulgaris* L. *Plant and Soil* **35**: 215–223.

Guzman P and Ecker J R. 1990. Exploiting the triple response of Arabidopsis to identify ethylene-related mutants. *The Plant Cell* **2**: 513–523.

Hirayama T, Kieber J J, Hirayama N, Kogan M, Guzman P, Nourizadeh S and Ecker J R. 1999. Responsive-to-antagonist1, a Menkes/Wilson disease – related copper transporter, is required for ethylene signaling in Arabidopsis. *Cell* **97**: 383–393.

Hua J and Meyerowitz E M. 1998. Ethylene responses are negatively regulated by a receptor gene family in *Arabidopsis thaliana*. *Cell* **94**: 261–271.

Huang Y, Li H, Hutchison C E, Laskey J and Kieber J J. 2003. Biochemical and functional analysis of CTR1, a protein kinase that negatively regulates ethylene signaling in Arabidopsis. *The Plant Journal* **33**: 221–233.

Jing H C, Schippers J H, Hille J and Dijkwel P P. 2005. Ethylene-induced leaf senescence depends on age-related changes and OLD genes in Arabidopsis. *Journal of Experimental Botany* **56**: 2915–2923.

Johnson P R and Ecker J R. 1998. The ethylene gas signal transduction pathway: a molecular perspective. *Annual Review of Genetics* **32**: 227–254.

Kazan K. 2015. Diverse roles of jasmonates and ethylene in abiotic stress tolerance. *Trends in Plant Science* **20**: 219–229.

KeÇpczyński J and KeÇpczyńska E. 1997. Ethylene in seed dormancy and germination. *Physiologia Plantarum* **101**: 720–726.

Kende H. 1993. Ethylene biosynthesis. *Annual Review of Plant Biology* **44**: 283–307.

Khan M S, Zaidi A, Wani P A and Oves M. 2009. Role of plant growth promoting rhizobacteria in the remediation of metal contaminated soils. *Environmental Chemistry Letters* **7**: 1–19.

Kidd F and West C. 1945. Respiratory activity and duration of life of apples gathered at different stages of development and subsequently maintained at a constant temperature. *Plant Physiology* **20**: 467.

Kieber J J, Rothenberg M, Roman G, Feldmann K A and Ecker J R. 1993. CTR1, a negative regulator of the ethylene response pathway in Arabidopsis, encodes a member of the raf family of protein kinases. *Cell* **72:** 427–441.

Klee H J. 2004. Ethylene signal transduction. Moving beyond Arabidopsis. *Plant Physiology* **135**: 660–667.

Lee K H and LaRue T A. 1992. Exogenous ethylene inhibits nodulation of Pisum sativum L. cv Sparkle. *Plant Physiology* **100**: 1759–1763.

Lim P O, Kim H J and Gil Nam H. 2005. Leaf senescence. *Annual Review of Plant Biology* **58**: 115–136.

Lin Z, Zhong S and Grierson D. 2009. Recent advances in ethylene research. *Journal of Experimental Botany* **60**: 3311–3336.

Liu Y and Zhang S. 2004. Phosphorylation of 1-aminocyclopropane-1-carboxylic acid synthase by MPK6, a stress-responsive mitogen-activated protein kinase, induces ethylene biosynthesis in Arabidopsis. *The Plant Cell* **16**: 3386–3399.

Llop-Tous I, Barry C S and Grierson D. 2000. Regulation of ethylene biosynthesis in response to pollination in tomato flowers. *Plant Physiology* **123**: 971–978.

Masucci J D and Schiefelbein J W. 1996. Hormones act downstream of TTG and GL2 to promote root hair outgrowth during epidermis development in the Arabidopsis root. *The Plant Cell* **8**: 1505–1517.

Matilla A J. 2000. Ethylene in seed formation and germination. *Seed Science Research* **10**: 111–126.

Matilla A J and Matilla-Vázquez M A. 2008. Involvement of ethylene in seed physiology. *Plant Science* **175**: 87–97.

Morgan P W and Drew M C. 1997. Ethylene and plant responses to stress. *Physiologia Plantarum* **100**: 620–630.

Mudge K W and Swanson B T. 1978. Effect of ethephon, indole butyric acid, and treatment solution pH on rooting and on ethylene levels within mung bean cuttings. *Plant Physiology* **61**: 271–273.

Müller R, Sisler E C and Serek M. 2000. Stress induced ethylene production, ethylene binding, and the response to the ethylene action inhibitor 1-MCP in miniature roses. *Scientia Horticulturae* **83**: 51–59.

Nakatsuka A, Murachi S, Okunishi H, Shiomi S, Nakano R, Kubo Y and Inaba A. 1998. Differential expression and internal feedback regulation of 1-aminocyclopropane-1-carboxylate synthase, 1-aminocyclopropane-1-carboxylate oxidase, and ethylene receptor genes in tomato fruit during development and ripening. *Plant Physiology* **118**: 1295–1305.

Neljubow D. 1901. Ueber die horizontale Nutation der Stengel von Pisum sativum und einiger anderer. *Pflanzen Beih. Bot. Zentralbl* **10**: 128–139.

O'Neill S D. 1997. Pollination regulation of flower development. *Annual Review of Plant Biology* **48**: 547–574.

Penmetsa R V and Cook D R. 1997. A legume ethylene-insensitive mutant hyperinfected by its rhizobial symbiont. *Science* **275**: 527–530.

Pitts R J, Cernac A and Estelle M. 1998. Auxin and ethylene promote root hair elongation in Arabidopsis. *The Plant Journal* **16**: 553–560.

Roman G, Lubarsky B, Kieber J J, Rothenberg M and Ecker J R. 1995. Genetic analysis of ethylene signal transduction in Arabidopsis thaliana: five novel mutant loci integrated into a stress response pathway. *Genetics* **139**: 1393–1409.

Salisbury F B and Ross C W. 1992. *Plant Physiology*. pp. 393–395. 4th edition. Wadsworth Publishing Company, Belmont, CA.

Schaller G E and Kieber J J. 2002. Ethylene. *The Arabidopsis Book*. https://doi.org/10.1199/0071.

Schiefelbein J W. 2000. Constructing a plant cell. The genetic control of root hair development. *Plant Physiology* **124**: 1525–1531.

Shi Y, Tian S, Hou L, Huang X, Zhang X, Guo H and Yang S. 2012. Ethylene signaling negatively regulates freezing tolerance by repressing expression of CBF and type-A ARR genes in Arabidopsis. *The Plant Cell* **24**: 2578–2595.

Shibuya K, Nagata M, Tanikawa N, Yoshioka T, Hashiba T and Satoh S. 2002. Comparison of mRNA levels of three ethylene receptors in senescing flowers of carnation (*Dianthus caryophyllus* L.). *Journal of Experimental Botany* **53**: 399–406.

Silva P O, Medina E F, Barros R S and Ribeiro D M. 2014. Germination of salt-stressed seeds as related to the ethylene biosynthesis ability in three Stylosanthes species. *Journal of Plant Physiology* **171**: 14–22.

Sińska I and Lewandowska U. 1991. Polyamines and ethylene in the removal of embryonal dormancy in apple seeds. *Physiologia Plantarum* **81**: 59–64.

Siriwitayawan G, Geneve R L and Downie A B. 2003. Seed germination of ethylene perception mutants of tomato and Arabidopsis. *Seed Science Research* **13**: 303–314.

Sun W, Van Montagu M and Verbruggen N. 2016. Small heat shock proteins and stress tolerance in plants. *Biochimicaet Biophysica Acta (BBA)-Gene Structure and Expression* **1577**: 1–9.

Tamimi S M and Timko M P. 2003. Effects of ethylene and inhibitors of ethylene synthesis and action on nodulation in common bean (Phaseolus vulgaris L.). *Plant and Soil* **257**: 125–131.

Tao J J, Chen H W, Ma B, Zhang W K, Chen S Y and Zhang J S. 2015. The role of ethylene in plants under salinity stress. *Frontiers in Plant Science* **6**: 1059.

Thomashow M F. 1999. Plant cold acclimation: freezing tolerance genes and regulatory mechanisms. *Annual Review of Plant Biology* **50**: 571–599.

Tripathi S K and Tuteja N. 2007. Integrated signaling in flower senescence: an overview. *Plant Signaling &Behavior* **2**: 437–445.

Wang K L C, Li H and Ecker J R. 2002. Ethylene biosynthesis and signaling networks. *The Plant Cell* **14**: S131–S151.

Wang W, Vinocur B and Altman A. 2002. Plant responses to drought, salinity and extreme temperatures: towards genetic engineering for stress tolerance. *Planta* **218**: 1–14.

Woeste K E and Kieber J J. 2000. A strong loss-of-function mutation in RAN1 results in constitutive activation of the ethylene response pathway as well as a rosette-lethal phenotype. *The Plant Cell* **12**: 443–455.

Woodson W R and Lawton K A. 1988. Ethylene-induced gene expression in carnation petals relationship to autocatalytic ethylene production and senescence. *Plant Physiology* **87**: 498–503.

Yang S F and Hoffman N E. 1984. Ethylene biosynthesis and its regulation in higher plants. *Annual Review of Plant Physiology* **35**: 155–189.

Zarembinski T I and Theologis A. 1994. Ethylene biosynthesis and action: a case of conservation. *Plant Molecular Biology*: 1579–1597. https://doi.org/10.1007/BF00016491.

Zegzouti H, Jones B, Frasse P, Marty C, Maitre B, Latché A and Bouzayen M. 1999. Ethylene-regulated gene expression in tomato fruit: characterization of novel ethylene-responsive and ripening-related genes isolated by differential display. *The Plant Journal* **18**: 589–600.

Zhang Z and Huang R. 2010. Enhanced tolerance to freezing in tobacco and tomato overexpressing transcription factor TERF2/LeERF2 is modulated by ethylene biosynthesis. *Plant Molecular Biology* **73**: 241–249.

Zimmerman P W and Hitchcock A E. 1933. Initiation and stimulation of adventitious roots caused by unsaturated hydrocarbon gases. *Contributions from Boyce Thompson Institutes* **5**: 351–369.

18 Brassinosteroids

18.1 INTRODUCTION

Brassinosteroids (BRs) are plant specific steroidal hormones, categorized by their polyhydroxylated sterol structure. First isolated from *Brassica napus* pollen, they are found in almost all the phyla of the plant such as alga, pteridophyte, gymnosperms, dicots and monocots (Grove et al., 1979; Bajguz, 2009). Brassinolide (BL) is the most bioactive plant steroids commonly known as brassinosteroids (BRs). They play various roles in plant growth and developmental processes by regulating several physiological functions, such as seed germination, cell elongation, cell division, senescence, vascular differentiation, reproduction, root development, photomorphogenesis and biotic and abiotic stress tolerance (Clouse and Sasse, 1998; Li and Chory, 1999; Rao et al., 2002; Sreeramulu et al., 2013). More than 70 BR intermediates, metabolites and their conjugates have been identified so far in various plants (Choe, 2006; Bajguz, 2007; Piotrowska and Bajguz, 2011; Sirhindi, 2013).

BRs are grouped according to the number of carbons in their structure into C_{27}, C_{28} or C_{29} BRs (Vardhini and Rao, 2003a, 2003b). However, 28-homobrassinolide (28-HomoBL) and 24-epibrassinolide (24-EpiBL) are the most bioactive and widely used in physiological and experimental studies. They elicit growth stimulation at nano- or micro-molar concentrations (Vardhini et al., 2006).

18.2 BIOSYNTHESIS OF BRASSINOSTEROIDS

Biosynthetic pathway of brassinolide starting from plant sterols was suggested by Mandava in 1988. 24-ethyl or a 24-ethylidene group (sitosterol, stigmasterol and isofucosterol) were the main sterols identified from *Phaseolus vulgaris* (Kim et al., 1988). Presence of similar BRs with a different carbon skeleton in the side chain (epibrassinolide, homobrassinolide, epicastasterone, (24S)-ethylbrassinone, norbrassinolide, brassinone) may be explained by the co-occurrence of campesterol as starting material for the biosynthesis of brassinolide together with other sterols, such as brassicasterol, b-sitosterol and 22-dehydrocholesterol (Khripach et al., 1999). Since BRs are a group of modified sterols, their biosynthetic pathways can be divided into two stages, sterol-specific and BR-specific pathways, i.e., squalene to campesterol (CR) and CR to BL, respectively (Choe, 2004). Endogenous BRs control the biosynthetic gene expression, which is involved in inactivating bioactive BR. Genes such as *DET2, DWF4, CPD, BR6ox1* and *ROT3* are feedbacks regulated in the BR-specific biosynthesis (Sun et al., 2013; Yu et al., 2011). Mevalonic acid (MVA), the starting molecule in the terpenoid pathway, is condensed and cyclized to produce 2,3-oxidosqualene. 2,3-oxidosqualene is further modified to form the major plant sterols such as sitosterol (SR) and CR. All of these parent sterols then serve as precursors of BL isologs, such as homo- or nor-BL (Choe, 2004). Sterols must be processed by the BR-specific pathway in order to form bioactive BRs. Generally, sterols are modified to have different functional groups, such as the saturation of a double bond at C5 and the formation of a 6-oxo group, addition of D-oriented vicinal hydroxyl groups at C-22 C-23, epimerization of a 3E-hydroxyl group to the 3D-configuration, addition of a 2D hydroxyl group, and a Baeyer-Villiger type oxidation in the B ring (Choe, 2004). The individual biosynthetic steps have been explained by metabolic tests using BR-overproducing cell lines of *Catharanthus roseus* (Rouleau et al., 1999). These authors developed BL-overproducing lines to overcome low biosynthetic activities in regular plant tissues or cell lines, which technically limits extraction and detection of BRs. Generally, biosynthesis of BR proceeds through multiple branched pathways and first branch occurs at CR and the second at campestanol (CN). CR can be either C-22 hydroxylated or C-5 reduced in bifurcated pathways that are termed the early and the late C-22 oxidation pathways (Choe et al., 2001). Additionally, CN proceeds to one of

DOI: 10.1201/9781003373889-23

two alternative pathways such as the early and the late C-6 oxidation pathways. Early C-6 oxidation pathway undergoes a two-step oxidation of the C-6 position at the CN stage, whereas in the late C-6 oxidation pathway, C-6 is oxidized at the second to last step (Fujioka et al., 2002). Bassinoids biosynthetic pathway in tomato.

18.3 DISTRIBUTION, REGULATION AND DEACTIVATION OF BRASSINOSTEROIDS

BRs are widely distributed and present in most plant tissues (Bajguz and Tretyn, 2003). However, the concentration of bioactive BRs (e.g., castasterone [CS] and brassinolide [BL]) varies with species and even with different parts in the same plant (Bancos et al., 2002; Bajguz and Tretyn, 2003; Shimada et al., 2003; Symons and Reid, 2004). Generally, reproductive organs such as pollen, seeds and fruits have the highest levels of bioactive BRs (Bajguz and Tretyn, 2003; Montoya et al., 2005; Symons et al., 2006; Nomura et al., 2007). In contrast, vegetative shoot tissues have much lower levels of bioactive BRs followed by root tissues with lowest levels (Bancos et al., 2002; Shimada et al., 2003; Symons and Reid, 2004). Generally, shoot tip (apex) and the young internodes is the main site for highest accumulation of bioactive BRs within the shoot tissues (Shimada et al., 2003; Symons and Reid, 2004).

Maintenance and regulation of endogenous BR level are crucial for various biological functions in plants (Tanaka et al., 2005). BR biosynthesis, transport and degradation are key mechanisms that maintain the endogenous level of BR in plant. Symons et al. (2007) reported that BR levels are not regulated by the long-distance transport around the plant because they produced *at original place in a particular tissue*, such as the fruit, roots, internodes and leaves. In addition, tissues with high BR levels also normally have higher transcript levels but this cannot be used for assessing BR levels (Shimada et al., 2003: Montoya et al., 2005; Symons et al., 2006; Nomura and Bishop, 2006). Furthermore, several steps in the BR biosynthesis pathway undergo feedback regulation in response to BR levels because BR synthesis is tightly controlled (Fujioka and Yokota, 2003; Nomura and Bishop, 2006). Therefore, endogenous BR levels in plant tissues are mainly regulated through the tissue-specific control of BR synthesis, catabolism and inactivation because they cannot be transported to long distance (Symons and Reid, 2004; Symons et al., 2008). Furthermore, excessive quantities of BRs are inactivated via various metabolic processes such as hydroxylation, glycosylation, sulfonation, demethylation, epimerization, esterification, dehydrogenation and structural changes to the steroidal skeleton and the side chain (Bajguz, 2007).

Brassinosteroid catabolism/metabolism involves various processes, such as acylation, sulfonation and glycosylation that play an important role in maintaining the optimum levels of bioactive BR in the cell (Saini et al., 2015). Several genes belonging to the family of BAHD acyltransferases, *brassinosteroid inactivator1* (*BIA1*), BAHD acyltransferases pizza (PIZ), bri1–5 enhanced 1 (BEN1), sulfotransferase 3 (BNST3) and *abnormal shoot-1* (*abs-1*) have been identified to be involved in BR acylation, sulfonation and inactivation (Marsolais et al., 2007; Yuan et al., 2007; Schneider et al., 2012). Glycotransferase enzymes UGT73C6 and its close homolog UGT73C5 catalyze the 23 *O*-glycosylation of CS and BL (Poppenberger et al., 2005). However, external environmental factors, such as salt and temperature stress, have also been shown to control BR biosynthesis (Maharjan and Choe, 2011; Sharma et al., 2013).

18.4 BRASSINOSTEROID SIGNALING

Various biochemical, genetic and proteomic methods have been used to understand the BR signaling pathway (Zhu et al., 2013; Fàbregas and Caño-Delgado, 2014; Belkhadir and Jaillais, 2015). BRs are perceived by the localized leucine-rich repeat receptors like kinase (LRR-RLK) BRI1 and its closely related homologs, such as BRI1-LIKE 1 (BRL1) and BRI1 LIKE 3 (BRL3) in the nano-molar range in the extracellular domain of the plasma membrane (Caño-Delgado et al., 2004). BRI1 and BRL1 have been reported to have similar basic skeleton with 25 and 24 units of LRRs, and they are involved in binding of BR (Saini et al., 2015). However, minor structural differences in the BRL1 and BRI1 indicate that BRI1 have less affinity compared to BRL1. Therefore, BR binds more strongly to BRL1 (She et al., 2013). Similarly, BRL2 (homolog of BRL1) have ability to bind to BR and also responsible for vascular development has (Ceserani et al., 2009). Although BRL2 with BRL1 showed high sequence identity, they also conserved Arg588 and Gly690 residues that are responsible for higher affinity of BRL1 for BL compared to BRL2 (Saini et al., 2015). She et al. (2013) reported that the occurrence of severely negatively charged residues located at the inner side of the binding pocket in BRL2 result in the alteration in its hydrophobicity to BRs.

BRI1 homodimer remains in inactive form through interaction with inhibitory protein BRI1 kinase inhibitor 1 (BKI1) as well as by auto-inhibitory function of its cytoplasmic kinase domain in the absence of BR (Wang and Chory, 2006; Jaillais et al., 2011). Simultaneously, active BR insensitive 2 (BIN2), a GSK3/Shaggy-like kinase, phosphorylates BES1/BZR2 family transcription factors and hinders their activity by protein degradation, reduced DNA binding and cytoplasmic retention by 14–3–3 proteins (Ye et al., 2010; Hao et al., 2013). BRI1 heteroligomerise by its co-receptor BRI1-associated receptor kinasel (BAK1), which is also known as somatic embryogenesis receptor-like kinase 3 (SERK3) are responsible for full activation of BR signaling (Nam and Li, 2002). Binding of BR to the 70-amino acids island domain of BRI1 activates a change in the receptor and results in the autophosphorylation of BRI1 kinase domain activation loop. It induces a partial kinase activity of BRI1 leading to the transphosphorylation at Tyr211 residue of BKI1, causing its dissociation from the membrane and thus allowing BRI1 to interact with BAK1 (Wang et al., 2005a, 2005b; Wang and Chory, 2006). Consequently, BR induces the conformational alterations in its receptor that are essential for BRI1-BAK1 interaction (Santiago et al., 2013; Sun et al., 2013). BKI1 promotes BRI1 signaling by binding to 14–3–3 protein and blocking their negative functions during BR signaling (Wang et al., 2011).

Using forster resonance energy transfer (FRET) and fluorescence lifetime imaging microscopy (FLIM) methods, it has been shown that significant amounts of BRI1-BAK1 hetero-oligomers are present at the cell surface even without BR (Bücherl et al., 2013). These authors assumed that preassembled BRI1-BAK1 after conjugation with ligand might undergo discrete rearrangements of their respective intracellular kinase domains, which is similar to signaling paradigms as observed in receptor tyrosine kinases in animals (RTKs) that undertake ligand-independent receptor dimerization followed by dimer reorganization upon ligand binding (Lemmon and Schlessinger, 2010; Bücherl et al., 2013). Activated BRI1 starts a cascade of phosphorylation events of its downstream plasma membrane bound receptors, such as cytoplasmic kinases (RLCKs), BR signaling kinases (BSKs) and constitutive differential growth 1 (CDG1) in order to transfer the signal from membrane receptors to cytoplasmic regulators of BR signaling (Tang et al., 2008; Kim et al., 2011; Sreeramulu et al., 2013).

BRI1 phosphorylates CGD1 and BSK1 at Ser-234 and Ser-230, subsequently, phosphorylate and activate BRI1-suppressorl (BSU1) as shown by mass-spectrometric analysis (Saini et al., 2015). BRI1 can activate BSU1 through BSK1 or CDG1, and BSK1-induced activation requires BRI1 kinase activity, while CDG1-induced activation does not require BRI1 but is enhanced by BRI1 (Saini et al., 2015). Therefore, either BSU1-CDG1 or BSUI-BSK is required for transducing the signal from the receptor kinase BRI1 to the GSK3-like kinase BIN2 (Kim et al., 2011). BSU1 formerly dephosphorylates GSK3 like kinase BR insensitive 2 (BIN2) to hinder its role and discharges the inhibitory effects on two master transcription factors (BZR1 and BZR2) of BR signaling (Wang et al., 2002; Yin et al., 2002; He et al., 2005; Kim and Wang, 2010). Finally, BZR1 and BES1 are dephosphorylated by protein phosphatase 2A (PP2A) and released from 14–3–3 proteins, which resulted in their nuclear localization to bind to the promoter of their target genes in order to regulate their gene expression (Sun et al., 2010; Tang et al., 2011; Yu et al., 2011; Wang et al., 2012b)

18.5 PHYSIOLOGICAL ROLES OF BRASSINOSTEROIDS

18.5.1 Growth Stimulation

BRs stimulate both cell division and cell elongation, and therefore, they are able to induce cell swelling, curvature and splitting of the second internode (Rao et al., 2002). By feeding BR-deficient mutants with BRs, such as cpd and dwf4, it was possible to restore their dwarf phenotype trait to wild-type in Arabidopsis (Szekeres et al., 1996; Azpiroz et al., 1998). Similarly, BR-deficient dwarf1 (brd1) mutant exhibits abnormal organization and polar elongation of leaf and stem cells and finally abnormal growth of organs in rice plant, which have deficiencies of BR C-6 oxidase (Hong et al., 2002). Genetic and molecular studies revealed that the BR-promoted cell elongation mostly depends on the expression of xyloglucan endotransglycosylases (XETs), which play an important role in integrating new xyloglucan into the growing cell wall (Yang et al., 2011a). Both BRU1 and TCH4 encode XET proteins that are highly and rapidly induced by BR treatment during early stages of elongation in soybean and Arabidopsis, respectively (Zurek and Clouse, 1994; Xu et al., 1995). Yang et al. (2011b) reported that BR-activated transcription factor BES1 could directly bind to the promoter regions of all cellulose synthase genes in Arabidopsis to enhance their expression and biomass accumulation, thereby making a significant contribution to cell elongation. It has been shown that 0.01 mg of brassinolide is sufficient to enhance elongation, curvature and splitting (Maeda, 1965) and is very effective in stimulating growth in young vegetative tissues (Sasse, 1991). Brassinosteroids promote epicotyl and hypocotyl elongation in several species (Mandava, 1988; Mayumi and Shibaoka, 1995; Clouse et al., 1996). It is reported that BRs regulate the expression of a number of genes functioning in auxin responses, especially auxin transport system in both dark and light-grown seedlings during hypocotyl elongation in Arabidopsis (Nemhauser et al., 2004). BRs can stimulate hypocotyl elongation in *Brassica chinensis* through increasing cell-wall relaxation (Wang et al., 2010). Besides this, they can also elongate peduncles of dicots as well as coleoptiles and mesocotyls of monocots (Mandava, 1988; Sasse, 1991; Clouse, 1996). Application of 24-epibrassinolide was found effective for enhancing root growth in *Pinus radiate* and homobrassinolide application increased the root biomass in sugar beet (Sasse, 1994; Schilling et al., 1991). Similarly, the use of brassinolide activates cell division in leaf protoplasts of Petunia hybrid and Chinese cabbage (Nakajima et al., 1996; Oh et al., 1998).

BRs promote shoot development in a number of plant species by promoting expression of genes that are responsible for cell elongation and wall extensibility, such as Arabidopsis, soybean, mung bean, azuki bean, pea, rice and tomato (Mandava, 1988; Zurek and Clouse, 1994; Horvath et al., 2003). Differentiation of vascular tissue can also be induced by applying BL exogenously in *Helianthus tuberosus* and *Zinnia elegans* (Clouse and Zurek, 1991; Fukuda, 1997).

18.5.2 Root Elongation and Development

Application of BRs at low concentrations was largely independent of auxin, and they do not alter the auxin transport inhibitor, 2,3,5-triidobenzoic acid (Müssig et al., 2003). Mouchel et al. (2006) reported that BRX encodes a plant-specific protein that is expected to regulate transcription. BRs and auxin might act synergistically to accelerate lateral root formation, which may be partially mediated by the patatin-related phospholipases A (Rietz et al., 2010).

18.5.3 Reproductive Development

Pollen is a rich source of endogenous BRs, and it has been shown that application of exogenous BL can promote pollen tube elongation (Grove et al., 1979; Hewitt et al., 1985). BR-deficient and BR-sensitive mutants, such as cpd, dwf4 and bri1, are male sterile or with significantly impaired male fertility (Szekeres et al., 1996; Kim et al., 2005). Previous studies have shown that exogenous application of BR and overexpression of BR synthetic genes can increase fruit size in cucumber and tomato, indicating that fruit development is also positively regulated by BRs (Montoya et al., 2005; Fu et al., 2008).

18.5.4 Flowering

The expression period of circadian rhythms of CCR2, CAB2 and CCA1 that control flowering time was altered by BRs (Hanano et al., 2006). Foliar application of brassinosteroids has been found effective in increasing the number of flowers in strawberry (Pipattanawong et al., 1996). However, application of brassinosteroids in autumn increased the number of flowers in grapes but reduced the flower number if applied in late winter (Yoshiok et al., 1990), indicating that external environmental factors such as temperature might control BR biosynthesis during the winter as reported (Sharma et al., 2013).

18.5.5 Germination

The application of brassinosteroids enhances seed germination in several species (Sasse et al., 1995; Jones-Held et al., 1996). BRs promote endosperm rupture and counteract the inhibitory effect of ABA-induced dormancy, leading to enhancement in germination (Steber and McCourt, 2001; Finch-Savage and Leubner-Metzger, 2006).

18.5.6 Senescence

Brassinolide enhanced senescence in Xanthium, Rumex, Cucumber and Mung bean (Mandava et al., 1981; He et al., 1996). BRs have been reported to regulate senescence via activated oxygen (ROS), since the actions of peroxidase, superoxide dismutase and catalase were altered by BR action, and malondialdehyde was significantly increased by BR treatment (Ding and Zhao, 1995).

18.5.7 Stress Tolerance

BRs play an important role in stress mitigation in plants that were exposed to various abiotic stresses, such as high temperature (Kurepin et al., 2008; Janeczko et al., 2011), chilling (Liu et al., 2011; Wang et al., 2014), salinity (Abbas et al., 2013), light (Wang et al., 2012a; Li et al., 2012a), drought (Li et al., 2012b; Mahesh et al., 2013), flooding (Lu et al., 2006; Liang and Liang, 2009), metals/metalloids (Arora et al., 2010a, 2010b; Ashraf et al., 2010; Bajguz, 2010), herbicides (Sharma et al., 2013), pesticides (Xia et al., 2006), insecticides (Xia et al., 2009b, 2011) and organic pollutants (Ahammed et al., 2012, 2013). BRs increase plant resistance to low temperature by maintaining chlorophyll contents and promoting plant growth, whereas at high temperature, they induce resistance by maintaining protein synthesis (Kulaeva et al., 1991).

REFERENCES

Abbas S, Latif H H and Elsherbiny E A. 2013. Effect of 24-epibrassinolide on the physiological and genetic changes on two varieties of pepper under salt stress conditions. *Pakistan Journal of Botany* **45**: 1273–1284.

Ahammed G J, Gao C J, Ogweno J O, Zhou Y H, Xia X J and Mao W H. 2012. Brassinosteroids induce plant tolerance against phenanthrene by enhancing degradation and detoxification in *Solanum lycopersicum* L. *Ecotoxicology and Environmental Safety* **80**: 28–36.

Ahammed G J, Ruan Y P, Zhou J, Xia X J, Shi K and Zhou Y H. 2013. Brassinosteroid alleviates polychlorinated biphenyls-induced oxidative stress by enhancing antioxidant enzymes activity in tomato. *Chemosphere* **90**: 2645–2653.

Arora P, Bhardwaj R and Kanwar M K. 2010a. 24-Epibrassinolide induced antioxidative defense system of *Brassica juncea* L. under Zn metal stress. *Physiology and Molecular Biology of Plants* **16**: 285–293.

Arora P, Bhardwaj R and Kanwar M K. 2010b. 24-Epibrassinoliide regulated diminution of Cr metal toxicity in *Brassica juncea* L. plants. *Brazilian Journal of Plant Physiology* **22**: 159–165.

Ashraf M, Akram N A, Arteca R N and Foolad M R. 2010. The physiological, biochemical and molecular roles of brassinosteroids and salicylic acid in plant processes and salt tolerance. *Critical Review in Plant Science* **29**: 162–190.

Azpiroz R, Wu Y, LoCascio J C and Feldmann K A. 1998. An Arabidopsis brassinosteroid-dependent mutant is blocked in cell elongation. *The Plant Cell* **10**: 219–230.

Bajguz A. 2007. Metabolism of brassinosteroids in plants. *Plant Physiology and Biochemistry* **45**: 95–107.

Bajguz A. 2009. Isolation and characterization of brassinosteroids from algal cultures of *Chlorella vulgaris* Beijerinck (Trebouxiophyceae). *Journal of Plant Physiology* **166**: 1946–1949.

Bajguz A. 2010. An enhancing effect of exogenous brassinolide on the growth and antioxidant activity in *Chlorella vulgaris* cultures under heavy metals stress. *Environmental and Experimental Botany* **68**: 175–179.

Bajguz A and Tretyn A. 2003. The chemical characteristic and distribution of brassinosteroids in plants. *Phytochemistry* **62**: 1027–1046.

Bancos S, Nomura T, Sato T, Molnar G, Bishop G J, Koncz C, Yokota T, Nagy F and Szekeres M. 2002. Regulation of transcript levels of the Arabidopsis cytochrome P450 genes involved in brassinosteroid biosynthesis. *Plant Physiology* **130:** 504–513.

Belkhadir Y and Jaillais Y. 2015. The molecular circuitry of brassinosteroid signaling. *New Phytologist* **206**: 522–540.

Bücherl C A, van Esse G W, Kruis A, Luchtenberg J, Westphal A H, Aker J and de Vries S C. 2013. Visualization of BRI1 and BAK1 (SERK3) membrane receptor heterooligomers during brassinosteroid signaling. *Plant Physiology* **162**: 1911–1925.

Caño-Delgado A, Yin Y, Yu C, Vafeados D, Mora-García S, Cheng J C and Chory J. 2004. BRL1 and BRL3 are novel brassinosteroid receptors that function in vascular differentiation in Arabidopsis. *Development* **131**: 5341–5351.

Ceserani T, Trofka A, Gandotra N and Nelson T. 2009. VH1/BRL2 receptor-like kinase interacts with vascular-specific adaptor proteins VIT and VIK to influence leaf venation. *The Plant Journal* **57**: 1000–1014.

Choe S. 2004. Brassinosteroid biosynthesis and metabolism. pp. 156–178. *Plant Hormones: Biosynthesis, Signal Transduction, Action* (Eds.) Davies P J. Kluwer Academic Publishers, Dordrecht, The Netherlands.

Choe S. 2006. Brassinosteroid biosynthesis and inactivation. *Physiologia Plantarum* **126**: 539–548.

Choe S, Fujioka S, Noguchi T, Takatsuto S, Yoshida S and Feldmann K A. 2001. Overexpression of DWARF4 in the brassinosteroid biosynthetic pathway results in increased vegetative growth and seed yield in Arabidopsis. *The Plant Journal* **26**: 573–582.

Clouse S D. 1996. Molecular genetic studies confirm the role of brassinosteroids in plant growth and development. *The Plant Journal* **10**: 1–8.

Clouse S D, Langford M and McMorris T C. 1996. A brassinosteroid-insensitive mutant in Arabidopsis thaliana exhibits multiple defects in growth and development. *Plant Physiology* **111**: 671–678.

Clouse S D and Sasse J M. 1998. Brassinosteroids: essential regulators of plant growth and development. *Annual Review of Plant Biology* **49**: 427–451.

Clouse S D and Zurek D. 1991. Molecular analysis of brassinolide action in plant growth and development. pp 122–140. *Brassinosteroids: Chemistry, Bioactivity and Applications* (Eds.) Cutler HG, Yokota T and Adam G. American Chemical Society, Washington, DC.

Ding W M and Zhao Y J. 1995. Effect of epi-BR on activity of peroxidase and soluble protein content of cucumber cotyledon. *Acta Phytophysiologica Sinica* **21**: 259–264.

Fàbregas N and Caño-Delgado A I. 2014. Turning on the microscope turret: a new view for the study of brassinosteroid signaling in plant development. *Physiologia Plantarum* **151**: 172–183.

Finch-Savage W E and Leubner-Metzger G. 2006. Seed dormancy and the control of germination. *New Phytologist* **171**: 501–523.

Fu L, Xu B T, Xu X R, Gan R Y, Zhang Y, Xia E Q and Li H B. 2008. Antioxidant capacities and total phenolic contents of 62 fruits. *Food Chemistry* **129**: 345–350.

Fujioka S, Takatsuto S and Yoshida S. 2002. An early C-22 oxidation branch in the brassinosteroid biosynthetic pathway. *Plant Physiology* **130**: 930–939.

Fujioka S and Yokota T. 2003. Biosynthesis and metabolism of brassinosteroids. *Annual Review of Plant Biology* **54**: 137–164.

Fukuda H. 1997. Tracheary element differentiation. *The Plant Cell* **9**: 1147.

Grove M D, Spencer G F, Rohwedder W K, Mandava N, Worley J F, Warthen J D and Cook J C. 1979. Brassinolide, a plant growth-promoting steroid isolated from Brassica napus pollen. *Nature* **81**: 216–217.

Hanano S, Domagalska M A, Nagy F and Davis S J. 2006. Multiple phytohormones influence distinct parameters of the plant circadian clock. *Genes to Cells* **11**: 1381–1392.

Hao J, Yin Y and Fei S Z. 2013. Brassinosteroid signaling network: implications on yield and stress tolerance. *Plant Cell Reports* **32**: 1017–1030.

He C J, Morgan P W and Drew M C. 1996. Transduction of an ethylene signal is required for cell death and lysis in the root cortex of maize during aerenchyma formation induced by hypoxia. *Plant Physiology* **112**: 463–472.

He J-X, Gendron J M, Sun Y, Gampala S S, Gendron N, Sun C Q and Wang Z-Y. 2005. BZR1 is a transcriptional repressor with dual roles in brassinosteroid homeostasis and growth responses. *Science* **307**: 1634–1638.

Hewitt F R, Hough T, O'Neill P, Sasse J M, Williams E G and Rowan K S. 1985. Effect of brassinolide and other growth regulators on the germination and growth of pollen tubes of *Prunus avium* using a multiple hanging-drop assay. *Australian Journal of Plant Physiology* **12**: 201–211.

Hong Z, Ueguchi-Tanaka M, Shimizu-Sato S, Inukai Y, Fujioka S, Shimada Y and Uozu S. 2002. Loss-of-function of a rice brassinosteroid biosynthetic enzyme, C-6 oxidase, prevents the organized arrangement and polar elongation of cells in the leaves and stem. *The Plant Journal* **32**: 495–508.

Horvath D P, Anderson J V, Chao W S and Foley M E. 2003. Knowing when to grow: signals regulating bud dormancy. *Trends in Plant Science* **8**: 534–540.

Jaillais Y, Hothorn M, Belkhadir Y, Dabi T, Nimchuk Z L, Meyerowitz E M and Chory J. 2011. Tyrosine phosphorylation controls brassinosteroid receptor activation by triggering membrane release of its kinase inhibitor. *Genes & Development* **25**: 232–237.

Janeczko A, Oklestkova J, Pociecha E, Koscielniak J and Mirek M. 2011. Physiological effects and transport of 24-epibrassinolide in heat-stressed barley. *Acta Physiologia Plantarum* **33**: 1249–1259.

Jones-Held S, VanDoren M and Lockwood T. 1996. Brassinolide application to Lepidium sativum seeds and the effects on seedling growth. *Journal of Plant Growth Regulation* **15:** 63–67.

Khripach V A, Zhabinskii V N and Malevannaya N N. 1999. Recent advances in brassinosteroids study and application. *ChemInform* **30**: 26–36.

Kim S K, Akihisa T, Tamura T, Matsumoto T, Yokota T and Takahashi N. 1988. 24-methylene-25-methylcholesterol in Phaseolus vulgaris seed: structural relation to brassinosteroids. *Phytochemistry* **27**: 629–631.

Kim T W, Guan S, Burlingame A L and Wang Z Y. 2011. The CDG1 kinase mediates brassinosteroid signal transduction from BRI1 receptor kinase to BSU1 phosphatase and GSK3-like kinase BIN2. *Molecular Cell* **43**: 561–571.

Kim T W, Hwang J Y, Kim Y S, Joo S H, Chang S C, Lee J S and Kim S K. 2005. Arabidopsis CYP85A2, a cytochrome P450, mediates the Baeyer-Villiger oxidation of castasterone to brassinolide in brassinosteroid biosynthesis. *The Plant Cell* **17**: 2397–2412.

Kim T W and Wang Z Y. 2010. Brassinosteroid signal transduction from receptor kinases to transcription factors. *Annual Review of Plant Biology* **61**: 681–704.

Kulaeva O N, Burkhanova E A, Fedina A B, Khokhlova V A, Bokebayeva G A, Vorbrodt H M and Adam G. 1991. Effect of brassinosteroids on protein synthesis and plant-cell ultrastructure under stress conditions. *ACS Symposium Series* **474**. https://doi.org/10.1021/bk-1991-0474.ch012.

Kurepin L V, Qaderi M M, Back T G, Reid D M and Pharis R P. 2008. A rapid effect of applied brassinolide on abscisic acid concentrations in *Brassica napus* leaf tissue subjected to short-term heat stress. *Plant Growth Regulation* **55**: 165–167.

Lemmon M A and Schlessinger J. 2010. Cell signaling by receptor tyrosine kinases. *Cell* **141**: 1117–1134.

Li J and Chory J. 1999. Brassinosteroid actions in plants. *Journal of Experimental Botany* **50**: 275–282.

Li Y H, Liu Y J, Xu X L, Jin M, An L Z and Zhang H. 2012b. Effect of 24-epibrassinolide on drought stress-induced changes in *Chorispora bungeana*. *Biologia Plantarum* **56**: 192–196.

Li Z Y, Xu Z S, He G Y, Yang G X, Chen M and Li L C. 2012a. A mutation in *Arabidopsis* BSK5 encoding a brassinosteroid-signaling kinase protein affects responses to salinity and abscisic acid. *Biochemistry Biophysical Research Communications* **426**: 522–527.

Liang J and Liang Y. 2009. Effects of plant growth substances on water-logging resistance of oilseed rape seedling. *Xinan Shifan Daxue Xuebao, Ziran Kexueban* **34**: 58–62.

Liu Y J, Jiang H F, Zhao Z G and An L Z. 2011. Abscisic acid is involved in brassinosteroids-induced chilling tolerance in the suspension cultured cells from *Chorispora bungeana*. *Journal of Plant Physiology* **168**: 853–862.

Lu X, Chen Y, Gong W and Chen Y. 2006. Effect of brassinolide on the seedling growth and waterlogging resistance of soybean. *Chinese Agricultural Science Bulletin* **23**: 37–38.

Maeda E. 1965. Rate of lamina inclination in excised rice leaves. *Physiologia Plantarum* **18**: 813–827.

Maharjan P M and Choe S. 2011. High temperature stimulates DWARF4 (DWF4) expression to increase hypocotyl elongation in Arabidopsis. *Journal of Plant Biology* **54**: 425.

Mahesh B, Parshavaneni B, Ramakrishna B and Rao S S R. 2013. Effect of brassinosteroids on germination and seedling growth of radish (*Raphanus sativus* L.) under PEG-6000 induced water stress. *American Journal of Plant Science* **4**: 2305–2313.

Mandava N B. 1988. Plant growth-promoting brassinosteroids. *Annual Review of Plant Physiology and Plant Molecular Biology* **39**: 23–52.

Mandava N B, Sasse J M and Yopp J H. 1981. Brassinolide, a growth-promoting steroidal lactone. *Physiologia Plantarum* **53**: 453–461.

Marsolais F, Boyd J, Paredes Y, Schinas A M, Garcia M, Elzein S and Varin L. 2007. Molecular and biochemical characterization of two brassinosteroid sulfotransferases from Arabidopsis, AtST4a (At2g14920) and AtST1 (At2g03760). *Planta* **225**: 1233.

Mayumi K and Shibaoka H. 1995. A possible double role for brassinolide in the reorientation of cortical microtubules in the epidermal cells of azuki bean epicotyls. *Plant and Cell Physiology* **36**: 173–181.

Montoya T, Nomura T, Yokota T, Farrar K, Harrison K, Jones J G and Bishop G J. 2005. Patterns of Dwarf expression and brassinosteroid accumulation in tomato reveal the importance of brassinosteroid synthesis during fruit development. *The Plant Journal* **42**: 262–269.

Mouchel C F, Osmont K S and Hardtke C S. 2006. BRX mediates feedback between brassinosteroid levels and auxin signalling in root growth. *Nature* **443**: 458–461.

Müssig C, Shin G H and Altmann T. 2003. Brassinosteroids promote root growth in Arabidopsis. *Plant Physiology* **133**: 1261–1271.

Nakajima K, Yamanaka Y, Nakae K, Kojima H, Ichiba M, Kiuchi N and Hirano T. 1996. A central role for Stat3 in IL-6-induced regulation of growth and differentiation in M1 leukemia cells. *The EMBO Journal* **15**: 3651.

Nam K H and Li J. 2002. BRI1/BAK1, a receptor kinase pair mediating brassinosteroid signaling. *Cell* **110**: 203–212.

Nemhauser J L, Mockler T C and Chory J. 2004. Interdependency of brassinosteroid and auxin signaling in Arabidopsis. *PLoS Biol* **2**: e258.

Nomura T and Bishop G J. 2006. Cytochrome P450s in plant steroid hormone synthesis and metabolism. *Phytochemistry Reviews* **5**: 421–432.

Nomura T, Ueno M, Yamada Y, Takatsuto S, Takeuchi Y and Yokota T. 2007. Roles of brassinosteroids and related mRNAs in pea seed growth and germination. *Plant Physiology* **143**: 1680–1688.

Oh P, McIntosh D P and Schnitzer J E. 1998. Dynamin at the neck of caveolae mediates their budding to form transport vesicles by GTP-driven fission from the plasma membrane of endothelium. *The Journal of Cell Biology* **141**: 101–114.

Piotrowska A and Bajguz A. 2011. Conjugates of abscisic acid, brassinosteroids, ethylene, gibberellins, and jasmonates. *Phytochemistry* **72**: 2097–2112.

Pipattanawong N, Fujishige N, Yamane K and Ogata R. 1996. Effects of brassinosteroid on vegetative and reproductive growth in two day-neutral strawberries. *Journal of the Japanese Society for Horticultural Science* **65**: 651–654.

Poppenberger B, Fujioka S, Soeno K, George G L, Vaistij F E, Hiranuma S and Bowles D. 2005. The UGT73C5 of *Arabidopsis thaliana* glucosylates brassinosteroids. *Proceedings of the National Academy of Sciences, United States of America* **102**: 15253–15258.

Rao S S R, Vardhini B V, Sujatha E and Anuradha S. 2002. Brassinosteroids-A new class of phytohormones. *Current Science* **82**: 1239–1245.

Rietz S, Dermendjiev G, Oppermann E, Tafesse F G, Effendi Y, Holk A, Parker J E, Teige M and Scherer G F. 2010. Roles of Arabidopsis patatin-related phospholipases a in root development are related to auxin responses and phosphate deficiency. *Molecular Plant* **3**: 524–538.

Rouleau M, Marsolais F, Richard M, Nicolle L, Voigt B, Adam G and Varin L. 1999. Inactivation of brassinosteroid biological activity by a salicylate-inducible steroid sulfotransferase from Brassica napus. *Journal of Biological Chemistry* **274**: 20925–20930.

Saini S, Sharma I and Pati P K. 2015. Versatile roles of brassinosteroid in plants in the context of its homoeostasis, signaling and crosstalks. *Frontiers in Plant Science* **6**: 950.

Santiago J, Henzler C and Hothorn M. 2013. Molecular mechanism for plant steroid receptor activation by somatic embryogenesis co-receptor kinases. *Science* **341**: 889–892.

Sasse J M. 1991. The case for brassinosteroids as endogenous plant hormones. *American Chemical Society* **474**: 158–166.

Sasse J M. 1994. Brassinosteroids and roots. pp. 228–232. *Proceedings-Plant Growth Regulator Society of America-Annual Meeting*. Volume 21. Plant Growth Regulator Society of America, Alexandria.

Sasse J M, Smith R and Hudson I. 1995. Effect of 24-epibrassinolide on germination of seeds of *Eucalyptus camaldulensis* in saline conditions. *Plant Growth Regulator Society of America* **22**: 136–141.

Schilling G, Schiller C and Otto S. 1991. Influence of brassinosteroids on organ relations and enzyme activities of sugar-beet plants. *American Chemical Society* **474**: Chapter 18.

Schneider K, Breuer C, Kawamura A, Jikumaru Y, Hanada A, Fujioka S and Yamaguchi S. 2012. Arabidopsis PIZZA has the capacity to acylate brassinosteroids. *PLOS ONE* **7**: e46805.

Sharma I, Bhardwaj R and Pati P K. 2013. Stress modulation response of 24-epibrassinolide against imidacloprid in an elite indica rice variety Pusa Basmati-1. *Pesticide Biochemistry Physiology* **105**: 144–153.

She J, Han Z, Zhou B and Chai J. 2013. Structural basis for differential recognition of brassinolide by its receptors. *Protein & Cell* **4**(6): 475–482.

Shimada Y, Goda H, Nakamura A, Takatsuto S, Fujioka S and Yoshida S. 2003. Organ-specific expression of brassinosteroid-biosynthetic genes and distribution of endogenous brassinosteroids in Arabidopsis. *Plant Physiology* **131**: 287–297.

Sirhindi G. 2013. Brassinosteroids: biosynthesis and role in growth, development, and thermotolerance responses. pp. 309–329. *Molecular Stress Physiology of Plants* (Eds.) Rout G R and Das A B. Springer, Delhi, India.

Sreeramulu S, Mostizky Y, Sunitha S, Shani E, Nahum H, Salomon D and Sessa G. 2013. BSKs are partially redundant positive regulators of brassinosteroid signaling in Arabidopsis. *The Plant Journal* **74**: 905–919.

Steber C M and McCourt P. 2001. A role for brassinosteroids in germination in Arabidopsis. *Plant Physiology* **125**(2): 763–769.

Sun Y, Fan X Y, Cao D M, Tang W, He K, Zhu J Y and Patil S. 2010. Integration of brassinosteroid signal transduction with the transcription network for plant growth regulation in Arabidopsis. *Developmental Cell* **19**: 765–777.

Sun Y, Han Z, Tang J, Hu Z, Chai C, Zhou B and Chai J. 2013. Structure reveals that BAK1 as a co-receptor recognizes the BRI1-bound brassinolide. *Cell Research*. https://doi.org/10.1038/cr.2013.131.

Symons G M, Davies C, Shavrukov Y, Dry I B, Reid J B and Thomas M R. 2006. Grapes on steroids. Brassinosteroids are involved in grape berry ripening. *Plant Physiol*ogy **140**: 150–158.

Symons G M and Reid J B. 2004. Brassinosteroids do not undergo long-distance transport in pea. Implications for the regulation of endogenous brassinosteroid levels. *Plant Physiology* **135**: 2196–2206.

Symons G M, Ross J J, Jager C E and Reid J B. 2007. Brassinosteroid transport. *Journal of Experimental Botany* **59**: 17–24.

Symons G M, Smith J J, Nomura T, Davies N W, Yokota T and Reid J B. 2008. The hormonal regulation of de-etiolation. *Planta* **227**: 1115–1125.

Szekeres M, Nemeth K, Kalman Z, Mathur J, Kauschmann A, Altmann T, Redei G P, Nagy F, Schell J and Koncz C. 1996. Brassinosteroids rescue the deficiency of CYP90, a cytochrome P450, controlling cell elongation and de-etiolation in *Arabidopsis*. *Cell* **85**: 171–182.

Tanaka K, Asami T, Yoshida S, Nakamura Y, Matsuo T and Okamoto S. 2005. Brassinosteroid homeostasis in Arabidopsis is ensured by feedback expressions of multiple genes involved in its metabolism. *Plant Physiology* **138**: 1117–1125.

Tang W, Kim T W, Oses-Prieto J A, Sun Y, Deng Z, Zhu S, Wang R, Burlingame A L and Wang Z Y. 2008. BSKs mediate signal transduction from the receptor kinase BRI1 in Arabidopsis. *Science* **321:** 557–560.

Tang W, Yuan M, Wang R, Yang Y, Wang C, Oses-Prieto J A and Gendron J M. 2011. PP2A activates brassinosteroid-responsive gene expression and plant growth by dephosphorylating BZR1. *Nature Cell Biology* **13**: 124–131.

Vardhini B V, Anuradha S and Rao S S R. 2006. Brassinosteroids – a great potential to improve crop productivity. *Indian Journal of Plant Physiology* **11**: 1–12.

Vardhini B V and Rao S S R. 2003a. Amelioration of water stress by brassinosteroids on seed germination and seedling growth of three varieties of sorghum. *Plant Growth Regulation* **41**: 25–31.

Vardhini B V and Rao S S R. 2003b. Influence of brassinosteroids on growth, yield, metabolite content and enzyme activities of tomato (*Lycopersicon esculentum* Mill.). *Proceedings of the National Academy of Sciences, India Section B: Biological Sciences* **73**: 307–315.

Wang D, Shi Q, Wang X, Wei M, Hu J, Liu J and Yang F. 2010. Influence of cow manure vermicompost on the growth, metabolite contents, and antioxidant activities of Chinese cabbage (*Brassica campestris* ssp. chinensis). *Biology and Fertility of Soils* **46**: 689–696.

Wang H, Yang C, Zhang C, Wang N, Lu D, Wang J and Wang X. 2011. Dual role of BKI1 and 14–3–3s in brassinosteroid signaling to link receptor with transcription factors. *Developmental Cell* **21**: 825–834.

Wang Q, Ding T, Gao L, Pang J and Yang N. 2012a. Effect of brassinolide on chilling injury of green bell pepper in storage. *Scientia Horticulturae* **144**: 195–200.

Wang X H and Chory J. 2006. Brassinosteroids regulate dissociation of BKI1, a negative regulator of BRI1 signaling, from the plasma membrane. *Science* **313**: 1118–1122.

Wang X H, Goshe M B, Soderblom E J, Phinney B S, Kuchar J A, Li J and Clouse S D. 2005a. Identification and functional analysis of in vivo phosphorylation sites of the Arabidopsis Brassinosteroid-Insensitive1 receptor kinase. *The Plant Cell* **17**: 1685–1703.

Wang X H, Li X, Meisenhelder J, Hunter T, Yoshida S, Asami T and Chory J. 2005b. Autoregulation and homodimerization are involved in the activation of the plant steroid receptor BRI1. *Developmental Cell* **8**: 855–865.

Wang X H, Shu C, Li H Y, Hu X Q and Wang Y X. 2014. Effects of 0.01% brassinolide solution application on yield of rice and its resistance to autumn low-temperature damage. *Acta Agriculturae Jiangxi* **26**: 36–38.

Wang Z Y, Bai M Y, Oh E and Zhu J Y. 2012b. Brassinosteroid signaling network and regulation of photomorphogenesis. *Annual Review of Genetics* **46**: 701–724.

Wang Z Y, Nakano T, Gendron J, He J, Chen M, Vafeados D and Chory J. 2002. Nuclear-localized BZR1 mediates brassinosteroid-induced growth and feedback suppression of brassinosteroid biosynthesis. *Developmental Cell* **2**: 505–513.

Xia X H, Huang Y Y, Wang L, Huang L F, Yu Y L and Zhou Y H. 2006. Pesticides-induced depression of photosynthesis was alleviated by 24-epibrassinolide pretreatment in *Cucumis sativus* L. *Pesticide Biochemistry and Physiology* **86**: 42–48..

Xia X J, Wang Y J, Zhau Y H, Tao Y, Mao W H and Shi K. 2009b. Reactive oxygen species are involved in brassinosteroid-induced stress tolerance in cucumber. *Plant Physiology* **150**: 801–814.

Xia X J, Zhou Y H, Ding J, Shi K, Asami T and Chen Z. 2011. Induction of systemic stress tolerance by brassinosteroid in *Cucumis sativus*. *New Phytologist* **191**: 706–720.

Xu W, Purugganan M M, Polisensky D H, Antosiewicz D M, Fry S C and Braam J. 1995. Arabidopsis TCH4, regulated by hormones and the environment, encodes a xyloglucan endotransglycosylase. *The Plant Cell* **7**: 1555–1567.

Yang C J, Lee S H, Goddard M E and Visscher P M. 2011a. GCTA: a tool for genome-wide complex trait analysis. *The American Journal of Human Genetics* **88**: 76–82.

Yang C J, Zhang C, Lu Y N, Jin J Q and Wang X L. 2011b. The mechanisms of brassinosteroids' action: from signal transduction to plant development. *Molecular Plant* **4**: 588–600.

Ye Q, Zhu W, Li L, Zhang S, Yin Y, Ma H and Wang X. 2010. Brassinosteroids control male fertility by regulating the expression of key genes involved in Arabidopsis anther and pollen development. *Proceedings of the National Academy of Sciences* **107**: 6100–6105.

Yin Y, Wang Z Y, Mora-Garcia S, Li J, Yoshida S, Asami T and Chory J. 2002. BES1 accumulates in the nucleus in response to brassinosteroids to regulate gene expression and promote stem elongation. *Cell* **109**: 181–191.

Yoshiok T, Nesumi H and Ito Y. 1990. Pleiotropic influences of brassinoides in fruit crops. *Plant Growth Regulation* **59**: 44–45.

Yu X, Li L, Zola J, Aluru M, Ye H, Foudree A, Guo H, Anderson S, Aluru S, Liu P, Rodermel S and Yin Y. 2011. A brassinosteroid transcriptional network revealed by genome-wide identification of BESI target genes in *Arabidopsis thaliana*. *Plant Journal* **65**: 634–646.

Yuan G F, Jia C G, Li Z, Sun B, Zhang L P, Liu N and Wang Q M. 2007. Effect of brassinosteroids on drought resistance and abscisic acid concentration in tomato under water stress. *Scientia Horticulturae* **126**: 103–108.

Zhu Y, Zuo M, Liang Y, Jiang M, Zhang J, Scheller H V and Zhang A. 2013. MAP65–1a positively regulates H2O2 amplification and enhances brassinosteroid-induced antioxidant defence in maize. *Journal of Experimental Botany* **64**: 3787–3802.

Zurek D M and Clouse S D. 1994. Molecular cloning and characterization of a brassinosteroid-regulated gene from elongating soybean (*Glycine max* L.) epicotyls. *Plant Physiology* **104**: 161–170.

19 Jasmonates and Other Bioregulators

19.1 INTRODUCTION

Oxylipins, the oxidation products of unsaturated fatty acids, represent a highly diverse group of substances that are implicated in a number of developmental processes and stress responses in plants (Püssa et al., 2009). Jasmonate (JAs), such as jasmonic acid (JA) and its derivatives (methyl jasmonate [MeJA], cis-jasmone, jasmonoyl isoleucine [JA-Ile], jasmonoyl ACC [JA-ACC]) and several other metabolites are well studied examples of plant oxylipins. Jasmonic acid was first isolated from fungus cultures of *Lasiodiplodia theobromae* (Demole et al., 1962). It has been reported that JAs are involved in various plant development processes such as trichome development, vegetative growth, cell cycle regulation, senescence and anthocyanin biosynthesis regulation and also respond to various biotic and abiotic stresses (Creelman and Mullet, 1997; Wasternack, 2007; Howe and Jander, 2008; Browse, 2009; Avanci et al., 2010; Pauwels and Goossens, 2011). JAs have been found in all higher plants; however, its level in plant tissues varies among tissue types, development and external stimuli (Vick and Zimmerman, 1987; Hamberg and Gardner, 1992; Sembdner and Parthier, 1993).

19.2 JASMONATE BIOSYNTHESIS

JA's biosynthesis is initiated from polyunsaturated fatty acids (PUFA), which are synthesized by one of the seven distinct branches of the lipoxygenase (LOX) pathway, the allene oxide synthase (AOS) branch (Feussner and Wasternack, 2002). The remaining six branches form other oxylipins including GLVs as well as epoxy-, hydroxy-, keto- or ether PUFA and epoxyhydroxy-PUFA (Feussner and Wasternack, 2002). Only 13-hydroperoxide from α-linolenic acid (18:3, α-LeA) can be utilized by the AOS branch for JA production in the oxylipin biosynthesis (Yan et al., 2013). JAs and MeJA are produced from α-linolinic acid (α-LeA) that is localized in chloroplast membranes (Wasternack and Kombrink, 2010). Jasmonate synthesis is initiated by lipoxygenases (LOXs), which catalyzes the region and stereo selective di-oxygenation of polyunsaturated fatty acids (Blée, 2002; Feussner and Wasternack, 2002; Howe and Schilmiller, 2002; Schaller and Stintzi, 2009; Wasternack, 2007). Linoleic and linolenic acids are oxygenated by specific LOX's at C-9 or C-13 to result in the corresponding (9S)- or (13S)-hydroperoxy-octadecadi (tri)enoic acids that is fed into at least seven alternative pathways and subsequently form a large variety of oxylipins (Blée, 2002; Feussner and Wasternack, 2002). Formation of α-linolenic acid (α-LeA) from lipids occurs through the action of phospholipases, with an intermediate compound being 13-hydroperoxy-9,11,15-octadecatrienoic acid (13-HPOT), which is formed by the addition of oxygen molecule to α-LeA, and the reaction is stimulated by 13-lipoxygenase (LOX). Subsequently, 13-HPOT is oxidized to allene oxide by allene oxide synthase (AOS) and then converted into 12-oxo phytodienoic acid (12-OPDA) by the enzyme allene oxide cyclase (AOC) (Ahmad et al., 2016). Both AOS and AOC are reported to be present in plastids, and they act in concert (Gfeller et al., 2010; Wasternack and Hause, 2013). AOS, which belong to the cytochrome P450 family, is a potential catalyst for JA biosynthesis (Schaller and Stintzi, 2009). 12-OPDA is the final product of metabolic pathway of JA biosynthesis, which occurs in the chloroplast and undertakes three cycles of β-oxidation in the peroxisomes (Delker et al., 2006; Wasternack, 2007; Bussell et al., 2014). JA formation from 12-OPDA with different metabolic conversions forming distinct JAs and MeJA is formed by the methylation reaction in the presence of JA methyl transferase (Ahmad et al., 2016). AOC is encoded by either one gene (tomato) or multiple gene family in *Arabidopsis* (Wasternack and Hause, 2013).

 DOI: 10.1201/9781003373889-24

Jasmonates are synthesized from lipid elements and perceived by protein receptors that activate signal transduction pathway (Wasternack and Hause, 2013). The JA-deficient mutant *Coronatine-insensitive1* (*Coi1*) was identified from *Arabidopsis*, which is responsible for inhibition of root length (Kazan and Manners, 2008). These authors observed that when *Coi1* was treated with JA and coronatine, COR is secreted by *Pseudomonas syringae*. It is a phytotoxin, which is structurally and functionally similar to JAs, *Coi1* mutants showed reduction in root length. It has been reported that *Coi1* mutants are not effective in pest and pathogen resistance, wound response, secondary metabolite biosynthesis and fertility because these responses in plants are jasmonate-dependent (Ahmad et al., 2016). *jar1* (Jasmonate insensitive 1) and *jai1* (*Jasmonate resistant1*) mutants are reported to suppress effectiveness of externally applied JAs in *Arabidopsis*. *JAR1* encodes a JA amino acid synthetase and assists JA link to isoleucine thus, showing its role as a signaling molecule by suppressing root growth (Staswick, 2008; Wasternack and Kombrink, 2010). Jai1 is a sterile mutant of tomato and is resistant to JA, which showed loss of seed maturation control, glandular trichomes appearance on leaves, unripened fruits and sepals indicating JA-signaling pathway in glandular trichomes of tomato defense responses (Li et al., 2004).

The *Coi1* gene codes for a protein having 16 leucine-rich repeats and F-box motif (Browse, 2009; Avanci et al., 2010). F-box protein forms multi protein structures that serve as receptors (Avanci et al., 2010). *Coi1* participate in eliminating repressors of the JA transduction and these repressors are mediated by SCF complexes (Xie et al., 1998; Jiang et al., 2013; Wasternack and Hause, 2013). SCF complex is made up of SKP1, F-box and Cullin proteins, which controls cell cycle regulation by activating ubiquitination of the cell cycle proteins and their degradation by the 26S proteasome (Kazan and Manners, 2008). However, until now there is no evidence available regarding the repressor that can provide indications about the relationship between entire transcriptome and the SCF^{Coi1} complex regulated by JA (Browse, 2009). JAI3 (jasmonate insensitive3) protein that was included in JAZ protein family is a primary repressor of gene expression in the JAs signaling (Chini et al., 2007). JAI3 and other JAZ proteins have been reported to interact with Coi1 and application of JA degrades proteasome 26S due to the activity of Coi1 (Chung et al., 2010). Furthermore, JAI3/JAZ3 and JAZ1 are reported to act as repressors of the JA signaling (Robson et al., 2010). Additionally, SCF^{Coi1}-dependent ubiquitination is required for the activation of JA responsive genes (Chini et al., 2007; Kazan and Manners, 2008; Wager and Browse, 2012).

19.3 VEGETATIVE SINKS AND PROTEIN STORAGE

Since jasmonate levels are high in vegetative tissues, it acts as protein storage in plants (Creelman and Mullet, 1997). In soybean, jasmonate levels are higher in axes, plumules and the hypocotyl hook in the cell elongation zone as compared to the nonelongating portion of stems and roots (Creelman and Mullet, 1995). Theses authors have also reported that JA levels are higher in young growing leaves of soybean compared to the older fully expanded leaves. Developing reproductive structures, such as pods, has higher JA levels than the seeds. Jasmonate has been proposed to play a role in the formation of tubers, a special type of vegetative sink (Pelacho and Mingo-Castel, 1991; Koda et al., 1992). Later, it was discovered that genes, which encode vegetative storage proteins (VSPs), are regulated by jasmonate confirming its role in protein storage (Anderson, 1988; Staswick, 1994). Vegetative storage proteins (VSPs) were first described in soybean, which is accumulated in the vacuoles of paraveinal mesophyll and bundle sheath cells that surround veins in soybean leaves (Franceschi and Giaquinta, 1983; Wittenbach, 1983). Further study showed that VSPs only accumulate in pods and other developing reproductive structures but did not accumulate in seeds, suggesting that the VSPs represent temporary deposits of amino acids derived from disassembly of Rubicon and other leaf proteins that are being mobilized for seed development (Staswick, 1989).

19.4 PHYSIOLOGICAL ROLES OF JASMONATE

19.4.1 Seed Germination

Jasmonate stimulates the germination of dormant seeds and germination of dormant embryos in apple by increasing the alkaline lipase activity, which might promote the mobilization of lipid reserves to provide sugars needed for seedling growth (Ranjan and Lewak, 1992). Inhibitors of lipoxygenase inhibit embryo germination and JA partially reverse inhibitor action (Creelman and Mullet, 1997). Level of JA increase after imbibition and further decline with seedling development, which has been related to seed reserve mobilization and therefore, may be a consequence rather than a trigger of germination (Creelman and Mullet, 1995). Increase in JAs causes lipid peroxidation to damage membrane s and therefore, it enhances the germination (Ranjan and Lewak, 1992). Jasmonate-insensitive mutants (jin4 and jarl) have shown increased sensitivity to ABA inhibition of germination and therefore, it is assumed that JA may stimulate seed germination by decreasing sensitivity to ABA (Staswick et al., 1991; Berger et al., 1996).

19.4.2 Flower and Fruit Development

Jasmonate level is usually high in developing reproductive tissues, such as flowers, fruits and seeds (Creelman and Mullet, 1997). Jasmonate and related volatile fatty acid derivatives have been reported to be involved in insect attraction to facilitate pollen dispersal. It has been reported in tomato and apple that fruit ripening can be stimulated by jasmonate through the production of ethylene (Czapski et al., 1992). These authors speculated that jasmonate levels might gradually increase in developing fruits leading to enhanced ethylene synthesis and subsequent promotion of fruit ripening. *Arabidopsis thaliana* with AtVsp showed high expression in flowers and developing fruit whereas, AtVspv proteins were absent in flowers of *coil* (JA-insensitive mutants), which could be induced by treating the plants with JA (Benedetti et al., 1995; Berger et al., 1996). Similarly, fatty acid mutants of *A. thaliana* that lack LA and JA and the *coil* (JA-insensitive mutant) fail to produce viable pollen unless supplied with JA (McConn et al., 1997). Pollen development, stamen elongation and the pollen release timing are all affected by JAs (Liechti and Farmer, 2006). Exogenous spray of JAs significantly increases the production of floral nectar in *Brassica napus* (Radhika et al., 2010). JA also regulates nectar secretion in flowers as it causes secretion of extra floral nectar (Heil, 2015).

19.4.3 Nodulation

JAs have been shown to inhibit the nodule formation in *Medicago truncatula* and *Lotus japonicas* (Sun et al., 2006; Nakagawa and Kawaguchi, 2006).

19.4.4 Senescence

Involvement of JA in leaf senescence has been described by several researchers (Taylor and Whitelaw, 2001; Seltmann et al., 2010). *Arabidopsis* mutants, which have reduced levels of JAs, *such as allene oxide synthase (aos), oxophytodienoate-reductase 3* (*opr3*) and *coronatine insensitive 1 (coil)*, which is insensitive to JAs, display temporal shifts in the onset of natural and dark-induced senescence (He et al., 2002; Castillo and León, 2008; Schommer et al., 2008; Danisman et al., 2012). He et al. (2002) reported that exogenous application of JA on wild-type *Arabidopsis* promoted leaf senescence and induced the expression of several *SAGs*, including JA biosynthesis genes. Under natural conditions, a gradual increase in the transcript levels of JA biosynthesis genes such as AOS, *LOX1, LOX3, LOX4, OPR1* and *OPR3* was observed in senescing leaves (He et al., 2002). Exogenous application of MeJA or JA has been found to be effective in enhancing senescence in *Artemisia absinthium* or *Arabidopsis* (Ueda and Kato, 1980; He et al., 2002)

19.4.5 Role of Jasmonates in Abiotic Stress

Exogenous application of jasmonates enhances drought resistance by increasing the antioxidative capacity of plants under water stress (Bandurska et al., 2003). JAs play an important role in signaling drought-induced antioxidant responses including the ascorbate metabolism (Li et al., 2004). Therefore, its application protects plants from drought-induced oxidative damage by enhancing the activity of antioxidant enzymes (Nafie et al., 2011). Application of MeJA improves water stress tolerance by regulating the growth, proliferation rate, proline accumulation, chlorophyll levels, tissue water status, oxidative stress and membrane lipid peroxidation in *Musa species* (Mahmood et al., 2012). Accumulation of JAs during drought stress plays a positive regulatory role in stomatal closure (Gehring et al., 1997; Suhita et al., 2004).

Foliar application of MeJA has been found effective in alleviating salinity stress symptoms in soybean seedlings (Yoon et al., 2009). Similarly, exogenous application of jasmonate can rescue growth in salt-sensitive cell lines. The salt stress response is modulated by JA-signaling components, such as JAZ proteins in grapevine (Ismail et al., 2012). JA biosynthesis mutant (fad3/7/8) and JA-signaling mutant (jarl) have been shown a greater sensitivity to ozone (Rao et al., 2000). MeJA can alleviate chilling injury by inducing the production of cryo-protective agents, proteinase inhibitors, polyamines and ABA; as well as by lowering the activity of LOX's and antioxidants (González-Aguilar et al., 2000; Cao et al., 2009; Zhao et al., 2013). JA biosynthesis genes, such as *OsDAD1, OsLOX2, OsAOC, OsAOS1, OsAOS2, OsOPR1* and *OsOPR7*, and JA signaling genes, such as *OsJAR1, OsbHLH148* and *OsCOI1a*, showed upregulation after cold exposure in rice seedlings (Du et al., 2013).

19.4.6 Plant Defense against Pathogens

Jasmonates is a key regulator that plays a crucial role in plant defense responses to pathogens (Pieterse et al., 2012). Furthermore, JAs activate gene-encoding protease inhibitors that help protect plants from insect damage (Johnson et al., 1989). The expression of gene that encodes the antifungal proteins, such as thionin, osmotin, PDF and the ribosome-inactivating protein, RIP, are activated by JAs (Becker and Apel, 1992; Chaudhry et al., 1994; Xu et al., 1994; Howe et al., 1996; Penninckx et al., 1996). JA induces genes involved in the synthesis of phytoalexin (Chs, Paland HMGR) and phenolics (polyphenol oxidase), both of which are involved in plant defense systems (Creelman et al., 1992; Choi et al., 1994; Doares et al., 1995).

Application of JA has shown a reduced number of *Frankliniella occidentalis* (thrips), flea beetles and aphids, which were ascribed to an increase in the activities of polyphenol oxidase and proteinase inhibitors (Thaler et al., 2001). Potato plants treated with JA shown increased resistance against *Phytophthora infestans* (Choi et al., 1994). JAs also increase plant's resistance to root pests like fungus gnat (*Bradysia impatiens*) in *Arabidopsis* and spinach (McConn et al., 1997; Schmelz et al., 2002).

REFERENCES

Ahmad P, Rasool S, Gul A, Sheikh S A, Akram N A, Ashraf M and Gucel S. 2016. Jasmonates: multifunctional roles in stress tolerance. *Frontiers in Plant Science* **7**: 12–21.

Anderson J M. 1988. Jasmonic acid-dependent increases in the level of specific polypeptides in soybean suspension cultures and seedlings. *Journal of Plant Growth Regulation* **7**: 203–211.

Avanci N C, Luche D D, Goldman G H and Goldman M H S. 2010. Jasmonates are phytohormones with multiple functions, including plant defense and reproduction. *Genetics and Molecular Research* **9**: 484–505.

Bandurska H, Stroiński A and Kubiś J. 2003. The effect of jasmonic acid on the accumulation of ABA, proline and spermidine and its influence on membrane injury under water deficit in two barley genotypes. *Acta Physiologiae Plantarum* **25**(3): 279–285.

Becker W and Apel K. 1992. Isolation and characterization of a cDNA clone encoding a novel jasmonate-induced protein of barley (Hordeum vulgare L.). *Plant Molecular Biology* **19**: 1065–1067.

Benedetti C E, Xie D and Turner J G. 1995. COI1-dependent expression of an Arabidopsis vegetative storage protein in flowers and siliques and in response to coronatine or methyl jasmonate. *Plant Physiology* **109**: 567–572.

Berger S, Bell E and Mullet J E. 1996. Two methyl jasmonate-insensitive mutants show altered expression of AtVsp in response to methyl jasmonate and wounding. *Plant Physiology* **111**: 525–531.

Blée E. 2002. Impact of phyto-oxylipins in plant defense. *Trends in Plant Science* **7**: 315–322.

Browse J. 2009. Jasmonate passes muster: a receptor and targets for the defense hormone. *Annual Review of Plant Biology* **60**: 183–205.

Bussell J D, Reichelt M, Wiszniewski A A, Gershenzon J and Smith S M. 2014. Peroxisomal ATP-binding cassette transporter COMATOSE and the multifunctional protein ABNORMAL INFLORESCENCE MERISTEM are required for the production of benzoylated metabolites in Arabidopsis seeds. *Plant Physiology* **164**: 48–54.

Cao S, Zheng Y, Wang K, Jin P and Rui H. 2009. Methyl jasmonate reduces chilling injury and enhances antioxidant enzyme activity in postharvest loquat fruit. *Food Chemistry* **115**: 1458–1463.

Castillo M C and León J. 2008. Expression of the β-oxidation gene 3-ketoacyl-CoA thiolase 2 (KAT2) is required for the timely onset of natural and dark-induced leaf senescence in Arabidopsis. *Journal of Experimental Botany* **59:** 2171–2179.

Chaudhry B, Müller-Uri F, Cameron-Mills V, Gough S, Simpson D, Skriver K and Mundy J. 1994. The barley 60 kDa jasmonate-induced protein (JIP60) is a novel ribosome-inactivating protein. *The Plant Journal* **6**: 815–824.

Chini A, Fonseca S, Fernandez G, Adie B, Chico J M, Lorenzo O and Micol J L. 2007. The JAZ family of repressors is the missing link in jasmonate signalling. *Nature* **448**: 666–671.

Choi D, Bostock R M, Avdiushko S and Hildebrand D F. 1994. Lipid-derived signals that discriminate wound-and pathogen-responsive isoprenoid pathways in plants: methyl jasmonate and the fungal elicitor arachidonic acid induce different 3-hydroxy-3-methylglutaryl-coenzyme A reductase genes and antimicrobial isoprenoids in Solanum tuberosum L. *Proceedings of the National Academy of Sciences* **91**: 2329–2333.

Chung H S, Cooke T F, DePew C L, Patel L C, Ogawa N, Kobayashi Y and Howe G A. 2010. Alternative splicing expands the repertoire of dominant JAZ repressors of jasmonate signaling. *The Plant Journal* **63**: 613–622.

Creelman R A and Mullet J E. 1995. Jasmonic acid distribution and action in plants: regulation during development and response to biotic and abiotic stress. *Proceedings of the National Academy of Sciences* **92**: 4114–4119.

Creelman R A and Mullet J E. 1997. Biosynthesis and action of jasmonates in plants. *Annual Review of Plant Biology* **48**: 355–381.

Creelman R A, Tierney M L and Mullet J E. 1992. Jasmonic acid/methyl jasmonate accumulate in wounded soybean hypocotyls and modulate wound gene expression. *Proceedings of the National Academy of Sciences* **89**: 4938–4941.

Czapski J, Horbowicz M and Saniewski M. 1992. The effect of methyl jasmonate on free fatty acids content in ripening tomato fruits. *Biologia Plantarum* **34**: 71.

Danisman S, Van der Wal F, Dhondt S, Waites R, de Folter S, Bimbo A and Angenent G C. 2012. Arabidopsis class I and class II TCP transcription factors regulate jasmonic acid metabolism and leaf development antagonistically. *Plant Physiology* **159**: 1511–1523.

Delker C, Stenzel I, Hause B, Miersch O, Feussner I and Wasternack C. 2006. Jasmonate biosynthesis in Arabidopsis thaliana-enzymes, products, regulation. *Plant Biology* **8**: 297–306.

Demole E, Lederer E and Mercier D. 1962. Isolement et détermination de la structure du jasmonate de méthyle, constituant odorant caractéristique de l'essence de jasmin. *Helvetica Chimica Acta* **45**: 675–685.

Doares S H, Narváez-Vásquez J, Conconi A and Ryan C A. 1995. Salicylic acid inhibits synthesis of proteinase inhibitors in tomato leaves induced by system in and jasmonic acid. *Plant Physiology* **108**: 1741–1746.

Du H, Liu H and Xiong L. 2013. Endogenous auxin and jasmonic acid levels are differentially modulated by abiotic stresses in rice. *Frontiers in Plant Science* **4**: 397.

Feussner I and Wasternack C. 2002. The lipoxygenase pathway. *Annual Review of Plant Biology* **53**: 275–297.

Franceschi V R and Giaquinta R T. 1983. The paraveinal mesophyll of soybean leaves in relation to assimilate transfer and compartmentation. I. Ultrastructure and histochemistry during vegetative development. *Planta* **157**: 411–421.

Gehring C A, Theimer T C, Whitham T G and Keim P. 1997. Ectomycorrhizal fungal community structure of pinyon pines growing in two environmental extremes. *Ecology* **79**: 1562–1572.

Gfeller A, Baerenfaller K, Loscos J, Chételat A, Baginsky S and Farmer E E. 2010. Jasmonate controls polypeptide patterning in undamaged tissue in wounded Arabidopsis leaves. *Plant Physiology* **156**: 1797–1807.

González-Aguilar G A, Gayosso L, Cruz R, Fortiz J, Báez R and Wang C Y. 2000. Polyamines induced by hot water treatments reduce chilling injury and decay in pepper fruit. *Postharvest Biology and Technology* **18**: 19–26.

Hamberg M and Gardner H W. 1992. Oxylipin pathway to jasmonates: biochemistry and biological significance. *Biochimica et Biophysica Acta (BBA)-Lipids and Lipid Metabolism* **1165**: 1–18.

He Y, Fukushige H, Hildebrand D F and Gan S. 2002. Evidence supporting a role of jasmonic acid in Arabidopsis leaf senescence. *Plant Physiology* **128**: 876–884.

Heil M. 2015. Extrafloral nectar at the plant-insect interface: a spotlight on chemical ecology, phenotypic plasticity, and food webs. *Annual Review of Entomology* **60**: 213–232.

Howe G A and Jander G. 2008. Plant immunity to insect herbivores. *Annual Review of Plant Biology* **59**: 41–66.

Howe G A, Lightner J and Ryan C A. 1996. An octadecanoid pathway mutant (JL5) of tomato is compromised in signaling for defense against insect attack. *The Plant Cell* **8**: 2067–2077.

Howe G A and Schilmiller A L. 2002. Oxylipin metabolism in response to stress. *Current Opinion in Plant Biology* **5**: 230–236.

Ismail A, Riemann M and Nick P. 2012. The jasmonate pathway mediates salt tolerance in grapevines. *Journal of Experimental Botany* **63**: 2127–2139.

Jiang S, Yao J, Ma K W, Zhou H, Song J, He S Y and Ma W. 2013. Bacterial effector activates jasmonate signaling by directly targeting JAZ transcriptional repressors. *PLoS Pathogen* **9**: e1003715.

Johnson R, Narvaez J, An G and Ryan C. 1989. Expression of proteinase inhibitors I and II in transgenic tobacco plants: effects on natural defense against Manduca sexta larvae. *Proceedings of the National Academy of Sciences* **86**: 9871–9875.

Kazan K and Manners J M. 2008. Jasmonate signaling: toward an integrated view. *Plant Physiology* **146**: 1459–1468.

Koda Y, Kikuta Y, Kitahara T, Nishi T and Mori K. 1992. Comparisons of various biological activities of stereoisomers of methyl jasmonate. *Phytochemistry* **31**: 1111–1114.

Li C, Schilmiller A L, Liu G, Lee G I, Jayanty S, Sageman C and Howe G A. 2004. Role of β-oxidation in jasmonate biosynthesis and systemic wound signaling in tomato. *The Plant Cell* **17**: 971–986.

Liechti R and Farmer E E. 2006. Jasmonate biochemical pathway. *Science's STKE* **322**: 1–3.

Mahmood M, Bidabadi S S, Ghobadi C and Gray D J. 2012. Effect of methyl jasmonate treatments on alleviation of polyethylene glycol-mediated water stress in banana (*Musa acuminata* cv. 'Berangan', AAA) shoot tip cultures. *Plant Growth Regulation* **68**: 161–169.

McConn M, Creelman R A, Bell E and Mullet J E. 1997. Jasmonate is essential for insect defense in Arabidopsis. *Proceedings of the National Academy of Sciences* **94**: 5473–5477.

Nafie E, Hathout T, Mokadem A and Shyma A. 2011. Jasmonic acid elicits oxidative defense and detoxification systems in Cucumis melo L. cells. *Brazilian Journal of Plant Physiology* **23**: 161–174.

Nakagawa T and Kawaguchi M. 2006. Shoot-applied MeJA suppresses root nodulation in Lotus japonicus. *Plant and Cell Physiology* **47**: 176–180.

Pauwels L and Goossens A. 2011. The JAZ proteins: a crucial interface in the jasmonate signaling cascade. *The Plant Cell* **23**: 3089–3100.

Pelacho A M and Mingo-Castel A M. 1991. Jasmonic acid induces tuberization of potato stolons cultured in vitro. *Plant Physiology* **97**: 1253–1255.

Penninckx I A, Eggermont K, Terras F R, Thomma B P, De Samblanx G W, Buchala A and Broekaert W F. 1996. Pathogen-induced systemic activation of a plant defensin gene in Arabidopsis follows a salicylic acid-independent pathway. *The Plant Cell* **8**(12): 2309–2323.

Pieterse C M, Van der Does D, Zamioudis C, Leon-Reyes A and Van Wees S C. 2012. Hormonal modulation of plant immunity. *Annual Review of Cell and Developmental Biology* **28**: 489–521.

Püssa T, Raudsepp P, Toomik P, Pällin R, Mäeorg U, Kuusik S and Rei M. 2009. A study of oxidation products of free polyunsaturated fatty acids in mechanically deboned meat. *Journal of Food Composition and Analysis* **22**: 307–314.

Radhika V, Kost C, Boland W and Heil M. 2010. The role of jasmonates in floral nectar secretion. *PLOS ONE* **5**: e9265.

Ranjan R and Lewak S. 1992. Jasmonic acid promotes germination and lipase activity in non-stratified apple embryos. *Physiologia Plantarum* **86**: 335–339.

Rao M V, Lee H I, Creelman R A, Mullet J E and Davis K R. 2000. Jasmonic acid signaling modulates ozone-induced hypersensitive cell death. *The Plant Cell* **12**: 1633–1646.

Robson F, Okamoto H, Patrick E, Harris S R, Wasternack C, Brearley C and Turner J G. 2010. Jasmonate and phytochrome A signaling in Arabidopsis wound and shade responses are integrated through JAZ1 stability. *The Plant Cell* **22**: 1143–1160.

Schaller A and Stintzi A. 2009. Enzymes in jasmonate biosynthesis – structure, function, regulation. *Phytochemistry* **70**: 1532–1538.

Schmelz E A, Grebenok R J, Ohnmeiss T E and Bowers W S. 2002. Interactions between Spinacia oleracea and Bradysia impatiens: a role for phytoecdysteroids. *Archives of Insect Biochemistry and Physiology* **51**: 204–221.

Schommer C, Palatnik J F, Aggarwal P, Chételat A, Cubas P, Farmer E E and Weigel D. 2008. Control of jasmonate biosynthesis and senescence by miR319 targets. *PLoS Bio*logy **6**: e230.

Seltmann M A, Stingl N E, Lautenschlaeger J K, Krischke M, Mueller M J and Berger S. 2010. Differential impact of lipoxygenase 2 and jasmonates on natural and stress-induced senescence in Arabidopsis. *Plant Physiology* **152**: 1940–1950.

Sembdner G A P B and Parthier B. 1993. The biochemistry and the physiological and molecular actions of jasmonates. *Annual Review of Plant Biology* **44**: 569–589.

Staswick P E. 1989. Developmental regulation and the influence of plant sinks on vegetative storage protein gene expression in soybean leaves. *Plant Physiology* **89**: 309–315.

Staswick P E. 1994. Storage proteins of vegetative plant tissues. *Annual Review of Plant Biology* **45**: 303–322.

Staswick P E. 2008. JAZing up jasmonate signaling. *Trends in Plant Science* **13**: 66–71.

Staswick P E, Huang J F and Rhee Y. 1991. Nitrogen and methyl jasmonate induction of soybean vegetative storage protein genes. *Plant Physiology* **96**: 130–136.

Suhita D, Raghavendra A S, Kwak J M and Vavasseur A. 2004. Cytoplasmic alkalization precedes reactive oxygen species production during methyl jasmonate-and abscisic acid-induced stomatal closure. *Plant Physiology* **134**: 1536–1545.

Sun J, Cardoza V, Mitchell D M, Bright L, Oldroyd G and Harris J M. 2006. Crosstalk between jasmonic acid, ethylene and Nod factor signaling allows integration of diverse inputs for regulation of nodulation. *The Plant Journal* **46**: 961–970.

Taylor J E and Whitelaw C A. 2001. Signals in abscission. *New Phytologist* **151**: 323–340.

Thaler J S, Stout M J, Karban R and Duffey S S. 2001. Jasmonate-mediated induced plant resistance affects a community of herbivores. *Ecological Entomology* **26**: 312–324.

Ueda J and Kato J. 1980. Isolation and identification of a senescence-promoting substance from wormwood (*Artemisia absinthium* L.). *Plant Physiology* **66**: 246–249.

Vick B A and Zimmerman D C. 1987. Oxidative systems for modification of fatty acids: the lipoxygenase pathway. *The Biochemistry of Plants* **9**: 53–90.

Wager A and Browse J. 2012. Social network, JAZ protein interactions expand our knowledge of jasmonate signaling Front. *Plant Physiology* **3**: 41.

Wasternack C. 2007. Jasmonates: an update on biosynthesis, signal transduction and action in plant stress response, growth and development. *Annals of Botany* **100**: 681–697.

Wasternack C and Hause B. 2013. Jasmonates: biosynthesis, perception, signal transduction and action in plant stress response, growth and development. An update to the 2007 review in Annals of Botany. *Annals of Botany* **111**: 1021–1058.

Wasternack C and Kombrink E. 2010. Jasmonates: structural requirements for lipid-derived signals active in plant stress responses and development. *ACS Chemical Biology* **5**: 63–77.

Wittenbach V A. 1983. Purification and characterization of a soybean leaf storage glycoprotein. *Plant Physiology* **73**: 125–129.

Xie D X, Feys B F, James S, Nieto-Rostro M and Turner J G. 1998. COI1: an Arabidopsis gene required for jasmonate-regulated defense and fertility. *Science* **280**: 1091–1094.

Xu Y I, Chang P F L, Liu D, Narasimhan M, Raghothama K G, Hasegawa P M and Bressan R A. 1994. Plant defense genes are synergistically induced by ethylene and methyl jasmonate. *The Plant Cell* **6**: 1077–1085.

Yan Y, Jiang W, Spinetti T, Tardivel A, Castillo R, Bourquin C and Zhou R. 2013. Omega-3 fatty acids prevent inflammation and metabolic disorder through inhibition of NLRP3 inflammasome activation. *Immunity* **38**: 1154–1163.

Yoon J Y, Hamayun M, Lee S K and Lee I J. 2009. Methyl jasmonate alleviated salinity stress in soybean. *Journal of Crop Science and Biotechnology* **12**: 63–68.

Zhao M L, Wang J N, Shan W, Fan J G, Kuang J F, Wu K Q and Lu W J. 2013. Induction of jasmonate signalling regulators MaMYC2s and their physical interactions with MaICE1 in methyl jasmonate-induced chilling tolerance in banana fruit. *Plant, Cell & Environment* **36**: 30–51.

Part VI

Hormonal Regulation of Physiological and Developmental Processes in Plants

The hormonal regulation of plant growth and development dominated and shaped experimental plant physiology for much of the 20th century. The modern molecular and genetic tools and techniques provide excellent opportunities for understanding many intricate details of the mechanism of hormone action and how they regulate basic plant processes. The publication of a whole-plant genome of Arabidopsis, tomato and other model plants along with the production of large numbers of mutants in these model plants heralded a new era marked by explosion of knowledge regarding metabolism, receptors and mechanisms, based on comparisons with better-understood organisms, reconstruction of proteins and, of course, on manipulation of the genome and its products. However, it is essential to recognize, at the same time, that knowledge of molecular events alone cannot provide complete answers to all questions, especially those at the whole organism or community scale.

This part, consisting of three chapters, explores the conceptual frameworks, reviews specific issues relating to the evidence relating to hormone sensitivity and reviews the full range of roles for plant hormones in regulating various physiological processes. These chapters also pinpoint relevant hypotheses, experiments conducted to elucidate the role of different plant phytohormones and the interpretation of results within a holistic context. It also highlights the shortcomings and future perspectives in advancing our understanding of these processes.

DOI: 10.1201/9781003373889-25

20 Hormonal Regulation of Physiological Processes

20.1 INTRODUCTION

Higher plants are extremely sensitive to their surrounding and are capable of responding to the constant changes in the surrounding by adjusting their internal growth and developmental programs. Being sessile organisms, plants respond to several environmental and endogenous signals which together with the intrinsic genetic program determine their form. Plants' response to these external as well as internal signals is mediated by a group of structurally and chemically diverse molecules known as plant hormones, plant growth substances or phytohormones. The concept of plant hormones originated from the classical experiment conducted by Charles Darwin and his son Francis on phototropism in the 1880s. They were able to demonstrate that when oat seedlings were exposed to a directional light source, a signal from the plant apex traveled to lower parts of the seedling that resulted in their bending toward the light source. This signal was subsequently identified as indoleacetic acid (IAA), the first known plant hormone. Since then, several leading research groups across the globe have attempted to understand the mechanisms by which this pattern of growth and development in plant is regulated. During the course of these investigations, they identified five groups of phytohormones, considered as "classical" phytohormones including auxin, cytokinin, gibberellins (GA), abscisic acid (ABA) and ethylene (Santner et al., 2009). Recently, the role of additional phytohormones, such as brassinosteroids (BR), jasmonic acid (JA), salicylic acid (SA), polyamines, strigolactones (SL), nitric oxide (NO) and peptide hormones, in plant growth and development has been demonstrated in various plants (Santner et al., 2009). These studies have demonstrated numerous, diverse dramatic effects of hormones on plant growth and development. Hormones that are active at very low concentration (10^{-6} M or less) act by binding tightly to specific "receptor" molecules, like protein molecule.

Previous studies mainly focused on physiological responses, biosynthetic pathways and transport mechanisms of phytohormones; however, owing to the recent advances in plant genetic, molecular and biochemical techniques, our knowledge of phytohormones, particularly mechanism of their action, has advanced substantially, and many of the genes involved in hormone perception and signaling pathways have been identified and characterized (Dharmasiri et al., 2013). Many known growth and developmental responses to hormones are mediated by gene expression (Santner et al., 2009). These responses involve mechanisms that may or may not require gene regulation, although non-gene regulatory hormonal responses have not been characterized extensively (Dharmasiri et al., 2013). This chapter focuses on the perception and signaling of various phytohormones and their role in various developmental processes.

20.2 PERCEPTION OF HORMONAL SIGNALS

Hormones are known to control the expression of genes by regulating the abundance of two types of gene regulatory proteins, transcriptional factors (TFs) and transcriptional repressors. They regulate either by degradation or by altering the activities of regulatory proteins through post-translational modifications (Dharmasiri et al., 2013). Most of the plant hormonal response pathways involve ubiquitin-mediated protein degradation (Dharmasiri et al., 2013). Ubiquitin (Ub), a small protein present in all eukaryotes, is specifically and covalently attached to other proteins through a process called ubiquitination (Dharmasiri et al., 2013). This either changes the properties of proteins or destroys

DOI: 10.1201/9781003373889-26

the Ub-tagged proteins through 26S proteasome (Glickman and Ciechanover, 2002). Ub-mediated protein degradation has been shown to be fundamental to many cellular processes (Vierstra, 2009), self-incompatibility (Zhang et al., 2009), control of shoot apical meristem (Di Giacomo et al., 2013) and chloroplast biogenesis (Huang et al., 2013). According to Dharmasiri and Estelle (2002), the ubiquitination process involves three sequential enzymes, Ub-activating enzyme (E1), Ub-conjugating enzyme (E2) and Ub-ligase (E3). These authors have also shown that E1 first activates Ub by coupling with ATP to form E1-Ub by linking C-terminal glycine of Ub to a cysteine of E1 through a thiolester bond. Then, Ub is transferred to a conserved cysteine in E2 and is finally transferred to the target protein through E3 Ub-ligase to recruit the target protein (Smalle and Vierstra, 2004). In *Arabidopsis*, there are two E1s, 37 E2s and more than 1,400 E3s (Vierstra, 2009). Although several groups of E3 ubiquitin ligases (RING [REALLY INTERESTING NEW GENE], U-box, HECT [HOMOLOGY TO E6-AP C TERMINAL], CUL3-BTB [CULLIN3-BROAD-COMPLEX/TRAMTRACK/BRIC-A-BRAC], CUL4-DDB1 [CULLIN4-DNA DAMAGE BINDING1], SCF [SKP1CULLIN1 F-BOX] and APC [ANAPHASE PROMOTING COMPLEX]) are known to exist in plants, only SCF, RING and CUL3-BTB E3 ligases play a direct role in biosynthetic pathways and responses of some hormones and strigolactone (Santner and Estelle, 2010). Dharmasiri and Estelle (2004) have shown that SCF type E3 ligase is a multi-subunit protein complex made up of SKP1, CULLIN1 (or CULLIN2), an F-box protein and RBX1. Target proteins to be ubiquitinated interact with the F-box protein while RBX1 protein recruits the Ub-conjugated E2 enzyme (Ub-E2). RBX1, CUL1 and SKP1 are shared by many different SCF complexes while the F-box protein is specific for each SCF complex, thus giving the specificity to the complex.

Hormones use SCF-dependent ubiquitin-proteasome pathway to regulate hormone-specific gene expression (Santner and Estelle, 2009, 2010). At least some ABA responses are likely to be regulated through ubiquitin-proteasome pathway using RING E3 ligases (Dharmasiri et al., 2013). In this system, RING finger motif recruits the Ub-E2 (Santner and Estelle, 2010). SA, on the other hand, uses CUL3-BTB type E3 ligase to regulate the abundance of a regulatory protein (Boatwright and Pajerowskaâ Mukhtar, 2013). Similar to SCF E3 ligases, CUL3-BTB E3s are multi-subunit complexes but use CULLIN3a or 3b as the scaffold protein while BTB protein recruits the target protein.

Auxins (Indole 3-acetic acid or IAA) are intensely studied because of their involvement in several developmental processes from embryogenesis to senescence and from root tip to shoot tip (Woodward and Bartel, 2005; Abel and Theologis, 2010). According to Friml and Jones (2010), auxin signal may be perceived inside the nucleus, extracellular matrix and ER, thus determining numerous downstream responses, mainly the rapid regulation of gene expression and cell expansion. Effendi and Scherer (2011) suggested a link between the role of ABP1 (AUXIN BINDING PROTEIN1) as a receptor in auxin signaling either on ER or on the outer surface of the plasma membrane and rapid cell expansion, epidermal cell pattern formation and auxin transport. Auxin signaling is a classic example of the involvement of SCF E3 ligase complex in perception of a phytohormone inside the nucleus. In the absence of auxin, gene expression is inhibited by transcriptional repressors of auxin responsive genes (Auxin/Indole-3-acetic acid protein – Aux/IAA). Szemenyei and coworkers have shown that Aux/IAA proteins bind to auxin response factors (ARFs) via the scaffolding function of TOPLESS1 (TPL1) protein and block the transcription of auxin responsive genes (Szemenyei et al., 2008). In contrast, in the presence of auxin, the F-box protein TIR1 (TRANSPORTER INHIBITOR RESPONSE1) of the SCFTIR1 complex interacts with Aux/IAA transcriptional repressors, promoting their ubiquitination and degradation of Aux/IAAs (Gray et al., 2001). Destruction of Aux/IAA releases the repression on ARF transcription factors, allowing the transcription of auxin response (Gray et al., 2001).

Contrary to the initial belief that auxin directly or indirectly promotes the modification of Aux/IAAs or an interacting protein, it has now been shown that auxin binds directly to the F-box protein TIR1 (receptor) and enhances rapid degradation of Aux/IAAs in *Arabidopsis* (Dharmasiri et al., 2005a, 2005b; Kepinski and Leyser, 2004, 2005). Further investigations revealed that five other members of the TIR1 protein family (AFB1–5, AUXIN SIGNALING F-BOX1-5) also function as auxin receptors (Greenham et al., 2011). In-depth studies have shown that Auxin signal is perceived together by TIR1/AFB with Aux/IAA proteins functioning as co-receptors (Tan et al., 2007; Villalobos et al., 2012). These authors have shown that auxins act as glue and enhance interaction between TIR1/AFB with Aux/IAA proteins.

20.2.1 Gibberellins (GAs)

GAs are perceived by a nuclear-localized soluble receptor, GID1 (GIBBERELLIN INSENSITIVE DWARF1) (Griffiths et al., 2006; Nakajima et al., 2006). A functional GID1 is required for GA responses (Griffiths et al., 2006; Murase et al., 2008; Shimada et al., 2008). Interaction of GA with the binding pocket of GID1 probably causes the lid to fold back, allowing it to interact with another group of proteins called DELLAs. According to the current model, DELLAs interact with GID1 in response to GA, and the complete GID1/GA/DELLA complex, in turn, interacts with corresponding SCFGID2/SLY1 complex. The GID2 (GIBBERELLIN INSENSITIVE DWARF2) and SLY1 (SLEEPY1) are F-box proteins in rice and *Arabidopsis*, respectively (McGinnis et al., 2003; Sasaki et al., 2003).

20.2.2 Cytokinin

According to Dharmasiri et al. (2013), cytokinin is perceived by endoplasmic reticulum localized AHK (ARABIDOPSIS HISTIDINE KINASE) receptors. Three AHK receptors, AHK2, AHK3 and AHK4/WOL1 (WOODENLEG1)/CRE1 (CYTOKININ RESPONSE1) have been identified in Arabidopsis (Hwang et al., 2012). Cytokinin receptors have an extracellular cytokinin-binding CHASE (CYCLASE/HIS KINASE-ASSOCIATED SENSING EXTRACELLULAR) domain, HIS KINASE domain and cytoplasmic receiver domain (To and Kieber, 2008). Following cytokinin binding to CHASE domain, AHKs autophosphorylate the conserved HIS residue in the HIS KINASE domain. Subsequently, the phosphate group is transferred to a conserved Asp residue in the receiver domain of AHKs and then transferred to AHPs (HISTIDINE PHOSPHOTRANSFER PROTEINS) in the cytoplasm (Argueso et al., 2010; Perilli et al., 2010; Hwang et al., 2012). Five AHPs (AHP1–5) involved in cytokinin signaling in *Arabidopsis* can actively shuttle between nucleus and cytoplasm independent of their phosphorylation status and cellular cytokinin level (Hutchison et al., 2006; Punwani et al., 2010). In the nucleus, phosphorylated AHPs can phosphorylate a conserved Asp residue in a group transcription factor called ARR Type A–C (ARABIDOPSIS RESPONSE REGULATORS) (To and Kieber, 2008; Kieber and Schaller, 2010). Phosphorylated Type-B ARs activate the transcription of cytokinin responsive genes including Type-A ARRs and CRFs (CYTOKININ RESPONSE FACTORS) (Rashotte et al., 2006; To and Kieber, 2008). Phosphorylation stabilizes Type-A ARRs and activates a negative feedback loop in cytokinin signaling (To et al., 2007).

20.2.3 Abscisic Acid (ABA)

ABA is involved in stress responses as well as in regulating leaf size, internode length, seed dormancy, bud dormancy, embryo and seed development and reproduction (Lee and Luan, 2012). According to Guo et al. (2011), ABA may be perceived by multiple receptors at various cellular locations. The ABA can bind to chloroplast envelope-localized (ChlH/ABAR [H SUBUNIT OF THE CHLOROPLAST Mg2+-CHELATASE]/[ABA RECEPTOR]) or plasma-membrane-localized (GTG1/GTG2 [GPCR-TYPE G PROTEIN1 AND 2]) receptors (Shen et al., 2006; Pandey et al., 2009). ChlH/ABAR binds with ABA (Shen et al., 2006) and promotes its interaction with WRKY transcriptional factors, WRKY40, 18 and 60 (Shang et al., 2010). At low ABA levels, WRKY40 represses key ABA-responsive genes. Therefore, the binding of ChlH/ABAR with WRKYs keeps them in the cytoplasm relieving the repression of ABA-responsive genes such as ABI5 and DREB2A (Shang et al., 2010). Currently, the best-established ABA perception happens via PYR/PYL/RCAR receptors. Binding of ABA to PYR/PYL/RCAR changes its conformation to facilitate the binding of PP2C (TYPE 2C PROTEIN PHOSPHATASE) phosphatase and inactivates it (Melcher et al., 2009; Miyazono et al., 2009; Park et al., 2009). When the cellular ABA levels are low, PP2Cs dephosphorylate SnRK2s (SNF1-RELATED PROTEIN KINASES) (Fujii et al., 2009; Umezawa et al., 2009) and prevent SnRK2s from phosphorylating its target proteins such as b-ZIP transcription factors and ion channels, which leads to repressed ABA signaling (Kobayashi et al., 2005; Geiger et al., 2009). The b-ZIP transcription factors are ABA-responsive element (ABRE) binding factors (AREBs) such as AREB1, 2, 3 and ABI5 (Uno et al., 2000; Furihata et al., 2006). When PP2Cs are inactivated in the presence of ABA, SnRKs get activated by autophosphorylation and lead to phosphorylation of b-ZIP transcription factors to regulate ABA-responsive gene activation (Fujii et al., 2009). SLAC1 and KAT1 are examples of two ion channels regulated by SnRK2s where both are involved in ABA-regulated stomatal movements (Pilot et al., 2001; Vahisalu et al., 2008).

Although it may not play a central role, ubiquitin-proteasome-mediated protein degradation has been implicated at least in some ABA responses. Of the several RING E3 ligases linked to ABA responses, AIP2 (ABI3 INTERACTING PROTEIN2) and KEG (KEEP ON GOING) RING E3 ligases regulate the abundance of ABI3 and ABI5 transcriptional factors, respectively, in response to ABA. While ABA protects ABI5 from KEG-mediated ubiquitination and destruction through an unknown mechanism, ABA enhances AIP2-mediated ubiquitination and degradation of ABI3 probably by enhancing the AIP2 expression (Santner and Estelle, 2010). Nevertheless, intermediate steps between ABA perception and these E3 ligases are not yet clear.

20.2.4 Ethylene

Ethylene is perceived through membrane-bound receptors located in the endoplasmic reticulum (ER). In *Arabidopsis*, there are five members of ethylene receptor proteins known as ETR1 (ETHYLENE RESPONSE1), ETR2, ERS1 (ETHYLENE RESPONSE SENSOR1), ERS2 and EIN4 (ETHYLENE INSENSITIVE4) (Yoo et al., 2009). They are divided into two sub-families (I and II) depending on the number of transmembrane domains (TMDs). Several scientists have studied these receptors and their function in ethylene signaling (Kendrick and Chang, 2008; Hua and Meyerowitz, 1998; Qu et al., 2007; Ju and Chang, 2012; Hall et al., 2000; Gao et al., 2008; Ji and Guo, 2013). According to the current model, activated ethylene receptors interact with CTR1 (CONSTITUTIVE TRIPLE RESPONSE1) in the absence of ethylene leading to the phosphorylation of EIN2 (ETHYLENE INSENSITIVE2). This will prevent the cleavage of EIN2-CEND and its subsequent translocation to the nucleus. As a result, EIN3 (ETHYLENE INSENSITIVE3) and EIL1 (ETHYLENE INSENSITIVE3-LIKE1) interact with EBF1/EBF2 (EIN3 BINDING F-BOX PROTEIN1) of the $SCF^{EBF1/2}$ leading to ubiquitination of EIN3/EIL1 and their subsequent degradation through 26S proteasome. Thus, in the absence of ethylene, ethylene-induced gene expression is inhibited due to the low abundance of EIN3/EIL1. Binding of ethylene to receptors inhibits the phosphorylation of CTR1, resulting in dephosphorylation of EIN2, thus enhancing its CEND cleavage and subsequent transport of CEND to the nucleus. The EIN2-CEND interferes with EIN3/EIL1 ubiquitination and degradation perhaps either by downregulating EBF1 and EBF2 expression or through an unknown mechanism, thus leading to the accumulation of EIN3/EIL1 transcription factors, resulting in ethylene-induced gene transcription (An et al., 2010).

20.2.5 Brassinosteroid

Brassinosteroid (BR) by binding to plasma-membrane-localized LRR (LEUCINE-RICH REPEAT) receptor like kinase BRI1 (BRASSINOSTEROID INSENSITIVE1) triggers its signaling (Kim et al., 2010). While BRI1 functions as the major BR receptor, its homologs BRL1 (BRI1 LIKE1) and BRL3 can also perceive BR (Yang et al., 2011). BKI1 (BRI1 KINASE INHIBITOR1) inhibits the kinase activity of BRI1 at low levels of cellular BRs by binding to the C-terminal region of BRI1 (Wang et al., 2001). At high levels of BR, binding of BR causes conformational changes in BRI1 that dissociates BKI1 and undergoes autophosphorylation (Wang et al., 2001; Jaillais et al., 2011). This autophosphorylation promotes the hetero-oligomerization of BRI1 with BAK1 (BRI1-ASSOCIATED RECEPTOR KINASE1) and transphosphorylation of BAK1 and BRI1 (Kim et al., 2010). BAK1 belongs to a small subfamily of receptor-like kinases called SERKs (SOMATIC EMBRYOGENESIS RECEPTOR KINASES). Among them, two SERKs, SERK1 and BKK1 (BAK1-LIKE1/SERK4), act redundantly with BAK1 (Kim et al., 2009). The perceived BR signal is transduced through phosphorylation of BSKs (BR SIGNALING KINASES) by BRI1 (Clouse, 2011). Phosphorylated BSKs then bind and activate protein phosphatase BSU1 (BRI1 SUPPRESSOR1) (Kim et al., 2009). The activated BSU1 dephosphorylates and inactivates a GSK3/Shaggy-like kinase, BIN2 (BRASSINOSTEROID INSENSITIVE2). This results in the accumulation of unphosphorylated, active BR transcription factors, BES1 (BRI1-EMS SUPPRESSOR1) and BZR1 (BRASSINAZOLE RESISTANT1) (Kim et al., 2009). BES1 and BZR1 are bHLH (BASIC HELIX-LOOP-HELIX) like transcription factors. BZR1 and BES1 bind to promoters containing BR-response elements (BRREs) and E-BOX motifs and regulate their transcription (Clouse, 2011; Yu et al., 2011).

20.2.6 Jasmonates

The mechanisms of perception and signal transduction of jasmonate are similar to that of auxin (Dharmasiri et al., 2013). The F-Box protein COI1 (CORONATINE INSENSITIVE1) and the transcriptional repressor JAZ (JASMONATE-ZIM DOMAIN) form the cofactor in the presence of physiologically active jasmonate-Ile (Chini et al., 2007; Thines et al., 2007). While Aux/IAA proteins interact directly with common co-repressor (TPL), JAZ proteins interact with TPL via the adaptor protein NINJA (NOVEL INTERACTOR OF JAZ) (Pauwels et al., 2010). Interaction of JAZ with SCFCOI1 leads to ubiquitination and subsequent degradation of JAZs by 26S proteasome complex to release the MYB/MYC transcription factors which induces jasmonate-sensitive gene transcription (Perez and Goossens, 2013).

20.3 HORMONAL CONTROL OF CELL DIVISION

Plants are unique in their ability to maintain permanent stem cell populations and form new organs and continue to grow for entire duration of their life span. Phytohormones, such as auxin, gibberellin, cytokinin, abscisic acid, brassinosteroid, ethylene, jasmonate and strigolactone, govern and coordinate various developmental processes needed to sustain continuous growth. As described in the earlier sections, each of these hormones is perceived and transduced by separate pathways; however, it has become evident in recent years that hormones do not act only in a linear pathway, rather are interconnected by a complex network of interactions and feedback circuits that determine the final outcome of the individual hormone actions. The role played by each of the hormones in controlling cell division, cell growth and cell differentiation is described in this section.

20.3.1 Auxin

Classical experiments with exogenously applied auxin provided ample evidences for auxin's role in stimulating cell division in both intact plant tissues and cultured cells (Davies, 1995). Auxin acts on multiple targets to control cell proliferation. In Arabidopsis, auxin has been shown to control the progression from G_1 to G_2 by inducing the expression of one of the cyclin-dependent kinases (CDKA1) (Hemerly et al., 1993; Ferreira et al., 1994; Inagaki and Umeda, 2011). Auxin has also been shown to regulate cyclin genes, such as *CYCB1* and *CYCA2* (Roudier et al., 2003), and promote G_1 to S progression by regulating the stability of E2FB protein (Magyar et al., 2005).

20.3.2 Cytokinin

Cytokinins (CKs) are mainly of two types, *trans-zeatin* and *N*6-(Δ2-isopentenyl) adenine (Hirose et al., 2008). CKs promote cell differentiation in the transition zone (TZ) of the root but do not inhibit cell proliferation in the proximal meristem (Dello Ioio et al., 2007). Recently, evidences have shown that CKs are involved in the control of cell division in the QC by downregulating the expression of an auxin influx carrier, *LAX2*. This indicates that CKs negatively control auxin maxima and QC specifications that are associated with enhanced cell division in the QC.

20.3.3 Gibberellin

Since gibberellin (GA) was first discovered, >130 GAs have been identified. However, only a few (e.g., GA1, GA3 and GA4) are bioactive, but their individual functions remain elusive (Yamaguchi, 2008; Shani et al., 2013). GAs regulate stem elongation, germination, dormancy, flowering, sex expression and senescence of leaves and fruit primarily by promoting cell growth (Davière and Achard, 2013). However, Achard et al. (2009) showed that GAs also control cell proliferation in the proximal meristem (PM) of *Arabidopsis* roots by degrading DELLAs and suppressing expression of CDK inhibitors. By promoting cell elongation, GAs enable the adjacent cells to elongate and divide, leading to enlargement of the root meristem (Takatsuka and Umeda, 2014; Ubeda-Tomás et al., 2009).

20.3.4 Brassinosteroid (BR)

BRs control cell expansion, cell elongation and vascular differentiation, pollen elongation and acceleration of senescence (Clouse, 2002). BRs promote cell proliferation in a manner similar to CKs; however, the mode of BR action is rather complicated at the root apex (Hu et al., 2000). BRs function in promoting QC divisions by positively controlling the expression of *ETHYLENE RESPONSE FACTOR 115* (*ERF115*), which encodes a rate-limiting factor of QC divisions (Heyman et al., 2013).

20.3.5 Ethylene

Ethylene signaling plays a major role in plant development, such as fruit ripening, senescence of leaves and flowers, and seedling triple response (Lin et al., 2009). Ortega-Martínez et al. (2007) showed that ethylene promotes cell division in the QC of *Arabidopsis* roots. Ethylene-induced division of QC cells is accompanied by QC-specific gene expression, and the surrounding stem cells remain undifferentiated (Ortega-Martínez et al., 2007). Ethylene has no dramatic effect on stem cell fate and cell division in the PM (Růžička et al., 2007).

20.3.6 Abscisic Acid (ABA)

ABA mediates plants' responses to environmental stresses, such as cold, salt stress, osmotic stress and pathogen attack (Melcher et al., 2010). ABA is generally recognized as a growth inhibitor (Zeevaart and Creelman, 1988; Finkelstein and Gibson, 2002). There are few reports dealing with mechanism underlying ABA-mediated inhibition of cell division. Wang et al. (1998) demonstrated that ABA inhibits cell division by inducing the expression of the *Arabidopsis* CDK inhibitor KRP1. In addition, ABA seems to inhibit cell division in the PM by modulating the expression of cyclins and CDK inhibitors, but not CDKs themselves. ABA is also associated with maintenance of the quiescent state of QC cells (Zhang et al., 2010). Since ABA is induced by external stresses, it might contribute to the maintenance of the SCN by preserving undifferentiated stem cells under stressful conditions (Takatsuka and Umeda, 2014).

20.3.7 Strigolactone

Strigolactones have a role in the development of plant shoots and roots (Ruyter-Spira et al., 2011). These authors have shown that balanced auxin–strigolactone interactions are crucial for controlling stem root growth.

20.4 HORMONAL CONTROL OF APICAL DOMINANCE

Apical dominance is the process by which the shoot apex exerts its control over the outgrowth of lateral buds further down the stem (Cline, 1994). Also known as "correlative inhibition," apical dominance is a "developmental correlation" where one organ of the plant impacts the growth of the other (Hillman, 1984; Sachs, 1991). In an herbaceous plant, as the apical shoot meristem continues to grow, the lateral buds arise in the leaf axils (Fahn, 1990). These lateral buds become dormant shortly after their formation or after a brief period of growth. Although the elongation of lateral buds is arrested, they carry out all essential metabolic processes (Stafstrom, 1995). These inhibited or quiescent buds may be fairly rudimentary or well formed (Hillman, 1984). Apical dominance is released and one or more of these lower axillary buds start to elongate within a few hours to days following the removal of the apical bud or decapitation (Tamas, 1987). Brown et al. (1967) used the term "apical control" for woody plants to describe the physiological conditions that give rise to the overall shape and form of the tree crown via various branching patterns. Both "apical dominance" and "apical control" are being used by several authors to explain this phenomenon (Martin, 1987; Loreti and Pisani, 1990; Wilson, 1990), but the former term is usually used in a narrower sense with respect to herbaceous plants to specify control exerted by the shoot apex over axillary bud outgrowth. In a broader sense, it also refers to the influence of shoot apex on the orientation and growth of lateral branches and leaves, rhizomes, stolons, and tubers and leaf abscission (Hillman, 1984). Apical control typically refers to the influence of the main growing point on all branches of a perennial plant (Martin, 1987), or as Bollmark et al. (1995) described it, "the influence of the top of the tree on the branches lower down." For instance, the pine tree has a single dominant central leader shoot and strong apical control that produces an attenuated-shaped crown. On the other hand, oak tree does not have a single dominant bud and several lateral buds grow out to produce more or less a rounded shape. Even the individual shoots may have weak apical dominance and the lateral buds of the current year grow to varying degrees during the first season, but resume active growth when the growing point is removed, damaged or bent. Hence, when used with respect to trees, apical dominance refers to the control of lateral bud outgrowth by the apex in an individual branch during the current season's growth. A number of hormones (ethylene, gibberellins and auxin) and other organic compounds (sugar) have been investigated for their role in apical dominance (Blake et al., 1980; Cline, 1997; Cline et al., 2001; Cline and Oh, 2006; Mason et al., 2014).

20.4.1 Quantitative Characterization of Apical Dominance

Descriptive (qualitative or subjective) evidences may not always be sufficient to understand the underlying cellular, biochemical and molecular processes involved in both maintenance and release of apical dominance (Cline, 1997). In many cases, quantitative data or objective evidences on outgrowth of lateral buds are required to make comparison among species or treatments used to overcome apical dominance. The most commonly used measure to detect release of apical dominance is to quantify outgrowth (i.e., length and diameter of buds or fresh weight) of lateral shoots following the treatment of the shoot apex. It is also critical to precisely measure or quantify the cellular, biochemical and molecular events both before and after applying the treatments. Periodic measurements of these processes and bud growth may be carried out to provide definitive answers.

20.4.2 Theories of Apical Dominance

From the time Thimann and coworkers demonstrated the role of endogenous hormones on bud growth in *Vicia faba*, a number of theories regarding apical dominance have been introduced to describe the role of plant hormone auxin (indole-3-acetic acid, IAA) that moves down the stem from the shoot tip (Thimann et al., 1934; Domagalska and Leyser, 2011; Gomez-Roldan et al., 2008; Dun et al., 2012; Ferguson and Beveridge, 2009). Depletion of IAA from the stem after decapitation was believed to induce the growth of lateral buds, and hence, the role of auxins was considered as a key factor in all apical dominance models (Domagalska and Leyser, 2011; Gomez-Roldan et al., 2008; Dun et al., 2012; Ferguson and Beveridge, 2009). In the auxin transport canalization-based model, axillary buds are thought to remain dormant as long as continuous flow of auxin from shoot tip was maintained (Domagalska and Leyser, 2011; Prusinkiewicz et al., 2009; Crawford et al., 2010; Bennett and Leyser, 2006). Other researchers have put forward a second messenger theory to describe the role of apically derived auxin (Dun et al., 2012; Brewer et al., 2009, 2013). According to these authors, auxin produced in the shoot tip inhibits axillary bud growth indirectly by inhibiting cytokinin production and/or promoting strigolactone synthesis. Unlike auxin, cytokinin and strigolactone are thought to move into the axillary bud to promote or inhibit bud growth, respectively. Dun and coworkers found that the exogenous application of strigolactones and cytokinin regulates axillary bud growth (Dun et al., 2012). Additionally, the application of auxin to the cut surface of the decapitated stump did not inhibit the initial bud growth in both *Pisum sativum* and *Arabidopsis thaliana*; although it did inhibit bud growth at the later stages (Crawford et al., 2010; Beveridge et al., 2000; Morris et al., 2005; Cline, 1996; Cline et al., 2001; Thimann, 1937). Auxin depletion caused by decapitation, stem girdling or auxin transport inhibitors did not always promote bud outgrowth (Morris et al., 2005; Cline, 1996; Dun et al., 2012). Auxin was also ineffective in inhibiting bud growth after decapitation in some species and under high light conditions (Cline, 1996). Furthermore, when the shoot tip and bud were separated by a long distance, bud growth after decapitation occurred even before any reduction in IAA level in the stem adjacent to the bud could be observed (Dun et al., 2012; Morris et al., 2005; Renton et al., 2012). Mason et al. (2014) and Morris et al. (2005) also showed that the auxin transport inhibitor, naphthylphthalamic acid, reduced auxin content in the upper zone of the stem but had no effect on early bud growth. Moreover, exogenous auxin did not prevent bud growth in decapitated plants for almost 24 h after treatment (Morris et al., 2005; Cline et al., 2001). Therefore, it is highly unlikely that IAA directly or indirectly regulates the early outgrowth of lateral buds in decapitated stems (Mason et al., 2014). These observations have ruled out auxin to be the sole regulator due to lack of correlation with early axillary bud growth. This has kicked off a debate among researchers as to how auxin inhibits the growth of those buds (Booker et al., 2003; Mason et al., 2014; Cline et al., 2001; Sachs and Thimann, 1967). In 1924, the nutritive hypothesis proposed that access to plant nutrients is the major factor regulating axillary bud growth (Loeb, 1924; Phillips, 1975; Mason et al., 2014). Various theories of apical dominance are discussed in this section.

20.4.2.1 Plant System for Studying Apical Dominance

The ideal plant system for investigating cellular, molecular and biochemical processes during apical dominance release is the one in which the apical dominance is moderately strong with rapidly growing shoot, responds quickly to decapitation and only highest lateral bud outgrow in response to decapitation with short transport distance (Cline, 1997). *Pisum sativum* and Japanese morning glory *(Ipomoea nil)* are considered ideal systems for studying apical dominance release (Cline, 1997;

Wickson and Thimann, 1958; Beveridge et al., 1994; Stafstrom, 1995; Cline and Riley, 1984). On the other hand, *Coleus* plant is not particularly suitable because the apical dominance is very weak and decapitation produces only a slight response.

20.4.3 Apical Dominance Hypotheses

Since the first observation by Thimann in the 1930s on the role of hormones in inhibiting bud development, hundreds of studies have been conducted to understand the cellular, biochemical and molecular processes involved in release of apical dominance and grow out of lateral buds in various species. These studies have given rise to at least four hypotheses on regulatory mechanism of apical dominance and lateral bud outgrowth. These include the classical hypothesis, auxin transport hypothesis, bud transition hypothesis and nutritive hypothesis. These hypotheses have been developed based on studies on intact, decapitated or *in vitro* systems (Dun et al., 2006). The main hypotheses have been addressed here.

Classical hypothesis: Using the decapitation experiments, several scientists have demonstrated that auxin content regulates shoot branching by influencing the levels, transport, and/or action of other signals, such as cytokinin and shoot multiplication signal (SMS) that are responsible for inhibition of bud outgrowth (Snow, 1937; Hall and Hillman, 1975; Morris, 1977; Bangerth, 1989; Sachs and Thimann, 1967; Li et al., 1995; Foo et al., 2005). Decapitation leads to increase in the levels of endogenous cytokinins in the stems and/or xylem sap of many legumes (Bangerth, 1994; Li et al., 1995). It also increases the delivery of cytokinins to axillary buds (Turnbull et al., 1997; Tanaka et al., 2006). Therefore, cytokinins biosynthesized in the stem may be transported into the axillary buds after decapitation. Nordstrom and coworkers have found auxin to affect cytokinin biosynthesis, although it is not clear if cytokinin levels increase just before outgrowth of axillary buds (Nordstrom et al., 2004; Turnbull et al., 1997).

It is also possible that cytokinin is not always a limiting factor preventing bud outgrowth, or other factors may affect the competency for buds to respond to cytokinin. Auxin supply from the shoot tip was found to influence the production of SMS in pea stems (Foo et al., 2005) because transcript levels of RMS1 that control SMS synthesis in pea were slightly elevated after the exogenous application of auxin to intact or decapitated plants.

Auxin transport hypothesis: According to this hypothesis, it is the auxin transport and not the level that regulates axillary bud outgrowth (Bangerth, 1989; Li and Bangerth, 1999; Leyser, 2005). The establishment of an auxin transport stream is needed for bud outgrowth. Auxin in the shoot tip of the main stem is transported in a basipetal direction. The transport stream of the main stem is suggested to limit the flow of auxin from the axillary bud, thus not allowing the axillary bud to establish its own auxin transport stream. Therefore, axillary bud is unable to outgrow due to enhanced auxin levels. In this case, auxin plays a dual role, i.e., sustaining initial development of lateral buds and inhibiting further growth (Bennett et al., 2006).

To fully test the auxin transport hypothesis of shoot branching, the following three main points require further investigations:

- Whether the quantity of endogenous auxin transported in the polar auxin stream is at, or near, maximal capacity in nonbranching, wild-type plants?
- Whether the inability to transport additional auxin in the stem limits the flow of auxin from the axillary bud. This will help elucidate whether the auxin transport stream of a nonbranching plant is indeed unable to incorporate additional auxin synthesized by the bud?
- Whether an enhanced flow of auxin from axillary buds occurs at a critical time or stage of bud outgrowth?

Bud transition hypothesis: The bud transition hypothesis proposes that a bud may stay in any of the three stages: stage of dormancy, stage of transition or stage of sustained growth (Stafstrom and Sussex, 1992; Devitt and Stafstrom, 1995; Cline, 1997; Napoli et al., 1999; Shimizu-Sato and Mori, 2001; Morris et al., 2005; Beveridge, 2006). Here, dormancy is used to describe the extremely low or negligible growth rate of the axillary bud despite the fact that these buds are metabolically active. Stafstrom and Sussex (1992) found the expression of a ribosomal protein gene (rpL27) to be increased prior to visible growth in buds of decapitated pea plants. However, while some buds continued with sustained growth, this progression was halted in other buds in the same shoot. These buds presumably reentered a stage of dormancy, reflected by a reduction in rpL27 expression. This and other similar studies (e.g., Madoka and Mori, 2000) support the presence of a transition stage or stages where axillary buds are more receptive than dormant buds to signals that stimulate outgrowth, yet remain able to revert to a dormant state. These researchers suggested that a variety of factors including the stage of whole-plant ontogenetic development, the particular node at which the bud arises, the age of the bud, genotype, light, temperature and photoperiod may be involved in determining the developmental stage of a bud (Stafstrom, 1995; Beveridge et al., 2003; Horvath et al., 2003).

Weberling (1989) proposed the existence of three morphological zones that influence plant growth and development: the enrichment zone, representing the region where inflorescences develop; the inhibition zone, adjacent to the enrichment zone where there is little bud outgrowth; and the innovation zone at the base of the plant, where buds either remain dormant or develop into axillary branches that more or less phenocopy the main shoot (for review, see Napoli et al., 1999). They proposed that these morphological zones influence the responsiveness of axillary buds to signals such as cytokinin (King and Van Staden, 1988), auxin (Morris et al., 2005) and SMS (Beveridge et al., 2003). Given that, in different situations, nongrowing buds have different responses to various treatments that may stimulate bud outgrowth, it is useful to consider that bud outgrowth consists of several stages and that particular signals may act at some, but not in all of these stages. Beveridge and coworkers suggested that auxin acts to inhibit continued growth of buds that are in a transition stage. These results indicate that auxin depletion alone, directly or indirectly, does not induce initial bud outgrowth, indicating that other factors influence the competency to respond to changes in auxin. The results of Morris et al. (2005) indicate that decapitation causes a rapidly propagated signal that triggers a dormant bud to enter a transition stage. In this transition stage, the bud starts preliminary growth only.

Feedback regulation is commonly involved in maintaining homeostasis in the system. In terms of shoot branching, lateral bud outgrowth is balanced with the growth of other plant parts, particularly other shoots. The feedback mechanisms identified thus far, for shoot branching, involve auxin, cytokinin and SMS (Beveridge, 2006).

Nutritive hypothesis: Some studies have drawn upon concepts of the nutritive hypothesis that suggest apical dominance is maintained largely by the sugar demand of the shoot tip, which limits the availability of sugars to axillary buds. After shoot tip is removed, sugars accumulate in axillary buds within a time frame consistent with bud growth. The decapitation response can also be mimicked by supplying exogenous sugars to intact plants. With these findings, the previously mentioned disparities between auxin content and bud outgrowth can be explained in terms of sugar availability to the buds. Decapitation removes apical sugar demand and rapidly increases sucrose availability to axillary buds. This is sufficient to cause bud release, regardless of the auxin status of the adjacent stem. The role of auxin is prominent in the later stages of branch growth (Morris et al., 2005; Bangerth, 1989) rather than during the initial bud release stage. Possessing both the rapid sucrose-based response and the longer-term auxin response could be advantageous to the plant. This provides an important advantage in terms of interplant competition and the relative speed at which the plant can recover from the loss of its shoot tip. However, if all buds were to continue to grow, the plant phenotype would be drastically altered from the initially apically dominant state to an overly bushy phenotype. Consequently, auxin plays a role in determining which buds will continue to grow out, by functioning with the other plant hormones, cytokinins and strigolactones, to either promote the progression of growing axillary buds into branches or force them back into dormancy

(Stafstrom et al., 1998). Unlike the sucrose effect, this hormonal effect is substantially dependent on bud position because auxin is depleted in a basipetal gradient, and buds growing in branches will become additional sources of auxin. Consequently, in contrast to hormone models (Bennett and Leyser, 2006; Crawford et al., 2010; Brewer et al., 2009, 2013; Prusinkiewicz et al., 2009), another outcome of this combined sugar demand and hormone signaling model of shoot branching is the ease with which the decapitation of plants with strong apical dominance (no branches) can lead to a new architecture of basal and aerial branches (Beveridge, 2000; Cline and Sadeski, 2002). In addition to providing rapid responses to changes in sink demand and regulating the number and position of branches, the involvement of sugars, and not simply hormones, could also prevent excessive bud outgrowth under poor growing conditions.

20.5 HORMONAL CONTROL OF SEED DORMANCY

The seed is an important stage in the life cycle of higher plants that ensures continuance and survival as a species. It is the dispersal unit that is able to survive the period between seed maturation and the establishment of the next generation as a seedling after it has germinated. For this purpose, the seed, mainly in a dry state, is well equipped to sustain extended periods of unfavorable conditions. To optimize germination over time, the seed enters a dormant state, which prevents pre-harvest germination as well. Numerous studies have been performed to better understand how germination is controlled by various environmental factors and chemicals. However, still very little is known about the process by which the embryo emerges from the seed to complete germination and how embryo emergence is blocked in dormant seeds (Bewley, 1997).

Seed development comprises two major phases: embryo development and seed maturation. Embryogenesis, which is a morphogenesis phase, starts with the formation of a single-cell zygote and ends in the heart stage when all embryo structures have been formed (Mayer et al., 1991). It is followed by a growth phase during which the embryo fills the seed sac (Goldberg et al., 1994). At the end of the embryo growth phase, cell division is arrested in the embryo (Raz et al., 2001). Hereafter, the seed, containing a fully developed embryo, undergoes maturation during which food reserves accumulate and dormancy and desiccation tolerance develops (Goldberg et al., 1994). Several detailed studies have been published on the developmental (Baud et al., 2002), metabolic (Fait et al., 2006), proteomic (Gallardo et al., 2001, 2002; Chibani et al., 2006) and gene expression (Girke et al., 2000; Nakabayashi et al., 2005; Finch-Savage et al., 2007) changes during seed development in dry seeds and subsequent germination. Seed development has been extensively studied using mutant's defective in various aspects of the process. Mutants affected in the morphogenesis phase result in lethality or have a seedling phenotype (Mayer et al., 1991; Meinke, 1995). In seed germination mutants, properties of germination and dormancy are affected, which sometimes are accompanied by pleiotropic effects that are specific for maturation, such as desiccation intolerance (Goldberg et al., 1994; Koornneef and Karssen, 1994).

20.6 HORMONAL CONTROL OF TUBERIZATION

Tuberization is strongly influenced by day length with short days (SDs) generally promoting tuber formation, although response to photoperiod varies among different genotypes (Ewing, 1995). While commercial potato varieties are relatively insensitive to day-length conditions, wild species like *Solanum demissum* or some lines of *Solanum tuberosum* ssp andigena, are totally dependent on photoperiod for tuber induction. Ewing and Struick (1992) have shown that the wild species tuberize under SD conditions and long days (LDs) or SD supplemented with a night break (SD + NB) inhibit tuberization. The length of the night period determines the tuberization response of these species. Batutis and Ewing (1982) have demonstrated that the light stable phytochrome PHYB is the major photoreceptor involved in the regulation of tuber induction. Grafting experiments have demonstrated that the photoreceptor controls the synthesis of a graft-transmissible inhibitory signal that is produced under LDs (Jackson et al., 1998). Since GA activity was shown to decrease under SD conditions, Ewing (1995) and Kumar and Wareing (1974) postulated that the graft-transmissible signal contains GA. This was further substantiated by the fact that treatments with the inhibitors of GA biosynthesis (ancymidol and paclobutrazol) promote tuberization in *S. tuberosum* ssp Andigena plants under noninductive LD conditions (Jackson and Prat, 1996).

20.7 HORMONAL CONTROL OF FLOWER INDUCTION

Plants have unique developmental mechanisms that allow them to transition into flowering in response to external signals. The key structure that provides this transition is the shoot apical meristem (SAM), which gives rise to the aerial structures of the plant, such as leaves, stems and, ultimately, flowers. The timing of the transition from vegetative growth to flowering is very fundamental to agriculture, horticulture and plant breeding because flowering is the beginning of sexual reproduction process (Bernier et al., 1993). Several research groups have attempted to understand this transition process and have produced a large volume of data on this topic (Bernier et al., 1981b; Halevy, 1985–1989; Bernier, 1988). In plants, this transition to flowering takes place in response to environmental cues, such as photoperiod, temperature and water availability. Plants in which flowering is not dependent on photoperiod or temperature are called "autonomous-flowering." However, these plants flower in response to irradiance. The environmental signals are perceived by mature leaves (photoperiod and irradiance), shoot apex (low or vernalizing temperature), all parts (temperature) and roots (water availability) (Bernier et al., 1993). Photoperiod influences initiation of flowering in some plants, whereas temperature, especially low temperature, induces flowering in others. However, these factors interact very closely with genotype to control flowering, and under certain circumstances, transition to flowering can be initiated by other factors. For instance, in *Pharbitis nil* (short-day plant) and *Silene armeria* (long-day plant), flowering can be induced under noninductive photoperiod by low temperature, irradiance, reducing water availability, root pruning or controlling nutrition (Bernier et al., 1993). Similarly, in mango, citrus and lychee, flower induction under subtropical conditions is governed by chilling temperatures from passing cold fronts during winter-spring months, but under tropical climatic conditions, the age of the last flush is the dominant factor regulating flowering (Davenport, 1993, 2000, 2003, 2007). Stems must have had sufficient rest (about four to five months) before they can be induced to flower in the absence of chilling temperatures. Mild water stress or low nitrogen fertility can replace the requirement for extended rest period. The age of the previous flush modifies the cool-temperature-induced floral response, with older stems more likely to produce flowers than younger stems. In some plants, vernalization, gibberellic acid and an increase in the proportion of far-red light in the light source can substitute for one another in promoting the transition to flowering (Martínez-Zapater and Somerville, 1990; Bagnall, 1992). This means that alternative pathways of flower induction may exist in these plants. Since different environmental signals are perceived by different plant parts, a close interaction between plant parts is required to initiate transition of vegetative buds into reproductive buds (Bernier et al., 1993). Although floral initiation has been the topic of intense investigations by several research groups over the last century, it is not fully clear how various factors regulate events or processes connected with flower induction. For instance, in citrus species, no single indispensable factor that stimulates flower induction has been identified. Similarly, in mango, although a number of factors (low temperature, proper balance of endogenous hormones, especially auxin from leaves and cytokinins from roots, a combination of age-regulated vegetative promoter and temperature-regulated florigenic promoter) are known to promote flowering, no threshold inductive temperature has been identified or how different management practices regulate production of vegetative and reproductive shoots is fully understood (Davenport, 1993, 2007). In protea, the presence of developing inflorescences or possible floral inhibiting factors from the previous flowering season is suggested to inhibit inflorescence initiation (Hoffman, 2006). Although flowering in papaya is not affected both photoperiod (Nakasone and Storey, 1955) and temperature, sex expression is altered by environmental cues, PGRs and other factors. Kozlowski and coworkers reported that flowering in most woody perennials does not appear to be under photoperiodic control, thereby suggesting a weak interaction between temperature and photoperiod (Kozlowski et al., 1991). However, these aspects need to be investigated further to elucidate the exact flowering mechanism in different fruit crops.

20.7.1 Theories of Flower Induction

Three major theories have been proposed to explain the chemical control of the transition to flowering.

- Florigen/antiflorigen concept
- Nutrient diversion hypothesis
- The model of multifactorial control

20.7.1.1 Florigen/Antiflorigen Concept

Julius Sachs (1865) is considered the father of the flower hormone concept, although it was Chailakhyan in 1936 who coined the word "florigen" to refer to a hypothetical flowering hormone consisting of two components (i.e., gibberellin and anthesin), which are involved in flowering. Of these, gibberellins play a key role in flowering of LDPs, whereas anthesin promotes flowering in short-day plants (SDPs). Bernier (1988) and Zeevaart (1982) suggested that the florigen/antiflorigen is simple, specific and universal hormone synthesized continuously in the leaves and transported to meristems where it acts to initiate flowering. However, these substances are yet to be isolated and identified (Singh, 1959, 1961; Zeevaart, 1982; Nunez-Elisea and Davenport, 1991; Davenport and Nunez-Elisea, 1997; Kulkarni, 1991; Davenport et al., 1995; Aukerman and Amasino, 1998). Lang (1980) suggested that all photoperiodic plants may contain similar antiflorigen. In plants that contain both florigen and antiflorigen, flower initiation is believed to be regulated by proper balance between these two factors at the meristem level. In mango, occurrence of flowering and non-flowering shoots on the same tree and inability of immature shoots to flower along with the mature shoots have been reported to be caused by the balance between the promoter from the mature leaves and inhibitor from the immature shoots (Singh, 1961). The exogenous application of gibberellins inhibits flowering, whereas inhibitory action of endogenous gibberellins can be overcome with anti-gibberellins such as paclobutrazol (Singh, 1961).

20.7.1.2 Nutrient Diversion Model

Despite several reports of extracts with flower-inducing activity and numerous attempts by several research groups to extract florigen, it remained a physiological concept rather than a chemical entity. In the meantime, T'SE (1975) and Sachs (1977) obtained clear evidence in "San Diego" bougainvillea that day length, cytokinin and selective defoliation treatments increased assimilate supply to the inflorescence axis before transition to flowering occurs. Therefore, they hypothesized that the reproductive development is constrained by assimilate supply and treatments that increase the supply of assimilate to apical meristems promote flowering (Sachs, 1977). This hypothesis was later called the nutrient diversion model (Ramina et al., 1979). According to this model, the inductive treatments result in an increase in the amount of assimilates moving to the apical meristem, where it induces flowering (Sachs and Hackett, 1983; Bernier, 1988). Specifically, the source (leaves)–sink (apical meristems) relationships within the plant regulates flowering as evidenced by higher concentration of assimilates in apical meristems under inductive conditions than those under noninductive conditions (Ramina et al., 1979; Sachs and Hackett, 1983). Several research groups have demonstrated greater accumulation and metabolism of sucrose, protein and amino acids along with an almost fourfold increase in the specific activities of amylase and protease in the bark tissues of ten-year-old flowering mango trees as compared with three-year-old, non-flowering trees. Therefore, they suggested that sugars may play a regulatory role in floral induction in mango (Singh, 1960; Chacko and Ananthanarayanan, 1982; Rameshwar, 1989; Goldschmidt et al., 1985; Goldsmith and Golomb, 1982) and other species (Sachs, 1977; Mishra and Dhillon, 1978; Ramina et al., 1979; Sachs et al., 1979; Bernier et al., 1981a; Sachs and Hackett, 1983). Sachs et al. (1979) proposed the theory of photoassimilate diversion to the apical bud for the carbohydrate-regulated flowering. Flowering promoters are believed to be agents that promote assimilate supply to the meristem. Some PGRs alter assimilate distribution in lower plants as is the case in *Pharbitis nil* where cytokinins promotes flowering by redistributing photosynthetic products to meristems (Ogawa and King, 1980). Sachs and Hackett (1983) reported that many factors controlling flower initiation also influence photosynthesis and/or assimilate availability. Further evidence to this theory came from studies in which sucrose or glucose applications promoted flowering in several photosensitive herbaceous plants, sometimes even under noninductive conditions (Bodson and Bernier, 1985). Carbohydrate levels have been suggested as a limiting factor for flower formation in citrus (Ogaki et al., 1963; Goldschmidt and Golomb, 1982) as well as in other fruit trees (Harley et al., 1942; Worley, 1979; Monselise and Goldschmidt, 1982).

In citrus and plums, gibberellins inhibit flowering, whereas carbohydrates promote flowering (Goldschmidt et al., 1985; Bradley and Crane, 1960). Goldsmith and coworkers, therefore, concluded that gibberellins and "carbohydrates" operate as independent regulatory factors in the flowering in these species.

20.7.1.3 Multifactorial Model

The view that assimilates are the only important component in directing the transition to flowering was replaced by the multifactorial control model, which states that a number of promoters and inhibitors including phytohormones and assimilates are involved in controlling the developmental transition to flowering (Bernier et al., 1981a; Bernier 1988). According to this model, flowering can occur only when the limiting factors are present at the apex in the appropriate concentrations and at the right times. This model attempted to account for the diversity of flowering responses by proposing that different factors could be limiting for flowering in plants with divergent genetic backgrounds and/or under particular environmental conditions. Genetic analysis of flowering time in pea, cereals and Arabidopsis also supports the idea that the transition to flowering is under multifactorial control (Snape et al., 1996; Weller et al., 1997, 2009; Koornneef et al., 1998).

20.7.2 Genetic Studies of Flowering

Flower development is a multistep process beginning with response of plants to environmental signal (photoperiod or vernalizing temperature), followed by the change of meristem identity, termination of the stem cell activity, organ polarity control, organ-identity determination and organogenesis (Gan et al., 2013). According to these authors, flowering is regulated by a closely linked genetic pathways involving transcription factors and signaling molecules, catalytic enzymes and structural proteins. Epigenetic regulators are key to the initiation and maintenance of the genetic pathways by controlling gene expression through chromosomes. Four response pathways have been proposed based on genetic studies in Arabidopsis. These include the photoperiod, vernalization, autonomous and gibberellin (GA) flowering response pathways (Mouradov et al., 2002; Périlleux and Bernier, 2002; Komeda, 2004; Corbesier and Coupland, 2005). Two genes central to long-day (LD)-induced flowering in Arabidopsis have been identified (CONSTANS [CO] and FLOWERING LOCUS T [FT]). The former (CO) encodes a nuclear zinc-finger protein, which in response to LD induces transcription of FT, encoding an RAF-kinase-inhibitor-like protein (Golembeski et al., 2014). Both genes are not expressed in the shoot apex. Expression from meristem-specific promoters of CO does not promote flowering, but early flowering is induced in plants in short day (SD) when FT is overexpressed in the meristem. An et al. (2004) and Ayre and Turgeon (2004) have shown that expression of CO in the phloem is sufficient to generate a phloem-mobile signal in Arabidopsis. An et al. (2004) further speculated FT protein to be either the mobile signal itself or FT controls the synthesis of a mobile substance that induces flowering.

In the vernalization pathway, flowering in cold-requiring accessions of Arabidopsis is controlled by the MADS-box gene, FLOWERING LOCUS C (FLC) and FT. FLC protein represses FT in the leaf. Prolonged exposure to vernalizing temperatures reduces FLC expression, thus inducing FT, which then acts as a flowering stimulus. FLC expression also inhibits response of shoot apex to the FT signal. The vernalizing temperature first induces the production of the FT signal in leaves followed by the response of apical meristem to the signal (Searle et al., 2006).

Huang and coworkers (2005) suggested that the FT mRNA is the mobile stimulus, which they considered florigen itself or at least a part of it in Arabidopsis. They concluded that FT mRNA produced in the leaf and transported to the apical meristem is the limiting factor for flowering. They correlated FT mRNA with flower formation in apical meristem in Arabidopsis. In some species, production of florigen appears to continue even after the plants are transferred to noninductive conditions, and induced leaves remain effective donors in grafting experiments three months after they had been moved from SD to LD. This phenomenon, called indirect induction or nonlocalized induction, suggests that florigen has self-perpetuating properties (Zeevaart, 1976). Zeevaart (2006) also suggested positive feedback whereby FT, once induced, further enhances its own expression, both in the donor leaf and in the apical meristem. He further suggested that, regardless of environmental cues involved in floral induction, florigen was the end product and that the CO/FT combination in the flowering response pathway is central to flowering in all plant species (Zeevaart, 2006). Therefore, the differences between SDPs and LDPs appear to reside in how the genes in the flowering pathway function and are regulated. It remains to be shown, of course, that FT is the universal systemic transmissible signal (mRNA or protein) that is required for flowering.

Flowering in tomato (*Solanum lycopersicum*), a day-neutral plant (DNP), was shown to be induced by a transmissible signal, generated by the ortholog of FT, SINGLE-FLOWER TRUSS (SFT) (Lifschitz et al., 2014). They showed that overexpression of FT or SFT induced early flowering in day-neutral tobacco (*Nicotiana tabacum*) and tomato. Unlike herbaceous plants, trees flower only after a long juvenile phase lasting many years. Böhlenius and coworkers reported that expression of FT is also a prerequisite for flowering in trees. Ectopic expression of an FT ortholog drastically shortened the juvenile phase and resulted in early flowering in aspen (*Populus* spp.). Moreover, expression of the FT ortholog increased with age of the trees (Böhlenius et al., 2006). Hsu and coworkers also reported that a critical level of FT2 expression was necessary before flowering can occur in juvenile *Populus deltoides*. This suggests that either FT or a product of FT is the flower-inducing signal (Hsu et al., 2006). Thus, the failure of apical meristems to induce flowering is related to insufficient production of florigen and/or rapid decay of florigen in the receptor.

20.7.3 Antiflorigen or Flowering Inhibitor

There is strong evidence that noninduced leaves can inhibit flowering (Zeevaart, 2006). Using ^{14}C labeled photoassimilate, King and Zeevaart (1973) explained the inhibiting effects in terms of interference in florigen movement to target receptor buds. Zeevaart (2006) also provided evidence for involvement of mobile signal in inhibiting flowering. Hanzawa and coworkers have shown that the TFL1 has homology with FT, and change of a single amino acid can convert TFL1 as a repressor of flowering to an activator of flowering (Hanzawa et al., 2005). However, at present, there is no gene function that can specifically be associated with transmissible inhibitor (Zeevaart, 2006).

20.8 HORMONAL CONTROL OF FLOWERING

The effects of major classes of hormones on flowering are explained in this section.

AUXINS

Auxin has long been postulated as a key regulator of flower development, but the exact role of auxin in flower development was not well defined until recently (Nemhauser et al., 1998; Cheng and Zhao, 2007; Okada et al., 1991). A link between auxin and flower development was first established when the auxin transport mutant pin1 was isolated and characterized (Okada et al., 1991; Galweiler et al., 1998). Auxin is considered both a promoter and an inhibitor of flower initiation in a number of crops, although inhibition is far more widespread than promotion (Salisbury, 1955). Normally, auxin promotes flowering at low concentrations and act as inhibiter at high concentrations. Inhibitory effect of high doses may be more to do with general growth inhibition (Jacobs, 1985) or an auxin-dependent ethylene biosynthesis (Krekule et al., 1985). The presence of auxin at optimal concentration is required if flowers are to be formed in some species (Tran Thanh Van, 1980). Chen (1987) reported that the concentration of indoleacetic acid (IAA) diffused from developing vegetative shoot tips was much higher compared to the concentrations of diffused auxin from bud rest, early panicle development and full flowering stages. However, Lal and Rani (1977) reported no measurable differences in free IAA between shoot tips sampled during "on" and "off" years. Naphthalene acetic acid (NAA) has been reported to inhibit bud break of mango in Egypt (Bakr et al., 1981). However, Singh (1961) and Pandey and Narwadkar (1984) observed no inhibition of bud break after application of auxin to mangoes.

Although there is overwhelming evidence for the role of auxin in floral development, there has been no direct evidence until recently because of the lack of in-depth understanding of auxin biosynthesis (Nemhauser et al., 1998; Cheng and Zhao, 2007). Recent studies on the involvement of the YUCCA (YUC) family of flavin monooxygenases in auxin biosynthesis provided unambiguous evidence that auxin is necessary for the formation of flowers (Zhao et al., 2001; Tobena-Santamaria et al., 2002; Cheng et al., 2006; Yamaguchi et al., 2013). The auxin biosynthetic YUC genes were found to have profound effects on flower development as they affect all aspects of flower development (Cheng et al., 2006). These authors have reported that YUC genes have unique yet overlapping roles in flower development. In general, fewer and smaller flowers are formed when more YUC genes are compromised. Yamaguchi and coworkers have identified three genes whose expression is mediated by auxin-activated transcription factor, MONOPTEROS (MP), and jointly regulate the initiation of flower primordial. These genes encode known regulators of flower development, LEAFY (LFY), that specifies floral fate and two AINTEGUMENTA-LIKE/PLETHORA transcription factors, which are key regulators of floral growth (Yamaguchi et al., 2013). These authors also revealed a mechanistic link between flower primordial initiation and subsequent steps in flower morphogenesis, in which LFY provides direct feedback to auxin pathway (Yamaguchi et al., 2013).

Disruption of auxin biosynthesis and its polar transport or signaling was shown not to prevent the transition from vegetative growth to reproductive growth, but converting inflorescence meristem into flower meristem does require normal auxin polar transport (Cheng and Zhao, 2007; Zafari et al., 2015). Auxin determines whether flower primordia are formed or not. As these researchers have noted, auxin is also essential for specifying floral organs and determining the pattern formation within a floral organ. Thus, auxin clearly plays an essential role in various aspects of flower development, although the underlying mechanisms have not yet been fully understood. Cheng and Zhao (2007) concluded that auxin is a key regulator for the fate of several founder cells in flower primordia.

GIBBERELLINS

GAs have been extensively studied for their role in flowering in LD and cold-requiring plants grown under noninductive conditions (Bernier et al., 1981a, 1981b: Lang, 1965: Zeevaart, 1976). The role of GAs in controlling flower initiation is rather a controversial matter (Bernier et al., 1981a; Heide et al., 1987). In some rosette LDP as well as in plants requiring cold treatment, exogenous application of GAs promotes flowering. In contrast, it does not induce flowering in SDPs grown under noninductive LD conditions. GA has been shown to inhibit flower initiation in some perennial crops like citrus, apple and the SDP strawberry (Bernier, 1988). The effectiveness of various GAs varies with species, time of application and growing conditions (Kinet et al., 1985; Bernier et al., 1981a; Lang, 1965).

The growth retardants CCC and diaminozide enhance flower formation in lemons (Bodson and Remade, 1987; Bruffert et al., 1985), oranges (Chailakhyan et al., 1977) and mangoes (Suryanarayana and Rao, 1977; Kurian and Lyer, 1993a, 1993b). In contrast, flowering in grape involves the formation of anlagen (undifferentiated primordia), differentiation of inflorescence primordia and, ultimately, the formation of flowers (Srinivasan and Mullins, 1980, 1981). GA was shown to enhance the induction of the first step leading to the differentiation into tendrils or inflorescences, while it inhibits the differentiation of inflorescences from anlagen (second step), thus promoting the development of vegetative tendrils. There have been numerous studies on the inhibitory effects of GA_3 on mango flowering (Singh, 1961; Kachru et al., 1971, 1972; Rajput and Singh, 1989; Nunez-Elisea and Davenport, 1991; Osthuyse, 1991). Gibberellic acid treatment of mangoes has been recommended in the Canary Islands to delay flowering until after the danger of winter frost has passed (Galan-Sauco, 1990). In the subtropics of Australia, it is used to prevent flowering in newly planted trees during the spring so that the full growing period can be utilized for vegetative growth, thereby hastening orchard establishment (Davenport and Nunez-Elisea, 1997). Response to GA_3 varies among cultivars, growing conditions and timing of application (Turnbull et al., 1996). Plant response to GA_3 also varies from year to year so that concentrations that are effective in delaying flowering in one year may not have any effect in the following season (Turnbull et al., 1996). Conversely, it may be so effective that it delays flowering beyond the inductive window, resulting in a vegetative flush (Kachru et al., 1971, 1972; Davenport and Nunez-Elisea, 1997). The variable response to applied GAs may be directly related to endogenous levels of the active gibberellin in buds at the time of application, as well as to the differential sensitivity of buds to GA_3, depending on their position (apical vs. axillary) or age (Nunez-Elisea and Davenport, 1991).

20.8.1 Cytokinins

Since their discovery, cytokinins have been shown to have effects on several physiological and developmental processes, including leaf senescence, nutrient mobilization, apical dominance, the formation and activity of shoot apical meristems and floral development. Physiological evidence suggests that they have a role in floral transition (D'Aloia et al., 2011). These authors have shown that the application of N^6-benzylaminopurine (BAP) promotes flowering of Arabidopsis plants grown in noninductive short days. They also proposed a mechanistic basis for the role of cytokinin in flowering. According to them, the response to cytokinin does not require *FLOWERING LOCUS T (FT)* but require *TSF* (a paralogue of TWIN SISTER OF FT) and *SOC1 (SUPPRESSER OF OVEREXPRESSION OF CONSTAN1).*

Exogenous cytokinins both promote and inhibit flower initiation in a variety of species, although promotive effects are much more frequent than inhibitory ones (Bernier et al., 1981b, 1981a). Cytokinin is an absolute requirement for flowering of explants from adult *Passiflora* plants, but juvenile explants do not flower even in its presence (Scorza and Janick, 1980). Chen (1987) described precocious bud break and flowering of mango shoots in response to an early October application of 100 ppm 6-Benzylaminopurine (BA). Full flowering was observed one month following the application compared with flowering three months later in non-treated trees. Similar results were reported by Nunez-Elisea et al. (1990) using the synthetic cytokinin-like thiadiazuron.

20.8.1.1 Ethylene

Continuous exposure of trees to smoke has been utilized to stimulate flowering of mango and a number of plants. However, only the branches which have attained sufficient age or "ripeness to flower" respond to the smoke treatment (Bueno and Valmayor, 1974). Rodriguez (1932), investigating smoke-induced flowering of pineapple, first proposed that ethylene generated by burning material may stimulate flowering. Numerous investigations since then have shown ethephon to be an effective floral promoter in some mango cultivars under specific conditions (Das et al., 1989). Indirect support for this hypothesis is also provided by Lopez and coworkers, who reported that KNO_3 stimulated flowering of mango is mediated by increased levels of endogenous ethylene (Lopez et al., 1984). Mosqueda-Vazquez and Avila-Resendiz (1985) reported that the efficacy of KNO_3 was negated by $CoCl_2$ and $AgNO_3$ compounds which inhibit the synthesis and action of ethylene, respectively, when sprayed 1–4 h after the KNO_3 application. Nunez-Elisea and Davenport (1991; Davenport and Nunez-Elisea, 1997) reported elevated ethylene production in mango shoots in response to ethephon sprays without an accompanying floral response. The role of ethylene in floral induction is controversial since it has opposite effects in different plant species. Ethylene has been found to inhibit or delay flower initiation in a number of species. Moreover, in *Xanthium*, a short-day plant, some are inhibited by ethylene treatment, while other plants of the same species are induced to flower by ethylene under long days (De Greef et al., 1989). Furthermore, the inhibitory effect of ethylene on flowering *Pharbitis nil* depends on the ABA level (Wilmowicz et al., 2008).

20.8.1.2 Abscisic Acid (ABA)

Exogenous application of ABA inhibits flowering in several SDPSs and LDPs grown in inductive conditions, and response to endogenous ABA is not always consistent with the day length. Bernier (1988) suggested that ABA, produced more or less constantly irrespective of day length, acts as a general inhibitor while its impact is overcome under inductive conditions or increased amounts of promoters. It might be the increased production of ethylene and translocation of food material to shoots that creates ideal conditions for flower induction.

20.9 HORMONAL CONTROL OF ABIOTIC STRESS

Abiotic stresses, such as drought, salinity, cold and heat, negatively impact plant growth and development, which ultimately affect the yield (Tuteja, 2007). They are the leading causes of crop losses worldwide (Tuteja, 2007; Qin et al., 2011; Bailey-Serres et al., 2012; Paleg and Blumwald, 2011). According to various studies, abiotic stresses trigger many physiological, biochemical and molecular responses that impact various cellular processes in plants (Wang et al., 2001, 2003; Hasanuzzaman et al., 2013). Among various phytohormones, abscisic acid (ABA) has been studied the most as the central regulator of abiotic stress resistance, especially the water stress in plants (Finkelstein, 2013; Yamaguchi-Shinozaki and Shinozaki, 2006; Wani and Kumar, 2015; Paleg and Blumwald, 2011; Santner and Estelle, 2009; Wang et al., 2009), enabling plants to cope with different stresses. ABA also plays an important role, allowing plants to adapt to cold temperatures. Both increased biosynthesis and the exogenous application of ABA improve cold tolerance in plants (Xue-Xuan et al., 2010). Under harsh environmental conditions, the level of ABA increases via ABA biosynthesis, and it often rapidly alters gene expression by inducing or preventing the degradation of transcriptional regulators via the ubiquitin–proteasome system (Santner and Estelle, 2010). The ABA binds to its receptor to initiate signal transduction leading to cellular responses to stresses (Ng et al., 2014). Therefore, ABA is also called a stress hormone (Mehrotra et al., 2014).

Plants develop various mechanisms to combat stress challenges, not always sufficient to fully protect them. ABA regulates various physiological processes from stomatal opening to protein storage and provides adaptation to drought, salt and cold stresses. Therefore, several strategies like abscisic acid (ABA) phytohormone engineering have been used to produce abiotic stress tolerance crop plants.

Initially, it was thought that ABA had a significant role in the process of fruit abscission (Ohkuma et al., 1963; Cracker and Abeles, 1969). However, later intensive studies showed that ABA was not directly involved in abscission process. The precise role of ABA in abscising organs is to promote senescence and stress responses, the processes preceding abscission (Finkelstein, 2013). ABA biosynthesis is significantly increased under drought or salinity stress conditions, stimulating stomatal closure, change in gene expression and adaptive physiological responses (Seki et al., 2002; Yamaguchi-Shinozaki and Shinozaki, 2006; Shinozaki and Yamaguchi-Shinozaki, 2007; Cutler et al., 2010; Kim et al., 2010; Sah et al., 2016). ABA also plays an important role in many cellular processes including seed development, dormancy, germination, vegetative growth (Xiong and Zhu, 2003; Finkelstein et al., 2008) and modulation of root architecture (Harris, 2015).

Since the discovery of ABA, several efforts have been devoted to understanding how ABA is synthesized under stress conditions. For ABA perception and signaling, two breakthroughs were achieved in 2009 by the discoveries of the soluble ABA receptor proteins and the core signaling complexes that perceive ABA and transmit cues to subsequent molecular events (Fujii et al., 2009; Ma et al., 2009; Park et al., 2009. Recently, Park et al. (2015) engineered ABA receptor using agrochemicals, thus offering new possibilities for the better development of the crop. A better understanding of ABA regulatory mechanisms will eventually contribute to the development of engineered stress-tolerant crop plants, the primary goal of plant molecular biologists.

Although ABA is the most studied stress-responsive hormone, other hormones, especially cytokinins, brassinosteroids (BR), Jasmonates and auxins, also have a direct or indirect role to play during environmental stress (Paleg and Blumwald, 2011). Cytokinin (CK) acts as an antagonist to ABA because exposure of plants to water stress results in decreased levels of CK. Genes-encoding proteins associated with CK signaling pathways were found to be differentially affected by various abiotic stresses (Argueso et al., 2009). BR has also been reported to induce the expression of stress-related genes, leading to the maintenance of photosynthetic activity, the activation of antioxidant enzymes, the accumulation of osmoprotectants and the induction of other hormone responses (Divi and Krishna, 2009). Interrelations between the different plant hormones results in synergetic or antagonistic interactions that are important in plants' response to abiotic stress (Wang et al., 2009).

REFERENCES

Abel S and Theologis A. 2010. Odyssey of auxin. *Cold Spring Harbor Perspectives in Biology* **2**: a004572.

Achard P, Gusty A, Cheminant S, Alioua M, Dhondt S, Coppens F, Beemster G T and Genschik P. 2009. Gibberellin signaling controls cell proliferation rate in *Arabidopsis*. *Current Biology* **19**: 1188–1193.

An F, Zhao Q, Ji Y, Li W, Jiang Z, Yu X, Zhang C, Han Y, He W, Liu Y, Zhang S, Ecker J R and Guo H. 2010. Ethylene induced stabilization of ETHYLENE INSENSITIVE3 and EIN3-LIKE1 is mediated by proteasomal degradation of EIN3 binding F-Box 1 and 2 that requires EIN2 in Arabidopsis. *The Plant Cell* **22**: 2384–2401.

An H L, Roussot C, Suárez-López P, Corbesier L, Vincent C, Piñeiro M, Hepworth S, Mouradov A, Justin S, Turnbull C and Coupland G. 2004. CONSTANS acts in the phloem to regulate a systemic signal that induces photoperiodic flowering of Arabidopsis. *Development* **131**: 3615–3626.

Argueso C T, Ferreira F J and Kieber J J. 2009. Environmental perception avenues: the interaction of cytokinin and environmental response pathways. *Plant, Cell and Environment* **32**: 1147–1160.

Argueso C T, Raines T and Kieber J J. 2010. Cytokinin signaling and transcriptional networks. *Current Opinion in Plant Biology* **13**: 533–539.

Aukerman M J and Amasino R M. 1998. Floral induction and florigen. *Cell* **93**: 491–494.

Ayre B G and Turgeon R. 2004. Graft transmission of a floral stimulant derived from CONSTANS. *Plant Physiology* **135**: 2271–2278.

Bagnall D J. 1992. Control of flowering in *Arabidopsis thaliana* by light, vernalisation, and gibberellins. *Australian Journal of Plant Physiology* **19**: 401–409.

Bailey-Serres J, Lee S C and Brinton E. 2012. Waterproofing crops: effective flooding survival strategies. *Plant Physiology* **160**: 1698–1709.

Bakr El, Abdalla K M, Meligi M A and Ismail A. 1981. Floral differentiation in mango as affected by growth regulators, ringing and defoliation. *Egyptian Journal of Horticulture* **8:** 161–166.

Bangerth F. 1989. Dominance among fruits/sinks and the search for a correlative signal. *Physiologia Plantarum* **76**: 608–614.

Bangerth F. 1994. Response of cytokinin concentration in the xylem exudate of bean (Phaseolus vulgaris L.) plants to decapitation and auxin treatment, and relationship to apical dominance. *Planta* **194**: 439–444.

Batutis E J and Ewing E E. 1982. Far-reversal of red-light effect during long-night induction of potato (Solanum tuberosum L.) tuberization. *Plant Physiology* **69**: 672–674.

Baud S, Boutin J P, Miquel M, Lepiniec L and Rochat C. 2002. An integrated overview of seed development in Arabidopsis thaliana ecotype Ws. *Plant Physiology and Biochemistry* **40:** 151–160.

Bennett T and Leyser O. 2006. Something on the side: axillary meristems and plan development. *Plant Molecular Biology* **60**: 843–854.

Bennett T, Sieberer T, Willett B, Booker J, Luschnig C and Leyser O. 2006. The Arabidopsis MAX pathway controls shoot branching by regulating auxin transport. *Current Opinion in Biology* **16**: 553–563.

Bernier G. 1988. The control of floral evocation and morphogenesis. *Annual Review Plant Physiology and Plant Molecular Biology* **39**(1): 175–219.

Bernier G, Havelange A, Houssa C, Petitjean A and Lejeune P. 1993. Physiological signals that induce flowering. *The Plant Cell* **5**(10): 1147–1155.

Bernier G, Kinet J-M and Sachs R M. 1981a. *The Physiology of Flowering*. Volume I. CRC Press, Boca Raton.

Bernier G, Kinet J-M and Sachs R M. 1981b. *The Physiology of Flowering*. Volume II. CRC Press, Boca Raton.

Beveridge C A. 2000. Long-distance signalling and a mutational analysis of branching in pea. *Plant Growth Regulation* **32**: 193–203.

Beveridge C A. 2006. Axillary bud outgrowth: sending a message. *Current Opinion in Plant Biology* **9**: 35–40.

Beveridge C A, Ross J and Murfet I. 1994. Branching mutant *rms-2* in *Pisum sativum*: grafting studies and endogenous indole-3-acetic acid levels. *Plant Physiology* **104**: 953–959.

Beveridge C A, Symons G M and Turnbull C G N. 2000. Auxin inhibition of decapitation induced branching is dependent on graft-transmissible signals regulated by genes Rms1 and Rms2. *Plant Physiology* **123**: 689–698.

Beveridge C A, Weller J L, Singer S R and Hofer J M I. 2003. Axillary meristem development. Budding relationships between networks controlling flowering, branching, and photoperiod responsiveness. *Plant Physiology* **131**: 927–934.

Bewley J D. 1997. Seed germination and dormancy. *Plant Cell* **9:** 1055–1066.

Blake T, Pharis R and Reid D. 1980. Ethylene, gibberellins, auxin and apical control of branch angle in conifer, *Cupressus arizonica. Planta* **148**: 64–68.

Boatwright J L and Mukhtar K P. 2013. Salicylic acid: an old hormone up to new tricks. *Molecular Plant Pathology*. https://doi.org/10.1111/mpp.12035.

Bodson M and Bernier G. 1985. Is flowering controlled by the assimilate level? *Physiology of Vegetable Crops* **23:** 491–501.

Bodson M and Remade B. 1987. Distribution of assimilates from various source-leaves during the floral transition of *Sinapis alba* L. *Physiology of Vegetable Crops* **25:** 341–350.

Böhlenius H, Huang T, Charbonnel-Campaa L, Brunner A M, Jansson S, Strauss S H and Nilsson O. 2006. CO/FT regulatory module controls timing of flowering and seasonal growth cessation in trees. *Science* **312**: 1040–1043.

Bollmark M, Chen H, Moritz T and Eliasson L. 1995. Relations between cytokinin level, bud development and apical control in Norway spruce, *Picea abies. Physiologia Plantarum* **95**: 563–568.

Booker J, Chatfield S and Leyser O. 2003. Auxin acts in xylem-associated or medullary cells to mediate apical dominance. *Plant Cell* **15**: 495–507.

Bradley M V and Crane J C. 1960. Gibberellin-induced inhibition of bud development in some species of plums. *Science* **131**: 825–826.

Brewer P B, Dun E A, Ferguson B J, Rameau C and Beveridge C A. 2009. Strigolactone acts downstream of auxin to regulate bud outgrowth in pea and Arabidopsis. *Plant Physiology* **150**(1): 482–493.

Brewer P B, Koltai H and Beveridge C A. 2013. Diverse roles of strigolactones in plant development. *Molecular Plant* **6**: 18–28.

Brown C, Mcalpine R and Kormanik P. 1967. Apical dominance and form in woody plants: a reappraisal. *American Journal of Botany* **54**: 153–162.

Bruffert J, Fontaine D and Imhoff C. 1985. *Anangallis Anrensis*. pp. 434–449.

Bueno P B and Valmayor R V. 1974. Potassium nitrate: key to mango flowering. *Agricultruiest Los Banos* **13**: 4–16.

Chacko E K and Ananthanarayanan T V. 1982. Accumulation of reserve substances in *Mangifera indica* L. during flower initiation. *Zeitschrift fur Pflanzenphysiologie* **106:** 281–285.

Chailakhyan M Kh, Grigor'eva N Y and Lozhnikova V N. 1977. Effect of leaf extracts from flowering tobacco plants on the flowering of *Chenopodium rubrum* shoots and seedlings. *Doklady Akademii Nauk SSR* **236**: 773–776.

Chen W S. 1987. Endogenous growth substances in relation to shoot growth and flower bud development of mango. *Journal of American Society of Horticultural Science* **112:** 360–363.

Cheng Y, Dai X and Zhao Y. 2006. Auxin biosynthesis by the YUCCA flavin monooxygenases controls the formation of floral organs and vascular tissues in Arabidopsis. *Genes and Development* **20**: 1790–1799.

Cheng Y and Zhao Y. 2007. A role for auxin in flower development. *Journal of Integrative Plant Biology* **49**: 99–104.

Chibani K, li-Rachedi S, Job C, Job D, Jullien M and Grappin P. 2006. Proteomic analysis of seed dormancy in Arabidopsis. *Plant Physiology* **142:** 1493–1510.

Chini A, Fonseca S, Fernandez G, Adie B, Chico J M, Lorenzo O, Garcia-Casado G, Lopez-Vidriero I, Lozano F M and Ponce M R. 2007. The JAZ family of repressors is the missing link in jasmonate signalling. *Nature* **448**: 666–671.

Cline M G. 1994. The role of hormones in apical dominance. New approaches to an old problem in plant development. *Physiologia Plantarum* **90**: 230–237.

Cline M G. 1996. Exogenous auxin effects on lateral bud outgrowth in decapitated shoots. *Annals of Botany* **78**: 255–266.

Cline M G. 1997. Concept and terminology of apical dominance. *American Journal of Botany* **84**: 1064–1069.

Cline M G, Chatfield S P and Leyser O. 2001. NAA restores apical dominance in the axr3–1 mutant of Arabidopsis thaliana. *Annals of Botany, London* **87**: 61–65.

Cline M G and Oh C. 2006. A reappraisal of the role of abscisic acid and its interaction with auxin in apical dominance. *Annals of Botany* **98**: 891–897.

Cline M G and Riley L. 1984. The presentation time for shoot inversion release of apical dominance in *Pharbitis nil*. *Annals of Botany* **53**: 897–900.

Cline M G and Sadeski K. 2002. Is auxin the repressor signal of branch growth in apical control? *American Journal of Botany* **89**: 1764–1771.

Clouse S D. 2002. Brassinosteroid signaling: novel downstream components emerge. *Current Biology* **12**: 485–487.

Clouse S D. 2011. Brassinosteroid signal transduction: from receptor kinase activation to transcriptional networks regulating plant development. *Plant Cell* **23**: 1219–1230.

Corbesier L and Coupland G. 2005. Photoperiodic flowering of Arabidopsis: integrating genetic and physiological approaches to characterization of the floral stimulus. *Plant Cell and Environment* **28**: 54–66.

Cracker L E A and Abeles F B. 1969. Abscission: role of abscisic acid. *Plant Physiology* **44**: 1144–1149.

Crawford S, Shinohara N, Sieberer T, Williamson L, George G, Hepworth J, Müller D, Domagalska M A and Leyser O. 2010. Strigolactones enhance competition between shoot branches by dampening auxin transport. *Development* **137**(17): 2905–2913.

Cutler S R, Rodriguez P L, Finkelstein R R and Abrams S R. 2010. Abscisic acid: emergence of a core signaling network. *Annual Review of Plant Biology* **61**: 651–679.

D'Aloia M, Bonhomme D, Bouché F, Tamseddak K, Ormenese S, Torti S, Coupland G and Périlleux C. 2011. Cytokinin promotes flowering in Arabidopsis via transcriptional activation of the *FT* paralogue *TSF*. *The Plant Journal*. https://doi.org/10.1111/j.1365-313X.2011.04482.x.

Das G C, Sahoo S C and Ray D P. 1989. Studies on effect of gibberellic acid and urea either alone or in combination on the growth and flowering behaviour of some 'on' and 'off' year shoots in Langra mango. *Acta Horticulturae* **231**: 495–499.

Davenport T L. 1993. Floral manipulation in mangos. pp. 54–60. *Proceedings of the Conference on Mango in Hawaii* (Eds.) Chia L E and Evans D O. Cooperative Extension Service, University of Hawaii, Honolulu, HI.

Davenport T L. 2000. Processes influencing floral initiation and bloom: the role of phytohormones in a conceptual flowering model. *Horticulture Technology* **10**: 733–739.

Davenport T L. 2003. Management of flowering in three tropical and subtropical fruit tree species. *HortScience* **38**: 1331–1335.

Davenport T L. 2007. Reproductive physiology of mango. *Brazilian Journal of Plant Physiology* **19**: 363–376.

Davenport T L and Nunez-Elisea R. 1997. Reproductive physiology. pp. 66–146. *The Mango* (Ed.) Litz R. CAB International, Wallingford.

Davenport T L, Ramos L and Alani A. 1995. Evidence for a transmissible florgenic promoter in mango. p. 159. *Proceedings 22th Annual Meeting of the Plant-growth Regulator Society of America.* Plant-growth Regulator Society of America, Alexandria.

Davière J M and Achard P. 2013. Gibberellin signaling in plants. *Development* **140**: 1147–1151.

Davies P J. 1995. The plant hormones: their nature, occurrence and functions. pp. 1–12. *Plant Hormones: Physiology, Biochemistry and Molecular Biology* (Ed.) Davies P J. Kluwer Academic Publishers, Dordrecht, The Netherlands.

De Greef J A, De Proft M P, Mekers O, Van Dijck R, Jacobs L and Philippe L. 1989. Floral Induction of Bromeliads by Ethylene. pp. 313–322. *Biochemical and Physiological Aspects of Ethylene Production in Lower and Higher Plants. Advances in Agricultural Biotechnology.* Volume 26. Diepenbeek, Belgium, Springer, Netherlands.

Dello Ioio R, Linhares F S, Scacchi E, Casamitjana-Martinez E, Heidstra R, Costantino P and Sabatini S. 2007. Cytokinins determine arabidopsis root-meristem size by controlling cell differentiation. *Current Biology* **17**: 678–682.

Devitt M L and Stafstrom J P. 1995. Cell cycle regulation during growth dormancy cycles in pea axillary buds. *Plant Molecular Biology* **29**: 255–265.

Dharmasiri N, Dharmasiri S and Estelle M. 2005a. The F-box protein TIR1 is an auxin receptor. *Nature* **435**: 441–445.

Dharmasiri N, Dharmasiri S, Weijers D, Lechner E, Yamada M, Hobbie L, Ehrismann J, Jurgens G and Estelle M. 2005b. Plant development is regulated by a family of auxin receptor F box proteins. *Developmental Cell* **9**: 109–119.

Dharmasiri S and Estelle M. 2002. The role of regulated protein degradation in auxin response. *Plant Molecular Biology* **49**: 401–409.

Dharmasiri S and Estelle M. 2004. Auxin signaling and regulated protein degradation. *Trends in Plant Science* **9**: 302–308.

Dharmasiri S, Jayaweera T and Dharmasiri N. 2013. Plant hormone signalling: current perspectives on perception and mechanism of action. *Ceylon Journal of Science (biological Science)* **42**: 1–17.

Di Giacomo E, Serino G and Frugis G. 2013. Emerging role of the ubiquitin proteasome system in the control of shoot apical meristem function F. *Journal of Integrative Plant Biology* **55**: 7–20.

Divi U K and Krishna P. 2009. Brassinosteroid: a biotechnological target for enhancing crop yield and stress tolerance. *New Biotechnology* **26**: 131–136.

Domagalska M A and Leyser O. 2011. Signal integration in the control of shoot branching. *Nature Reviews Molecular Cell Biology* **12**: 211–221.

Dun E A, de Saint Germain A, Rameau C and Beveridge C A. 2012. Antagonistic action of strigolactone and cytokinin in bud outgrowth control. *Plant Physiology* **158**: 487–498.

Dun E A, Ferguson B J and Beveridge C A. 2006. Apical dominance and shoot branching. Divergent opinions or divergent mechanisms? *Plant Physiology* **142**: 812–819.

Effendi Y and Scherer G N F E. 2011. AUXIN BINDING-PROTEIN1 (ABP1), a receptor to regulate auxin transport and early auxin genes in an interlocking system with PIN proteins and the receptor TIR1. *Plant Signaling and Behavior* **6**: 1101–1103.

Ewing E E. 1995. The role of hormones in potato (*Solanum tuberosum*, L.) tuberization. pp. 698–724. *Plant Hormones: Physiology, Biochemistry and Molecular Biology* (Ed.) Davies P J. Kluwer Academic, Dordrecht.

Ewing E E and Struik P C. 1992. Tuber formation in potato: induction, initiation and growth. *Horticulture Review* **14**: 89–197.

Fahn A. 1990. *Plant Anatomy*. 4th edition. Pergamon Press, New York.

Fait A, Angelovici R, Less H, Ohad I, Urbanczyk-Wochniak E, Fernie A R and Galili G. 2006. Arabidopsis seed development and germination is associated with temporally distinct metabolic switches. *Plant Physiology* **142:** 839–854.

Ferguson B J and Beveridge C A. 2009. Roles for auxin, cytokinin, and strigolactone in regulating shoot branching. *Plant Physiology* **149**: 1929–1944.

Ferreira P C, Hemerly A S, Engler J D, van Montagu M, Engler G and Inzé D. 1994. Developmental expression of the *arabidopsis* cyclin gene cyc1At. *The Plant Cell* **6**: 1763–1774.

Finch-Savage W E, Cadman C S, Toorop P E, Lynn J R and Hilhorst H W. 2007. Seed dormancy release in Arabidopsis Cvi by dry after-ripening, low temperature, nitrate and light shows common quantitative patterns of gene expression directed by environmentally specific sensing. *Plant Journal* **51:** 60–78.

Finkelstein R R. 2013. Abscisic acid synthesis and response. *Arabidopsis Book* **11**: e0166.

Finkelstein R R and Gibson S I. 2002. ABA and sugar interactions regulating development: 'cross-talk' or 'voices in a crowd'? *Current Opinion in Plant Biology* **5**: 26–32.

Finkelstein R R, Reeves W, Ariizumi T and Steber C. 2008. Molecular aspects of seed dormancy. *Annual Review of Plant Biology* **59**: 387–415.

Foo E, Buillier E, Goussot M, Foucher F, Rameau C and Beveridge C A. 2005. The branching gene RAMOSUS1 mediates interactions among two novel signals and auxin in pea. *Plant Cell* **17**: 464–474.

Friml J Ã and Jones A R. 2010. Endoplasmic reticulum: the rising compartment in auxin biology. *Plant Physiology* **154**(2): 458–462.

Fujii H, Chinnusamy V, Rodrigues A, Rubio S, Antoni R, Park S Y, Cutler S R, Sheen J, Rodriguez P L and Zhu J K. 2009. In vitro reconstitution of an abscisic acid signaling pathway. *Nature* **462**: 660–664.

Furihata T, Maruyama K, Fujita Y, Umezawa T, Yoshida R, Shinozaki K and Yamaguchi-Shinozaki K. 2006. Abscisic acid-dependent multisite phosphorylation regulates the activity of a transcription activator AREB1. *Proceedings of the National Academy of Sciences of the USA* **103**: 1988–1993.

Galan-Sauco V. 1990. *Los Frutales Tropicales en los Sub tropicos Aguacate, Mango, Litchi, Longan*. Ediciones Mundi-Prensa, Madrid.

Gallardo K, Job C, Groot S P C, Puype M, Demol H, Vandekerckhove J and Job D. 2001. Proteomic analysis of Arabidopsis seed germination and priming. *Plant Physiology* **126:** 835–848.

Gallardo K, Job C, Groot S P C, Puype M, Demol H, Vandekerckhove J and Job D. 2002. Importance of methionine biosynthesis for Arabidopsis seed germination and seedling growth. *Physiologia Plantarum* **116:** 238–247.

Galweiler L, Guan C, Muller A, Wisman E, Mendgen K, Yephremov A and Palme K. 1998. Regulation of polar auxin transport by AtPIN1 in Arabidopsis vascular tissue. *Science* **282**: 2226–2230.

Gan E S, Huang J and Ito T. 2013. Functional roles of histone modification, chromatin remodeling and microRNAs in Arabidopsis flower development. *International Review of Cell and Molecular Biology* **305**: 115–161.

Gao Z, Wen C-K, Binder B M, Chen Y-F, Chang J, Chiang Y-H, Kerris R J, Chang C and Schaller G E. 2008. Heteromeric interactions among ethylene receptors mediate signaling in Arabidopsis. *Journal of Biological Chemistry* **283**: 23801–23810.

Geiger D, Scherzer S, Mumm P, Stange A, Marten I, Bauer H, Ache P, Matschi S, Liese A, Al-Rasheid K A, Romeis T and Hedrich R. 2009. Activity of guard cell anion channel SLAC1 is controlled by drought-stress signaling kinase-phosphatase pair. *Proceedings of the National Academy of Sciences of the USA* **106**: 21425–21430.

Girke T, Todd J, Ruuska S, White J, Benning C and Ohlrogge J. 2000. Microarray analysis of developing Arabidopsis seeds. *Plant Physiology* **124:** 1570–1581.

Glickman M H and Ciechanover A. 2002. The ubiquitin-proteasome proteolytic pathway: destruction for the sake of construction. *Physiological Reviews* **82**: 373–428.

Goldberg R B, Paiva G and Yadegari R. 1994. Plant embryogenesis: zygote to seed. *Science* **266:** 605–614.

Goldschmidt E E, Aschkenazi N, Herzano Y, Schaffer A A and Monselise S P. 1985. A role for carbohydrate levels in the control of flowering in citrus. *Scientia Horticulturae* **26**: 159–166.

Goldschmidt E E and Golomb A. 1982. The carbohydrate balance of alternate-bearing citrus trees and the significance of reserves for flowering and fruiting. *Journal of American Society Horticultural Science* **107**: 206–208.

Golembeski G E, Kinmonth-Schultz H A, Song Y H and Imaizumi T. 2014. Photoperiodic flowering regulation in *Arabidopsis thaliana. Advances in Botanical Research* **72**: 1–28.

Gomez-Roldan V, Fermas S, Brewer P B, Puech-Pagès V, Dun E A, Pillot Jean-Paul, Letisse F, Matusova R, Danoun S, Portais Jean-Charles, Bouwmeester H, Bécard G, Beveridge C A, Rameau C and Rochange S F. 2008. Strigolactone inhibition of shoot branching. *Nature* **455**: 189–194.

Gray W M, Kepinski S, Rouse D, Leyser O and Estelle M. 2001. Auxin regulates SCFTIR1-dependent degradation of AUX/IAA proteins. *Nature* **414**: 271–276.

Greenham K, Santner A, Castillejo C, Mooney S, Sairanen I, Ljung K and Estelle M. 2011. The AFB4 auxin receptor is a negative regulator of auxin signaling in seedlings. *Current Biology* **21**: 520–525.

Griffiths J, Murase K, Rieu I, Zentella R, Zhang Z-L, Powers S J, Gong F, Phillips A L, Hedden P and Sun T P. 2006. Genetic characterization and functional analysis of the GID1 gibberellin receptors in Arabidopsis. *The Plant Cell Online* **18**: 3399–3414.

Guo J, Yang X, Weston D J and Chen J G. 2011. Abscisic acid receptors: past, present and future. *Journal of Integrative Plant Biology* **53**: 469–479.

Halevy A H. 1985–1989. *Hand Book of Flowering*. Volume I–VI. CRC Press, Boca Raton.

Hall A E, Findell J L, Schaller G E, Sisler E C and Bleecker A B. 2000. Ethylene perception by the ERS1 protein in Arabidopsis. *Plant Physiology* **123**: 1449–1458.

Hall S M and Hillman J R. 1975. Correlative inhibition of lateral bud growth in *Phaseolus vulgaris* L. Timing of bud growth following decapitation. *Planta* **123**: 137–143.

Hanzawa Y, Money T and Bradley D. 2005. A single amino acid converts a repressor to an activator of flowering. *Proceedings of National Academy of Sciences of the Unites States of America* **101**: 7748–7753.

Harley C P, Magness J R, Masure M P, Fletcher L A and Degmon E S. 1942. Investigations on the cause and control of biennial bearing of apple trees. *United States Department of Agriculture Technical Bulletin* **792**.

Harris J M. 2015. Abscisic acid: hidden architect of root system structure. *Plants* **4**: 548–573.

Hasanuzzaman M, Nahar K, Alam M M, Roychowdhury R and Fujita M. 2013. Physiological, biochemical, and molecular mechanisms of heat stress tolerance in plants. *International Journal of Molecular Science* **14**: 9643–9684.

Heide M, Bush M G and Evans L T. 1987. Inhibitory and promotive effects of gibberellic acid on floral Initiation and development in *Poa pratensis and Bromus inermis*. *Physiologia Plantarum* **69**: 342–350.

Hemerly A S, Ferreira P, de Almeida Engler J, Van Montagu M, Engler G and Inzé D. 1993. cdc2a expression in *Arabidopsis* is linked with competence for cell division. *The Plant Cell* **5**: 1711–1723.

Heyman J, Cools T, Vandenbussche F, Heyndrickx K S, Van Leene J, Vercauteren I, Vanderauwera S, Vandepoele K, De Jaeger G, Van Der Straeten D and De Veylder L. 2013. ERF115 controls root quiescent center cell division and stem cell replenishment. *Science* **342**: 860–863.

Hillman J R. 1984. Apical dominance. pp. 127–148. *Advanced Plant Physiology* (Ed.) Wilkins M. Pitman, London.

Hirose N, Takei K, Kuroha T, Kamada-Nobusada T, Hayashi H and Sakakibara H. 2008. Regulation of cytokinin biosynthesis, compartmentalization and translocation. *Journal of Experimental Botany* **59**: 75–83.

Hoffman E W. 2006. *Flower Initiation and Development of Protea cv. Carnival*. Dissertation Presented for the Degree of Doctor of Philosophy (Agric.) at the University of Stellenbosch, South Africa.

Horvath D P, Anderson J V, Chao W S and Foley M R. 2003. Knowing when to grow: signals regulating bud dormancy. *Trends in Plant Science* **8**: 534–540.

Hsu C Y, Liu Y, Luthe D S and Yuceer C. 2006. Poplar FT2 shortens the juvenile phase and promotes seasonal flowering. *Plant Cell* **18**: 1846–1861.

Hu Y, Bao F and Li J. 2000. Promotive effect of brassinosteroids on cell division involves a distinct CycD3-induction pathway in *Arabidopsis*. *The Plant Journal* **24**: 693–701.

Hua J and Meyerowitz E M. 1998. Ethylene responses are negatively regulated by a receptor gene family in *Arabidopsis thaliana*. *Cell* **94**: 261–271.

Huang T, Böhlenius H, Eriksson S, Parcy F and Nilsson O. 2005. The mRNA of the Arabidopsis gene FT moves from leaf to shoot apex and induces flowering. *Science* **309**: 1694–1696.

Huang W, Ling Q and Jarvis P. 2013. The ubiquitin-proteasome system regulates chloroplast biogenesis. *Communicative and Integrative Biology* **6**: e23001.

Hutchison C E, Li J, Argueso C, Gonzalez M, Lee E, Lewis M W, Maxwell B B, Perdue T D, Schaller G E, Alonso J M, Ecker J R and Kieber J J. 2006. The Arabidopsis histidine phosphotransfer proteins are redundant positive regulators of cytokinin signaling. *Plant Cell* **18**: 3073–3087.

Hwang I, Sheen J and Muller B. 2012. Cytokinin signaling networks. *Annual Review of Plant Biology* **63**: 353–380.

Inagaki S and Umeda M. 2011. Cell-cycle control and plant development. *International Review of Cell and Molecular Biology* **291**: 227–261.

Jackson S D, James P, Prat S and Thomas B. 1998. Phytochrome B affects the levels of a graft-transmissible signal involved in tuberization. *Plant Physiology* **117**: 29–32.

Jackson S D and Prat S. 1996. Control of tuberization in potato by gibberellins and phytochrome B. *Physiologia Plantarum* **98**: 407–412.

Jacobs W P. 1985. The role of auxin in inductive phenomena. *Biologia. Plantarum* **27:** 303–309.

Jaillais Y, Hothorn M, Belkhadir Y, Dabi T, Nimchuk Z L, Meyerowitz E M and Chory J. 2011. Tyrosine phosphorylation controls brassinosteroid receptor activation by triggering membrane release of its kinase inhibitor. *Genes and Development* **25**: 232–237.

Ji Y and Guo H. 2013. From endoplasmic reticulum (ER) to nucleus: EIN2 bridges the gap in ethylene signaling. *Molecular Plant***6**: 11–14.

Ju C and Chang C. 2012. Advances in ethylene signaling: protein complexes at the endoplasmic reticulum membrane. *AoB Plants Pls*: 031. https://doi.org/10.1093/aobpla/pls031.

Kachru R B, Singh R N and Chacko E K. 1971. Inhibition of flowering in mango *(Mangifera indica* L.) by gibberellic acid. *HortScience* **6:** 140–141.

Kachru R B, Singh R N and Chacko E K. 1972. Inhibition of flowering in *Mangifera indica* L by gibberellic acid. *Acta Horticulturae* **24:** 206–209.

Kendrick M D and Chang C. 2008. Ethylene signaling: new levels of complexity and regulation. *Current Opinion in Plant Biology* **11**: 479.

Kepinski S and Leyser O. 2004. Auxin-induced SCFTIR1-Aux/IAA interaction involves stable modification of the SCFTIR1 complex. *Proceedings of the National Academy of Sciences of the USA* **101**: 12381–12386.

Kepinski S and Leyser O. 2005. The Arabidopsis F-box protein TIR1 is an auxin receptor. *Nature* **435**: 446–451.

Kieber J J and Schaller G E. 2010. The perception of cytokinin: a story 50 years in the making. *Plant Physiology* **154**: 487–492.

Kim J S, Huang T Y and Bokoch G M. 2009. Reactive oxygen species regulate a slingshot-cofilin activation pathway. *Molecular Biology of Cell***20**: 2650–2660.

Kim T H, Böhmer M, Hu H, Nishimura N and Schroeder J I. 2010. Guard cell signal transduction network: advances in understanding abscisic acid, CO_2, and Ca_2C Signaling. *Annual Review of Plant Biology* **61**: 561–591.

Kinet J M. 1993. Environmental, chemical, and genetic control of flowering. *Horticultural Reviews* **15**: 279–334.

Kinet J M, Sachs R M and Bernier G. 1985. *The Physiology of Flowering*. Volume III. CRC Press, Boca Raton.

King R A and Van Staden J. 1988. Differential responses of buds along the shoot of Pisum sativum to isopentyladenine and zeatin application. *Plant Physiology and Biochemistry* **26**: 253–259.

King R W and Zeevaart J A. 1973. Floral stimulus movement in Perilla and flower inhibition caused by noninduced leaves. *Plant Physiology* **51**: 727–738.

Kobayashi Y, Murata M, Minami H, Yamamoto S, Kagaya Y, Hobo T, Yamamoto A and Hattori T. 2005. Abscisic acid-activated SNRK2 protein kinases function in the gene-regulation pathway of ABA signal transduction by phosphorylating ABA response element-binding factors. *The Plant Journal* **44**: 939–949.

Komeda Y. 2004. Genetic regulation of time to flower in Arabidopsis thaliana. *Annual Review of Plant Biology* **55**: 521–535.

Koornneef M, Alonso-Blanco C, Peeters A J M and Soppe W. 1998. Genetic control of flowering time in *Arabidopsis*. *Annual Review of Plant Physiology and Plant Molecular Biology* **49**: 345–370.

Koornneef M and Karssen C M. 1994. Seed dormancy and germination. pp. 313–334. *Arabidopsis* (Eds.) Meyerowitz E M and Somerville C R. Cold Spring Harbor Laboratory Press, Cold Spring Harbor, NY.

Kozlowski T T, Kramer P J and Pallardy S G. 1991. *Physiological Ecology of Wordy Plants*. Academic Press, San Diego.

Krekule J J, Pavlova L, Souckova D and Machackova I. 1985. Auxin in flowering of short-day and long-day *Chenoing* species. *Biologia Plantarum* **27**: 310–317.

Kulkarni V J. 1991. Physiology of flowering in mango studied by grafting. *Acta Horticulturae* **291**: 95–104.

Kumar D and Wareing P F. 1974. Studies on tuberization of *Solanum andigena*. *New Phytologist* **73**: 833–840.

Kurian R M and Iyer C P A. 1993a. Chemical regulation of tree size in mango *(Mangifera indica* L.) cv. Alphonso. II. Effects of growth retardants on flowering and fruit set. *Journal of Horticultural Science* **68:** 355–360.

Kurian R M and Iyer C P A. 1993b. Chemical regulation of tree size in mango *(Mangifera indica* L.) cv. Alphonso. III. Effects of growth retardants on yield and quality of fruits. *Journal of Horticultural Science* **68:** 361–364.

Lal K and Rani S. 1977. Auxins of mango shoot tip and their significance in flowering. *Pantnagar Journal of Research* **2:** 31–35.

Lang A. 1965. Physiology of flower initiation. pp. 1380–1536. *Encyclopedia of Plant Physiology*. Volume 15 (Ed.) Ruhland W H. Springer, Berlin, Heidelberg, New York.

Lang A. 1980. Inhibition of flowering in long day plants. pp. 310–322. *Plant Growth Substances* (Ed.) Skoog F. Springer-Verlag, Berlin.

Lee S C and Luan S. 2012. ABA signal transduction at the crossroad of biotic and abiotic stress responses. *Plant, Cell and Environment* **35**: 53–60.

Leyser O. 2005. The fall and rise of apical dominance. *Current Opinion in Genetics and Development* **15**: 468–471.

Li C J and Bangerth F. 1999. Autoinhibition of indoleacetic acid transport in the shoots of two-branched pea (Pisum sativum) plants and its relationship to correlative dominance. *Physiologia Plantarum* **106**: 415–420.

Li C J, Herrera G J and Bangerth F. 1995. Effect of apex excision and replacement by 1-naphthylacetic acid on cytokinin concentration and apical dominance in pea plants. *Physiologia Plantarum* **94**: 465–469.

Lifschitz E, Ayre B G and Eshed Y. 2014. Florigen and anti-florigen – a systemic mechanism for coordinating growth and termination in flowering plants. *Frontiers in Plant Science* **16**. https://doi.org/10.3389/fpls.2014.00465.

Lin Z, Zhong S and Grierson D. 2009. Recent advances in ethylene research. *Journal of Experimental Botany* **60**: 3311–3336.

Loeb J. 1924. *Regeneration from a Physico-Chemical Viewpoint*. McGraw-Hill, New York.

Lopez M R, Mosqueda-Vazquex R and Pimiento B E. 1984. El nitrato de potasio como promoter de etileno endogeno y la induction floral del mango (*Mangifera indica* L.) cv. Manila. *Resumenes del Congreso Nacional de Rtogenetica, Institute Technologias Agropegurio, Numero 20 de Aguascalientes. Ags.* 100 p. Universidad Autónoma, Chapingo, México.

Loreti F and Pisani D P. 1990. Structural manipulation for improved performance in woody plants. *HortScience* **25**: 64–70.

Ma Y, Szostkiewicz I, Korte A, Moes D, Yang Y, Christmann A and Grill E. 2009. Regulators of PP2C phosphatase activity function as abscisic acid sensors. *Science* **324**: 1064–1068.

Madoka Y and Mori H. 2000. Acropetal disappearance of PsAD1 protein in pea axillary buds after the release of apical dominance. *Plant Cell Phy*siology **41**: 556–564.

Magyar Z, De Veylder L, Atanassova A, Bakó L, Inzé D and Bögre L. 2005. The role of the *Arabidopsis* E2FB transcription factor in regulating auxin-dependent cell division. *The Plant Cell* **17**: 2527–2541.

Martin G. 1987. Apical dominance. *HortScience* **22**: 8244–8833.

Martínez-Zapater J M and Somerville C R. 1990. Effect of light quality and vernalization on late-flowering mutants of *Arabidopsis thaliana*. *Plant Physiology* **92**: 770–776.

Mason M G, Ross J J, Babst B A, Wienclaw B and Beveridge C A. 2014. Sugar demand, not auxin, is the initial regulator of apical dominance. *PNAS* **111**: 6092–6097.

Mayer U, Ruiz R A T, Berleth T, Misera S and Jürgens G. 1991. Mutations affecting body organization in the Arabidopsis embryo. *Nature* **353**: 402–407.

McGinnis K M, Thomas S G, Soule J D, Strader L C, Zale J M, Sun T and Steber C M. 2003. The Arabidopsis SLEEPY1 gene encodes a putative F-box subunit of an SCF E3 ubiquitin ligase. *The Plant Cell Online* **15**: 1120–1130.

Mehrotra R, Bhalothia P, Bansal P, Basantani M K, Bharti V and Mehrotra S. 2014. Abscisic acid and abiotic stress tolerance – different tiers of regulation. *Journal of Plant Physiology* **171**: 486–496.

Meinke D W. 1995. Molecular genetics of plant embryogenesis. *Annual Review of Plant Physiology and Plant Molecular Biology* **46**: 369–394.

Melcher K, Ng L M, Zhou X E, Soon F F, Xu Y, Suino-Powell K M, Park S Y, Weiner J J, Fujii H, Chinnusamy V, Kovach A, Li J, Wang Y, Li J, Peterson F C, Jensen D R, Yong E L, Volkman B F, Cutler S R, Zhu J K and Xu H E. 2009. A gate-latch-lock mechanism for hormone signaling by abscisic acid receptors. *Nature* **462**: 602–608.

Melcher K, Zhou X E and Xu H E. 2010. Thirsty plants and beyond: structural mechanisms of abscisic acid perception and signaling. *Current Opinion in Structural Biology* **20**: 722–729.

Mishra K A and Dhillon B S. 1978. Carbohydrates and mineral composition of leaves in relation to fruit-bud differentiation in 'Langra' mango. *Indian Journal of Agricultural Science* **47:** 46–50.

Miyazono K, Miyakawa T, Sawano Y, Kubota K, Kang H J, Asano A, Miyauchi Y, Takahashi M, Zhi Y, Fujita Y, Yoshida T, Kodaira K S, Yamaguchi-Shinozaki K and Tanokura M. 2009. Structural basis of abscisic acid signaling. *Nature* **462**: 609–614.

Monselise S P and Goldschmidt E E. 1982. Alternate bearing in fruit trees. *Horticulture Review* **4**: 128–173.

Morris D A. 1977. Transport of exogenous auxin in two-branched dwarf pea seedlings (*Pisum sativum* L.). *Planta* **136**: 91–96.

Morris S E, Cox M C H, Ross J J, Krisantini S and Beveridge C A. 2005. Auxin dynamics after decapitation are not correlated with the initial growth of axillary buds. *Plant Physiology* **138**: 1665–1672.

Mosqueda-Vazquez R and Avila-Resendiz C. 1985. Floral induction of Mango with KNO3 applications and its inhibition by AgNO3 or CoCl2 application. *Horticultura Mexicana* **1**: 93–101.

Mouradov A, Cremer F and Coupland G. 2002. Control of flowering time: interacting pathways as a basis for diversity. *Plant Cell* **14**: S111–S130.

Murase K, Hirano Y, Sun T-P and Hakoshima T. 2008. Gibberellin-induced DELLA recognition by the gibberellin receptor GID1. *Nature* **456**: 459–463.

Nakabayashi K, Okamoto M, Koshiba T, Kamiya Y and Nambara E. 2005. Genome-wide profiling of stored mRNA in Arabidopsis thaliana seed germination: epigenetic and genetic regulation of transcription in seed. *Plant Journal* **41:** 697–709.

Nakajima M, Shimada A, Takashi Y, Kim Y C, Park S H, Ueguchi-Tanaka M, Suzuki H, Katoh E, Iuchi S and Kobayashi M. 2006. Identification and characterization of Arabidopsis gibberellin receptors. *The Plant Journal* **46**: 880–889.

Nakasone H V and Storey W B. 1955. Studies on the inheritance of fruiting height of *Carica papaya* L. *American Society for Horticultural Science Proceedings* **66:** 168–182.

Napoli C A, Beveridge C A and Snowden K C. 1999. Reevaluating concepts of apical dominance and the control of axillary bud outgrowth. *Current Topics Developmental Biology* **44**: 127–169.

Nemhauser J L, Zambryski P C and Roe J L. 1998. Auxin signaling in *Arabidopsis* flower development? *Current Opinion in Plant Biology* **1:** 531–535.

Ng L M, Melcher K, Teh B T and Xu H E. 2014. Abscisic acid perception and signaling: structural mechanisms and applications. *Acta Pharmacologia Sinica* **35**: 567–584.

Nordstrom A, Tarkowski P, Tarkowska D, Norbaek R, Astot C, Dolezal K and Sandberg G. 2004. Auxin regulation of cytokinin biosynthesis in *Arabidopsis thaliana*: a factor of potential importance for auxin-cytokinin regulated development. *Proceedings National Academy of Sciences, USA* **101**: 8039–8044.

Nunez-Elisea R, Caldeir M L and Davenport T L. 1990. Thidiazuron effects on growth Initiation and expression in mango *(Mangifera indica* L.). *HortScience* **25**: 1167.

Nunez-Elisea R and Davenport T L. 1991. Effect of duration of low temperature treatment on flowering of containerized 'Tommy Atkins' mango. pp. 39–41. In *Proceedings of the Plant Growth Regulator Society of America (USA).* Boston, MA.

Nunez-Elisea R and Davenport T L. 1991. Flowering of 'Keitt' mango in response to deblossoming and gibberellic acid. *Proceedings of Florida State Horticultural Society* **104:** 41–43.

Ogaki C, Fujita K and Ito H. 1963. Investigations on the cause and control of alternate bearing in Unshiu orange trees. IV. Nitrogen and carbohydrate contents in the shoots as related to blossoming and fruiting. *Journal of Japanese Society of Horticultural Science* **32**: 157–167.

Ogawa Y and King R W. 1980. Flowering in seedlings of *Pharbitis nil* induced by benzyl adenine applied under a non-inductive day length. *Plant Cell Physiology* **21**: 1109–1116.

Ohkuma K, Lyon J L, Addicott F T and Smith O E. 1963. Abscisin II, an abscission-accelerating substance from young cotton fruit. *Science* **142**: 1592–1593.

Okada K, Ueda J, Komaki M K, Bell C J and Shimura Y. 1991. Requirement of the auxin polar transport system in early stages of Arabidopsis floral bud formation. *Plant Cell* **3**: 677–684.

Ortega-Martínez O, Pernas M, Carol R J and Dolan L. 2007. Ethylene modulates stem cell division in the *Arabidopsis thaliana* root. *Science* **317**: 507–510.

Osthuyse S A. 1991. Effect of promotion on flowering and lateral branching of young mango trees in the nursery. *South African Mango Growers' Association Yearbook* **11**: 54.

Paleg Z and Blumwald E. 2011. Hormonal balance and abiotic stress tolerance in crop plants. *Current Opinion in Plant Biology* **14**: 290–295.

Pandey R M and Narwadkar P R. 1984. Studies on the induction of growth and flowering in Dashehari mango. *Indian Journal of Horticulture* **41:** 171–176.

Pandey S, Nelson D C and Assmann S M. 2009. Two novel GPCR-type G proteins are abscisic acid receptors in Arabidopsis. *Cell* **136**: 136–148.

Park I K, Mundy-Bosse B, Whitman S P, Zhang X, Warner S L, Bearss D J and Caligiuri M A. 2015. Receptor tyrosine kinase Axl is required for resistance of leukemic cells to FLT3-targeted therapy in acute myeloid leukemia. *Leukemia* **29**: 2382–2389.

Park S Y, Fung P, Nishimura N, Jensen D R, Fujii H, Zhao Y, Lumba S, Santiago J, Rodrigues A, Chow T F, Alfred S E, Bonetta D, Finkelstein R, Provart N J, Desveaux D, Rodriguez P L, McCourt P, Zhu J K, Schroeder J I, Volkman B F and Cutler S R. 2009. Abscisic acid inhibits type 2C protein phosphatases via the PYR/PYL family of START proteins. *Science* **324**: 1068–1071.

Pauwels L, Barbero G Fn, Geerinck J, Tilleman S, Grunewald W, Parez A C, Chico J M, Bossche R V, Sewell J and Gil E. 2010. NINJA connects the co-repressor TOPLESS to jasmonate signaling. *Nature* **464**: 788–791.

Perez A C L and Goossens A. 2013. Jasmonate signalling: a copycat of auxin signalling? *Plant, Cell and Environment.* https://doi.org/10.1111/pce.12121.

Périlleux C and Bernier G. 2002. The control of flowering: Do genetical and physiological approaches converge? pp. 1–32. *Plant Reproduction. Annual Plant Reviews.* Volume 6 (Eds.) O'Neill S D and Roberts J A. Sheffield Academic Press, Sheffield.

Perilli S, Moubayidin L and Sabatini S. 2010. The molecular basis of cytokinin function. *Current Opinion in Plant Biology* **13**: 21–26.

Phillips I D J. 1975. Apical dominance. *Annual Review of Plant Physiology and Plant Molecular Biology* **26**: 341–367.

Pilot G, Lacombe B, Gaymard F, Cherel I, Boucherez J, Thibaud J B and Sentenac H. 2001. Guard cell inward K+ channel activity in arabidopsis involves expression of the twin channel subunits KAT1 and KAT2. *The Journal of Biological Chemistry* **276**: 3215–3221.

Prusinkiewicz P, Crawford S, Smith R S, Ljung K, Bennet T, Ongaro V and Leyser O. 2009. Control of bud activation by an auxin transport switch. *Proceedings of National Academy of Science, USA* **106**: 17431–17436.

Punwani J A, Hutchison C E, Schaller G E and Kieber J J. 2010. The subcellular distribution of the Arabidopsis histidine phosphotransfer proteins is independent of cytokinin signaling. *The Plant Journal* **62**: 473–482.

Qin F, Shinozaki K and Yamaguchi-Shinozaki K. 2011. Achievements and challenges in understanding plant abiotic stress responses and tolerance. *Plant Cell Physiology* **52**: 1569–1582.

Qu R, Li X and Luo Y. 2007. Wind-dragged corolla enhances self-pollination: a new mechanism of delayed self-pollination. *Annals of Botany* **100**: 1155–1164.

Rajput C B S and Singh J N. 1989. Effects of urea and GA, sprays on the growth, flowering and fruiting characters of mango. *Acta Horticulturae* **231**: 301–305.

Rameshwar A. 1989. Mango flowering – stress induced. *Acta Horticulturae* **231:** 433–439.

Ramina A, Hackett W P and Sachs R M. 1979. Flowering in bougainvillea-function of assimilate supply and nutrient diversion. *Plant Physiology* **64**: 810–813.

Rashotte A M, Mason M G, Hutchison C E, Ferreira F J, Schaller G E and Kieber J J. 2006. A subset of Arabidopsis AP2 transcription factors mediates cytokinin responses in concert with a two-component pathway. *Proceedings of the National Academy of Sciences of the USA* **103**: 11081–11085.

Raz V, Bergervoet J H and Koornneef M. 2001. Sequential steps for developmental arrest in Arabidopsis seeds. *Development* **128**: 243–252.

Renton M, Hanan J, Ferguson B J and Beveridge C A. 2012. Models of long-distance transport: how is carrier-dependent auxin transport regulated in the stem? *New Phytologist* **194**: 704–715.

Rodriguez A G. 1932. Influence of smoke and ethylene on the fruiting of pineapple (*Ananas sativus* Schult.). *Journal of Agricultural University, Puerto Rico* **16:** 5–18.

Roudier F, Fedorova E, Lebris M, Lecomte P, Gyorgyey J, Vaubert D, Horvath G, Abad P, Kondorosi A and Kondorosi E. 2003. The *Medicago* species A2-type cyclin is auxin regulated and involved in meristem formation but dispensable for endoreduplication-associated developmental programs. *Plant Physiology* **131**: 1091–1103.

Ruyter-Spira C, Kohlen W, Charnikhova T, van Zeijl A, van Bezouwen L, de Ruijter N, Cardoso C, Lopez-Raez J A, Matusova R, Bours R, Verstappen F and Bouwmeester H. 2011. Physiological effects of the synthetic strigolactone analog GR24 on root system architecture in *Arabidopsis*: another belowground role for strigolactones? *Plant Physiology* **155**: 721–734.

Růžička K, Ljung K, Vanneste S, Podhorská R, Beeckman T, Friml J and Benková E. 2007. Ethylene regulates root growth through effects on auxin biosynthesis and transport-dependent auxin distribution. *The Plant Cell* **19**: 2197–2212.

Sachs J. 1865. Wirkung des lichtes auf die blütenbilding unter vermittlung der laubblätter. *Botany. Ztg.* **23**: 117–121; 125–131; 133–139.

Sachs R M. 1977. Nutrient diversion: a hypothesis to explain the chemical control of flowering. *HortScience* **12**: 220–222.

Sachs R M and Hackett W P. 1983. Source-sink relationships and flowering. pp. 263–272. *Strategies of Plant Reproduction* (Ed.) Meudt W J. Beltsville Symposia in Agricultural Research. Allanheld, Osmun Publishing, Totowa, NJ.

Sachs R M, Hackett W P, Ramina A and Maloo C. 1979. Photosynthetic assimilation and nutrient diversion as controlling factors in flower initiation in bougainvillea (San Diego Red) and *Nicotiana tabacum cv.* Wis. 38. pp. 95–102. *Photosynthesis and Plant Development* (Eds.) Marcella R, Lysters H and van Poucke M. W. Junk, The Hague.

Sachs T. 1991. *Pattern Formation in Plant Tissues*. Cambridge University Press, New York.

Sachs T and Thimann K V. 1967. The role of auxins and cytokinins in the release of buds from dominance. *American Journal of Botany* **54**(1): 136–144.

Sah S K, Reddy K R and Jiaxu Li. 2016. Abscisic acid and abiotic stress tolerance in crop plants. *Frontiers in Plant Science* **7**: 571–597.

Salisbury F B. 1955. The dual role of auxin in flowering. *Plant Physiology* **30**: 327–334.

Santner A, Calderon-Villalobos L I A and Estelle M. 2009. Plant hormones are versatile chemical regulators of plant growth. *Nature Chemical Biology* **5**: 301–307.

Santner A and Estelle M. 2009. Recent advances and emerging trends in plant hormone signaling. *Nature* **459**: 1071–1078.

Santner A and Estelle M. 2010. The ubiquitin proteasome system regulates plant hormone signaling. *The Plant Journal* **61**: 1029–1040.

Sasaki A, Itoh H, Gomi K, Ueguchi-Tanaka M, Ishiyama K, Kobayashi M, Jeong D-H, An G, Kitano H and Ashikari M. 2003. Accumulation of phosphorylated repressor for gibberellin signaling in an F-box mutant. *Science Signaling* **299**: 1896–1898.

Scorza R and Janick J. 1980. In-vitro flowering of Passiflora suberosa L. *Journal of American Society of Horticultural Science* **105**: 892–897.

Searle I, He Y, Turck F, Vincent C, Fornara F, Kro¨ ber S, Amasino R A and Coupland G. 2006. The transcription factor FLC confers a flowering response to vernalization by repressing meristem competence and systemic signaling in Arabidopsis. *Genes Development* **20**: 898–912.

Seki M, Ishida J, Narusaka M, Fujita M, Nanjo T, Umezawa T, Kamiya A, Nakajima M, Enju A, Sakurai T, Satou M, Akiyama K, Yamaguchi-Shinozaki K, Carninci P, Kawai J, Hayashizaki Y and Shinozaki K. 2002. Monitoring the expression pattern of around 7,000 Arabidopsis genes under ABA treatments using a full-length cDNA microarray. *Functional and Integrative Genomics* **2**: 282–291.

Shang Y, Yan L, Liu Z Q, Cao Z, Mei C, Xin Q, Wu F Q, Wang X F, Du S Y, Jiang T, Zhang X F, Zhao R, Sun H L, Liu R, Yu Y T and Zhang D P. 2010. The Mg-chelatase H subunit of *Arabidopsis* antagonizes a group of WRKY transcription repressors to relieve ABA-responsive genes of inhibition. *Plant Cell* **22**: 1909–1935.

Shani E, Weinstain R, Zhang Y, Castillejo C, Kaiserli E, Chory J, Tsien R Y and Estelle M. 2013. Gibberellins accumulate in the elongating endodermal cells of *Arabidopsis* root. *Proceedings of the National Academy of Sciences, USA* **110**: 4834–4839.

Shen Y Y, Wang X F, Wu F Q, Du S Y, Cao Z, Shang Y, Wang X L, Peng C C, Yu X C, Zhu S Y, Fan R C, Xu Y H and Zhang D P. 2006. The Mg-chelatase H subunit is an abscisic acid receptor. *Nature* **443**: 823–826.

Shimada A, Ueguchi-Tanaka M, Nakatsu T, Nakajima M, Naoe Y, Ohmiya H, Kato H and Matsuoka M. 2008. Structural basis for gibberellin recognition by its receptor GID1. *Nature* **456**: 520–523.

Shimizu-Sato S and Mori H. 2001. Control of outgrowth and dormancy in axillary buds. *Plant Physiology* **127**: 1405–1413.

Shinozaki K and Yamaguchi-Shinozaki K. 2007. Gene networks involved in drought stress response and tolerance. *Journal of Experimental Botany* **58**: 221–227.

Singh L B. 1960. *The Mango Botany, Cultivation and Utilization*. Leonard Hill, London.

Singh L B. 1961. Biennial bearing in mango – Effect of gibberellic acid and maleic hydrazide. *Horticultural Advances* **5:** 96–106.

Singh R N. 1959. Movement of flowering substances in mango leaves. *Horticultural Advances* **3**: 20–27.

Smalle J and Vierstra R D. 2004. The ubiquitin 26S proteasome proteolytic pathway. *Annual Review of Plant Biology* **55**: 555–590.

Snape J W, Quarrie S A and Laurie D A. 1996. Comparative mapping and its use for the genetic analysis of agronomic characters in wheat. *Euphytica* **89**: 27–31.

Snow R. 1937. On the nature of correlative inhibition. *New Phytologist* **36**: 283–300.

Srinivasan C and Mullins M C. 1980. Effect of temperature and growth regulators on formation of anlagen, tendrils and inflorescences in *Vitis vinifera*. *Annals of Botany* **45**: 439–446.

Srinivasan C and Mullins M C. 1981. Physiology of flowering in the grape vine – A review. *American Journal of Enology and Viticulture* **32**: 47–63.

Stafstrom J P. 1995. Developmental potential of shoot buds. pp. 257–279. *Plant Stems: Physiology and Functional Morphology* (Ed.) Gartner B. Academic Press, San Diego, CA.

Stafstrom J P, Ripley B D, Devitt M L and Drake B. 1998. Dormancy-associated gene expression in pea axillary buds. Cloning and expression of PsDRM1 and PsDRM2. *Planta* **205**: 547–552.

Stafstrom J P and Sussex I M. 1992. Expression of a ribosomal protein gene in axillary buds of pea seedlings. *Plant Physiology* **100**: 1494–1502.

Suryanarayana V and Rao V N M. 1977. Ascorbic acid changes in shoots of mango cv. Mulgoa as affected by growth retardants in relation to flowering. *Indian Journal of Plant Physiology* **20**: 88–90.

Szemenyei H, Hannon M and Long J A. 2008. TOPLESS mediates auxin-dependent transcriptional repression during Arabidopsis embryogenesis. *Science Signaling* **319**: 1384.

Takatsuka H and Umeda M. 2014. Hormonal control of cell division and elongation along differentiation trajectories in roots. *Journal of Experimental Botany* **65**: 2633–2643.

Tamas I. 1987. Hormonal regulation of apical dominance. pp. 393–410. *Plant Hormones and Their Role in Plant Growth and Development* (Ed.) Davies P. Martinus Nijhoff, Boston, MA.

Tan X, Calderon-Villalobos L I A, Sharon M, Zheng C, Robinson C V, Estelle M and Zheng N. 2007. Mechanism of auxin perception by the TIR1 ubiquitin ligase. *Nature* **446**: 640–645.

Tanaka M, Takei K, Kojima M, Sakakibara H and Mori H. 2006. Auxin controls local cytokinin biosynthesis in the nodal stem in apical dominance. *Plant Journal* **45**: 1028–1036.

Thimann K V. 1937. On the nature of inhibitions caused by auxin. *American Journal of Botany* **24**: 407–412.

Thimann K V, Skoog F and Kerckhoff W G. 1934. On the inhibition of bud development and other functions of growth substance in *Vicia faba*. *Proceedings of Royal Society of London – Series B Biological Science* **114**: 317–339.

Thines B, Katsir L, Melotto M, Niu Y, Mandaokar A, Liu G, Nomura K, He S Y and Howe G A. 2007. JAZ repressor proteins are targets of the SCFCOI1 complex during jasmonate signalling. *Nature* **448**: 661–665.

To J P, Deruere J, Maxwell B B, Morris V F, Hutchison C E, Ferreira F J, Schaller G E and Kieber J J. 2007. Cytokinin regulates type-A Arabidopsis response regulator activity and protein stability via two-component phosphorelay. *Plant Cell* **19**: 3901–3914.

To J P and Kieber J J. 2008. Cytokinin signaling: two-components and more. *Trends in Plant Science* **13**(2): 85–92.

Tobena-Santamaria R, Bliek M, Ljung K, Sandberg G, Mol J N, Souer E and Koes R. 2002. FLOOZY of petunia is a flavin monooxygenase-like protein required for the specification of leaf and flower architecture. *Genes and Development* **16**: 753–763.

Tran Thanh Van K. 1980. Control of morphogenesis by inherent and exogenously. *Cytotoxicity* **11A**(Supplement): 175–194.

T'SE A T. 1975. *Inflorescence Development in Bougainvillea "San Diego Red" in Relation to Assimilate Accumulation and Diversion*. PhD Thesis. University of California, Davis.

Turnbull G C N, Anderson K L and Winston E C. 1996. Influence of gibberellin treatment on flowering and fruiting patterns mango. *Australian Journal of Experimental Agriculture* **36**: 603–611.

Turnbull C G N, Raymond M A A, Dodd I C and Morris S E. 1997. Rapid increases in cytokinin concentration in lateral buds of chickpea (*Cicer arietinum* L.) during release of apical dominance. *Planta* **202**: 271–277.

Tuteja N. 2007. Abscisic and abiotic stress signaling. *Plant Signaling and Behaviour* **2**: 135–138.

Ubeda-Tomás S, Federici F, Casimiro I, Beemster G T S, Bhalerao R, Swarup R, Doerner P, Haseloff J and Bennett M J. 2009. Gibberellin signaling in the endodermis controls Arabidopsis root meristem size. *Current Biology* **19**: 1194–1199.

Umezawa T, Sugiyama N, Mizoguchi M, Hayashi S, Myouga F, Yamaguchi-Shinozaki K, Ishihama Y, Hirayama T and Shinozaki K. 2009. Type 2C protein phosphatases directly regulate abscisic acid-activated protein kinases in Arabidopsis. *Proceedings of the National Academy of Sciences of the USA* **106**: 17588–17593.

Uno Y, Furihata T, Abe H, Yoshida R, Shinozaki K and Yamaguchi-Shinozaki K. 2000. Arabidopsis basic leucine zipper transcription factors involved in an abscisic acid-dependent signal transduction pathway under drought and high-salinity conditions. *Proceedings of the National Academy of Sciences of the USA* **97**: 11632–11637.

Vahisalu T, Kollist H, Wang Y F, Nishimura N, Chan W Y, Valerio G, Lamminmaki A, Brosche M, Moldau H, Desikan R, Schroeder J I and Kangasjarvi J. 2008. SLAC1 is required for plant guard cell S-type anion channel function in stomatal signaling. *Nature* **452**: 487–491.

Vierstra R D. 2009. The ubiquitin-26S proteasome system at the nexus of plant biology. *Nature Reviews Molecular Cell Biology* **10**: 385–397.

Villalobos L I A C N, Lee S, De Oliveira C, Ivetac A, Brandt W, Armitage L, Sheard L B, Tan X, Parry G and Mao H. 2012. A combinatorial TIR1/AFB-Aux/IAA co-receptor system for differential sensing of auxin. *Nature Chemical Biology* **8**: 477–485.

Wang H, Qi Q, Schorr P, Cutler A J, Crosby W L and Fowke L C. 1998. ICK1, a cyclin-dependent protein kinase inhibitor from *Arabidopsis thaliana* interacts with both Cdc2a and CycD3, and its expression is induced by abscisic acid. *The Plant Journal* **15**: 501–510.

Wang L, Wang Z, Xu Y, Joo S-H, Kim S-K, Xue Z, Xu Z, Wang Z and Chong K. 2009. OsGSR1 is involved in crosstalk between gibberellins and brassinosteroids in rice. *The Plant Journal* **57**: 498–510.

Wang W X, Vinocur B and Altman A. 2003. Plant responses to drought, salinity and extreme temperatures: towards genetic engineering for stress tolerance. *Planta* **218**: 1–14.

Wang W X, Vinocur B, Shoseyov O and Altman A. 2001. Biotechnology of plant osmotic stress tolerance and physiological and molecular considerations. *Acta Horticulturae* **560**: 285–292.

Wani S H and Kumar V. 2015. Plant stress tolerance: engineering ABA: a potent phytohormone. *Transcriptomics* **3**: 1000113. https://doi.org/10.4172/2329-8936.1000113.

Weberling F. 1989. Structure and evolutionary tendencies of inflorescences in the Leguminosae. *Monographs in Systematic Botany from the Missouri Botanical Garden* **29**: 35–58.

Weller J L, Hecht V, Liew L C, Sussmilch F C, Wenden B, Knowles C L, Vander Schoor J K, et al. 2009. Update on the genetic control of flowering in garden pea. *Journal of Experimental Botany* **60**: 2493–2499.

Weller J L, Reid J B, Taylor S A and Murfet I C. 1997. The genetic control of flowering in pea. *Trends in Plant Science* **2**: 412–418.

Wickson M and Thimann K. 1958. The antagonism of auxin and kinetin in apical dominance. *Physiologia Plantarum* **11**: 62–74.

Wilmowicz E, Kesy J and Kopcewicz J. 2008. Ethylene and ABA interactions in the regulation of flower induction in Pharbitis nil. *Journal of Plant Physiology* **165**: 1917–1928.

Wilson B. 1990. The development of tree form. *HortScience* **25**: 52–54.

Woodward A W and Bartel B. 2005. Auxin: regulation, action, and interaction. *Annals of Botany* **95**: 707–735.

Worley R E. 1979. Fall defoliation date and seasonal carbohydrate concentration of pecan wood tissue. *Journal American Society for Horticultural Science* **104**: 195–199.

Xiong L and Zhu J K. 2003. Regulation of abscisic acid biosynthesis. *Plant Physiology* **133**: 29–36.

Xue-Xuan X, Hong-Bo S, Yuan-Yuan M, Gang X, Jun-Na S, Dong-Gang G and Cheng-Jiang R. 2010. Biotechnological implications from abscisic acid (ABA) roles in cold stress and leaf senescence as an important signal for improving plant sustainable survival under abiotic-stressed conditions. *Critical Reviews in Biotechnology* **30**: 222–230.

Yamaguchi N, Wu M F, Winter C M, Berns M C, Nole-Wilson S, Yamaguchi A, Coupland G, Krizek B A and Wagner D. 2013. A molecular framework for auxin-mediated initiation of flower primordial. *Developmental Cell* **24**: 271–282.

Yamaguchi S. 2008. Gibberellin metabolism and its regulation. *Annual Review of Plant Biology* **59**: 225–251.

Yamaguchi-Shinozaki K and Shinozaki K. 2006. Transcriptional regulatory networks in cellular responses and tolerance to dehydration and cold stresses. *Annual Review Plant Biology* **57**: 781–803.

Yang C J, Zhang C, Lu Y N, Jin J Q and Wang X L. 2011. The mechanisms of brassinosteroids' action: from signal transduction to plant development. *Molecular Plant* **4**: 588–600.

Yoo S D, Cho Y and Sheen J. 2009. Emerging connections in the ethylene signaling network. *Trends in Plant Science* **14**: 270.

Yu X, Li L, Zola J, Aluru M, Ye H, Foudree A, Guo H, Anderson S, Aluru S, Liu P, Rodermel S and Yin Y. 2011. A brassinosteroid transcriptional network revealed by genome-wide identification of BESI target genes in *Arabidopsis thaliana. The Plant Journal* **65**: 634–646.

Zafari M, Malekzadeh A and Jahani R. 2015. Effects of Plant hormone auxin in flower development. *Agriculture Science Developments* **4**: 27–20.

Zeevaart J A D. 1976. Physiology of flower formation. *Annual Review of Plant Physiology* **27**: 321–348.

Zeevaart J A D. 1982. Transmission of the floral stimulus from a long-short-day plant, *Bryophyllum diagremontianum*, to the short-long-day plant *Echeveria harmsii. Annals of Botany, London* **49**: 549–552.

Zeevaart J A D. 2006. Florigen coming of age after 70 years – Current perspective essay. *The Plant Cell* **18**: 1783–1789.

Zeevaart J A D and Creelman R A. 1988. Metabolism and physiology of abscisic acid. *Annual Review of Plant Physiology and Plant Molecular Biology* **39**: 439–473.

Zhang L, Nebane N M, Wennerberg K, Li Y, Neubauer V, Hobrath J V, McKellip S, Rasmussen L, Shindo N and Sosa M. 2010. A high-throughput screen for chemical inhibitors of exocytic transport in yeast. *Chembiochem***11**: 1291–1301.

Zhang Y, Zhao Z and Xue Y. 2009. Roles of proteolysis in plant self-incompatibility. *Annual Review of Plant Biology* **60**: 21–42.

Zhao Y, Christensen S K, Fankhauser C, Cashman J R, Cohen J D, Weigel D and Chory J. 2001. A role for flavin monooxygenase-like enzymes in auxin biosynthesis. *Science* **291**: 306–330.

21 Hormonal Control of Fruit Setting, Growth and Ripening

21.1 INTRODUCTION

Botanically, a fruit is the seed-bearing structure in flowering plants (angiosperms) formed from the fertilized ovary. They are the end product of a matured ovary containing a single seed as in cereal grains (rice, wheat, barley, rye) or a fleshy succulent structure as in common fruits such as bananas, apples, water melon and strawberries. Fruits serve as the means by which angiosperms disseminate seeds. Humans and animals by using edible fruits to fulfill their nutrition needs help in seed dispersal. In that sense, fruits account for a substantial portion of the world's agricultural output. Common fruits such as bananas, apples, oranges, grapes, lemons and strawberries are regarded as the fleshy seed-bearing structures that are sweet or sour in taste when ripe and not edible when raw or immature. Botanically, fruits also include many structures that are not commonly termed as fruits, e.g., grains like wheat, corn, and many common vegetables like tomatoes, beans, peas, cucumbers, squash, peppers, eggplant, okra and sweet corn, as well as the fungal spores (Lewis, 2002; Desai, 2004).

Several processes take place as the fruits ripen to become edible and senesce; these changes may take place while the fruits are still attached to the parent plant or after their harvest. Fruits such as tomato, banana and avocado are harvested at their mature green stage and are inedible until they ripe. In contrast, other fruits like orange, strawberry, grape and boysenberry need to stay on the tree or vine until they become ready to eat in order to develop their desired eating quality. The major physiological changes taking place during the fruit ripening may be collectively characterized as follows (Atkinson et al., 2016):

- changes in carbohydrate metabolism resulting in increased sweetness owing to the accumulation of total and reducing sugars;
- changes in color and flavor due to the formation of aroma volatiles;
- textural changes leading to softness;
- accumulation of organic acids associated with the development of flavor.

These changes in the ripened fruit make them attractive to be consumed by man and animals helping in the dispersal of seeds for the next generation. The lignified pits and seeds encased in the fibrous core may be discarded after eating the flesh while the smaller seeds may pass through the animals' digestive system to be deposited with their excrement. Broadly the fruits are distinguished into two classes, viz., the climacteric and the non-climacteric ones, based on the patterns of respiration during the process of ripening. Kidd and West (1938) first showed that fruits, such as apples, are characterized by their increase in the rate of respiration during ripening. This finding led to classification of fruits into climacteric and non-climacteric based on the patterns of respiration (Table 21.1).

In this chapter, only the horticulturally classified fleshy fruits are primarily considered for detailed discussion on the regulation of fruit formation, setting, fruit growth, development, fruit drop and ripening.

DOI: 10.1201/9781003373889-27

TABLE 21.1
Some Commonly Known Climacteric and Non-Climacteric Fruits

Climacteric fruits	Non-climacteric fruits
Apple	Beans
Avocado	Beats
Banana	Brussels sprout
Cantaloupe	Cauliflower
Fig	Cherry
Guava	Citrus (lemon, lime, orange, tangerine, sweet lime, etc.)
Mango	Cucumber
Melon	Grape
Nectarine	Loquat
Papaya	Lychee
Passionfruit	Okra
Pear	Olive
Persimmon	Pepper
Tomato	Pineapple
	Pomegranate
	Rambutan
	Strawberry
	Watermelon

21.2 HORMONAL REGULATION OF FRUIT FORMATION, SETTING AND GROWTH

21.2.1 Regulation by Ethylene

The role of ethylene in plant growth has been the subject of primary interest of the plant physiologists. Ethylene has been known to inhibit cell division, DNA synthesis and growth in the meristems of roots, stems and axillary buds without influencing the RNA synthesis. Apical dominance often is broken when ethylene is removed, apparently because the gas inhibits polar auxin transport irreversibly, thereby reducing the shoot's auxin content just as if the apex has been removed. A similar mechanism may underlie ethylene-induced release of buds, tubers, root initials and seeds from dormancy. Ethylene often inhibits cell expansion within 15 min and delays differentiation, which will make the previously expanding cells eventually grow to enormous size. Tropistic responses are inhibited when ethylene reversibly and rapidly prevents lateral auxin transport. In most of these cases and some other instances, ethylene action is mimicked by the auxin application, since the latter induces ethylene formation. Regulation by ethylene extends to abscission to flower formation and fading, and to fruit growth and ripening. Production of ethylene is controlled by auxin and by red light, auxin acting to induce a labile enzyme needed for ethylene synthesis and red light to repress ethylene production (Stanley, 1973).

Ethylene has also been found to regulate root cap size and cell differentiation in maize, because its addition or the inhibition of its synthesis affected root cap development. Ethylene also appeared to regulate the auxin redistribution system residing in the root cap (Ponce et al., 2005). Flower formation is initiated by the specifically differentiated buds and flower-bud initiation is a necessary precursor to fruit formation. It was postulated that flower-bud initiation depends on the balance between carbohydrates and nitrogenous protein compounds in developing tissues (C/N ratio);

during the following years, it is established that, in all plants, flower-bud initiation and subsequent fruit formation are controlled by growth regulators (GRs). The biosynthesis of GRs and the balance among them are, in turn, controlled by the environmental forces, especially temperature and light (Grierson, 2001). The gibberellins were made commercially available facilitating scientists to conduct investigations to inhibit flower-bud formation and induction of parthenocarpy in a wide range of horticultural plant crops. Now tracheal sap has become a convenient source of naturally occurring GRs for research on the role of GRs in bud initiation. The abscisic acid (ABA) also has been known to be involved both in flower-bud formation and fruit development. GRs appear to control messenger RNA (mRNA) required to synthesize the necessary enzymes *de novo* for fruit development after the anthesis. According to Grierson (2001), much research is involved using exogenous application of both natural and synthetic GRs to manipulate bud differentiation and fruit development. However, flower-bud initiation and subsequent entire cycle of flowering and fruiting can be regulated solely by the intelligent manipulation of temperature and light. Saltveit (1983) developed a remarkable practice of 39-week cycle, repeated at 4-week intervals, dark and lighted cold rooms and greenhouse facilities to provide a continuous supply of three varieties of container grown apples for year-round harvesting.

Saltveit (1999) has demonstrated numerous effects of ethylene on growth, development and storage life of many fruits, vegetables and ornamental crops at μl^{-1} concentrations. Harvested fruits and vegetables may be intentionally or unintentionally exposed to biologically active levels of ethylene, and both endogenous and exogenous sources of ethylene contribute to its biological activity. Ethylene synthesis and sensitivity are enhanced during certain stages of plant growth and development, as well as by a number of biotic and abiotic stresses. Exposure to ethylene may occur advertently in storage or transit, from atmospheric pollution or from ethylene produced by adjacent crops. Intentional exposures are primarily used to ripen the fruits uniformly. The detrimental or the beneficial effects of ethylene on fruit quality center on altering or accelerating the natural processes of development, ripening and senescence. The beneficial effects of ethylene, however, differ in both degree and direction. Care must, therefore, be taken to insure that crops sensitive to ethylene are only exposed to the desired atmosphere. A number of techniques to control the effects of ethylene in relation to their application with commercially important fruits and vegetables, giving the examples of general and specific beneficial and detrimental effects of ethylene of these products, have been discussed (Saltveit, 1999).

With the increasing public demand of various kinds of seedless fruits, parthenocarpy has become a highly valuable phenomenon in many fruit crops, though pollination is essential for normal fruit set. In case of citrus fruits, pollination is considered unnecessary or even undesirable because it increases the number of seeds in the "seedless" varieties before the introduction of crosses such as the tangelos (tangerine × grapefruit). Such apparently fruitful varieties when grown in small trial plots were found to be almost completely barren when planted in large-scale multi-hectare blocks. Thus, for some varieties like Orlando tangelo, pollination by some other variety was necessary as is the case in apples and pears. Parthenocarpy is desirable in crops like table grapes because seedlessness fetch higher prices than seeded grapes. In Thompson Seedless grapes, fruit set depends on GRs involved in pollen tube development although the pollen tube does not reach ovule to fertilize the ovule (Grierson, 2001). Fertilization is often crucial for fruit set and pericarp development (Atkinson et al., 2016). As the fertilized ovules develop into seeds, this influence on pericarp growth continues where production of hormones by the endosperm and developing embryo promotes pericarp growth.

Ethylene is usually regarded as the "ageing" hormone owing to its vital role in accelerating such developmental processes as ripening, senescence and abscission; however, ethylene regulates many other aspects of growth and development throughout the life cycle of the plant (Schaller, 2012). Multiple mechanisms have already been identified by which transcriptional output from ethylene signaling pathway can be tailored to meet the needs of any specific developmental pathway. Schaller (2012) specially cited the report of Lumba et al. (2012) on how vegetative transitions are regulated through the effect of the transcription factor *FUSCA3* on ethylene-controlled gene expression, providing an elegant example of how hormonal control can be integrated into a developmental pathway.

Ethylene production is closely associated with fruit ripening as it regulates and coordinates different aspects of the ripening process, such as the development of color, flavor, aroma and texture. Typically, fruits generate barely detectable amounts of ethylene until ripening when there is a burst of its production, especially in the climacteric types. A major rise in ethylene production may take place before, just after or close to the respiratory peak in the climacteric fruits such as apple, banana, avocado, fig, mango, papaya, passion fruit, pear and tomato. As with the respiratory rise, the levels of ethylene produced in these fruits may vary widely between the species (Klee and Giovannoni, 2011).

The involvement of ethylene biosynthesis and signaling in fruit set and early fruit development in zucchini squash has been reported (Martinez et al., 2013). The results demonstrated that a reduction of ethylene production or signaling in the zucchini flower was able to induce fruit set and early fruit development and, therefore, that ethylene was actively involved in fruit set and early fruit development. Auxin and TIBA treatments inducing fruit set and early fruit development in this species also inhibited ethylene production and the expression of ethylene biosynthesis and response genes. The authors have presented a model that discusses the crosstalk between ethylene and auxin in the control of fruit set and early fruit development in zucchini squash.

21.2.2 Regulation by Other Growth Hormones

De Jong et al. (2009) noted a positive correlation between the number of seeds in the fruit and the fruit size. The significance of seeds as sources of hormones for initiation and stimulation of fruit growth is implied by fruits' response to exogenous hormones in parthenocarpic systems. Parthenocarpy can be achieved by applying auxins and gibberellins to the unfertilized embryo or by employing auxin transport inhibitors such as chloroflurenol to prevent loss of auxin from embryos so as to exceed the threshold level for pericarp response. Parthenocarpic studies in tomato and cucumber have shown that high auxin levels enhance embryo cell division, and the latter is more critical than the subsequent cell expansion in determining the fruit size (Atkinson et al., 2016). Such results imply a cooperative mode of action where gibberellins combine with auxins to initiate cell division. Gillaspy et al. (1993) observed that seed cytokinins and cell division are similarly related because tomato seeds accumulate cytokinins that subsequently influence cell division in the surrounding pericarp tissue. In a critical review to analyze the role of silver nitrate ($AgNO_3$) in modulating plant growth and development, Vinod Kumar et al. (2009) observed that the basic studies on ethylene have opened new vistas for applied research in the area of micro-propagation, somatic embryogenesis, *in vitro* flowering, growth promotion, fruit ripening and sex expression. Silver nitrate has been shown to be a very potent inhibitor of ethylene action and is widely used in plant tissue culture. Certain properties of silver nitrate such as easy availability, solubility in water, specificity and stability make it very useful for various applications in exploiting plant growth regulation and morphogenesis *in vitro* and *in vivo*. Silver ion-mediated responses seem to be involved in polyamines, ethylene and calcium-mediated pathways, which play a critical role in regulating physiological processes, including morphogenesis. The molecular basis for regulation of morphogenesis under the influence of silver nitrate is completely lacking. The authors have reviewed silver nitrate-mediated *in vitro* and *in vivo* studies with an emphasis on fundamental and applied aspects of plant growth modulation under the influence of silver nitrate.

The interdependence between seed development and fruit growth shows up in final fruit size. Parthenocarpic fruit have reduced auxin content and are often smaller than the wild-type fruit. In apples, seed numbers frequently correlate with fruit growth or with fruit shape and size. In kiwifruit, inadequate pollination results in distortion, emerging a curvilinear relationship between seed number and fruit weight, and a similar response is observed when young seeds are surgically removed from immature strawberry fruit, resulting in corresponding distortion in fruit development. Atkinson et al. (2016) concluded that despite ample evidence that natural control of fruit shape is primarily exerted by plant hormones originating from seeds and stimulating growth to varying degrees, this is not true for all fruits. In banana, fertile seeds actually suppress development of the fleshy pulp, an anomalous case where failure of fertilization allows an ovary to grow. In tomato and kiwifruit marrow, spatial seed distribution is dictated by the ovary shape, and the seed distribution, in turn, influences the pericarp growth, thus; the fruit size and shape become a function of initial ovary shape and the subsequent fertilization and seed development.

21.3 HORMONAL REGULATION OF FRUIT DROP AND RIPENING

21.3.1 Regulation of Fruit Drop

Plant organs include fruits as well as other parts like leaves, petals, buds and bud scales. Old branches and roots are shed from the parent plants owing to two distinct processes: general attrition and abscission (Sexton, 2001). While the general attrition is responsible for the detachment of dead or dying tissues like bark or old branches and roots through mechanical forces such as wind or differential growth of the stem rupture of an inherently week region of tissue, abscission is involved in the loss of plant parts including the fruits. Abscission is an active metabolic process resulting in the weakening of anything from 1 to 20 rows of cells in genetically programmed abscission zones. The breaking of cell wall is a vital element in the loss of plants' structural integrity. Unlike attrition abscission is under the precise internal hormonal control (Sexton, 2001).

Researchers in the 1970s noticed that use of IAA to induce parthenocarpic development of fruits also delayed fruit abscission. As a result, it was generally hypothesized that abscission resulted from a reduction in the amounts of auxin in the abscission zone caused by a reduced auxin supply from the senescing distal organ (Addicott, 1970). The hormonal regulations of senescence and abscission have been discussed in detail in Chapter 22.

21.3.2 Regulation of Fruit Ripening

The process of fruit ripening is closely associated with phenomenon of senescence. Senescence of a plant organ is often defined as the final stage in its ontogeny during which a series of essentially irreversible or deteriorative events occurs leading to cellular breakdown and death. Fruit ripening, on the other hand, refers to changes occurring during early stages of senescence of fruits, which make them fit for consumption (or acceptable to eat). Such changes typically include changes in color, texture, taste and flavor (aroma) of the fruit. From the botanical point of view, however, fruit ripening means that the seeds are ready for dispersal, and the attractive colors, sweetness or tasty juicy pulp and characteristic aroma of the ripened fleshy fruit might be related to this function. Siddika et al. (2013) reported that the ripened small or ivy gourd (*Coccinia cordifolia* L.) fruits had the highest amounts of protein, total and reducing sugars, β-carotene, vitamins B_1, B_2 and C, whereas the matured and immature fruits were rich in minerals.

21.3.3 Role of Ethylene in Regulation of Ripening

Burg and Burg (1962) summarized internal ethylene contents and ethylene production in a variety of fruits including some which had previously been described as devoid of ethylene, e.g., the mango, pineapple, orange and lemon. Mangoes were found to have a normal pattern of ethylene evolution coincident with their respiratory climacteric. The applied ethylene hastens mango ripening even after the climacteric state has begun; this effect is related to the log of the gas concentration and becomes apparent within 15 min. The ratio between the internal and external ethylene gradient and the rate of ethylene production varied by less than a factor of ten among fruits studied and can be conservatively estimated as 2 ppm per μl/kg/h. Direct measurements of ethylene have indicated that accumulation of a stimulatory concentration of ethylene precedes the onset of the climacteric in both mangoes and bananas, and a similar conclusion can be directly derived for several other fruits.

Ethylene as an endogenous plant hormone induces a variety of plant growth and developmental changes from germination to senescence, epinasty, abscission and fruit ripening as well as its participation in several biotic and abiotic stresses (Abeles et al., 1992). The control of these processes by ethylene involves regulation of ethylene biosynthesis and perception followed by signaling transcription factors regulating ethylene-responsive genes. Ethylene-induced gene expression was studied in systems in which the biosynthesis of ethylene is stimulated during developmental processes such as ripening of fruits, senescence of flower petals or during pathogen infection (Ohme-Takagi et al., 2000). The functional analysis of the promoters of these genes showed that the ethylene-responsive *cis*-elements of fruit ripening genes and senescence genes differed from that of defense genes whose expression is induced by ethylene in response to pathogen infection.

Latest developments in fruit development and ripening have recently been reviewed by White (2002). Considerable progress has been made in identifying genes controlling the development of dry dehiscent fruits in the model plant species *Arabidopsis thaliana*. In a model fleshy fruit system (tomato), major efforts have been made to dissect biochemical and genetic regulatory cascades controlling ripening by describing intermediates of the ethylene-signaling cascade, potential crosstalk between ethylene and auxin signals, and the role of ethylene-independent signals. Recent isolation of the *NOR* and *LeMADS-RIN* genes that participate in ethylene-independent signaling in tomato, and the discovery of a homolog of the *RIN* gene expressed in a non-climacteric strawberry fruit suggests that common regulatory cascade may operate in all fruits. The transcriptional profiling during the development and ripening of both climacteric and non-climacteric fruit has supported these observations. A number of novel genes involved in the biochemistry of fruit development and ripening have been identified. The use of phylogenies based on chloroplast gene sequences has allowed an insight into the evolution of fruit forms and fruit biochemistry, which may be used to manipulate commercial fruit species for the benefit of the consumer.

Alexander and Grierson (2002) elucidated the ripening mechanisms in climacteric tomato fruit with particular attention to the role of specific isotherms of ACC oxidase in controlling ethylene synthesis during and subsequent autocatalytic phase of ethylene production during ripening. Recent information on the structure and role of six different putative ethylene receptors in tomato has been discussed including evidences supporting the receptor inhibition model of ripening, possible differences in histidine kinase activity between receptors, and the role of receptor LeETR4 in ripening.

Molecular and genetic analysis of fruit development and ripening of fleshy fruits in recent years has resulted in significant advancements in knowledge with respect to understanding ethylene biosynthesis and response, cell-wall metabolism as well as the environmental cues that impact fruit ripening. Using tomato model system with well-characterized ripening has recently resulted in non-ripening fruits, for analyzing fruit ripening and development. Moore et al. (2002) demonstrated that the exogenous application of ethylene did not restore ripening but induced the expression of ethylene-regulated ripening genes, suggesting that both mutations block necessary aspects of ripening outside the realm of ethylene's influence. Both mutations, therefore, represented genes upstream of ethylene control and additional non-ethylene-mediated aspects of ripening. Both genes have now been isolated through positional cloning strategies, and it was shown that ripening is regulated, in part, by an MADS-box transcription factor at the *rin* locus. These developmental tools for tomato genomics have expanded the potential of the tomato system to elucidate genetic regulatory components impacting fruit development, ripening and fruit nutritional quality.

Chaves and de Mello-Farias (2006) reviewed the involvement of ethylene in fruit ripening since its role in fruit ripening was established in the 1930s. They reported that presently, there is a large amount of information on ethylene and fruit ripening, much of which was gathered at a time when gas was used for illumination. Time has moved on and we are living in the "genomic era" in which the role of the phytohormone ethylene on the development and fruit-ripening processes may be very important in developing newer strategies on fruit and vegetable conservation and in obtaining biotechnological products of high added value by genetically engineering commercial plant varieties.

In contrast, the non-climacteric fruits, such as blueberry, cherry, citrus, cucumber, grape, pineapple and strawberry, do not exhibit sharp changes; their respiration rate either remains almost constant or may show a steady decline until senescence intervenes, with little or no increase in ethylene production. Paul et al. (2012) have recently shown that such distinction is less clear cut. In some climacteric fruits, ripening changes may occur independent of ethylene, whereas some non-climacteric fruits pass through ethylene-dependent changes during ripening. With the development of more sensitive ethylene measuring devices, many non-climacteric fruits appear to show an increase in ethylene evolution at previously undetectable levels upon ripening. Barry and Giovannoni (2007) suggested that regulation of fruit ripening is ubiquitous, and the climacteric and non-climacteric behavior are more accurately envisaged as the extremes of a continuum of responses with the acquisition of sensitivity to ethylene playing role (Johnston et al., 2009). According to Paul et al. (2012), the sensitivity model is supported by the observation that unique climacteric and non-climacteric fruits both increase their respiration rate when exposed to exogenous ethylene. In kiwifruit, a hybrid ripening pattern is seen, with most of the ripening changes occurring in the absence of any detectable rise in ethylene and CO_2 production; a climacteric response occurs only toward the end of ripening. The exposure to exogenous ethylene, however, promotes ripening of kiwifruit.

Ethylene metabolism has remained the main focus for the biochemical research into fruit ripening; its pathway of biosynthesis involves the conversion of Methionine, a sulfur-containing amino acid to SAM (*S*-adenosylmethionine) through the action of SAM synthase, which is then converted into ACC (1-aminocyclopropane-1-carboxylic acid) through the action of ACC synthase (ACS), and ACC is finally converted into ethylene through the action of ACC oxidase (ACO) (Atkinson et al., 2016).

In the climacteric fruit, ethylene biosynthesis occurs at very low and basal levels during fruit development prior to ripening. This ethylene production is auto-inhibitory and is known as System1 ethylene. At the ripening initiation, there is a change in ethylene biosynthesis regulation, which becomes auto-stimulatory involving the induction of specific ripening-related ACO and ACS genes different from those that are responsible for System1 ethylene (Barry and Giovannoni, 2007). Ethylene production increases greatly owing to the autocatalytic regulation, known as System2 ethylene. Barry et al. (2000) showed that at the point at which the fruit becomes competent to ripen, there is a transition from System1 to System2 ethylene that may be regulated by developmental gene such as *RIN*.

Transgenic research with a number of fruit types has yielded wealth of knowledge regarding the steps involved in the control of fruit ripening, Barry and Giovannoni (2007) demonstrated that tomato fruit with reduced levels of the ripening associated ACC oxidase or ACC synthase or with depletion of ACC levels employing a bacterial ACC deaminase, developed and grew normally but their ripening was delayed or almost completely prevented, depending upon whether the fruit was attached to the plant or detached. However, gassing of the ethylene-suppressed fruit with exogenous ethylene caused normal ripening to resume. Experiments with melon and apple showed that effects of the suppression of ethylene biosynthesis depended on the fruit species and the extent of the suppression level. The development of color, softening, accumulation of sugars and organic acids and production of aroma volatiles differed with species and ethylene levels. For example, in antisense ACO melon fruit, rind degreening, softening and accumulation of organic acids were sensitive to the levels of ethylene, but flesh pigmentation was ethylene independent. It was inferred that ethylene does not control ripening as once believed and that additional regulation by other hormones as well as certain developmental controlled factors is involved.

According to Johnston et al. (2009), fruit-ripening process comprises a series of parallel processes involving both the ethylene-independent and ethylene-dependent pathways, the latter one requiring different sensitivities to ethylene to proceed, leading to a model whereby ethylene acts as a modulator to coordinate ripening in a developmentally choreographed pattern. Another important factor in the regulation of fruit ripening is the way in which plants perceive ethylene and the signal transduction pathway that leads to ethylene response (Atkinson et al., 2016). Ethylene is perceived by receptors that are negative regulators of the signaling pathway. In the absence of ethylene, the receptors actively suppress ethylene responses, but when these receptors bind ethylene, they undergo a conformational change leading to the removal of suppression allowing de-repression of the signaling pathway. Ethylene receptors are multigene families (six genes in tomato) encoding two types of closely related proteins; one subfamily with a histidine kinase domain and the other with a serine/threonine kinase function (Atkinson et al., 2016). In Arabidopsis, the receptors seem to act redundantly since removal of any one by mutation does not cause complete insensitivity to ethylene. In tomato, Kevany et al. (2007) noted that suppression of either of two receptor genes caused an early-ripening phenotype. The significance of receptors in tomato fruit ripening has also been shown by the semi-dormant mutant Never-ripe (*Nr*), the fruits of which do not ripen. This was due to a mutation in the *ETR3* receptor, impairing the fruit in its ability to receive ethylene. Antisense inhibition of this mutant gene restored normal ripening in the *Nr* mutant (Hackett et al., 2000). Kevany et al. (2007) reported that in tomato, the turnover of receptors (degradation of existing receptor proteins and the synthesis of new ones) controls the timing of ripening.

21.3.4 Ethylene Inhibitors

In recent years, the application of ethylene inhibitors such as 1-methylcyclopropene (1-MCP) is normally being employed to retard the rates of fruit ripening in certain cultivars of apples. Watkins (2008) provided an overview of 1-MCP trials conducted in several horticultural crops. The success of using 1-MCP to retard fruit ripening depends on the physiological state of the fruit, especially on the natural rate of replacement of the ethylene receptors that are blocked by the ethylene inhibitors such as 1-MCP. Because the binding of the latter to existing ethylene receptors is irreversible, a single exposure of 1-MCP will delay ripening for many days depending upon the rate of synthesis of new receptors. Although there has been rapid uptake of 1-MCP use for commercial storage of certain apple cultivars, this treatment was found ineffective for other varieties. This is associated with the way in which apple fruits ripen, involving only a limited softening with their firmness retention and crispness being the crucial quality component. The latter contrasts with the physiology of other fruit species whose ripening involves the fruit softening coordinated with changes in flavor and color. While 1-MCP extends storage life of fruits such as avocado, pear and banana, achieving uniform ripening afterwards may be difficult in these fruits (Watkins, 2008). Johnston et al. (2009) reasoned those certain aspects of fruit ripening such as softening, flavor and color have varying degrees of sensitivities to ethylene, which are influenced differently by the partial suppression of ethylene perception and climacteric, resulting in poorer flavor and color. In case of peach fruit, inhibition of ripening was found to be rapidly overcome, and repeated exposure to 1-MCP were required which were commercially not feasible. In all the varieties of fruits, carefully optimized fruit's stage of maturity, the concentration of 1-MCP, frequency of exposure and storage temperature were crucial factors in the blocking of ethylene action to extend the storage life of fruits (Atkinson et al., 2016). Bitter pit, superficial scald, core flush and water core of apples have been described in detail illustrating the present understanding of postharvest the problems in fruit crops.

21.3.5 Regulation of Ripening by Other Growth Hormones

Several other growth hormones also appear to play important roles in fruit ripening; especially the declining levels of auxin and increasing levels of abscisic acid (ABA) may control the ripening process in non-climacteric fruits, such as grape and strawberry. ABA may play a role in controlling the onset of ethylene production in climacteric fruits. The research work uncovering the interaction between auxin, abscisic acid and ethylene in the regulation of fruit ripening and senescence is an emerging area of research (McAtee et al., 2013)

There is a marked accumulation of ABA in fruit tissues during ripening. In climacteric fruits such as avocado and pear, the level of ABA is constant during maturation but rises rapidly during ripening and coincides with rise in ethylene production during ripening. Adato et al. (1976) observed a threefold increase in free ABA level during ripening of detached avocado and pear at 19°C to a maximum level of about 7,000 μg/kg fresh weight.

Lahav and Gottreich (1984) made an attempt to review and compile the available literature on the postharvest physiology of banana fruit. The growth regulators reviewed included auxins, gibberellins, cytokinins, ethephon, CCC, Alar, ABA, morphactin and Triiodobenzoic acid (TIBA). None of these is used routinely. While chlorophenoxy and naphthalene compounds are definitely harmful, others are worthy of further investigations. The effects of four growth hormones, namely, indoleacetic acid (IAA, 10 ppm), gibberellins (GA_3, 50 ppm), cytokinin (kinetin, 2 ppm) and abscisic acid (ABA) on the ripening of three varieties of bananas (*M. paradisiaca* L.) stored at 20° C were evaluated as judged from the parameters such as physical and chemical changes (firmness, pulp/peel ratio, units of Losecke's color chart, total chlorophyll, total yellow [xanthophylls and carotene] pigments, starch, total sugars, acidity, ascorbic acid, cellulose and hemicelluloses) and five hydrolytic and oxidative enzymes (α-amylase, starch phosphorylase, acid phosphatase, peroxidase and catalase). The results indicated that the postharvest dip of bananas in the aqueous solutions of indoleacetic acid and abscisic acid significantly hastened the rate of banana ripening while the fruits dipped in solutions of GA_3 and kinetin slowed down ripening. The activities of all the five enzymes enhanced substantially during the storage of fruits at 20°C and the potential storage life of bananas significantly decreased with an advance in the stage of maturity on plant (Desai and Deshpande, 1975, 1978a, 1978b, 1979; Salunkhe and Desai, 1984a; Sharma and Singh, 2000).

The ripening of banana fruits could be controlled chemically to enhance their shelf life. Desai and Deshpande (1978b) reported that a postharvest dip of banana (*Musa paradisiaca* L.) fruits in the aqueous solutions of abscisic acid and indoleacetic acid significantly hastened ripening, judging from increases in total sugars, acidity, ascorbic acid and units of von Losecke's color chart during storage at 20°C. The treatment of bananas with gibberellic acid and kinetin, on the contrary, retarded banana ripening as indicated by higher values for firmness, starch, cellulose and hemicellulose. The results obtained suggested that ripening of banana can be controlled chemically. The differential effects of GRs on banana ripening were interpreted through their probable action on five hydrolytic and oxidative enzymes (Desai and Deshpande, 1979).

The postharvest treatment of papaya fruit with GA_3, vitamin K, silver nitrate, kinetin and cobalt chloride were found to extend the shelf life of papayas appreciably without any adverse effects on palatability (Mehta et al., 1986). The levels of succinate and malate dehydrogenase activity were suppressed during the postharvest ripening of papaya fruits treated with the above chemicals, indicating the lower levels of mitochondrial respiratory activity denoting the retardation of ripening in the climacteric papaya fruit. The postharvest dip treatment of "Alphonso" mango with cycocel, alar, gibberellic acid and menadione bisulfite significantly retarded ripening, as judged by the number of ripe fruits, during storage at 28°C. However, after cycocel treatment, retardation of ripening was observed in the early part of storage only. The value for the Brix/acid ratio was low and that for firmness was high in fruits treated with gibberellic acid and menadione bisulfite, indicating their unripe character. The results suggest that mango ripening can be controlled by the use of chemicals.

Venkatesan and Tamilmani (2010) investigated the effects of ethrel on the ripening of the off-season fruits of *Mangifera indica* L. var. Neelum. The control fruits were kept in the laboratory under the natural conditions while the experimental fruits were treated with different concentrations of ethrel (100, 200 and 300 ppm). In the control fruits, partial ripening led to incomplete metabolic changes, which did not alter the presence of sourness in the fruits, rendering them unfit to eat. On the other hand, the fruits treated with different concentrations (100, 200 and 300 ppm) of ethrel ripened on 13th, 11th and 9th days, respectively, after the treatment. The color changed from greenish to yellow and the fruits were palatable. In the control fruits, partial ripening led to incomplete metabolic changes, which did not alter the presence of sourness in the fruits. These studies were carried out using the peel and the pulp of fruit tissues individually. The phenols decreased during ripening, in both the treated and control fruits. On the other hand, the activity of peroxidase, polyphenoloxidase and catalase increased. Among the different levels of the ethrel tried, 200 ppm was found to be optimum for the ripening of the off-season mango fruit.

The storage life of most fruits is being successfully extended commercially by reducing the storage temperature to retard the fruit metabolism and the development of fungal rots. The basic effects of lower temperatures on fruits can also be supplemented by modified (MA) or controlled atmosphere (CA) storage wherein the oxygen concentration of the storage environment is reduced to less than 1% and concentration of carbon dioxide is enhanced to a maximum of 10% along with the use of sub-atmospheric techniques (Salunkhe and Desai, 1984a, 1984b; Sharma and Singh, 2000; Atkinson et al., 2016). The ways in which all these techniques impact the fruit essentially depend on the physiological state and stage of harvest maturity of the fruit. The commercial requirements of the fruit, such as whether the fruit is required to be stored for short or long periods will dictate the appropriate physiological stage of fruits' harvest maturity.

Under Fe deficiency, strategy I (non-germinaceous) plants upregulate the expression of many acquisition genes and develop morphological changes in their roots (Romera et al., 2011). The regulation of these responses is not completely known, but since the 1980s, different results suggest a role for auxin, ethylene and, more recently, nitric oxide. The upregulation of Fe acquisition genes does not depend solely on these hormones that would act as activators but also on some other signals, probably phloem Fe, that would act as inhibitors. It is not known which of the hormones considered is the last activator of the Fe acquisition genes, but some results suggest that auxin acts upstream of ethylene and NO and that, perhaps, ethylene is the last activator.

Johnston et al. (2009) showed that antisense suppression of *Malus domestica 1-AMINO-CYCLOPROPANE-CARBOXYLASE OXIDASE* (*MdACO1*) resulted in fruit with an ethylene production sufficiently low to be able to assess ripening in the absence of ethylene. The exposure of these fruits to different concentrations of exogenous ethylene showed that flesh softening, volatile biosynthesis and starch degradation had differing ethylene sensitivity and dependency. Early-ripening events such as the conversion of starch to sugars showed a low dependency for ethylene, but a high sensitivity to low concentrations of ethylene (0.01 $\mu l\ l^{-1}$). By contrast, later ripening events such as flesh softening and ester volatile production showed a high dependency for ethylene but were less sensitive to low concentrations (needing 0.1 $\mu l\ l^{-1}$ for a response). A sustained exposure to ethylene was required to maintain ripening, indicating that the role of ethylene may go beyond that of ripening initiation. These results suggest a conceptual model for the control of individual ripening characters in apple, based on both ethylene dependency and sensitivity.

21.3.6 Transgenic Breeding Strategies

The newer technologies, such as transgenic breeding strategies and genetic engineering are being employed successfully to investigate gene function and to improve different aspects of fruit quality as well as the production. Barry and Giovannoni (2007) suppressed ethylene biosynthesis to arrest fruit ripening by silencing either the ACS or ACO genes through genetic engineering. This technique is yet to find its commercial application. Suppression of cell wall-modifying enzyme, polygalacturonase, in tomato line has been found to extend fruits' shelf life. "Flavr Savr" became the first genetically engineered whole food to go on the market, though following the consumer resistance it had to be withdrawn a few years later. Papaya grown in Hawaii was devastated by ring spot virus; the transgenic modification has been successfully used to control this disorder in papaya through overexpression of a transgene of the virus coat protein that successfully interfered with the viral replication, providing the required resistance (Ferreira et al., 2002).

Dahmani-Mardas et al. (2010) employed a non-transgenic technique known as TILLING (targeting induced local lesions IN genomes) where the population of seeds or plants is chemically mutated at random followed by high-throughput screening to identify individuals where the target gene is affected and a desired trait has been improved. Though large populations are required for screening, individuals with reduced functionality of the encoded protein or even knockout in any non-essential gene can usually be obtained. This technology has been used successfully in melons to identify lines with a mutated gene and enhanced shelf life.

Wang et al. (2007) isolated two *MdERFs* (ethylene-response factors) from ripening apple fruit. The features of their conserved motifs indicated that *MdERF-1* and *MdERF-2* belonged to group VII and group IX categories in *Arabidopsis*, respectively. *MdERF-1* was expressed predominantly in ripening fruit, although a small degree of expression was also seen in non-fruit tissues. *MdFRFs-2*, however, was expressed exclusively in ripening fruit. The increased expression in ripening fruit was repressed by treatment with 1-methylcyclopropene (1-MCP), a potent antagonist of ethylene receptors. The results indicated that transcription is regulated positively by the ethylene signaling system.

Zou et al. (2014) investigated the relationship between gene expression related to ethylene and fruit ripening disorder in papaya caused by chilling injury. Papaya fruits were stored at 7°C and 12°C for 25 and 30 days, respectively, then treated with exogenous ethylene and followed by ripening at 25°C for 5 days. The symptoms of chilling injury were observed in fruit stored at 7°C for 20 days. Large differences in the expression levels of 22 genes involved in ethylene production were seen during 7°C storage with chilling injury. The genes with altered expression could be classified into the group of genes that were upregulated, including *ACS1/2/3, EIN2, EIN3s/EIL1, CTR1/2/3* and *ERF 1/3/4*, and the group of genes that were unregulated, including *ACO1/2*, *ERS* and *EBF1*. The results showed that pulp firmness had a significantly positive correlation with the expression of *ACS2, ACO1, CTR1/4, EIN3a/b* and *EBF1/2* in fruit without chilling injury. This positive correlation was changed to negative one in fruit after storage at 7°C for 25 days with chilling injury. The coloring index displayed significantly negative correlation with the expression levels of *ACS2, ACO1/2, CTR4, EIN3a/b* and *ERF3* in fruit after storage at 7°C for 25 days with chilling injury. This study suggested that these genes may play an important role in the abnormal softening and coloration with chilling injury in papaya.

21.4 DEVELOPMENTAL FACTORS

Ethylene is not the only factor regulating the fruit ripening. Atkinson et al. (2016) have described other developmental factors controlling ripening. A cold treatment can trigger fruit ripening in detached apple and kiwifruit, acting independently of ethylene or by increasing sensitivity to existing very low levels of ethylene (Tacken et al., 2010; Mworia et al., 2012). Although most ripening-associated traits appear to be regulated by the hormonal changes, there are a number of genes that control the developmental switch to ripening (Klee and Giovannoni, 2011). Some were identified in tomato in studies with ripening mutants which arose spontaneously during commercial tomato production, e.g., the *ripening-inhibitor* (*rin*), *Colorless non-ripening* (*Cnr*) and *non-ripening* (*nor*) mutants have fruits that fail to ripen though with different characteristics. Most of these do not ripen in response to exogenous ethylene treatment but some having ethylene-responsive genes will ripen by ethylene treatment. Yokotani et al. (2009) reported that transgenic tomato fruit with suppressed ethylene signaling pathway and treated with 1-MCP showed an ethylene-independent increase in the expression of ripening-related ACS genes and ethylene production. Atkinson et al. (2016) have dealt in detail several developmental factors that extend the storage life of fruits, including temperature and relative humidity, controlled and modified atmospheres, and sub-atmospheric or hypobaric storage.

REFERENCES

Abeles F B, Morgan P W and Saltveit Jr. M E. 1992. *Ethylene in Plant Biology*. 2nd edition. Academic Press, San Diego.

Adato I, Gazit S and Blumenfeld A. 1976. Relationship between changes in abscisic acid and ethylene production during ripening of avocado fruit. *Australian Journal of Plant Physiology* **3**: 555–558.

Addicott F T. 1970. Plant hormones in the control of abscission. *Biological Review* **45**: 485–524.

Alexander L and Grierson D. 2002. Ethylene biosynthesis and action in tomato: a model for climacteric fruit ripening. *Journal of Experimental Botany* **53**: 2039–2055.

Atkinson R G, Brummell D A, Burdon J N, Patterson K J and Schaffer R J. 2016. Fruit growth, ripening and post-harvest physiology. Chapter 11. pp. 1–50. *Plants in Action*. 2nd edition (Eds.) Passaarkli M. Australian Society of Plant Scientists, Australia, New Zealand Horticultural Society and New Zealand Institute of Agricultural and Horticultural Science, Macmillan Publishers, New York.

Barry C S and Giovannoni J J. 2007. Ethylene and fruit ripening. *Journal of Plant Growth Regulation* **26**: 143–159.

Barry C S, Llop-Tous M T and Grierson D. 2000. The regulation of 1-aminocyclopropane-1-carboxylic acid synthase gene expression during the transition from System 1 to System 2 ethylene synthesis in tomato. *Plant Physiology* **123**: 979–986.

Burg S P and Burg E A. 1962. Role of ethylene in fruit ripening. *Plant Physiology* **37**: 179–188.

Chaves A L S and de Mello-Farias P C. 2006. Ethylene and fruit ripening: from illumination gas to the control of gene expression more than a century of discoveries. *Genetic and Molecular Biology* **29**: 508–515.

Dahmani-Mardas F, Troadec C, Boualem A, Leveque S, Alsadon A A, Aldoss A A, Dogimont C and Bendahmane A. 2010. Engineering melon plants with improved fruit shelf life using the TILLING approach. *PLOS ONE* **5**: el 5776.

De Jong M, Mariaani C and Vriezea W H. 2009. The role of auxin and gibberellins in tomato fruit set. *Journal of Experimental Botany* **60**: 1523–1532.

Desai B B. 2004. *Handbook of Seeds*. 2nd revised edition. Marcel Dekker, New York.

Desai B B and Deshpande P B. 1975. Chemical transformations in three varieties of banana (*M. paradisiaca L.*) fruits stored at 20°C. *Mysore Journal of Agricultural Science* **9**: 634–643.

Desai B B and Deshpande P B. 1978a. Hydrolytic and oxidative enzymes during banana ripening. *Journal of Maharashtra Agricultural Universities* **9**: 24–26.

Desai B B and Deshpande P B. 1978b. Chemical control of ripening in banana. *Physiologia Plantarum* **44**: 238–240.

Desai B B and Deshpande P B. 1979. Influence of growth regulators on relative activities of some hydrolytic and oxidative enzymes during banana ripening. *Indian Journal of Plant Physiology* **22**: 189–190.

Ferreira S A, Pitz K Y and Manshardt R. 2002. Virus coat protein transgenic papaya provides practical control of papaya ringspot virus in Hawaii. *Plant Disease* **86**: 101–105.

Gillaspy G and Gruissem B H. 1993. Fruits: a developmental perspective. *Plant Cell* **5**: 1439–1451.

Grierson W. 2001. Fruit development, maturation, and ripening. pp. 143–159. *Handbook of Plant and Crop Physiology*. 2nd edition (Ed.) Pessarakli M. Marcel Dekker, New York.

Hackett R M, Ho C W, Lin Z, Humphrey C, Forte C, Fray R G and Grierson D. 2000. Antisense inhibition of the *Nr* gene restores normal ripening to the tomato *Never-ripe* mutant, consistent with the ethylene receptors-inhibition model. *Plant Physiology* **124**: 1079–1085.

Johnston J W, Gunaseelan K, Pidakala P, Wang M and Schaffer R J. 2009. Coordination of early and late ripening events in apples is regulated through differential sensitivities to ethylene. *Journal of Experimental Botany* **60**: 2689–2699.

Kevany B M, Tieman D M, Taylor M G, Cin V D and Klee H J. 2007. Ethylene receptor degradation controls the timing of ripening in tomato fruit. *Plant Journal* **51**: 458–467.

Kidd F and West C. 1938. The uptake of oxygen by apples. p. 105. *Report of Food Investment Board*. London.

Klee H J and Giovannoni J J. 2011. Genetics and control of tomato fruit ripening and quality attributes. *Annual Review of Genetics* **45**: 41–59.

Kumar V, Parvantam G and Gokare A R. 2009. $AgNO_3$-a potential regulator of ethylene activity and plant growth modulator. *Electronic Journal of Biotechnology* **12**: 1–15.

Lahav E. and Gottreich M. 1 984. The effect of growth hormones on bananas: a review. *Plant Growth Regulation* **2**: 15–30.

Lewis R A. 2002. *CRC Dictionary of Agricultural Sciences*. CRC Press, Boca Raton, FL.

Lumba S, Tsuchiro T, Delmas F, Hezky J, Provart N, Lu Q S M, McCourt P and Gazzamini S. 2012. The embryonic leaf identity gene FUSCA3 regulates vegetative phase transitions by negatively modulating ethylene regulated gene expression in Arabidopsis. *BMC Biology* **10**: 8.

Martinez C, Manzano S, Megias Z, Garrido D, Pico B and Jamilena M. 2013. Involvement of ethylene biosynthesis and signaling in fruit set and early fruit development in Zucchini squash (Cucurbita pepo L.). *BMC Plant Biology* **13**: 139.

McAtee P, Karim S, Schaffer R and David K. 2013. A dynamic interplay between phytohormones is required for fruit development, maturation, and ripening. *Frontiers in Plant Science* **4**: 79.

Mehta P M, Raj S S and Raju P S Jr. 1986. Influence of fruit ripening retardants on succinate and malate dehydrogenases in papaya fruit with emphasis on preservation. *Indian Journal of Horticultural Science* **43**: 169–173.

Moore S, Vrebalov J, Payton P and Giovannon J. 2002. Use of genomics tools to isolate key ripening genes and analyse fruit maturation in tomato. *Journal of Experimental Botany* **53**: 2023–2030.

Mworia E G, Yoshikawa T, Salikon N, Oda C, Asiche W O, Yokotani N, Abe D, Ushijima K, Nakano R and Kubo Y. 2012. Low-temperature modulated fruit ripening is independent of ethylene in 'Sanuki Gold' kiwifruit. *Journal of Experimental Botany* **63**: 963–971.

Ohme-Takagi M, Suzuki K and Shinshi H. 2000. Regulation of ethylene-induced transcription of defense genes. *Plant Cell Physiology* **41**: 1187–1192.

Paul V, Pandey R and Srivastava G C. 2012. The fading distinctions between classical patterns of ripening in climacteric and non-climacteric fruit and the ubiquity of ethylene: an overview. *Journal of Food Science and Technology* **49**: 1–21.

Ponce G, Barlow P W, Feldman L I and Cassab G I. 2005. Auxin and ethylene interactions control mitotic activity of the quiescent center, root cap size and pattern of cap cell differentiation in maize. *Plant Cell and Environment* **28**: 710–732.

Romera F J, Garcia M J, Alcantara E and Vicente R P. 2011. Latest findings about the interplay of auxin, ethylene and nitric oxide in the regulation of Fe deficiency responses by strategy I plants. *Plant Signaling and Behavior* **6**: 167–170.

Saltveit M E Jr. 1983. Year-round production of apple (Malus domestica) fruit from the container grown trees. *HortScience* **18**: 919–920.

Saltveit M E. 1999. Effect of ethylene on quality of fresh fruit and vegetables. *Postharvest Biology and Technology* **15**: 279–292.

Salunkhe D K and Desai B B. 1984a. *Postharvest Biotechnology of Fruits*. pp. 1–184. CRC Press, Boca Raton, FL.

Salunkhe D K and Desai B B. 1984b. *Postharvest Biotechnology of Vegetables*. pp 1–232. CRC Press, Boca Raton, FL.

Schaller G F. 2012. Ethylene and the regulation of plant development. *BMC Biology* **10**: 9.

Sexton R. 2001. Abscission. Chapter 10. pp. 205–227. *Handbook of Plant and Crop Physiology*. 2nd edition (Ed.) Pessarkli M. Marcel Dekker, Inc., New York.

Sharma R M and Singh R R. 2000. Harvesting post-harvest handling and physiology. pp. 94–147. *Post-Harvest Technology of Fruits and Vegetables* (Eds.) Verma L R and Joshi V K. Indus Publishing Company, New Delhi, India.

Siddika A, Ashraduzzaman Md, Khatun S, Khan M M H and Pervin F. 2013. Changes in the contents of chemical composition and activities of hydrolytic and oxidative enzymes of Coccinia cordifolia L. fruits. *Journal of Biological Science* **21**: 35–41.

Stanley S P. 1973. Ethylene in plant growth. *Proceedings of National Academy of Sciences, United States of America* **70**: 591–597.

Tacken E, Ireland H, Gunaseelan K, Karunairetnam S, Wang D and Schuitz K. 2010. The role of ethylene and cold temperature in the regulation of the apple POLYGALACTURONASE gene and fruit softening. *Plant Physiology* **153**: 294–305.

Venkatesan T and Tamilmani C. 2010. Effects of ethrel on phenolic changes during ripening of the off-season fruits of mango (Mangifera indica L. var. Neelam). *Current Botany* **1**: 22–28.

Wang A, Tan D, Takahashi A, Li T Z and Harada T. 2007. MdERFs, two ethylene-response factors involved in apple fruit ripening. *Journal of Experimental Botany* **58**: 3743–3748.

Watkins C B. 2008. Overview of 1-methyl cyclopropene trials for edible horticultural crops. *HortScience* **43**: 86–94.

White P J. 2002. Recent advances in fruit development and ripening: an overview. *Journal of Experimental Botany* **53**: 1995–2000.

Yokotani N, Nakona R, Imanishi S, Nagata M, Inaba A, Kubo Y, et al. 2009. Ripening associated ethylene biosynthesis in tomato fruit is autocatalytically and developmentally regulated. *Journal of Experimental Botany* **60:** 3433–3442.

Zou Y, Zhang L, Rao S, Zhu X, Ye L and Chen W. 2014. The relationship between the expression of ethylene-related genes and papaya fruit ripening disorder caused by chilling injury. *PLOS ONE* **12**: e116002. https://doi.org/10.1371/journal.pone.0116002.

22 Hormonal Regulation of Senescence and Abscission

22.1 INTRODUCTION

Senescence, derived from the Latin word "*senescere*" meaning "to grow old," refers to biological aging, leading to the gradual deterioration of function(s) of the part of the organ and/or the whole organism. Senescence can be either cellular senescence or senescence of the whole organism. It is commonly believed that cellular senescence underlies organismal senescence. The science of biological aging is "biogerontology." Leopold (1980) defined senescence as the endogenously controlled naturally occurring deteriorating changes leading to the death of plant cells, tissues, organs or organisms, whereas the term "aging" includes all the deteriorating changes occurring in time without reference to death as a consequence. While the aging is considered as the process happening during the entire life span of the organism, senescence is just the final developmental phase that culminates in death (Leopold, 1980; Sexton and Woolhouse, 1984; Nooden, 1988). Penarrubia and Moreno (2001) described the term "*apoptosis*," the programmed death of cells, as the intracellular process by which the cell promotes a set of genetically determined self-destructive activities, including the specific proteolysis and nuclear chromatin fragmentation. Gan and Amasino (1997) postulated that many features of stress response and senescence at the cellular level are carried out through the operation of programmed cell death.

Senescence has been regarded as the terminal differentiation of the natural developmental process usually taking place during the life cycle of an organ or organism, though a variety of environmental stresses and pathogenesis (biotic stress) may induce senescence at any stage of plants' life cycle (Thomas and Stoddart, 1980). The differential gene expression patterns in senescence processes induced by different factors do not usually result in significant variation at the physiological level, because all the senescence responses result from triggering of the same adaptive mechanisms (most aimed at the mobilization and transport of nutrients out of the decaying tissues) that are constitutively present in plants. These induced adaptive responses (senescence syndrome) are the hallmark of senescence irrespective of the circumstances such as aging, stress or pathogen attack. Far from being a chaotic breakdown, this degenerative process of senescence occurs as an orderly loss of functions and structures and comprises of an array of physiological and biochemical processes to remove the nutrients from the decaying tissues. This senescence syndrome includes the turnover of macromolecules and lipids transporting the mobilized nutrients out of the senescing structures toward other plant parts such as growing organs like young leaves, fruits or specialized storage tissues such as the bark of deciduous trees. Thus, senescence simultaneously promotes both the degradative and protective enzyme activities (Penarrubia and Moreno, 2001). Senescence of horticultural crop plants is of vital significance as it encompasses phenomena of economic importance occurring both in the field and during the storage and post-harvest handling of the products. Employing modern biotechnologies such as recombinant DNA technology it has been possible to control senescence genetically to manipulate the yield and quality of these commercially important crops.

22.2 ROLE OF HORMONES IN PLANT SENESCENCE

Several plant growth regulators have been known to play a significant role in senescence, some of which initiate the process through ethylene and abscisic acid production, whereas others like

DOI: 10.1201/9781003373889-28

cytokinins repress the process of senescence (Bruinsma, 1983; Buchanan et al., 2003; Chen et al., 2013): The role of hormones that are involved in plant senescence is explained in the following section.

22.2.1 Abscisic Acid

The abscisic acid (ABA) plays a key role in the developmental processes, such as seed germination and plant growth as well as in plants' response to the environment. The abscisic acid promotes the senescence and abscission of leaves, the levels of ABA increasing dramatically in leaves undergoing senescence. Some of the key ABA biosynthesis genes are known to be upregulated during the senescence processes, including NECD and aldehyde oxidase genes. It is postulated that ABA increases H_2O_2 levels either by production or by accumulation in leaves to accelerate senescence.

22.2.2 Ethylene

Ethylene, well known as the fruit ripening hormone, is also involved in the control of senescence of leaves and flowers. The levels of ethylene increase significantly during senescence. In *Arabidopsis* leaves, a gene, *EDR1*, is thought to regulate senescence by inducing ethylene synthesis.

22.2.3 Cytokinins

Cytokinins play a major role in the cellular proliferation as well as in shoot formation and branching. The cytokinins are known to delay the onset of senescence as indicated by a drop in the levels of cytokinins during senescence. The genes *SAG 12, AHK 3* and *IPT* are affected by cytokinins during senescence in Arabidopsis. The cytokinins influence senescence through regulation of primary metabolism and regulating sucrose levels.

22.2.4 Auxin

The levels of auxins increase during leaf senescence, and many genes involved in the auxin biosynthesis, such as tryptophan synthase, are upregulated during aging-dependent senescence. Other genes that are probably involved in the regulation of leaf senescence by auxins are *ARF2* in *Arabidopsis* at the transcriptional level whose content is increased during the senescence.

22.2.5 Jasmonic Acid (JA)

JA is an important factor regulating the onset of leaf senescence because a mutant that lacks a JA-biosynthetic enzyme gene delays leaf senescence and the application of JA to leaves accelerates senescence (He et al., 2002; Seltmann et al., 2010). This hormone is shown to reduce the photochemical efficiency of Photo System II and the levels of chlorophyll content of the detached leaves when applied exogenously. A gene thought to be involved with jasmonic acid in *Arabidopsis* is *Coronatine Insensitive1.*

22.2.6 Salicylic Acid

Salicylic acid has a role in plants' response to pathogens. It also plays a role in age-dependent leaf senescence with its level increasing dramatically by as much as four times in leaves undergoing senescence. In *Arabidopsis*, the genes associated with senescence are *SAG 12* and *PR1a*, which are shown to be undetectable when salicylic acid is not present in leaves. The transcriptome changes occurring through the presence of salicylic acid are similar to those that occur naturally in age-dependent senescence. Salicylic acid is also known to be associated with the pathogen-related cell death, and a similar role occurs in senescence-related cell death via the *hys 1* gene in *Arabidopsis*.

22.3 HORMONAL REGULATION OF SENESCENCE

22.3.1 Senescence and Fruit Ripening

The ripening of fruit is the physiological process, normally associated with senescence processes along with its metabolic features such as dismantling of chloroplastic components and other hydrolytic and oxidative changes in color, flavor, softness and taste. Fruit ripening has many specific metabolic characteristics whose fate as a physiological process is different from that of senescence. The development of organoleptic and physicochemical properties of fruit ripening eventually facilitates the spreading of seeds. Fruits, thus, act as a sink of nutrients during ripening while other senescing organs serve as a source (Nooden, 1988).

Penarrubia and Moreno (2001) categorized three types of factors that are known to influence the onset and progression of plant senescence:

- internal factors that are related to development and aging;
- environmental conditions, especially those causing plant stress;
- hormonal factors related to intercommunication with the rest of the plant.

Most plant growth regulators (phytohormones) influence senescence in plant organs either by promoting ethylene, abscisic acid and jasmonates or by producing cytokinins, auxins and gibberellins. The cytokinins and ethylene have been reported to exert greatest effect on plant senescence processes (Penarrubia and Moreno, 2001). Gan and Amasino (1996) noted a decline in cytokinin levels during natural leaf senescence and the exogenous application of cytokinins to excised leaves prevented senescence. Also, the delayed leaf senescence in tobacco plants expressing the maize homeobox gene, *knotted 1* under the control of promoter of a senescence-regulated gene (*SAG12*) is accompanied by the increased cytokinin content (Ori et al., 1999). The most conclusive effect of cytokinins in senescence has been the transformation of tobacco plants with an enzyme of cytokinin synthesis (the *Agrobacterium tumefaciens isopentenyl transferase* gene) under the *Sag 12* promoter. These plants show an autoregulatory production of ethylene, thereby avoiding the deleterious effects of overproduction and a concomitant delay of plant leaf senescence without any phenotypic alterations (Gan and Amasino, 1995).

According to Buchanan-Wollaston (1997), senescence is a complex, highly regulated, developmental phase in the life of a leaf, resulting in the coordinated degradation of macromolecules and the subsequent mobilization of components to other plant parts. The application of molecular biotechniques to the study of leaf senescence has, in recent years, enabled the isolation and characterization of a large range of cDNA clones representing genes that show increased expression in senescing leaves. The analysis of these genes and identification of the function of encoded proteins will allow a clear picture of the complex processes taking place during senescence to be assembled. Until now, the genes-encoding degradative enzymes such as proteases and nucleases, the enzymes of carbohydrate and lipid metabolism and those involved in nitrogen mobilization have been identified as the senescence enhancing genes. The combined action of several internal and external signals may be involved in the induction of senescence. The analysis of the regulatory mechanisms controlling the expression of senescence-inducing genes will allow the signaling pathways involved in the regulation of senescence to be elucidated. The research studies with transgenic plants and mutants are discovering the role played by cytokinins and ethylene in the regulation of leaf senescence.

22.3.2 Ethylene

Ethylene has been shown to accelerate the onset of senescence in some plant species (Penarrubia and Moreno, 2001). Garbic and Bleecker (1995) demonstrated that the *Arabidopsis* mutant etr 1–1, which is insensitive to ethylene owing to an inactive receptor, showed a delay of several days in the loss of chloroplast. Similarly, leaf senescence was transiently shown to be delayed in transgenic tomato plants with blocked ethylene synthesis due to antisense expression of the 1-amino cyclopropane-1-carboxylic acid (ACC) oxidase gene; however, once the senescence process has begun, the expression pattern of SAGs does not differ from that of the wild-type plants (John et al., 1995). These results indicate that ethylene influences senescence timing, and since the ethylene concentration enhances during the stress conditions, this could be a mechanism for adjusting the speed of plant response to the environment (Penarrubia and Moreno, 2001). The control of senescence appears to be exerted through regulation of gene expression acting at both the nuclear and the chloroplast genome level. Sexton and Woolhouse et al. (1984) postulated nuclear control of cell senescence according to the following facts:

- The mutation in the nuclear genome alters senescence syndrome.
- The chloroplast senescence is prevented by nucleation.
- The selective inhibitors of organelle RNA polymerases do not inhibit senescence.

Also, cycloheximide, the inhibitor of protein synthesis (80S cytoplasmic ribosomes), blocks a variety of senescence-related changes, although chloramphenicol (an inhibitor of 70S ribosomes) does not delay senescence. It has been suggested that the control of the nuclear-encoded chloroplast RNA polymerase could be the key step in this coordination (Penarrubia and Moreno, 2001).

Although the loss of chlorophyll is one of the most obvious changes during the senescence of green plant organs, several biochemical changes precede the chlorophyll loss, including a decline in photosynthetic capacity (Gepstein, 1988) and lowering of the protein content. Senescence can also be followed through measurements of oxidative damage, such as the extent of lipid peroxidation (Gutteridge and Halliwell, 1990). In the cases in which a hormone is directly implicated in senescence, the measurement of the hormonal levels may be a good approach to detect early syndrome of decay. In avocado and tomato fruits, Rowan et al. (1958) noticed a burst of ethylene preceded the onset of ripening. Pastori and Trippi (1993) monitored senescence in maize by determining the chlorophyll loss, lipid peroxidation and cell electrolyte leakage. According to Penarrubia and Moreno (2001), the gene expression analysis of selected markers may provide more sensitive and accurate monitoring of the senescence process. In view of the wealth of information being gathered through molecular analysis at the gene level, the understanding of the senescence regulation is expected to improve in the coming years, which will uncover the signaling mechanisms, clarifying the bonds between senescence and other related processes like stress responses and fruit ripening, extending the possibilities of genetic engineering of the senescence features in horticultural crops to improve their nutritional quality for the commercial benefit. Bruise resistant tomatoes or tomatoes that will not over ripen by delaying senescence have now been reported (Sheehy et al., 1988; Smith et al., 1988).

Uauy et al. (2006) and Waters et al. (2009) noted that leaf senescence occurs along with color changes in leaves and is an easily visible phenomenon in the life cycle of plants. Leaf senescence involves degradation of macromolecules, such as chlorophyll, carbohydrates, lipids, protein and nucleic acids, and contributes to the mobilization of these nutrients from the old leaves to the growing or storage tissues. The significance of efficient regulation of leaf senescence has been reported by the study on the domestication of cultivated wheat. Loci tightly linked to the enrichment of several important nutrients in cereal grains encode transcription factors (TFs) that regulate the onset of leaf senescence in ancestral wheat plants. The onset of leaf senescence is largely influenced by the age of the plant but is also affected by the changes in environmental conditions. Ethylene and other plant hormones accelerate or delay leaf senescence so that plants are better able to cope with adverse environmental changes to achieve better seed yields and biomass production (Buchanan-Wollaston et al., 2003).

22.3.3 Regulation of Senescence by Other Hormones

In addition to ethylene, JA and the developmental signals, other factors such as cytokinin, auxin and hydrogen peroxide are also involved in the regulation of leaf senescence in plants (Ellis et al., 2005; Lim et al., 2010; Smykowski et al., 2010; Yang et al., 2011; Wu et al., 2012; Jaradat et al., 2013). Several TFs that regulate the onset of leaf senescence under these additional signals have been reported. These signals integrate with ethylene to regulate the onset of leaf senescence, although there have been no reports of a direct interaction of ethylene with TFs acting downstream of these signals. Whether these signals act in an ethylene-independent manner during the onset of leaf senescence remains to be elucidated.

Recent research studies have focused on the roles of TFs and ethylene in the regulation of the onset of leaf senescence, emphasizing that regulation occurs at multiple levels downstream of the ethylene signaling pathway. The development of leaf is tightly linked to the onset of senescence; the mechanisms involved in these processes need further clarification. Jing et al. (2005) showed that the effect of ethylene on the stimulation of leaf senescence depends on the duration of ethylene exposure. Therefore, regulation of the appropriate duration of ethylene exposure could represent another candidate in the modulations of ethylene signals and the onset of leaf senescence. Further research efforts are needed to determine the mechanisms transforming the ethylene signal into the onset of leaf senescence.

22.4 ABSCISSION

Abscission (from Latin *ab*, meaning "away" and *scindere* meaning "to cut") is shedding of various parts of an organism, such as plants dropping their leaves, flowers, fruit or seeds. A plant may abscise a part either to discard a member that is no longer necessary, e.g., a leaf during autumn, a flower following fertilization or for the purpose of reproduction or as a means of defense against biotic/abiotic stresses such as infestation of gall aphids. Most deciduous plants drop their leaves by abscission before winter, whereas evergreen plants continuously abscise their leaves. Fruit drop is a form of abscission and a plant may abscise its fruit while still immature in order to conserve resources needed to bring other fruits to maturity. Plants may also abscise damaged leaves to conserve water or photosynthetic efficiency (Williams and Whitham, 1986).

22.4.1 Abscission Zones

Sexton (2001) described the shedding of plant parts such as leaves, flowers or fruit as a result of the weakening of abscission zones (AZs) which bisect the base of leaf, petioles, leaflets, petals, styles, flower buds, axillary buds, young fruit and the nodes of young stems. The AZs are not inherently weak and can bear 0.5 kg weight on a bean leaf AZ without rupture. However, after the induction of abscission, the same AZ will weaken so as to break at the slightest touch. Such progressive loss of structural integrity occurs over 72 h after an initial 18 h lag (Sexton et al., 1989). The leaf and fruit abscission normally takes up to three days, though the process can be extremely rapid; some petals and flowers bud are known to shed within 1 to 4 h after the inductive treatments (Roberts et al., 2000). Abscission occurs at genetically determined AZs where the induction of wall-degrading enzymes such as cellulase and polygalacturonase weakens a restricted band of cells called the separation layer. These separation layer cells are thought to be specially preprogrammed to respond to the inductive stimuli. The generation of the mechanical forces to facilitate separation of the loosened separation zone cells, followed by the rupture of xylem is an important component of the abscission process. After abscission, the wound is protected from microbial attack by the formation of chitinase and β-1, 3-glucanase, division and suberization of its surface cells, and blockage of the vascular traces (Sexton, 2001).

22.5 REGULATION OF ABSCISSION BY HORMONES

Induction of abscission is thought to be a normal and an integral part of the senescence program and is accompanied by the yellowing of leaves and fruit ripening. In most systems, this process can be accelerated to take place prematurely in the absence of senescence, e.g., pollination dramatically accelerates petal abscission. Leaf loss in temperate species accompanies senescence, which is induced by the environmental factors such as changes in photoperiod, low temperatures, as well as the drought. Leaf fall is often associated with lamina senescence, and water-stressed ivy plants shed leaves with the same chlorophyll content as those still attached to normal healthy plants.

Fruits are abscised at several distinct stages of development – a large number of immature fruits are shed, which appears to be a natural thinning process. However, during the growth phase of a fruit, abscission appears to be inhibited by the presence of developing seeds. Mature fruits are often induced to abscise as one of the terminal events in the ripening program. The manipulation of fruit abscission is of commercial importance in developing mechanical harvesting. After the demonstration that removal of the leaf blade causes the abscission of the remaining leaf stalk, it was hypothesized that reduction in the supply of photosynthates from the leaf caused the AZ cells to collapse and cause fracture. It was earlier shown that a tiny fragment of blade left attached to the petiole was enough to prevent abscission. Thus, it was thought unlikely that this small area of leaf was providing sufficient nutrients, it was suggested that the healthy blade probably produced a hormonal factor that prevented abscission (Sexton, 2001).

The discovery that auxin was produced by leaf blades led others to demonstrate that pollen rich in auxin applied to the cut end of a detached petiole delayed abscission (La Rue, 1936). Researchers using IAA to induce parthenocarpy in fruits reported that this treatment also delayed abscission in fruits. It was then postulated that the abscission resulted from a reduction in the amounts of auxin in the abscission zone was caused by a reduced auxin supply from the senescing distal organ (Addicott, 1970).

By the beginning of the 20th century, it was discovered that the traces of the gas used for illumination would cause the abscission of leaves, petals and fruits. The active component identified as ethylene was found to promote abscission at very low concentrations of 1–8 ppm. Years later, it was demonstrated that ethylene was synthesized by many plants. Milbrath et al. (1940) showed that ethylene produced by apples would defoliate roses. Auxin was in vogue at the time and two decades later the role of ethylene was established as a natural regulator of abscission in plants (Sexton, 2001). Osborne (1955) reported that the diffusates from senescent petioles contained a soluble factor that accelerated explants' abscission. Van Stevenick (1959) showed that the developing pods at the base of a lupine inflorescence stimulated abscission of the flowers above them; an abscission stimulator was also extracted from the young pods. Soon, a growth inhibitor was shown to be present in young cotton fruits approaching abscission (Ohkuma et al., 1963); the substance after purification was found to be abscisic acid, which accelerated abscission. Subsequently, the lupine pod extract was also shown to be the abscisic acid. Thus, ABA was accepted to be the potential regulator of abscission in plants (Addicott, 1982; Sexton, 2001).

22.5.1 Regulation of Abscission by IAA

After the establishment of the involvement of auxin in the inhibition of leaf abscission, attempts were made to measure its levels in naturally abscising systems. It was noticed that both the extractable and diffusible auxin levels dropped rapidly as the leaves yellowed approaching abscission (Addicott, 1970; Roberts and Osborne, 1981; Elkinawy, 1984). Similar correlations were observed between low auxin levels and fruit abscission. Carns (1966) evolved a simple model, which suggested that if the auxin levels in the AZ remained above a critical level, abscission was inhibited. The abscission-promoting factors such as ageing, frost damage and water stress were thought to lower the levels of free auxin in the zone. The rate of auxin transport from the distal organ was a major influence on IAA levels in the AZ, although the rates of synthesis and degradation were also implicated (Davenport et al., 1980).

The levels of auxin on the stem or proximal side of the AZ also influenced abscission, necessitating modifications of the simple auxin concentration theory. It was also demonstrated that the presence of young auxin-producing leaves on the stem seemed to accelerate loss of debladed petioles. The removal of the apical bud and young leaves delayed abscission below it, the effect of the apical bud being restored if replaced by a supply of IAA (Jacobs, 1979).

Based on the observation that applying auxin to the proximal side of an AZ accelerated abscission, Addicott et al. (1955) proposed the "gradient theory," which states that the direction of the auxin gradient across the AZ and not the absolute concentration of auxin is important. Whereas the auxin approaching the zone from the distal end inhibited abscission, that moving from the stem accelerated the process (Addicott, 1970). Guar and Leopold (1955) challenged the gradient theory of auxin based on their finding that very high levels of auxin applied to the stem side would often inhibit abscission. Abeles and Rubinstein (1964) demonstrated that the applications of auxin would promote the synthesis of ethylene, a potent accelerator of abscission. Thus, it was possible to attribute the accelerating effect of proximal auxin to enhanced ethylene production coupled to a failure of proximal IAA to reach the AZ before the abscission was under way. According to Abeles (1967), the proximally applied auxin would move to the zone much more slowly than would the distal applications, as its movement by diffusion would be opposed by basipetal auxin transport. The higher auxin concentrations applied proximally would diffuse to the zone more rapidly accounting for the inhibition sometimes observed. It has also been proposed that the auxin applied proximally may induce the synthesis of the ethylene precursor, aminocyclopropane carboxylic acid (ACC), which, in turn, diffuses to the zone to promote ethylene synthesis (Sexton, 2001). Although ethylene production by the application of proximal auxin explains the acceleration of abscission, this does not entirely account for the enhancement effect of the apex in Jacob's (1962) experiments, still resulting in the advocation of the gradient theory (Sexton, 2001).

Rubinstein and Leopold (1963) demonstrated that if auxin was added more than 12 h after deblading leaves, it accelerated abscission rather than preventing it. Thus, a view was put forth that explants were undergoing two-stage responses after excision: in the first stage, addition of auxin would inhibit abscission, prolonging the process, and in the second stage, the auxin levels fell, weakening the auxin acceleration. The auxin-induced ethylene formation was expected to explain the opposing effects of auxin in two stages. Rubinstein and Leopold (1963) proposed that in stage 1, the AZ cells were sensitive to auxin but insensitive to ethylene and in stage 2, the reverse applied. If auxin was added in the first stage, it promoted the duration of the ethylene-insensitive condition and if the auxin levels fell below a critical threshold, entering the second stage, the auxin became ineffective and the additional ethylene acceleration weakened. This hypothesis was further supported by the experiments that reduced the accelerating effect of auxin in the second stage by removing ethylene or by inhibiting its production (Abeles, 1967). It is accepted that the loss of responsiveness to auxin in the second stage is owing to the loss of auxin response. Jaffe and Goren (1979) noted that during the second stage, there was not only a reduction in IAA's ability to inhibit abscission but also evoked an H^+ ion efflux. It was speculated that these two diverse processes may become ineffective, having lost a common component of the response machinery such as a receptor. Possibly, the auxin is not simply reaching the separation zone cells if its additions are delayed owing to a decline in the auxin transport (Sexton, 2001). However, after abscission when the auxin is added to the cells of the fracture surface, it still appears to inhibit the production of the cell-wall-degrading enzymes, suggesting that at least one system is auxin sensitive (Osborne et al., 1985; Tucker et al., 1993). According to Sexton (2001), the role of auxin in the regulation of abscission has been eclipsed by the research work on ethylene though auxin (IAA) additions will completely prevent any effect of ethylene for extended periods of time. It was suggested that the IAA should be included in any model of the abscission control. Figure 32.1, based on the Sexton's unpublished data (cited by Sexton, 2001), depicts an illustrative balance model on the relative AZ concentrations of auxin on the right and ethylene on the left. If the left-hand side of the balance goes down, weakening starts, whereas if the right end is down, the process is inhibited. The ethylene and auxin concentrations are influenced by a variety of factors (shown at the top of the diagram). The balance may also be affected by the position of the fulcrum which can move from the center toward either end. Moving this to right represents a decrease in the sensitivity to auxin and an increase in sensitivity to ethylene and vice versa. Changing sensitivity may involve the level of receptors; some of the factors that change the latter are also illustrated (Sexton, 2001).

22.5.2 Regulation of Abscission by Ethylene

Ethylene induces shedding of several plant organs in a wide range of plant species (Sexton et al., 1985). The threshold concentrations required to induce such responses vary from 0.1 to 0.5 μl/l. Other chemical regulators such as the unsaturated hydrocarbons like propene, acetylene and butene also act as ethylene substitutes but are much less effective (Abeles and Gahagan, 1968). Some other analogues, such as 2,5-norbornadiene, are also competitive inhibitors (Sisler et al., 1985; McNicol et al., 1989).

The dramatic changes in the responsiveness of AZs to ethylene are evident by how distal auxin addition to fruit and leaf AZs renders them insensitive to ethylene. Halevy et al. (1984) demonstrated that unfertilized cyclamen flowers did not shed their corollas in ethylene, whereas the fertilized flowers did. Also, the styles of orange would not abscise in the presence of ethylene until the occurrence of fertilization (Goldschmidt and Leshem, 1971). Sexton et al. (1985) showed that the presence of both the ethylene gas and its receptors are needed for ethylene action to occur. The factors increasing the sensitivity, such as water deficiency, aging and ethylene per se, are believed to enhance the levels of the receptors, whereas the auxins reduced the same (Sexton et al., 1985; Sexton, 2001).

Hall (1952) had earlier put forward evidence to show that young leaves seemed to produce more ethylene than the older ones; therefore, ethylene was not widely regarded as the natural regulator of abscission. After the demonstration that auxin inhibited the accelerating effects of ethylene, the auxin/ethylene balance in the plant tissue was considered as an important aspect of abscission regulation by the growth regulators. In young leaves there was sufficient auxin to inhibit ethylene's abscission accelerating effect, unlike in the older leaves (Sexton, 2001).

According to Burg (1962), the concentrations of ethylene in plants were such that they could easily control abscission. The earlier claims that the ethylene was a natural regulator of abscission were based on the proof that increased ethylene production rates were correlated with abscission and were shown to exist in a wide variety of leaves, fruit and flowers, although Sexton et al. (1985) noted no such simple correlation to constitute as a proof of involvement of ethylene firmly in abscission, because there is a necessity to show that the endogenous ethylene concentrations increase above the threshold concentrations of 0.5–1.0 μl/l, that are needed to cause accelerated abscission if added exogenously (Sexton et al., 1985). It is difficult to extract enough gas from very tiny AZs to make such measurements. A strong correlation does exist between the ethylene production rates and its internal concentrations. Employing this relationship, Ben-Yehoshua and Aloni (1974) and Morgan and Durham (1980) showed that ethylene production rates of 3 to 5 μl/l of ethylene/kg/h were required to trigger abscission which was shown to be exceeded in many abscising systems (Sexton et al., 1985; Morgan et al., 1992; Sexton, 2001). Burdon and Sexton (1990), in their experiments with raspberry, found that these fruits are well suited to direct measurements of the ethylene gas concentrations in AZs, having 70 to 100 abscission zones enclosed within the fruit. The ethylene levels around these zones showed that the gas concentrations were less than the threshold level of 0.5 μl/l in green fruit, exceeding in ripening, abscising fruit.

Jackson et al. (1973) measured the rate of ethylene production in senescent bean leaves just prior to abscission followed by the application of (2-chlorethyl) phosphonic acid (CEPA) to younger petioles to generate similar amounts, the latter caused abscission. Several experiments of delayed abscission resulting from the reduced levels of internal ethylene in AZs, employing potassium permanganate or mercuric perchlorate to absorb the gas, have been reported (Jackson and Osborne, 1970). Using hypobaric or low-pressure atmospheres which increase diffusive loss has been more effective approach (Sexton et al., 1985). Aminoethoxyvinyl glycine (AVG), an inhibitor of ethylene biosynthesis also slows down natural abscission (Davenport and Manners, 1982). The inhibitors off ethylene action such as silver ions, which also inhibit ethylene responses such as fruit ripening and floral senescence, are equally effective in abscission (Beyer, 1976; Cameron and Reid, 1983; Ketring and Melouk, 1982; Reid, 1985). Though it is widely accepted that ethylene is involved in natural abscission, whether its action is direct or indirect remains to be elucidated. Tudela and Primo-Millo (1992) forwarded evidence that the movement of the ethylene precursor ACC in the xylem from water-stressed roots to the leaves may be one of the important mechanisms for inducing abscission. Ethylene lowers auxin concentrations by reducing its synthesis (Ernest and Valdovinos, 1971) and transport (Beyer, 1973) of the hormone and enhancing loss by conjugation and breakdown (Riov and Goren, 1979). Sexton et al. (1985) proposed that ethylene speeds abscission indirectly by reducing auxin levels. Beyer (1975) earlier had demonstrated that ethylene was actually involved in abscission both directly and indirectly. A system whereby the leaf blade and abscission zone could be exposed to ethylene independently was evolved. The abscission did not occur when the zone or leaf blade was treated alone; however, the lamina was shed when both the zone and blade were treated. The treatment of the leaf blade with ethylene decreased auxin transport down the petiole by over 80%. It was demonstrated that this effect could be mimicked by the inhibitors of auxin transport and could also be reversed if the supply of auxin from the blade was augmented (Beyer, 1975). The latter suggested that ethylene perhaps served indirectly in decreasing the auxin levels in the AZ by interfering with the movement of auxin from the leaf blade. The abscission could be hindered by impeding the auxin transport alone; ethylene's presence at the zone suggested a second direct role in abscission induction. Whether ethylene induces abscission or just accelerates is yet to be elucidated clearly. If ethylene is the inducer of abscission, removing it or interfering with ethylene action should not just slow down the process but arrest it completely. In a majority of such experiments, either the abscission does occur eventually or the observations were not continued long enough to distinguish between the process of stopping and slowing (Sexton, 2001).

22.5.3 Involvement of ABA in Abscission

After the isolation of ABA from young cotton fruit, a number of reports showed that increases in ABA were positively correlated with the abscission (Davis and Addicott, 1972; Addicott, 1970, 1982; Guinn, 1982a, 1982b). In contrast, a series of reports followed showing that there was no simple relationship between the endogenous ABA and the abscission of leaf (Davenport et al., 1977; Peterson et al., 1980), fruit or flower bud (Porter, 1977; Takeda and Crane, 1980). However, correlations do not either prove or disprove hormone's involvement, and it is commonly agreed that increases in tissue sensitivity to the hormone could induce abscission without a change in the ABA concentration (Sexton, 2001).

Among the hundreds of plants sprayed with ABA majority did not show response to abscission (Milborrow, 1974), contrasted with the reports of Addicott (1982). Abscisic acid is much more effective when applied to explants and is believed to act by promoting ethylene formation. Sagee et al. (1980) reported that ABA was ineffective in the presence of ethylene synthesis inhibitor, AVG. The current consensus seems to be that ABA is not directly involved in abscission (Sexton, 2001).

22.5.4 Involvement of Other Growth Regulators in Abscission

Although the gibberellins and cytokinins have been known to influence abscission, their effects are thought to be less important than that of the other potent hormones (Addicott, 1982; Sexton and Woolhouse, 1984). Kuang et al. (1992) reported that the cytokinins delay abscission, probably by delaying senescence indirectly. The observation that gibberellic acid accelerates abscission (Chatterjee and Leopold, 1964; Morgan, 1976), the ambiguity prevails as to whether or not the effect is mediated through ethylene (Abeles, 1967; Marynick, 1977). A few long-chain unsaturated fatty acids such as linolenic acid may also enhance abscission, though it is not clear whether these compounds are involved in the regulation of natural abscission; however, they do accumulate in certain senescent tissues (Sexton, 2001).

REFERENCES

Abeles F B. 1967. Mechanism of action of abscission accelerators. *Physiologia Plantarum* **20**: 442–454.

Abeles F B and Gahagan H E. 1968. III. Abscission: the role of ethylene analogues, carbon dioxide and oxygen. *Plant Physiology* **43**: 1255–1258.

Abeles F B and Rubinstein B. 1964. Regulation of ethylene evolution and leaf abscission by auxin. *Plant Physiology* **39**: 963–969.

Addicott F T. 1970. Plant hormones in the control of abscission. *Biological Review* **45**: 485–524.

Addicott F T. 1982. *Abscission*. University of California Press, Berkeley.

Addicott F T, Lynch R S and Carns H R. 1955. Auxin gradient theory of abscission regulation. *Science* **121**: 644–645.

Ben-Yehoshua S and Aloni B. 1974. Endogenous ethylene and abscission of orange leaf explants. *Botanical Gazette* **135**: 41–44.

Beyer E M. 1973. Abscission: support for a role of ethylene modification of auxin transport. *Plant Physiology* **52**: 1–5.

Beyer E M. 1975. Abscission: the initial effect of ethylene in the leaf blade. *Plant Physiology* **55**: 322–327.

Beyer E M. 1976. A potent inhibitor of ethylene action in plants. *Plant Physiology* **58**: 268–271.

Bruinsma J. 1983. Hormonal regulation of senescence, ageing, fading and ripening. pp. 141–163. *Postharvest Physiology and Crop Preservation* (Ed.) Liebermann M. Plenum Press, New York.

Buchanan-Wollaston V. 1997. The molecular basis of leaf senescence. *Journal of Experimental Botany* **48**: 181–199.

Buchanan-Wollaston V, Earl S, Harrison E, Mathas E, Navabpour S and Page T. 2003. The molecular analysis of leaf senescence – a genomics approach. *Plant Biotechnology* **1**: 3–22.

Burdon J N and Sexton R. 1990. The role of ethylene in shedding of raspberry fruit. *Annals of Botany* **66**: 111–120.

Burg S P. 1962. The physiology of ethylene formation. *Annual Review of Plant Physiology* **13**: 265–302.

Cameron A C and Reid M S. 1983. Use of silver thiosulfate to prevent flower abscission. From potted plants. *Scientia Horticulturae* **19**: 373–378.

Carns H R. 1966. Longevity of plant organs: internal factors controlling abscission. *Annual Review of Plant Physiology* **13**: 403–436.

Chatterjee S K and Leopold A C. 1964. Kinetin and gibberellins actions on abscission processes. *Plant Physiology* **39**: 334–337.

Chen X, Liu J, Lin G, Wang A, Wang Z and Lu G. 2013. Overexpression of AtWRKY28 and AtWRKy75 in *Arabidopsis* enhances resistance to oxalic acid and *Sclerotinia sclerotiorum*. *Plant Cell Reports* **32**: 1589–1599.

Davenport T L, Jordan W R and Morgan P W. 1977. Movement of and endogenous levels of abscisic acid during water stress-induced abscission in cotton seedlings. *Plant Physiology* **59**: 1165–1168.

Davenport T L and Manners M M. 1982. Nuceller senescence and ethylene production as they relate to avocado fruit abscission. *Journal of Experimental Botany* **33**: 815–825.

Davenport T L, Morgan P W and Jordan W R. 1980. The ability of petioles *to* transport auxin decreased with age (position) of the leaves. *Plant Physiology* **65**: 1023–1025.

Davis L A and Addicott F T. 1972. Abscisic Acid: correlations with abscission and with development in the cotton fruit. *Plant Physiology* **49**: 644–648.

Elkinawy M. 1984. Hormonal changes associated with leaf senescence in cotton (*Gossypium barbadense*). *Physiologia Plantarum* **62**: 593.

Ellis C M, Nagpal P, Young J C, Hagen G, Guilfoyle T J and Reed J W. 2005. Auxin respoense factor 1 and auxin response factor 2 regulate senescence and floral organ abscission in *Arabidopsis thaliana*. *Development* **132**: 4563–4574.

Ernest L C and Valdovinos J G. 1971. Regulation of auxin levels in *Coleus blumes* by ethylene. *Plant Physiology* **48**: 402–406.

Gan S and Amasino R M. 1995. Inhibition of leaf senescence by autoregulated production of cytokinin. *Science* **270**: 1986–1988.

Gan S and Amasino R M. 1996. Cytokinins in plant senescence: from spray and pray to clone and play. *Bioessays* **18**: 557–565.

Gan S and Amasino R M. 1997. Making sense of senescence. *Plant Physiology* **113**: 313–319.

Garbic V and Bleecker A B. 1995. Ethylene regulates the timing of leaf senescence in *Arabidopsis*. *Plant Journal* **8**: 595–602.

Gepstein S. 1988. Photosynthesis. pp. 85–109. *Senescence and Aging in Plants* (Eds.) Nooden L D and Leopold A C. Academic Press, London.

Goldschmidt E E and Leshem B. 1971. Style abscission in the citron (*Citrus medica* L.) and other citrus species: morphology, physiology, and chemical control with picloram. *American Journal of Botany* **58**: 14–23.

Guar B K and Leopold A C. 1955. The production of abscission by auxin. *Plant Physiology* **30**: 487–490.

Guinn G. 1982a. Abscisic acid and abscission of young cotton bolls in relation to water availability and boll load. *Crop Science* **22**: 580–583.

Guinn G. 1982b. Fruit age and changes in abscisic acid production and abscission rate of cotton fruits. *Plant Physiology* **69**: 349–352.

Gutteridge J M C and Halliwell B. 1990. The measurement and mechanism of lipid peroxidation in biological systems. *Trends in Biochemical Science* **15**: 129–135.

Halevy A H, Whitehead C S and Kofranek A M. 1984. Does pollination induce corolla abscission of cyclamen flowers by promoting ethylene production? *Plant Physiology* **75**: 1909–1093.

Hall W C. 1952. Evidence of the auxin-ethylene balance hypothesis of foliar abscission. *Botanical Gazette* **113**: 310–322.

He Y, Fukushige H, Hildebrand D F and Gan S. 2002. Evidence supporting a role of jasmonic acid in *Arabidopsis* leaf senescence. *Plant Physiology* **128**: 876–884.

Jackson M B, Hartley C and Osborne D J. 1973. Timing abscission in *Phaseolus vulgaris* L. controlling ethylene production and sensitivity to ethylene. *New Phytologist* **72**: 1251–1260.

Jackson M B and Osborne D J. 1970. Ethylene, the natural regulator of leaf abscission. *Nature* **225**: 1019–1022.

Jacobs W P. 1962. Longevity of plant organs: internal factors controlling abscission. *Annual Review of Plant Physiology* **13**(1): 403–436.

Jacobs W P. 1979. *Plant Hormones and Plant Development*. 6th edition. Cambridge University Press, Cambridge.

Jaffe M T and Goren R. 1979. Auxin and early stages of the abscission process of citrus leaf explants. *Botanical Gazette* **140**: 378–383.

Jaradat M R, Feurtado J A, Huang D, Lu Y and Cutler A J. 2013. Multiple roles of the transcription factor AtMYBRI/AtMYB44 in ABA signaling, stress responses and leaf senescence. *BMC Plant Biology* **13**: 192.

Jing H, Schippers J H M, Hille J and Dijkwel P P. 2005. Ethylene-induced leaf senescence depends on age-related changes and OLD genes in Arabidopsis. *Journal of Experimental Botany* **56**: 2915–2923.

John I, Drake R, Farrell A, Cooper W, Lee P, Horton P and Grierson D. 1995. Delayed leaf senescence in ethylene deficient ACC-oxidase antisense tomato plants – molecular and physiological analysis. *Plant Journal* **7**: 483–490.

Ketring D L and Melouk H A. 1982. Ethylene production and leaflet abscission of three peanut genotypes infected with *Cercospora archidicola* Hori. *Plant Physiology* **69**: 789–792.

Kuang A, Peterson C M and Dute R R. 1992. Leaf abscission in soybean-cytochemical and ultrastructural-changes following benzylaminopurine treatment. *Journal of Experimental Botany* **43**: 1611–1619.

Leopold A C. 1980. Aging and senescence in plant development. pp. 1–12. *Senescence in Plants*. CRC Press, Boca Raton, FL.

La Rue C D. 1936. The effect of auxin on the abscission of petioles. *Proceedings of National Academy of Sciences, United States of America* **22**: 254–259.

Lim P O, Lee I C, Kim J, Kim H J, Ryu J S, Woo H R and Nam H G. 2010. Auxin response factor 2 (ARF2) plays a major role in regulating auxin-mediated leaf longevity. *Journal of Experimental Botany* **61**: 1419–1430.

Marynick M C. 1977. Patterns of ethylene and carbon dioxide evolution during cotton explants abscission. *Plant Physiology* **59**: 484–489.

McNicol R J, Williamson B and Young K. 1989. Ethylene production by black current flowers infected by *Botrytis cinerea*. *Acta Horticulturae* **262**: 209–215.

Milborrow B V. 1974. The chemistry and physiology of abscisic acid. *Annual Review of Plant Physiology* **25**: 259–307.

Milbrath J A, Hansen J A and Hartman J A. 1940. Defoliation of rose plants with ethylene gas. *Science* **91**: 99–100.

Morgan P W. 1976. Gibberellic acid and indole acetic acid compete in ethylene promoted abscission. *Planta* **129**: 275–276.

Morgan P W and Durham J L. 1980. The physiology of ethylene formation. *Annual Review of Plant Physiology* **13**: 265–302.

Morgan P W, He C-J and Drew M C. 1992. Intact leaves exhibit a climacteric-like rise in ethylene production before abscission. *Plant Physiology* **100**: 1587–1590.

Nooden L D. 1988. The phenomena of senescence and aging. pp. 2–51. *Senescence and Aging in Plants* (Eds.) Nooden L D and Leopold A C. Academic Press, London.

Ohkuma K, Lyon J L and Addicott F T. 1963. Abscission II, an abscission accelerating substance from young cotton fruit. *Science* **142**: 1592–1593.

Ori N, Juarez M T, Jackson D, Yamaguchi J, Banowetz G M and Hake S. 1999. Leaf senescence is delayed in tobacco plants expressing the maize homeobox gene *knotted 1* under the control of a senescence-activated promoter. *Plant Cell* **11**: 1073–1080.

Osborne D J. 1955. Acceleration of abscission by a factor produced in senescent leaves. *Nature* **176**: 1166–1163.

Osborne D J, McManus M T and Webb J. 1985. *Ethylene and Plant Development*. p. 197. (Eds.) Roberts J A and Tucker J A. Butterworth, London.

Pastori G M and Trippi V S. 1993. Antioxidative protection in a drought-resistant maize strain during leaf senescence. *Physiologia Plantarum* **87**: 227–231.

Penarrubia L and Moreno J. 2001. Senescence in plants and crops. Chapter 9, pp. 181–203. *Handbook of Plant Crop Physiology*. 2nd edition (Ed.) Pessarkli M. Marcel Dekker, Inc., New York.

Peterson J C, Sacalis J N and Durkin D J. 1980. Promotion of leaf abscission in intact *Ficus benjamina* by exposure to water stress. *Journal of American Society of Horticultural Science* **105**: 788–792.

Porter N G. 1977. The role of abscisic acid in flower abscission of *Lupinus luteus*. *Physiologia Plantarum* **40**: 50–54.

Reid M S. 1985. Growth regulation and abscission in ornaments. *Acta Horticulturae* **167**: 57–68.

Riov J and Goren R. 1979. Effect of ethylene on auxin transport and metabolism in midrib sections in relation to leaf abscission of woody plants. *Plant Cell and Environment* **2**: 83–89.

Roberts J A and Osborne D J. 1981. Auxin and the control of ethylene production during the development and senescence of leaves and fruits. *Journal of Experimental Botany* **32**: 875–887.

Roberts J A, Whitelaw C A, Gonzalez-Carranza Z H and McManus M T. 2000. Cell separation processes in plants – models, mechanisms and manipulation. *Annals of Botany* **86**: 223–235.

Rowan K S, Pratt H K and Robertson R N. 1958. Relationship of high-energy phosphate content, protein synthesis and the climacteric rise in the respiration of ripening avocado and tomato fruits. *Australian Journal of Biological Science* **2**: 329–335.

Rubinstein B and Leopold A C. 1963. Analysis of the auxin control of bean leaf abscission. *Plant Physiology* **38**: 262–267.

Sagee O, Goren R and Riov J. 1980. Abscission of citrus leaf explants. *Plant Physiology* **66**: 750–753.

Seltmann M A, Stingl N E, Lautenschlaeger J K, Krischke M, Mueller M. and Berger S. 2010. Differential impact of lipoxygenase 2 and jasmonates on natural and stress-induced senescence in Arabidopsis. *Plant Physiology* **153**: 1940–1950.

Sexton R. 2001. Abscission. Chapter 10, pp. 205–227. *Handbook of Plant Crop Physiology*. 2nd edition (Ed.) Pessarkli M. Marcel Dekker, Inc., New York.

Sexton R, Lewis L N, Trewavas A J and Kelly P. 1985. Ethylene and abscission. pp. 173–196. *Ethylene and Plant Development* (Eds.) Roberts J A and Tucker G A. Butterworths, London.

Sexton R, Tucker M L, del Campilo E and Lewis L N. 1989. *Cell Separation in Plants*. p. 69. (Eds.) Osborne D J and Jackson M B. NATO ASI Series H35. Springer-Verlog, Berlin, Germany.

Sexton R and Woolhouse H W. 1984. Senescence and abscission. pp. 469–497. *Advanced Plant Physiology*. John Wiley & Sons, New York.

Sheehy R E, Kramer M K and Hiatt W R. 1988. Reduction of polygalacturonase activity in tomato fruit by antisense RNA. Proceedings of National Academy of Sciences, United States of America **85**: 8805–8809.

Sisler E C, Goren R and Huberman M. 1985. Effect of 2, 5-norbonadiene on abscission and ethylene production in citrus leaf explants. Physiologia Plantarum **63**: 114–120.

Smith C J S, Watson C F, Ray J, Bird C R. Morris P C, Schuch W and Grierson D. 1988. Antisense RNA inhibition of polygacturonase gene expression in ripening tomatoes. Nature **33**: 724–726.

Smykowski A, Zimmermann P and Zentgraf U. 2010. G-box Binding Factor 1 reduces CATALASE2 expression and regulates the onset of leaf senescence in Arabidopsis. Plant Physiology **153**: 1321–1331.

Takeda F and Crane J C. 1980. Abscisic acid in pistachio as related to inflorescence bud abscission. Journal of American Society of Horticultural Science **105**: 573–576.

Thomas H and Stoddart J L. 1980. Leaf senescence. *Annual Review of Plant Physiology* **31**: 83–111.

Tucker M J, Matters G L, Kohler S M, Kemmerer F C, Baird S L and Sexton R. 1993. *Cellular and Molecular Aspects of Plant Hormone Ethylene*. p. 265 (Eds.) Pech J C, Latche A and Balague C. Kluwer Academic Publisher, Boston.

Tudela D and Primo-Millo E. 1992. 1-Aminocyclopropane-1-carboxylic acid transported from roots to shoots promotes leaf abscission in Cleopatra mandarin (*Citrus reshni*, Hort. ex Tan) seedlings rehydrated after water stress. *Plant Physiology* **100**: 131–137.

Uauy C, Distelfeld A, Fahima T, Blechl A and Dubcovsky J. 2006. A NAC gene regulating senescence improves grain protein, zinc, and iron content in wheat. *Science* **314**: 1298–1301.

Van Stevenick R F M. 1959. Abscission-accelerators in lupines (Lupinus luteus L.). Nature **183**: 1246–1248.

Waters M T, Wang P, Korkaric M, Capper R G, Saunders N J and Langdale J A. 2009. GLK transcription factors coordinate expression of the photosynthetic apparatus in Arabidopsis. *Plant Cell* **21**: 1109–1128.

Williams A G and Whitham T G. 1986. Premature leaf abscission: an induced plant defense against gall aphids. *Ecology* **67**: 1619–1627.

Wu A, Allu A D, Ganapati P, Siddiqui H, Dortay H and Zanor M. 2012. JUNGBRUNNENI, a reactive oxygen species-responsive NAC transcription factor, regulates longevity in Arabidopsis. Plant Cell **24**: 482–506.

Yang S D, Seo P J, Yoon H K and Park C M. 2011. The Arabidopsis NAC transcription factor VNI2 integrates abscisic acid signals into leaf senescence via the COR/RD genes. Plant Cell **23**: 2155–2168.

Glossary of Terms

Abscisic Acid (ABA) – A plant hormone that regulates plant's response to various abiotic stresses. It also plays an important role in dormancy seeds and buds.

Abscission – Shedding of leaves, flowers, flower parts and fruit following formation of scar tissue in a plant, called the abscission zone.

Adenine – One of the four heterocyclic bases that make up DNA and RNA. In RNA and DNA, it is present as the ribonucleoside (called adenosine) and deoxyribonucleoside (called deoxyadenosine), respectively.

Adventitious – When applied to roots, it refers to roots arising from organ other than root like stem and leaf.

Amino Acid – These are that make up the proteins molecules and consist of a carboxylic acid group (COOH) and an amino group (NH_2) attached to a carbon atom. The central carbon atom may also carry other organic groups.

AMP – Adenosine monophosphate, an intermediate in energy metabolism.

Analogous – Molecules or compounds with similar structures or functions but not having evolved from a common ancestor.

Androecium – The male part of the flower.

Angiosperm – A plant belonging to *Angiospermae*, the taxonomic category encompassing all the true flowering plants and producing seeds enclosed on a capsule.

Anther – The part of the stamen bearing the pollen.

Apex – The end, top or peak of a structure. For example, the apical meristem is at the top of the stem, stigma is at the apex of the pistil.

Apical – Pertaining to the apex or top of a plant shoot or root.

Apical Dominance – Suppression of the lateral meristems by an apical meristem or main growing point.

Apical Meristem – Refers to an actively growing region of the main stem of a plant or root.

Apomixes – Type of reproduction in plants involving specialized generative tissues of flower without involving fertilization.

Asexual Reproduction – Reproduction that does not involve fusion of gametes, e.g., cutting, leaves, grafting, budding, parthenogenesis.

ATP (Adenosine 5′-triphosphate) – Complex biochemical molecules that stores biochemical energy and made up of the purine adenine attached to the sugar d-ribose, and with a polytriphosphate group attached to the other end of the ribose.

Auxin – A plant hormone that stimulates growth but suppresses development of meristems. Indoleacetic acid is a natural auxin.

Axil – The angle formed at appoint where the leaf (or the petiole) is attached to the stem. The buds or structures that are present at the axil are called axial buds.

Axis – The main shoot or stem of the plant.

Bulb – An underground storage structure comprised of a basal plate (compressed perennial stem) and numerous leaf scales, e.g., onion.

Calyx – The outermost whorl of flower.

Carbohydrate – A molecule made up of several carbon atoms connected in a chain and each bearing a hydrogen atom (H) and a hydroxyl group (OH), e.g., sugar and starch.

Carotenoid – Related to molecules similar to carotene.

Carpel – Female part of flower forming the fourth or inner whorl in flowers of angiosperms.

Chasmogamy – The unique feature of the flower referring to its opening to allow cross-pollination to occur.

Chlorophyll – The green pigment in green plants. It plays a key role in capturing the energy in light and provide it to the cell in a chemically usable form. Most common forms are Chlorophyll a and chlorophyll b.

Chloroplast – A major organelle in green plants containing the green pigment, chlorophyll. It is involved in photosynthesis.

Chromoplast – A pigment-containing carotenoid.

Chromosome – A long chain of DNA with its associated protein molecules.

Coleoptile – Protective sheath around epicotyl in grasses.

Corm – A modified stem that act as a storage structure, e.g., Crocus and gladiolus.

Corolla – Whorl of petals of a flower that forms the second layer from outside.

Cotyledon – The first embryonic leaf in a seed. While the monocots have one cotyledon, the dicots have two.

Cytokinesis – The process in which the cytoplasm of eukaryotic cell is divided in at the end of mitotic division.

Cytokinin – A plant hormone that stimulates cell division and promotes the growth of meristems. Benzyl-aminopurine is a common synthetic cytokinin.

Cytoplasm (or Cell Sap) – The cytosol (the gel-like substance enclosed within the cell membrane) and the organelles together is called cytoplasm.

Dehiscence – Spontaneous shedding of the ripe fruit to disperse the seeds. In pollen, the opening of the anther to release the pollen to expose them to pollinators.

Determinate – The size or growth of an organ is limited by cessation of meristematic activity during the year.

Development – It is the process by which structures originate and mature as a *plant* grows.

Dicot – A plant having two seedling leaves or cotyledons.

Differentiation – The process by which cells become specialized to perform particular functions.

DNA – Deoxyribonucleic acid, a polymer of deoxyribonucleotides that carries the hereditary information of the living cell.

Dormancy – A state of growth where the plant, plant parts or seed remains inactive.

Egg – The female reproductive structure.

Embryo – Earliest multicellular stage in the development of an organism following fertilization. The process by which embryos are formed is called embryogenesis.

Endoplasmic Reticulum – A network of channels in the cytoplasm of cells, to which ribosomes are attached while synthesizing proteins.

Endosperm – The part in the seed that stores food for germinating seeds.

Enzyme – A specialized molecule of RNA, mostly proteins that catalyzes chemical reactions without itself being consumed in the process.

Epicotyl – The embryonic shoot above the cotyledons. The shoot apical meristem is located here.

Epigeal Germination – For dicots, germination in which the cotyledons remain aboveground during germination.

Epigenetic – Pertaining to modifying the expression of genes by potentially reversible chemical modification of the DNA chains, specifically by methylation of cytosine bases.

Flower – The plant organ comprising various structures used in sexual reproduction.

Fruit – The structure that develops after fertilization. In develops from a carpel or aggregation of carpels.

Gamete – It is a haploid sexual cell: The male gametophyte is the pollen, and the egg cell is the female gametophyte.

Gene – It is the basic physical and functional unit of heredity.

Genome – The total set of all the genes of an organism.

Genomics – The study of genomes is called genomics.

Genotype – The genetic constitution of an individual.

Germination – The process wherein the embryo begins to elongate along its axis and proceeds to grow out breaking the seed wall outermost seed cover.

Gibberellic Acid – A plant hormone that stimulates cell elongation.

Golgi Apparatus – A structure in the cytoplasm where newly synthesized proteins are collected and sorted according to their usage.

Gynoecium – The female reproductive structure consisting of ovule, style and stigma.

Hormone – Are substances that regulate the cellular processes, often at a distance. They are active at low concentrations.

Hypocotyl – The stem of germinating seed below the cotyledon.

Hypogeal Germination – A type of germination in which the cotyledons remain belowground and never appear aboveground (e.g., lily).

Inflorescence – A reproductive shoot of a flowering plant, bearing one or more partial or complete flowers.

Internode – The portions of a shoot or stem between consecutive nodes.

Lateral Meristem – A meristem located on the side of a shoot, like in the leaf axil.

Leaf – A flattened vegetative shoot on a plant, a principal site of photosynthesis.

Meiosis – The process in sexual reproduction haploid daughter cells (gametes) are produced from diploid parent.

Membrane – A boundary layer around a cell or around organelles (e.g., mitochondria or chloroplast) within a cell, made up of a double layer of phospholipids.

Meristem – A growing point on a living plant that has the capacity to produce new cells and organs.

Messenger RNA or mRNA – Copies of DNA genes that code for proteins.

MicroRNA – Small RNA molecules (about 22 nucleotides long) that bind to the 3′ untranslated or rarely to the 5′ untranslated ends of mRNA to block or rarely to activate translation of the mRNA into protein.

Microtubule – A polymeric protein (i.e., composed of many identical protein molecules aggregated into long fibers) in cytoplasm, forming a scaffolding that participates in moving objects around inside the cell.

Mitochondrion – A structure found in eukaryotic cells that is surrounded by double membranes, which is the primary site of respiration in the cell.

Mitosis – The type of cell division in which two daughter cells that identical to each other and to their parent are produced.

Morphogenesis – Differentiation and growth of the structure of an organism (or a part of an organism).

Node – Position along a shoot where a leaf or other structure is attached.

Nucleotide – A molecule having either a purine or a pyrimidine base connected to a five-carbon sugar, either ribose or deoxyribose, and a phosphate group connected to the other end of the sugar moiety.

Nucleus of a cell – A structure found in living eukaryotic cells, delimited by a membrane and containing the chromosomes.

Organelle – A complex structure found within eukaryotic cells, and separated from the cytoplasm of the cell by a double membrane (e.g., chloroplast, mitochondria).

Ovary – The female reproductive structure that contains the ovule and forms the fruit after fertilization.

Ovule – The haploid female reproductive cell, located in the ovary.

Perianth – The flower structure that covers the calyx (sepals) and corolla (petals).

Pericarp – The structure developing from the wall of the ovary that protects or encloses the seed or seeds in an angiosperm.

Periclinal – Parallel to a surface. Cell division that takes place parallel to axis is called the periclinal division (opposite anticlinal division).

Petal – Second outmost whorl in flower. It is leaf-like often brightly colored structures to facilitate cross-pollination.

Petiole – the stem-like extension of the base of a leaf used for attachment of leaves to the stem.

Phenotype – The physical appearance and functioning of an individual.

Phosphorylation – Process where a phosphate molecules or group is added.

Photosynthesis – The process by which light energy is captured and used for converting carbon dioxide and water molecules into sugar.

Pistil – Female sexual organ of a flower, comprised of ovule style and stigma

Pollen – Male sexual structure that transmits the male gamete to the female stigma.

Polymerase – An enzyme that makes copies or transcriptions of DNA or RNA. RNA polymerases synthesize new RNA molecules using a DNA sequence as the pattern.

Promoter – A region of DNA extending 150 to 300 base pairs upstream from the transcription start site that contains binding sites for RNA polymerase and a number of proteins, transcription factors and related proteins that regulate the transcription of the adjacent gene.

Radicle – The original root that develops from the germinating seed.

Recalcitrant – Seeds that cannot be forced into a true dormant state.

Recombinant DNA – DNA produced by cutting and recombining DNA fragments from different sources. *Recombinant* is a cell that results from recombination of genes.

Rhizome – An underground modified stem growing horizontally.

RNA – Ribonucleic acid that is crucial to the functioning of living cells. Ribosomal RNA (rRNA) is central to the functioning of ribosomes. Messenger RNA (mRNA) carries the message from the DNA in the nucleus to the ribosome in the cytoplasm, where it is translated into the protein molecule by the ribosome.

Sepal – Outermost whorl of the flower. It is leaf-like structure that encloses the rest of the parts in a flower.

Sessile – Having a flower or leaf born directly on the stem or peduncle rather than on an elongated stalk.

Shoot – An aboveground portion of the plant consisting of stem, leaf, buds, branches.

Stem Cell – A pluripotent cell capable of continues division and being developmentally modified to serve a specialized function of the living organism.

Stigma – The female structure at the tip of the pistil.

Stomata – (Plural) Openings or pores on leaf surfaces that permit gaseous exchange between the interior of the leaf and the atmosphere.

Style – The structure that extends from the ovary and bears the stigma at its distal end.

Sugar – A carbohydrate molecule containing four to seven carbon atoms in a chain, and usually a terminal aldehyde group at one end of the chain or an internal carbonyl (keto) group.

Tetrad – Pollen grains in clusters of four.

Transcription – The process of creating a complementary RNA copy of a DNA sequence.

Transcription Factor – A protein that facilitates transcription of DNA into RNA.

Transcriptome – The entire set of transcribed genes in an organism.

Transfer RNA or tRNA – RNA molecules that transfer amino acids in a sequence-specific manner to growing protein chains at the ribosomes.

Translocation – In chromosomal translocation, refers to joining of non-adjacent segments of chromosomes.

Tuber – A perennial underground storage structure comprised of a swollen root with one or more shoot buds (potato, dahlia).

Vacuole – Bubble like structures within the cytoplasm and separated from the cytoplasm by a membrane used for storing organic compounds or molecules.

Vernalization – Cold treatment of plants to terminate dormancy or to induce flowering.

Zygote – The diploid structure resulting from the fusion of male and female gametes. This becomes the embryo after undergoing embryogenesis.

Author Index

C

D

E

F

G

H

I

J

K

N

O

P

Q

R

S

Subject Index

R

S

T

V

W

X

Z